T0142371

Advances in Intelligent Systems and Computing

Volume 690

Series editor

Janusz Kacprzyk, Polish Academy of Sciences, Warsaw, Poland
e-mail: kacprzyk@ibspan.waw.pl

About this Series

The series "Advances in Intelligent Systems and Computing" contains publications on theory, applications, and design methods of Intelligent Systems and Intelligent Computing. Virtually all disciplines such as engineering, natural sciences, computer and information science, ICT, economics, business, e-commerce, environment, healthcare, life science are covered. The list of topics spans all the areas of modern intelligent systems and computing.

The publications within "Advances in Intelligent Systems and Computing" are primarily textbooks and proceedings of important conferences, symposia and congresses. They cover significant recent developments in the field, both of a foundational and applicable character. An important characteristic feature of the series is the short publication time and world-wide distribution. This permits a rapid and broad dissemination of research results.

Advisory Board

Chairman

Nikhil R. Pal, Indian Statistical Institute, Kolkata, India
e-mail: nikhil@isical.ac.in

Members

Rafael Bello Perez, Universidad Central "Marta Abreu" de Las Villas, Santa Clara, Cuba
e-mail: rbellop@uclv.edu.cu

Emilio S. Corchado, University of Salamanca, Salamanca, Spain
e-mail: escorchado@usal.es

Hani Hagras, University of Essex, Colchester, UK
e-mail: hani@essex.ac.uk

László T. Kóczy, Széchenyi István University, Győr, Hungary
e-mail: koczy@sze.hu

Vladik Kreinovich, University of Texas at El Paso, El Paso, USA
e-mail: vladik@utep.edu

Chin-Teng Lin, National Chiao Tung University, Hsinchu, Taiwan
e-mail: ctlin@mail.nctu.edu.tw

Jie Lu, University of Technology, Sydney, Australia
e-mail: Jie.Lu@uts.edu.au

Patricia Melin, Tijuana Institute of Technology, Tijuana, Mexico
e-mail: epmelin@hafsamx.org

Nadia Nedjah, State University of Rio de Janeiro, Rio de Janeiro, Brazil
e-mail: nadia@eng.uerj.br

Ngoc Thanh Nguyen, Wroclaw University of Technology, Wroclaw, Poland
e-mail: Ngoc-Thanh.Nguyen@pwr.edu.pl

Jun Wang, The Chinese University of Hong Kong, Shatin, Hong Kong
e-mail: jwang@mae.cuhk.edu.hk

More information about this series at http://www.springer.com/series/11156

Feng Qiao · Srikanta Patnaik
John Wang
Editors

Recent Developments in Mechatronics and Intelligent Robotics

Proceedings of the International Conference on Mechatronics and Intelligent Robotics (ICMIR2017) - Volume 1

 Springer

Editors
Feng Qiao
Shenyang Jianzhu University
Shenyang
China

John Wang
Montclair State University
Montclair, NJ
USA

Srikanta Patnaik
SOA University
Bhubaneswar
India

ISSN 2194-5357   ISSN 2194-5365   (electronic)
Advances in Intelligent Systems and Computing
ISBN 978-3-319-65977-0   ISBN 978-3-319-65978-7   (eBook)
DOI 10.1007/978-3-319-65978-7

Library of Congress Control Number: 2017958551

Printed on acid-free paper

This Springer imprint is published by Springer Nature
The registered company is Springer International Publishing AG
The registered company address is: Gewerbestrasse 11, 6330 Cham, Switzerland

Preface

On behalf of the Organizing Committee I welcome the paper presenter and participants to the International Conference on Mechatronics and Intelligent Robotics (ICMIR2017) held at Kunming, China, during May 20–21, 2017. This annual conference is being organized each year by Interscience Research Network, an international professional body, in association with International Journal of Computational Vision and Robotics and International Journal of Simulation and Process Modelling, published by Inderscience Publishing House. I must welcome this year's General Chair Prof. Feng Qiao, Shenyang JianZhu University, Shenyang, China, for his generous contribution to ICMIR-2017. He has also contributed an issue of his journal International Journal of Simulation and Process Modelling to the selected papers of the conference.

Like every edition, this edition of ICMIR2017 was academically very rich and we had three eminent professors as keynote speakers namely Prof. John Wang, Dept. of Information Management & Business Analytics, School of Business Montclair State University, USA, Prof. Kevin Deng, Distinguished Professor and Executive Director of Automotive Research Institute, Jilin University, and Dr. Nilanjan Dey, Department of Information Technology, Techno India College of Technology, Kolkata, India.

There has been a rapid progress during last 5 (five) years. The domain covers various areas such as: robotic-assisted manufacturing; advanced mechanisms and robotics; systems modelling and analysis; instrumentation and device control; automation systems; intelligent sensing and control; medical robotics; and autonomous and complex systems. New technologies are constantly emerging, which are enabling applications in various domains and services. Intelligent Mechatronics and Robotics is no longer a functional area within the department of mechanical or electronics, but is an integral part of the manufacturing function of any organization. In the recent time, Intelligent Mechatronics and Robotics is probably the single most important facilitator of the manufacturing process. The result of research in this domain is now influencing the process of globalization, particularly in the productive, manufacturing and commercial spheres. Creating economic opportunities and contributing to monotony reduction is another thrust area for the emerging epoch of Intelligent Mechatronics and Robotics.

This edition of ICIMR covered the following areas but not limited to intelligent mechatronics, robotics and biomimetics, novel and unconventional mechatronic systems, modelling and control of mechatronics systems, elements, structures, mechanisms of micro- and nano-systems, sensors, wireless sensor networks and multi-sensor data fusion, biomedical and rehabilitation engineering, prosthetics and artificial organs, AI, neural networks and fuzzy logic in mechatronics and robotics, industrial automation, process control and networked control systems, telerobotics, human–computer interaction, human–robot interaction, artificial intelligence, bio-inspired robotics, control algorithms and control systems, design theories and principles, evolutional robotics, field robotics, force sensors, accelerometers, and other measuring devices, healthcare robotics, human–robot interaction, kinematics and dynamics analysis, manufacturing robotics, mathematical and computational methodologies in robotics, medical robotics, parallel robots and manipulators, robotic cognition and emotion, robotic perception and decision, sensor integration, fusion, and perception. This volume covers various articles covering the recent developments in the area of Intelligent Mechatronics and Robotics categorized into seven (7) tracks, such as:

1. Intelligent Systems
2. Intelligent Sensor & Actuator
3. Robotics
4. Mechatronics
5. Modelling & Simulation
6. Automation & Control and
7. Robot Vision

Srikanta Patnaik
Programme Chair: ICMIR-2017

Conference Organizing Committee

General Chair

Feng Qiao   Shenyang JianZhu University, Shenyang, China

Programme Chair

Srikanta Patnaik   SOA University, Bhubaneswar, Odisha, India

Organizing Chair

Zhengtao Yu   Kunming University of Science and Technology, Kunming, China

Technical Programme Committee

Xilong Qu	Hunan University of Finance and Economics, China
Yong Ma	Electronic Information School Wuhan University, China
Xiaokun Yang	University of Houston Clear Lake, China
Hao Wang	State Key Laboratory of Rolling and Automation, Northeastern University, China
Vladicescu Popentiu	Florin, City University, UK
Imran Memon	Zhejiang University, China
Guangzhi Qu	Oakland University, USA
V.S.S. Yadavalli	University of Pretoria, South Africa
Bruno Apolloni	Università degli Studi di Milano, Italy

Harry Bouwman	Delft University of Technology, Netherlands
Shyi-Ming Chen	National Taiwan University of Science and Technology, Taiwan
Yahaya Coulibaly	University Technology Malaysia, Malaysia
B.K. Das	Government of India, India
Joseph Davis	The University of Sydney, Australia
Arturo De La Escalera Hueso	Universidad Carlos III de Madrid, Spain
Ali Hessami	Vega Systems, UK
Yen-Tseng Hsu	National Taiwan University of Science and Technology, Taiwan
Lakhmi C. Jain	Bournemouth University, UK
Sanjay Jain	National University of Singapore, Singapore
Chidananda Khatua	Intel Corporation Inc., USA
Ayse Kiper	Middle East Technical University, Turkey
Ladislav J. Kohout	Florida State University, USA
Reza Langari	Texas A&M University, USA
Maode Ma	Nanyang Technological University, Singapore
N.P. Mahalik	California State University, Fresno, USA
Rabi N. Mahapatra	Texas A&M University, USA

Invited Speakers

John Wang	Department of Information Management & Business Analytics, School of Business Montclair State University, USA
Kevin Deng	Automotive Research Institute, Jilin University
Nilanjan Dey	Department of Information Technology, Techno India College of Technology, Kolkata, India

Acknowledgement

Like every year, this edition of ICMIR-2017 was also attended by more than 150 participants and 172 papers were shortlisted and published in this proceeding. The papers covered in this proceeding are the result of the efforts of the researchers working in various domains of Mechatronics and Intelligent Robotics. We are thankful to the authors and paper contributors of this volume.

We are thankful to the editor in chief and the anonymous review committee members of the Springer series on *Advances in Intelligent Systems and Computing* for their support to bring out the proceedings of 2017 International Conference on Mechatronics and Intelligent Robotics. It is noteworthy to mention here that this was really a big boost for us to continue this conference series.

We are thankful to our friends namely Prof. John Wang, from School of Business Montclair State University, USA, Prof. Kevin Deng, from Automotive Research Institute, Jilin University, and Dr. Nilanjan Dey, from Techno India College of Technology, Kolkata, India, for their keynote address. We are also thankful to the experts and reviewers who have worked for this volume despite the veil of their anonymity.

We are happy to announce here that next edition of the International Conference on Mechatronics and Intelligent Robotics (ICMIR2017) will be held at Kunming, China in association with Kunming University of Science and Technology, Kunming, China during last week of April 2018.

It was really a nice experience to interact with the scholars and researchers who came from various parts of China and outside China to participate the ICMIR-2017 conference. In addition to the academic participation and presentation, the participants must have enjoyed their stay, during the conference and sightseeing trip at Kunming.

I am sure that the readers shall get immense ideas and knowledge from this volume of AISC series volume on "Recent Developments in Mechatronics and Intelligent Robotics".

Contents

Intelligent Sensor and Actuatorr

Robotics

Intelligent Systems

Modified Chaos Particle Swarm Algorithm Using Interpolation of Zadoff-Chu Sequence

Wan Shaosong[1], Cao Jian[2(✉)], Li Gang[2], and Wang Ke[2]

[1] XiJing University, Xi'an 710123, Shaanxi, People's Republic of China
[2] Air Force Engineering University, Xi'an 710051, Shaanxi, People's Republic of China
cao_jian1972@163.com

Abstract. Motion detection from image sequences is the purpose of changes will be extracted from the background image. Moving target segmentations of regions objectives, track and understand later treatment is very important, since it will process only corresponds to regions of pixels in the image. However, because of the dynamic change of the background image, such as weather, light, shadow and chaos interference effects makes motion detection becomes a difficult task. In the monitoring system, a basic and crucial step is to determine the moving object in the video sequence, motion detection is directly related to the effect of moving target tracking and follow-up of target behavior recognition, is the key to the quality and practicality of the system as a whole. People always want a can be used to monitor a wide variety of environments, to meet the various requirements of the universal algorithm for moving object detection. However, the practical application is very difficult to meet such requirements, practical applications in a variety of environments thus far, but also from the algorithm complexity, reliability, and timeliness and other aspects into consideration.

Keywords: Algorithm · Optimization · Intelligent lock · Particle swarm · Design

1 Introduction

Moving target detection in image sequence that contains the motion information is properly processed, thereby removing the static background, detecting moving targets and relative movement information, and the integration of these information, are key parameters for the subsequent phases of the visual system provides reliable data source [1, 2]. Moving object detection principle is to preserve those characteristics information is important for visual inspection, and get rid of those useless for detecting moving objects to maximize the redundant information [3].

Motion detection at this stage there are four main methods: optical flow method, time difference, background subtraction method and classification of statistical models. Light flow method in camera movement of conditions Xia also can detection out independent of movement target, however most of light flow calculation quite complex, not can for real-time processing; time points detection method for movement environment has strong of since adaptability, but general cannot completely extraction out all related regional of pixel points, in movement entity in the easy produced empty phenomenon;

F. Qiao et al. (eds.), *Recent Developments in Mechatronics and Intelligent Robotics*,
Advances in Intelligent Systems and Computing 690, DOI 10.1007/978-3-319-65978-7_1

background removal method applies Yu fixed camera of case, it first for background established background model, through will current image frame and background model for compared.

Determines the brightness for a larger area, that is considered to be foreground area. This method is very fast, you can get the whole area, but for scene illumination and noise-sensitive, in the practical application of certain algorithms are needed to maintain and update the background model to changes in the environment; background statistical model adaptive selection of threshold, but complexity is often too high. An ideal target detection should be able to apply to a variety of environments. Should have a good target detection algorithm in all weather conditions should be robust; should be robust to changes in ambient light; to adapt to individual objects in a scene motion disturbances, such as shaking trees, water fluctuations; capable of handling, the large area of disorganized campaigns and target superimposed interference within the field of view.

People always want to get a perfect target detection algorithm can be applied to a variety of environments, but it is very difficult to solve the problem, because not only have to take into account the algorithm will try to adapt to a variety of environments, and unless there is a specific hardware support, otherwise they have to be in its complexity, reliability and real-time aspects of compromise.

2   Intelligent Lock Software

The software is shown as Fig. 1.

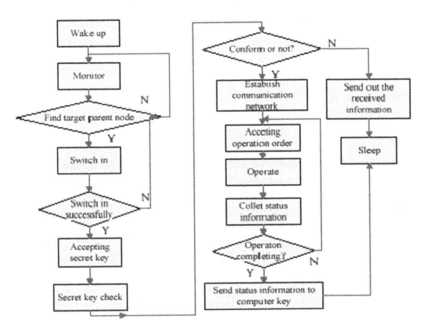

Fig. 1. Structure diagram of software

Algorithm flow chart follows as Fig. 2.

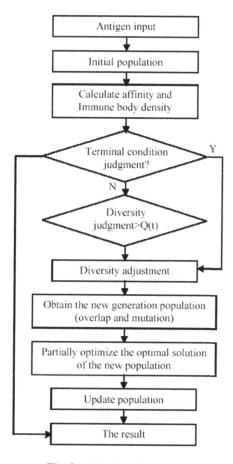

Fig. 2. Algorithm flow chart

Moving object detection is directly related to the effect of moving target tracking and follow-up of target recognition, which relates to the quality and practicality of the system as a whole. People always want a can be used to monitor a wide variety of environments, able to meet the various requirements of the universal algorithm for moving target detection. However, the practical application is very difficult to meet such requirements, because the actual application takes into account not only the algorithm can apply in a variety of environments, but also from the algorithm complexity, relia-bility, and real-time aspects of compromise into account.

3 Circuit of Hardware

Optical flow method has the advantage of optical flow not only carry the motion of moving objects, but also carries a rich information about the scene structure, it can not know any case information, detect moving objects, and can be used for movement of the camera, but most of the optical flow method is complex and time-consuming. Unless there is a specific hardware support, and it is difficult to achieve real-time detection; optical flow separation accuracy and correctness of the optical flow estimation are closely related. This optical flow separation on the noise is more sensitive and the athletic edge estimates are not precise, the calculation is more complex, estimates need to determine in advance the number of sports. So in order to get the best results will have to deal with motion estimation and segmentation.

Figure 3 reflects system framework.

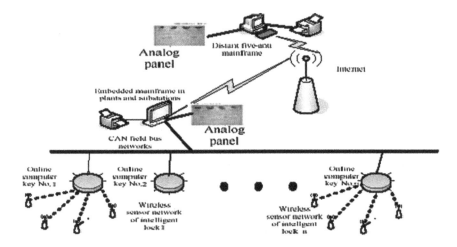

Fig. 3. Multi-layer system network

Based on background there and time points weighted of movement target detection method, although can extraction out full of target image, but in actual application in the still has many problem need solution, actual application in the a site "pure" of background image always not easy get of, a simple of practices is Dang scene in the no any target Shi collection a site image as background image, but with time of over, outside of light will changes, this will caused background image of changes, thus this used a site fixed background image of method, only suitable external conditions are good occasions. In order to achieve video surveillance for a long time, you need to update background image after a period of time. Because the background images will not be "fixed", so most researchers had abandoned the non-adaptive method of obtaining background, based on background subtraction methods over time, is the key to how the adaptive update background image. The hardware details are designed in Fig. 4.

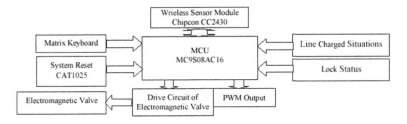

Fig. 4. Hardware structures

4 Comparisons Between New Ones and Original Ones

Post-processing stream of a pixel in a period of observation, statistics the brightness value of the pixel at that point, we can see that over a period of time, the brightness value is concentrated in a small area of. Result is shown in Fig. 5.

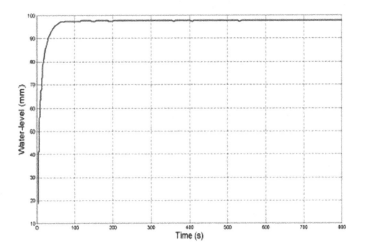

Fig. 5. System response graph

The comparison results are shown in Fig. 6.

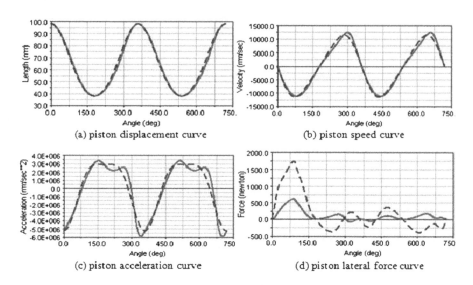

Fig. 6. Analyses and comparison

5   Summary

Fixed background is a simple solution, can detect objects in the scene. In addition, the scene background reference image reduces the effect of revealing background, improving the texture of moving target detection accuracy. Without the prospect of initial reference frames of moving object image is difficult to obtain for many systems, and fixed background can not update with the change of scene and cannot adapt to changes in lighting and backgrounds such as structure and content, not suitable for long-term monitoring applications. Therefore, how to automatically initialize and update the background model is an important topic.

References

1. Nasaroui, O., Gonzdez, F., Dasgupta, D.: The fuzzy artificial immune system. Fuzzy Syst. **1**(2), 711–716 (2002)
2. de Castro, L.N., Timmis, J.: Artificial Immune System: A Novel Computational Intelligence Approach. Springer, Heidelberg (2015)
3. Jiao, L., Wang, L.: A novel genetic algorithm based on immunity. IEEE Trans. Syst. Man Cybern. Part A Syst. Hum. **30**(5), 552–561 (2016)

Air Quality Index Prediction Using Error Back Propagation Algorithm and Improved Particle Swarm Optimization

Jia Xu[✉] and Lang Pei

College of Computer Science, Wuhan Qinchuan University, Wuhan, China
461406563@qq.com

Abstract. As the latest evaluation standards of air quality released by the State Environmental Protection Department, the Air Quality Index (AQI) is influenced by sulfur dioxide (SO_2), nitrogen dioxide (NO_2), particulate matter with particle size below 10 microns (PM10), particulate matter (PM2.5), carbon monoxide (CO) and ozone (O_3) in the air. The variation of AQI shows nonlinearity and complexity. In order to improve prediction accuracy, this paper proposes an air quality prediction model based on Error Back Propagation (BP) algorithm. The model is optimized by Particle Swarm Optimization (PSO) algorithm using dynamic inertia weight and experience particles. The experimental results show the improved PSO-BP model significantly reduces iteration time, effectively improves the prediction accuracy, and provides a new method for the AQI prediction.

Keywords: BP · PSO · Dynamic inertia weight

1 Introduction

With the development of social economy, cities' developments have been accelerated and the car ownership has increased. The contents of SO_2, NO_2, PM10, PM2.5 and O_3 in the atmosphere have increased gradually; Environmental pollution problems have been increasingly serious. As the State Environmental Protection Department released the latest air quality assessment standards in 2012, the air quality has become a major issue in relation to the future fate of mankind [1]. The study around this problem also came into being.

AQI is a single non-dimensional numerical form and is used to quantitatively describe the air quality. Its value looks seemingly disorderly, but the variation in a long time shows a certain rule. AQI is comprehensively influenced by SO_2, NO_2, PM10, PM2.5, CO and O_3; its value shows the characteristics of non-linear and abrupt changes. So AQI is a complicated nonlinear system. We can find the internal relation of influencing factors by the historical monitoring data, and then establish a prediction function to realize AQI prediction. Its principle is similar to the Artificial Neural Network (ANN). At present, the air quality forecasting application of ANN is still in the exploratory stage. The BP neural network is one of the most widely used neural network models and has the typical characteristics of neural networks [2]. However, the BP algorithm has slow

F. Qiao et al. (eds.), *Recent Developments in Mechatronics and Intelligent Robotics*,
Advances in Intelligent Systems and Computing 690, DOI 10.1007/978-3-319-65978-7_2

convergence and is easy to fall into a local minimum. To solve these problems, scholars have put forward many improved methods. For example, Li et al. [3] introduced a variable learning rate and an additional momentum into the BP algorithm to jump out of the local minimum of the error surface. But the training speed of this method is not very satisfactory. Zhang et al. [4] used the PSO algorithm to improve the learning strategies of the BP neural network. In this paper, the learning speed of BP algorithm is improved, but for the high dimension complex problem, the PSO algorithm is faced with the premature convergence problem.

This paper adopts the dynamic inertia weight and experience particles [5] to improve the standard PSO algorithm; it uses the improved algorithm to optimize the BP network learning strategy and then builds the PSO-BP prediction model to realize AQI simulation. The experimental results show that the improved PSO-BP algorithm not only could shorten the algorithm iteration time, but also could improve the convergence speed and the prediction accuracy.

2   Standard BP Neural Network

The BP neural network is a multilayer feed forward network with the one-way transmission. It is composed of input layer, hidden layer and output layer, its main characteristic is the signal forward propagation and the error back propagation [6]. The standard BP is a feed forward neural network with three layers topological structure and has only one hidden layer. The research shows if we select the suitable connection weights and the transfer function, a neural network with enough neurons and only one hidden layer could approximate any smooth, measurable function between the input and output [7]. Therefore, this paper uses the standard BP neural network as the network prototype.

The standard BP neural network adopts the sigmoid function to calculate the network output of each level. The sigmoid function is a non-decreasing continuous function; its value is a floating point number between -0.5 and 1.5. The function represents the state continuous neuron model and is processed very conveniently.

When the total error of input samples can't achieve the desired effect, the network enters the error back propagation stage. Using the formula (1), the BP algorithm calculates the weights between the hidden layer and the output layer. The weights can be used to inversely modify the weight matrix to achieve the optimization algorithm. The formula (1) is defined as

$$
\begin{cases}
\Delta w_{jk} = \eta \delta_k^y z_j = \eta \left(d_k - y_k \right) y_k \left(1 - y_k \right) z_j \\
\Delta v_{ij} = \eta \delta_j^z x_i = \eta \left(\sum_{k=1}^{m} \delta_k^y w_{jk} \right) z_j \left(1 - z_j \right) x_i
\end{cases}
\tag{1}
$$

Where Δw_{jk} and Δv_{ij} are the adjusting weights. δ_k and δ_j are the error signals of each level. η is a proportional coefficient and its value is a random number between 0 and 1. x_i is an input component. z_j is an output component of the hidden layer. y_k is an output component of the output layer. d_k is an expected output component.

3   Improved PSO Algorithm

The PSO algorithm [8, 9] is a kind of typical swarm intelligence algorithm, it can simulate the foraging behavior of birds in the nature to find an optimal solution through individuals collaborating and information sharing. In 1998, Shi and Eberhart in the academic paper "A Modified Particle Swarm Optimizer" [10] introduced an inertia weight into the evolution equation, thus the standard PSO algorithm was born.

The inertia weight is a very important parameter in the standard PSO algorithm. The larger the inertia weight is, the stronger the exploration ability is; the smaller the inertia weight is, the stronger development ability is [11]. In this paper, we have studied literatures [12–15], compared and analyzed the advantages and disadvantages of the linear inertia weight. At last, we linearly increase the value of the inertia weight before 1000 iterations, and then linearly decrease its value, so as to balance the exploration ability and development ability of the improved PSO algorithm. So we define the dynamic inertia weight as a function of the iteration time and the function is defined as

$$
w(k) = \begin{cases} 1 \times \dfrac{t}{MaxNum} + 0.25, & 0 \le \dfrac{t}{MaxNum} \le 0.5 \\[4mm] -1 \times \dfrac{t}{MaxNum} + 1.25, & 0.5 < \dfrac{t}{MaxNum} \le 1 \end{cases} \tag{2}
$$

where k is the current iteration. MaxMum is the maximum iteration time.

In the algorithm learning process, the fitness of each particle has showed the weakening "choice" behavior, [5] introduced experienced particles into the speed evolution equation to adjust the individual extreme and the global extreme, so as to improve the algorithm convergence speed and accuracy. The updated formulas of the individual extreme (3) and the global extreme (4) are defined as

$$
Pb_i'(t) = \begin{cases} Pb_i(t), & i < 2 \\ r_1 \times Pb_i(t) + r_2 \times Pb_m(t) + r_3 \times Pb_n(t), & i \ge 2 \end{cases} \tag{3}
$$

$$
Pgb'(t) = r_1 \times Pb_1(t) + r_2 \times Pb_2(t) + r_3 \times Pb_3(t) \tag{4}
$$

where Pb_i is the current individual extreme value. Pb_m and Pb_n are the experienced individual extreme values; they are randomly selected from previous ones in the same generation. Pb_i' is the updated individual extreme value. Pgb' is the updated global extreme value. Pb_1, Pb_2 and Pb_3 are three best individual extremes from the same generation. r_1, r_2 and r_3 are random values between -0.5 and 1.5, and $r_3 = 1 - r_1 - r_2$.

4   Simulation Design and Analysis

The simulation experiment selects 13 groups of Wuhan between May 1, 2016 and May 13, 2016 as the samples; they come from the China air quality on-line monitoring and analysis platform (http://www.aqistudy.cn/). PM2.5, PM10, CO, NO_2, O_3 and SO_2 are the network inputs, AQI is the target data. The first 12 groups are training samples, NO. 13 data is a test sample. Through experimental comparison, the network structure of improved PSO-BP model is 6-8-1; the population size is 20; the particle dimension is 65. The initial position component is a random number between −1 and 1. The current velocity component is a random number between −0.5 and 0.5. The maximum iterations number is 2000 and the minimum error is 0.001.

Firstly, the algorithm convergence analysis and the network output curve analysis are shown as Fig. 1(a) and (b).

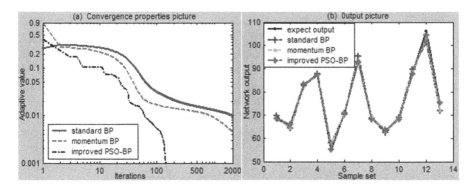

Fig. 1. (a) Convergence properties picture; (b) Output picture

By Fig. 1(a) and (b), the convergence speed of the standard BP algorithm and the momentum BP algorithm are slow in the late training; they can't reach the minimum convergence precision; the predicted values have the obvious deviation. The improved PSO-BP algorithm keeps a good convergence rate, it can reach the minimum error at 143 iterations; Its deviation values only show in No. 8, No. 9, No. 10 and No. 12. So the improved algorithm has the better effect of convergence and forecast than others.

Secondly, the relative error value comparison of three algorithms is shown in Table 1.

In Table 1, the deviation value of the improved PSO-BP algorithm in No. 8, No. 9, No. 10 and No. 12 is between 1.13 and 1.45, other values are all smaller than 0.9. So the algorithm doesn't have large deviation value. Its Ave_Error value is only 0.64%, which is smaller than other two algorithms. So the improved PSO-BP algorithm in this paper is obviously better than others.

Table 1.

NO.	Expected output	Standard BP		Momentum BP		Improved PSO-BP	
		Predicted value	Error (%)	Predicted value	Error (%)	Predicted value	Error (%)
1	68	69.801345	2.65	68.402343	0.59	68.568022	0.84
2	66	64.478980	2.30	64.435718	2.37	65.873440	0.19
3	83	82.555808	0.54	82.613787	0.47	83.318232	0.38
4	87	87.602480	0.69	87.068520	0.08	87.304945	0.35
5	55	58.106085	5.65	58.013594	5.48	55.155547	0.28
6	71	70.300322	0.99	70.347997	0.92	70.951919	0.07
7	93	95.401679	2.58	93.363184	0.39	92.745650	0.27
8	69	68.808340	0.28	69.869844	1.26	68.178084	1.19
9	63	62.557808	0.70	62.927736	0.11	63.714685	1.13
10	69	69.101081	0.15	68.885957	0.17	68.165413	1.21
11	88	89.685779	1.92	88.707819	0.80	87.670740	0.37
12	106	101.425017	4.32	103.009014	2.82	104.462997	1.45
13	75	71.917179	4.11	72.021966	3.97	75.448108	0.60
Ave_Error (%)		2.07		1.49		0.64	

5 Conclusion

In this paper, the dynamic inertia weight and experience particles are used in the improved PSO-BP algorithm to optimize network weights and thresholds. This integration method makes full use of the neural network learning ability and the global optimization of PSO algorithm. It provides a new method for predicting AQI. In the future, this paper will start from the nonlinear adjustment method of inertia weight, and further improve the global optimization of PSO algorithm.

References

1. Zhang, Y., Xiao, D., Zhao, Y.: A study of meteorological prediction with neural network based on time series. J. Wuhan Univ. Technol. **27**(2), 237–240 (2003)
2. Li, X.: Air quality forecasting based on GAB and fuzzy BP neural network. J. Hua zhong Univ. Sci. Technol. **41**(supp I), 63–65 (2013)
3. Li, Z., Zhou, B., Lin, N.: Classification of daily load characteristics curve and forecasting of short-term load based on fuzzy clustering and improved BP algorithm. Power Syst. Prot. Control **3**, 56–60 (2012)
4. Zhang, D., Han, S., Li, J., Nie, S.: BP algorithm based on improved particle swarm optimization. Comput. Simul. **2**, 147–150 (2011)
5. Jia, X., Yan, Y., Rui, Z.: Graduate enrollment prediction by an error back propagation algorithm based on the multi-experiential particle swarm optimization. In: 11th International Conference on Natural Computation (ICNC 2015), Zhangjiajie, China, pp. 1163–1168 (2015)
6. Lu, Y., Tang, D., Hao, X.: Productivity matching and quantitative prediction of coalbed methane wells based on BP neural network. Sci. China (Technol. Sci.) **54**(5), 1281–1286 (2011)

7. Hornik, K., Stinchcombe, M., White, H.: Multilayer feedforward networks are universal approximators. Neural Netw. **2**(5), 359–366 (1989)
8. Kennedy, J., Eberhart, R.: Particle swarm optimization. In: IEEE International Conference on Neural Networks, pp. 1942–1948 (1995)
9. Eberhart, R., Kennedy, J.: A new optimizer using particle swarm theory. In: 6th International Symposium on Micro Machine and Human Science, pp. 39–43 (1995)
10. Shi, Y., Eberhart, R.C.: A modified particle swarm optimizer. In: Proceedings of the IEEE Conference on Evolutionary Computation, Piscataway, NJ, pp. 69–73 (1998)
11. Tian, Y., Zhu, R., Xue, Q.: Research advances on inertia weight in particle swam optimization. Comput. Eng. Appl. **44**(23), 39–41 (2008)
12. Shi, Y., Eberhart, R.: Empirical study of particle swarm optimization. In: Proceedings of the 1999 Congress on Evolutionary Computation, pp. 1945–1950 (1999)
13. Fujimoto, R., Perumalla, K., Park, A.: Large-scale network simulation: how big? How fast? In: Proceedings of the 11th IEEE/ACM International Symposium on Modeling, Analysis and Simulation of Computer Telecommunications Systems, Orlando, Florida (2003)
14. Cui, H., Zhu, Q.: Convergence analysis and parameter selection in particle swarm optimization. Comput. Eng. Appl. **43**(23), 89–91 (2007)
15. Hu, J., Xu, J., Wang, J, Xu, T.: Research on particle swarm optimization with dynamic inertia weight. In: International Conference on Management and Service Science, Beijing, China (2009)

Picture Reconstruction Optimization Using Neural Networks

Yiheng Hu[⊠], Zhihua Hu, and Jin Chen

Shanghai Polytechnic University, Shanghai, China
yhhu@sspu.edu.cn

Abstract. The multilayer perceptron (MLP), as one of neural network types, is a function of one or more predictors which minimizes the error between the inputs and target variables. In this paper, the network architecture is designed and the optimal parameters are chosen. A novel method is proposed to add polynomial features to help get better results on the accuracy of reconstruction picture. The number of epochs for training is also an important fact for time consuming which has strong relationship with the avoid overtraining algorithm. The accuracy of the reconstruction work and the size of time consuming data are examined by experimental work.

Keywords: Neural network · Picture reconstruction · Multilayer perceptron · Optimization

1   Introduction

An MLP can be viewed as a logistic regression classifier which maps sets of input data into a set of appropriate outputs using a learnt non-linear transformation. This transformation projects the input data into hidden layers. A single hidden layer is sufficient to make MLPs universal, although there are substantial benefits to using many such hidden layers, i.e. the very premise of deep learning. The hidden layers compute the activations and then pass them the output layer. This type of network is trained with the back-propagation (BP) learning algorithm. BP is popular for incremental learning and it has been proven as the most common approximation technique. The BP neural network is a gradient descent algorithm by minimizing the error signal between actual output and the desire value [1].

This paper focuses on using BP algorithm to build a MLP classifier to reconstruct a 2-colored picture, as shown in Fig. 1. Firstly, the network architecture is designed and the optimal parameters are chosen. Then, according to the effectiveness and efficiency of engineering consideration, the optimal parameters are chosen, including: optimum sampling strategy, optimum method on avoid overtraining and optimal methods on increasing learning rate. From the whole experimental process, the results including the learning curve of the training process and the tabulation of the computed weights are obtained for analysis. The comparison and analysis of the results shows that this neural network is properly tuned to achieve a perfect compromise between effectiveness and efficiency.

© Springer International Publishing AG 2018

F. Qiao et al. (eds.), *Recent Developments in Mechatronics and Intelligent Robotics*,
Advances in Intelligent Systems and Computing 690, DOI 10.1007/978-3-319-65978-7_3

Fig. 1. Double-color cover.

2   Initial Architecture Design of the Network

There is a trade-off between effectiveness of the training network and the efficiency of the computational program. In order to explore the optimal method to solve the problem, a series of experiments are studied to achieve the optimum parameters.

Firstly, the choice of the number of hidden layers depend on the applications. Actually, for most problems, one hidden layer is sufficient. It is possible to introduce a great risk to coverage to the local minimum by using the two hidden layers to improve the model. So a preliminary three layer model with one hidden layer is built.

Then, the amount of neurons in inside layer is set by the type of pattern; for this project, the system must have 2 inputs because there are four pixels in the pattern. Similarly, the size of the output layer is determined by the amount of patterns which can be recognized; for this project, 1 output neurons are set. However, there are no fast rules for the amount selections of the hidden neuron and the network normally works well within a range of these variables. For this project, 21 hidden neurons are temporarily set and then it will be experimented for the best results. The weights of the network are set to random number from 0–1. The learning rate is initially set to 0.5. Before dealing with the BP algorithms, the input data should also be normalized, that is remove the mean value and divided by the maximum absolute value.

3   Optimal Sampling Strategy

In this section, we want to find out the best sampling strategy, which means that it has the best effectiveness and efficiency. Here we suppose that we are using the same neural network. We use difference strategies but sample the same size of data, that is 10000 points. Figure 2 illustrates the sampling results. In this project, the pixels around the edges and corners are much more important than other pixels. It can be noted, after about 250 times, the algorithm can be converged. Considering the reconstruction accuracy, exposure distance training data distribution is better than the other methods. One more thing we should take a focus is that although the uniform sampling strategy of MSE achieves quite low (about 0.02), the reconstruction accuracy is not good enough, which proves that MSE is not a good choice for this project (Table 1).

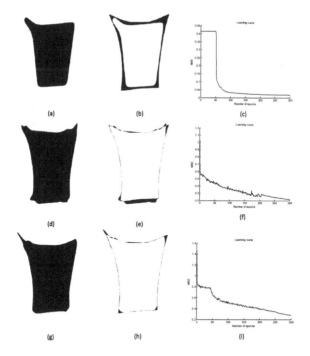

Fig. 2. Results for three sampling methods. (a) (d) (g) are reconstruction results, (b) (e) (h) are reconstruction error results. (c) (f) (i) are learning curves.

Table 1. Results of different sampling strategies.

Training data distribution	Data size	Reconstruction accuracy
Uniform	10000	0.9219
Linear	10000	0.9732
Exposure	10000	0.9894

4   Optimal Methods on Avoid Overtraining

An effective way of avoiding overtraining is called early-stopping method of training. As follows,

(1) The training data are divided into training set, validation set and test set. (in our project it is a **six-to-two-to-two** proportion.)
(2) Training on the training set only, and evaluating each sample error on the validation set. (in our project it is after every fifth epoch).
(3) Once the validation set error is higher than the previous check error, stop training.
(4) Using the weights of the network in the previous step as a result of the training run.

Unfortunately, this simple procedure is complex in practice, and the validation error may fluctuate during training. As shown in Fig. 3, the validation set error is not evolved smoothly enough.

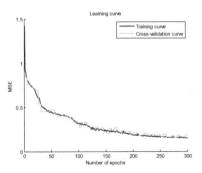

Fig. 3. Training with cross-validation.

It can be noticed that a proper stopping criterion is needed. Actually, we have lots of ways to avoid fluctuating in practice [2], like GL, UP, PQ, etc. Here we choose PQ, which is more robust and effective. The result is shown in Fig. 4.

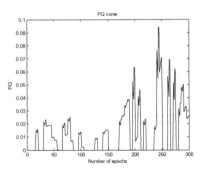

Fig. 4. Curve of PQ method.

It can noticed that, if we choose the constant to 0.08, the algorithm could stop at reasonable epoch. By using this PQ cross-validation method, the results are illustrated in Table 2.

Using cross-validation of PQ method, we can get relatively stable results to avoid over training. Although sometimes, like Run number 6 and 9, the epochs to stop is a little bit earlier that the reconstruction accuracy is not good enough comparing with other methods. We can add another condition for when to stop that the number of epochs should over 200.

Table 2. Results of cross-validation.

No. of run	No. of epochs	Test accuracy	Reconstruction accuracy
1	248	0.9430	0.9893
2	259	0.9271	0.9801
3	281	0.9610	0.9919
4	289	0.9518	0.9901
5	271	0.9405	0.9901
6	210	0.9401	0.9701
7	278	0.9523	0.9912
8	291	0.9491	0.9891
9	197	0.9312	0.9623
10	267	0.9563	0.9856

5 Optimal Methods on Increasing Learning Rate

5.1 Learning-Rate Annealing Schedules

We test the algorithm using different value of (100, 70, 50, 30, 5, 1). The Fig. 5(a) shows the curve of when take different values. The Fig. 5(b) shows the results of learning process.

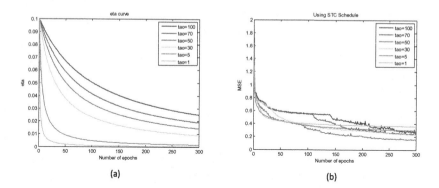

(a)   (b)

Fig. 5. Results of search-then-converge-schedule.

From the results, we find that when, the learning curve could convergence in shorter time and the MSE could be more lower.

5.2 Momentum Constant

Another simple method is shown as follows,

$$\Delta w_{ji}(n) = \eta \delta_j(n) y_i(n) + \alpha \cdot \Delta w_{ji}(n-1)$$

this is a positive number and it is called the momentum constant. In this experiment, different momentum constant $\alpha \in \{0, 0.1, 0.3, 0.5, 0.9\}$ are used to observe their effect on network convergence and accuracy. Simulated learning curves are obtained, as shown in Fig. 6. Apparently, 0.5 is the best choice for momentum constant.

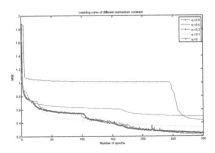

Fig. 6. Results of search-then-converge-schedule.

6   Conclusion

In this paper, pictures using neural network is reconstructed. In order to achieve the highest accuracy, we carefully chosen the optimal parameters of the neural network. (1) Optimum sampling strategy: exposure distance sampling; (2) Optimum method on avoid overtraining: Cross-validation with PQ methods, parameter in PQ is chosen 0.08; (3) Optimum methods on increasing learning rate: using STC schedule and momentum constant to increase learning rate with parameter chosen.

Acknowledgments. I would like to acknowledge the funding support of Control Theory and Control Engineering Discipline XXXPY1609 and the funding is A20NH1609B21-92.

References

1. Tuli, R.: Character recognition in neural networks using back propagation method. In: IACC IEEE 3rd International, pp. 593–599 (2013)
2. Prechelt, L., Orr, G.B.: Neural Networks: Tricks of the Trade. LNCS, pp. 53–67. Springer, Berlin. ISBN: 978-3-642-35289-8. Accessed 15 Dec 2013

An Architecture Design for Fighting Method Decision Based on Course of Action

Zhu Jiang[⊠], Wen ChuanHua, Xu Xin Wei, and Du Wei

NanJing Command College, Nanjing, China
`pearlriver_1981@163.com`

Abstract. Fighting method decision is becoming more and more intricate than ever in military area. Traditional problem-solving architectures and related methods, led to inaccuracy and hard to control, were out of time then. In this paper, a design of architecture including some military models, operating processes and mathematics expression methods, is proposed for choosing fighting methods. COA (Course of Action), a semantic bridge between fighting methods and automatic modeling bricks, plays an indispensable rule. A lot of emphases are mainly put on the formalization and semantic interpretation of COA so as to benefit designing and evaluating the fighting methods. Based on COA, a regular operating process is described, and correlative mathematics expressions are formulated. These technologies can be embodied in the fighting method design, to acquire precise effect and promote automatic degree of testing system. It can be widely used in platforms and agents in C2 (command and control) headquarter.

Keywords: War system · COA · Decision process · Semantic

1 Introduction

Choosing fighting methods is a dynamic decision problem. As Sun Tzu said "the solider works out his victory in relation to the foe when he is facing" [1], fighting methods are highly depend on several factors such as natural surroundings, operation goals, combat styles, weapon equipment, solider quality, battle situation and space-time condition.

In traditional qualitative analysis, there are lots of ambiguity and uncertainty in natural oral expressions. So, it can not be directly translated into a machine-readable knowledge, led to some handicaps of utilizing auto-generated forces. How to combine machine with human intelligence for perusing C2 (command and control) precision and accuracy [2] is an important issue.

Quantitative analyses overtake and incorporate the qualitative one. Firstly, formalization technologies are applied in standardizing inconsistent oral meanings. Secondly, decision problems involved should be formulated by quantitative norm. Thirdly, processes should be well-defined and regulated in a certain phase.

This paper propose an architecture. In Sect. 2, COA, acting as a bridge of commanders and engineers, is explained by a formal graph and semantic schema. In Sect. 3, a COA based design process and a two-level blueprint is put forward naturally.

© Springer International Publishing AG 2018
F. Qiao et al. (eds.), *Recent Developments in Mechatronics and Intelligent Robotics*,
Advances in Intelligent Systems and Computing 690, DOI 10.1007/978-3-319-65978-7_4

In Sect. 4, mathematic meanings and formulas are given for further digitization. In Sect. 5, experiment platform is designed.

2   A COA Semantic Model

COA concepts are brought here to represent action sequence mapping with a specific fighting method. Since COA is a refining knowledge of "5h" (when, where, who, why and how) behind an action, US army thinks highly of COA and glad to regulate decision by COA [3].

As Fig. 1 shows, the responsibility of military headquarters are developing COA, getting a feasible plan and perusing the peak of the efficiency.

Fig. 1. Decision process

Semantic technologies are utilized to express the implicit knowledge. Ontology provides an useful way to describe objects and their relationships [5], and OWL, provides rich sets to create ontology and markup information.

Figure 2 is an OWL graph which is machine readable and understandable [4]. With norms like "Relation, Act, Object, Task, Entity", an example of COA is shown in Fig. 3. In this graph, nodes represent tasks or acts, and links represent relations.

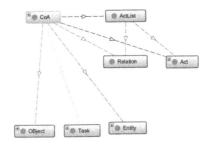

Fig. 2. COA semantics illustrated by OWL

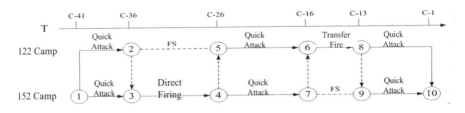

Fig. 3. An example of COA

Illustrated process is divided into several phases which are marked by time label such as C-41. Action sequence will be unfolded from the beginning to the end, and actions are well arranged in the timeline. Once been well defined and explained in programmatic and semantic domain, the military concept model is constructed easily.

3   A COA-Based Design Process

As Fig. 4 shows, accompanied with the military decision process, Fighting method decision not only contains designing basic fighting method (initial COA) in advance, but also contains designing of specific fighting methods (Optional COAs) according to conditions. Each divergence at decision points may lead to diversified COAs. In war game platform, COA, composed of key nodes, is the base line run through the top task tree to bottom action models. So, COA is the blueprint for event-driven war-gaming [6] and it stimulate changing states.

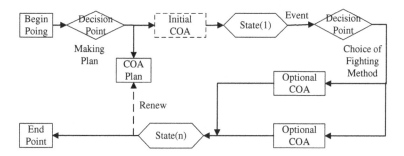

Fig. 4. Process of designing fighting method

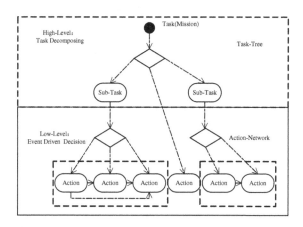

Fig. 5. The effect of the designing process

The effect of COA design is briefly explained in Fig. 5. In high level, task decomposing is done to create a task-tree (composed of more fine-grained sub-tasks at different levels). In low level, much concentration is put on action models. The two levels are merged together as a quality-to-quantity bridge.

4   Mathematic Expression of Decision Problem

Designing COAs is also an optimization problem from military operation perspective. Describing mathematically is the prerequisite for solve such a problem, for mathematic form provides algorithms for resource availability judgment and corruption test.

(1)   Mapping task to COA.

$$\text{Task} \xrightarrow{h_1} Sub\text{Task} \xrightarrow{h_2} COA, \quad h = h_1 * h_2 \tag{1}$$

h_1 is a division of task. h_2 is a mapping from Subtask to COA. $h_2 = C \times T \times P$, C is the capability constraints, T is time constraints, P is space constraints.

(2)   Expression of COA.

$$COA = (SEQ, Action^+) | (PAR, Action^+) | (OPT, Action^+) \tag{2}$$

SEQ (Sequential), PAR (Parallel), OPT (Optional)

(3)   State space of COA

$$\Omega = (A(t), G(A)) \tag{3}$$

Ω is the state space of COAs, composed of process sequence A(t) and datagram G(A). Supposing the combat task can be divided into several phases, $k = 1, 2, \cdots, K$, and the action in the phase is A_k, $k = 1, 2, \cdots, K$. So, $G(A_k)$ is the process datagram in phase k.

$\Omega_{\max} = A_1 \times A_2 \times \ldots \times A_k = \prod\limits_{i=1}^{k} A_i$ is the probably maximum value of Ω.

(4)   Benefit of COA

$$COA \xrightarrow{ER_1} Benefit \quad ER = \left\{ \begin{bmatrix} L_{i,1}^t \\ v_{i,1}^t \end{bmatrix} \begin{bmatrix} L_{i,2}^t \\ v_{i,2}^t \end{bmatrix} \cdots \begin{bmatrix} L_{i,j}^t \\ v_{i,j}^t \end{bmatrix} \cdots \begin{bmatrix} L_{i,n}^t \\ v_{i,n}^t \end{bmatrix} \right\} \tag{4}$$

i – unit with the order label i, $i \in ER$;

j – action with the lable j, $j \in COA$;

$L_{i,j}^t$ – in phase t, the capability of i to fulfill j action task.

$v_{i,j}^t$ – the capability level of $L_{i,j}^t$.

In different phases, performing actions need respective capabilities of entity. *Capability* $(G(A_k))$ is the resource requirement of actions, whose value can't surpass the local restraint or instant restraint of capability resource limit. Evaluating COA depend on what degree the fighting capabilities can be utilized. Decision point Ω^*, $(A^\bullet, G^*) \in \Omega$ is chosen with the best benefit.

5 Conclusion

After the architecture applied into COA war game testing bed [7], it proved to had several advantages:

(1) Practical. The complexity of fighting is simplified by layering and decomposing. When COA needs to be refined, the experimenter could adjust from different levels, and merely tackle at local level to limit effluence.
(2) Automatic. Interactions between elements of fighting were described with strict logic rules and translated into a mathematics format, that is good for constructing intelligent human interference and use machine to aid the headquarter.
(3) Intuitive. This work benefits grasping the main course of war, and untangle the complex relations between acts from a systematic view. It benefits supervising the element states in a war, intuitively reflecting the combat process. It can enhance the understanding of the fighting method.

Acknowledgements. This work is supported by Natural Science Foundation of China (71401177) and Postdoctoral fund (2016m603049).

References

1. Sun, W.: Master Sun's Art of War. Military Press, China
2. Saikia, A.V.M.A., VM, A.V.S.M.: The OODA cycle, netcentricity and effect based operation. Defence Manag. **18**(3), 1–7 (2008)
3. Vakas, D., et al.: Commander behavior and course of action selection in JWARS. In: Proceedings of the 10th CGF & BR Conference (2001)
4. Wang, D., Shen, R.: Modeling and robustness of knowledge network in supply chain. Trans. Tianjin Univ. **20**, 151–156 (2014)
5. Garstka, J.: Network centric operations conceptual framework version1.0 [EB/OL]. http://www.oft.osd.mil/library/library files/document 353, NCOCF Version1.0(FINAL).doc DDDAS Workshop Report. NSF. Sponsored Workshop on. DDDAS-Dynamic Data Driven Applications. Systems [EB/OL]. http://www.cise.nsf.gov/dddas. Accessed 18 Sept 2007
6. Zhu, J., Du, W., Liu, D.W.: Hyper-network multi agent model for military system and its use case. In: 2014 IEEE 7th Joint International Information Technology and Artificial Intelligence Conference (ITAIC 2014), 2014-12, Chongqing, China, pp. 358–363 (2014)
7. Michael, B.: JBOTS, Crocadile & TDSS: 3 conflict driven multi-agent system for education, experimentation, & decision support. In: The Australian Conference on Artificial Life ACAL2003, Canberra, pp. 1–12, December 2003

Trajectory Generation and Optimization of Slider Crank for Servo Press Based on Bezier Curve Model and Genetic Algorithm

Juxin Qu[✉]

School of Mechanical and Automotive Engineering, South China University of Technology, Guangzhou, China
qujuxin@126.com

Abstract. An optimal method for curve generation and optimization of the slider motion in slide-crank servo press is proposed and studied. The advantage of this method is the combination of Bezier curve model and genetic algorithm. Firstly, the input motion characteristics are defined by Bezier curves. Then the mathematical model based on the servo press motion is established. Afterwards, the output motion characteristics with different requirements are described. The motion curves are classified and summarized according to different constraint functions. At last, the simulation and experimental results are compared. Results show the design method meets the process requirements of the servo press successfully.

Keywords: Slider-crank mechanism · Motion trajectory · Bezier curve · Simulink

1 Introduction

Servo press is one kind of important equipment in metal forging and sheet stamping. There is no doubt that servo press possesses many advantages, for example, high precision, high adaptability and low noise. Slider-crank mechanism is one common driven type of servo press. With properly controlled servo motors, the speed and the stroke length of the slider can be adjusted on the basis of the technological requirements of the part processing. So servo press can perform different forming tasks such as deep drawing, flanging, and blanking tasks like punching, in which the slider moves according to a specified speed curve [1].

The obvious advantage of servo press is variable speed drive of the crank and the slider output motion characteristic can be performed in accordance with a desired trajectory. Most research has been done in the variable input speed mechanism. In the early stages, it has been addressed to cams [2–4] which are the simplest mechanisms used in driving systems. Later, linkages mechanisms, which consist of a number of links, were gained more and more focus and research. Professor Yan suggested a new method to acquire the required slider output motion by adjusting the speed of the crank basing on

© Springer International Publishing AG 2018
F. Qiao et al. (eds.), *Recent Developments in Mechatronics and Intelligent Robotics*,
Advances in Intelligent Systems and Computing 690, DOI 10.1007/978-3-319-65978-7_5

a rigid slider-crank mechanism [5]. More research was also about six-bar Watt-mechanism [6], four-bar linkages mechanism [7], and the Stevenson-type mechanism [8]. Their targets mainly focused on several aspects such as meeting the process requirement, minimizing the peak acceleration and reducing vibration and energy consumption.

In order to obtain the ideal motion curve, more and more research and design has been done on curve models and different optimization algorithms, for example, spline curve with polynomial curve fitting [2, 9], non-uniform B-spline with global optimum control point algorithm [10], and Bezier curves with recursive quadratic programming [8, 11]. Genetic Algorithms (GA) are commonly used to search problems and find the optimal solution [12, 13]. By combining the GA optimization method and Bezier curve model, the motion curves of the crank can be designed and optimized.

In this article, a new curve optimization method for designing motion curve of slider in a slider-crank servo press is proposed. Firstly, section one describes the design models for slider-crank mechanisms. Secondly, in section two, paper shows how the GA technique and the Bezier curve model can be used to obtain the optimal shape of a processing curve, specifically to minimize the peak value of acceleration for the slider. The motion curves are classified and summarized according to different constraint functions, and the displacement, velocity and acceleration are optimized to meet different requirements. Besides, the selection of the optimized variables is also discussed, the most appropriate number of optimization variables is obtained, which has seldom reported in other literatures so far. Then section three gives the experiment testing of the design examples. Finally, section four concludes this work. The results indicate that the obtained slider motion curves can effectively meet the requirements of different forming process by combining Bezier curve model with GA.

2   Bezier Curve Model of Slider-Crank Servo Press

In this research, the Bezier function is applied to directly describe the crank rotation angle as Eq. (1). Where θ_0, θ_1, ..., θ_i are the angular displacement representing the locations of the control points, $B_{i,n}(t)$ is Bernstein basis function, and t is the normalized time. Hence, the expressions of the crank velocity and acceleration trajectory can be derived in turn by differentiations as (3) and (4).

$$\theta(t) = \sum_{i=0}^{n} \theta_i B_{i,n}(t), \quad 0 \le t \le 1 \tag{1}$$

$$B_{i,n}(t) = \binom{n}{i} t^i (1-t)^{n-i}, \quad i = 0, 1, 2 \dots n \tag{2}$$

$$\omega(t) = n \sum_{i=0}^{n-1} (\theta_{i+1} - \theta_i) B_{i,n-1}(t), \quad 0 \le t \le 1 \tag{3}$$

$$\alpha(t) = n(n-1) \sum_{i=0}^{n-2} \left(\theta_{i+2} - 2\theta_{i+1} + \theta_i\right)B_{i,n-2}(t), \quad 0 \leq t \leq 1 \tag{4}$$

The relationship between actual time t_{ac} and t is expressed as $t = \frac{t_{ac}}{\tau}$, where τ is the period time of the crank for a cycle stroke. The actual crank rotation angle, speed, acceleration can be respectively written in terms of the parameter transformation as follows: $\theta_{ac} = \theta(t)$, $\omega_{ac} = \frac{\omega(t)}{\tau}$, $\alpha_{ac} = \frac{\alpha(t)}{\tau^2}$.

A typical crank-slide mechanism system is shown in Fig. 1. The bottom dead point position is defined as ordinate zero point. Locations of the control points $\theta_1, \theta_2, \ldots, \theta_{n-1}$ are the optimization variables, R is the length of the OA, λ is the ratio of radius of crankshaft and connecting rod length. Let θ denote the crank rotation angle at time t when the slide displacement is s. Based on kinematical analysis of slider-crank mechanism, the motion expressions of the slider are obtained as follows:

$$s = R\left[1 + \cos\theta + \frac{\lambda}{4}(1 - \cos2\theta)\right] \tag{5}$$

$$v = \dot{\theta}R\left(-\sin\theta + \frac{\lambda}{2}\sin2\theta\right) \tag{6}$$

$$a = \ddot{\theta}R\left(-\sin\theta + \frac{\lambda}{2}\sin2\theta\right) - \dot{\theta}^2R(\cos\theta - \lambda\cos2\theta) \tag{7}$$

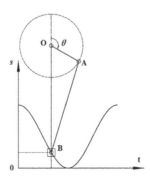

Fig. 1. The slider-crank model for servo press

3   Optimization

Process curve optimization is necessary to minimize energy cost and maximize efficiency. The aim of my work is to optimize the motion curves of slide with a given process requirement within capacity of servo press. The flowchart of the optimization procedure is shown in Fig. 2.

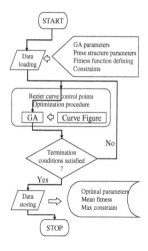

Fig. 2. Flow chart of the optimization procedure

3.1 Fitness Function

The precision of the products and the life of the mold can be reduced by the vibration. The main source of vibration is the inertial force which is related to the acceleration of the slider. The aim of optimization is to minimize the peak value of the acceleration. The expression is as Eq. (8). Here, x is variable array of the optimization procedure. The derived motion curves, Eqs. (1), (3) and (4), still contain undetermined control points $\theta_0, \theta_1, \ldots, \theta_n$. Due to the nature of Bezier curve, the designed curves of slider motion automatically meet the requirements of continuance except the points $t = 0$ and $t = 1$. So the boundary conditions of $\theta(t)$, $\omega(t)$ and $\alpha(t)$ must be defined specially. Based on the analysis above, the $\theta_0, \theta_{n-2}, \theta_{n-1}$ and θ_n can be derived from the Eq. (9). So, $(n - 3)$ variables need to optimize when there are $(n + 1)$ control points.

$$\underset{x}{\text{Min}} f(x) = \max|a(t)| \tag{8}$$

$$\begin{cases} \theta(0) = \theta(1) \\ \omega(0) = \omega(1) \\ \alpha(0) = \alpha(1) \end{cases} \Rightarrow \begin{cases} \theta_0 = 0 \\ \theta_{n-2} = 360 + \theta_2 - 4\theta_1 \\ \theta_{n-1} = 360 - \theta_1 \\ \theta_n = 360 \end{cases} \tag{9}$$

3.2 Constraint Function

Constraint function has defined the requirements of the output motion trajectory including displacement, velocity and acceleration which need to meet. In the different punching process, the constraint function has the specific expression. Several kinds of constraints are listed as follows.

(1) Required displacement
   In some of the stamping process, the displacement curve of the slider is required to pass through a specified point S_k at time t_k.

$$s(t_k) = S_k, \; k = 1, 2, 3, \ldots \tag{10}$$

(2) Constant output velocity
   In a specified period when the slide moves at a constant velocity, the following equality constraint must be contained. The output velocity v is approximately constant during period from t_a to t_b.

$$v(t_k) = V_a, \; a \le k \le b \tag{11}$$

(3) Restricted velocity
   In the formula, the output velocity v is not allowed to over a required value V during the period from t_m to t_n.

$$\left| v(t_k) \right| \le V, \, m \le k \le n \tag{12}$$

(4) Restricted acceleration
   In this equation, A is a required value and the output acceleration a is not allowed to over A during the period from t_p to t_q.

$$\left| a(t_k) \right| \le A, \, p \le k \le q \tag{13}$$

3.3   Result

3.3.1   Defining Number of Variables

Through analysis and calculation, on the premise of meeting the boundary continuity conditions in Eq. (9), the Bezier curve will become a straight line if the number of variables is zero. This state is the same as constant speed input of crank. The maximum acceleration of slide is 3.919 m/s^2, which is obtained through the theoretical calculation, and the value takes place at the normalized time point $t = 0.5$. In order to obtain the best results, different number of variables is selected for optimizing curve.

The higher order of curve, the greater amount of calculation, and it needs more complicated control system. Figure 3 shows choosing 7 variables for optimization is more appropriate in order to acquire lower acceleration peak because more variables cannot make the peak acceleration obviously decline.

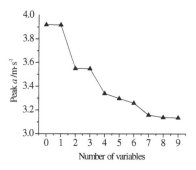

Fig. 3. Peak value of the slider acceleration

3.3.2 Required Displacement

In the stamping process, it usually demands the slider should have definite displacement at prescribed time point. According to the Eq. (10) and based on a certain technological requirements, the constraint of optimization problem can be defined as follows: $s(0.3) = 14$; $s(0.5) = 4.5$; $s(0.6) = 0$. The optimal variables are obtained as vector $x = [47.897, 107.489, 178.605, 305.663, 27.399, 89.848, 283.574]$ by a GA optimization project with programming required displacement constraint function subroutine. The corresponding slide motion characteristics are shown in Fig. 4. The displacement curve can exactly pass through the prescribed time points. The accurate results which come from MATLAB are $s_1 = (0.3, 14.2255)$, $s_2 = (0.5, 4.4659)$, $s_3 = (0.6, 0.0003)$, and the peak of the slider acceleration is 7.298 m/s^2. The value appears at $t = 0.07$, which is located in the initial stage of one stroke.

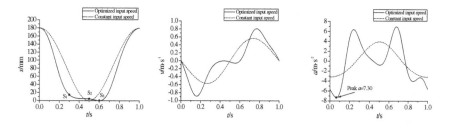

Fig. 4. The motion characteristics of slider - Required displacement

3.3.3 Constant Output Velocity

During forward stroke of a stamping process, it is assumed that constant velocity drawing is given at interval [0.2, 0.3]. According to the Eq. (11) and based on a certain technological requirements, the constraint of optimization problem is defined as $v(t_k) = v_{0.2}$, $0.2 \leq k \leq 0.3$. Through a GA optimization project with programming constant output velocity constraint function subroutine, the optimal variables are obtained as vector $x = [38.969, 84.33, 101.788, 109.035, 147.717, 222.156, 209.717]$. The corresponding slide motion characteristics are shown in Fig. 5. The accurate results at interval [0.2, 0.3] are obtained as follows: $v_{0.2} = -0.4210$, maximum velocity value is -0.4210, minimum velocity value is -0.4317, and the peak of the slider acceleration is 4.284 m/s^2.

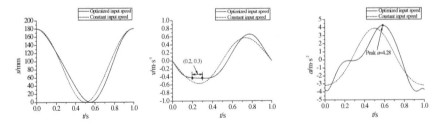

Fig. 5. The motion characteristics of slider - Constant output velocity

3.3.4   Restricted Velocity

The punch speed is very important in the forming process such as drawing and blanking. In a process, it requires that the slide velocity is smaller than a certain value in the processing interval. According to the Eq. (12), the constraint of optimization problem can be defined as follows: $|v(t_k)| \leq 0.15, 0.4 \leq k \leq 0.6$.

The optimal variables are obtained as vector $x = [46.808, 101.41, 105.324, 247.328, 150.650, 113.815, 191.177]$. The corresponding slide motion curves are presented in Fig. 6. The maximum velocity value is -0.1331 at interval [0.4, 0.6], and the peak of the slider acceleration is 5.991 m/s^2.

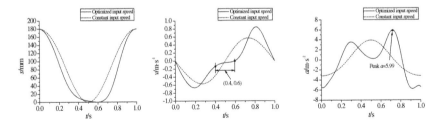

Fig. 6. The motion characteristics of slider- Restricted output velocity

3.3.5   Restricted Acceleration

Sometimes the process requires restricted slide acceleration in a specified interval. According to the Eq. (13), the constraint of optimization problem can be defined as follows: $|a(t_k)| \leq 0.4, 0.5 \leq k \leq 0.6$. Through a GA optimization project with programming restricted acceleration constraint function subroutine, the optimal variables are obtained as vector $x = [51.325, 134.171, 50.58, 236.489, 247.273, 108.046, 110.296]$. The corresponding slide motion characteristics are shown in Fig. 7. The maximum acceleration value is 0.3746 at interval [0.5, 0.6], and the peak of the slider acceleration is 7.802 m/s^2.

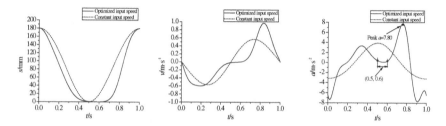

Fig. 7. The motion characteristics of slider - Restricted output acceleration

3.3.6 Multi-constraint Optimization

In practical production, the process requirements are usually multi-constraint and the slide displacement and velocity are restricted by time. According to the above discussion, the constraint of optimization problem can be defined as follows: $s(0.35) = 20$; $s(0.6) = 0$; $v(t_k) = v0.5, 0.5 \leq k \leq 0.55$.

The optimal variables are obtained as vector x = [37.276, 67.697, 159.025, 253.232, 52.156, 155.547, 209.816] by a GA optimization project with programming multi-constraint function subroutine. The motion trajectory characteristics of slide and crank are presented in Fig. 8. The result show s_1 = (0.35, 19.6504), s_2 = (0.6, 0.0091), v_a = (0.5, −0.0805), v_b = (0.55, −0.0769), the maximum offset value Δv = 0.0036 at interval [0.5, 0.55], and the peak of the slider acceleration is 6.233 m/s^2. The results indicate that the multi-constraint curve design can be obtained after optimization.

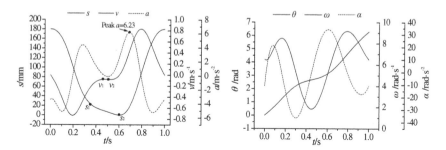

Fig. 8. The motion characteristics of slide and crank- Multi-constraint optimization

4 Experimental Testing

In order to prove the effectiveness of the proposed method, the experimental testing is performed on GPS-110 servo press with a data collection system as shown in Fig. 9. GPS-110 servo press uses DSP and FPGA as the chip of control system and has 'free curve' machining mode. The real-time data acquisition of slider motion is accomplished by an analog output photoelectric displacement sensor. Velocity and acceleration curves of slider are obtained by processing displacement data.

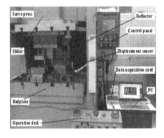

Fig. 9. The slider movement curve testing equipment

Figure 10 show the experimental results of curve optimization examples about different constraint. The testing motion curves satisfy the optimized ones well. The slider displacement is selected as the target in experiment testing and obtained directly, while the velocity and acceleration of slider are acquired after processing data. Therefore, the velocity and acceleration curves appear fluctuations compared with the displacement curves.

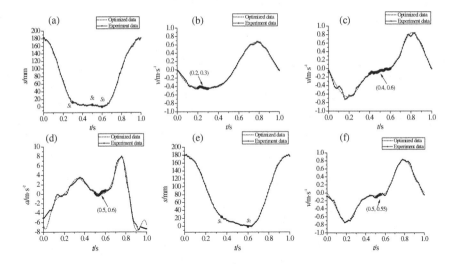

Fig. 10. Comparison of simulations and experiments

The prescribed points in constraint which required displacement are shown in Fig. 10a. Figure 10b and c show the prescribed velocity intervals of constraints which are constant output velocity and restricted velocity respectively. Figure 10d shows the prescribed acceleration interval in constraint which is restricted acceleration. The prescribed points and velocity interval in multi-constraint are shown in Fig. 10e and f.

5   Conclusion

This paper presents a novel concept of process curve design for slider-crank servo press. It is an optimal method which combines Genetic Algorithm with Bezier curve model. Different technological requirements and different constraints are considered. In all situations, the crank and slide motion trajectories are optimized, presented and discussed. Experiments are carried out on the GPS - 110 servo presses. Testing results accord with the numerical stimulation well. The results show that the combination of Bezier curve model and genetic algorithm is an effective curve design method for stamping process. The designed curves can satisfy the process requirements well. More importantly, the method will possess more flexibility and adaptability in production.

References

1. Osakada, K., Mori, K., Altan, T., Groche, P.: Mechanical servo press technology for metal forming. CIRP Ann. Manuf. Technol. **60**(2), 651–672 (2011)
2. Wu, L.L., Chang, W.T., Liu, C.H.: The design of varying-velocity translating cam mechanisms. Mech. Mach. Theory **42**(3), 352–364 (2007)
3. Yao, Y.A., Zhang, C., Yan, H.S.: Motion control of cam mechanisms. Mech. Mach. Theory **35**(4), 593–607 (2000)
4. Sun, C.Q., Ren, A.H., Sun, G.X.: Optimum design of motion curve of cam mechanism with lowest maximum acceleration. Appl. Mech. Mater. **86**, 666–669 (2011)
5. Yan, H.S., Chen, W.R.: On the output motion characteristics of variable input speed servo-controlled slider-crank mechanisms. Mech. Mach. Theory **35**(4), 541–561 (2000)
6. Yan, H.S., Chen, W.R.: A variable input speed approach for improving the output motion characteristics of Watt-type presses. Int. J. Mach. Tools Manuf **40**(5), 675–690 (2000)
7. Yan, H.S., Soong, R.C.: Kinematic and dynamic design of four-bar linkages by links counterweighing with variable input speed. Mech. Mach. Theory **36**(9), 1051–1071 (2001)
8. Yan, H.S., Chen, W.R.: Optimized kinematic properties for stevenson type presses with variable input speed approach. J. Mech. Des. **124**(2), 350–354 (2002)
9. Nguyen, V.T., Kim, D.J.: Flexible cam profile synthesis method using smoothing spline curves. Mech. Mach. Theory **42**(7), 825–838 (2007)
10. Wang, S.H., Hu, J.H., Wang, Y.K.: Electronic cam trajectory generation algorithm based on global optimal control points of non uniform b-spline. Appl. Mech. Mater. **418**, 219–224 (2013)
11. Kharal, A., Saleem, A.: Neural networks based airfoil generation for a given Cp using Bezier–PARSEC parameterization. Aerosp. Sci. Technol. **23**(1), 330–344 (2012)
12. Ziolkowski, M., Gratkowski, S.R.: Genetic algorithm and Bezier curves-based shape optimization of conducting shields for low-frequency magnetic fields. IEEE Trans. Magn. **44**(6), 1086–1089 (2008)
13. Shang, W.F., Zhao, S.D., Shen, Y.J.: A flexible tolerance genetic algorithm for optimal problems with nonlinear equality constraints. Adv. Eng. Inform. **23**(3), 253–264 (2009)

Double Half-Orientation Code and Nonlinear Matching Scheme for Palmprint Recognition

Chenghao Zhang, Weifeng Zhong[(✉)], Chunyu Zhang, and Xi Qin

School of Automation, Harbin University of Science and Technology, Harbin, China
910381219@qq.com

Abstract. The algorithm of palmprint recognition based on direction coding is an important algorithm supplement of identity authentication at this stage. In this paper, a palmprint recognition algorithm based on bi-directional coding and nonlinear matching is proposed. By modifying the original Gabor filter, this paper defines a group of "half-Gabor" filters, and adopts the "half-Gabor" filters so as to extract the double half-orientation code of the palmprint. Experimental results on the three categories of palmprint database show the high precision performance of the method proposed in this paper compared with the prior art method. The double-half Direction Codes are matched by Non-linear Matching Method. Moreover, the method is compared with Hamming distance matching method, which proves the superiority of this method in precision. The results show that the two algorithms are effective and accurate.

Keywords: Palmprint recognition · Nonlinear · Directional feature · Feature extraction

1 Introduction

Plamprint recognition, as a relatively novel and new biological recognition technology [1, 2], is concerned recently. Palmprint is the internal surface of a palm, and has not only the fingerprint features such as detail point, singular point and texture, but also the special authentication characteristics such as mainline and wrinkle. Therefore, the palmprint recognition algorithm has good implementability and strong reliability, etc.

Along with the continuous improvement of prior art, many persons have proposed the orientation coding method. For example, Zhang et al. have [3] researched and developed online palmprint recognition system; the palmprint is extracted by standard 2-D Gabor filterin order to propose different coding methods and obtain satisfactory effect. Zhang and Kong [4] put forward a relatively competitive code approach by extracting the predominating palmprint orientation, wherein a total of six differently-oriented Gabor filters are adopted for the convolution with the image of palmprint. We define the orientation of the maximum filter response as the predominating palmprint orientation.

This paper adopts new type of Double Half-orientation Code (hereinafter referred to as DHOC) method so as to extract the directional characteristics of the palmprint. Moreover, the similarity of palmprint is evaluated by combining with an effective nonlinear angle matching score. Compared with Hamming distance, the nonlinear

F. Qiao et al. (eds.), *Recent Developments in Mechatronics and Intelligent Robotics*,
Advances in Intelligent Systems and Computing 690, DOI 10.1007/978-3-319-65978-7_6

matching scheme is proven to have excellent performance. The experiment mainly focuses on three categories of palmprint databases, and the results indicate that the method proposed in this paper has excellent performance compared with prior art.

2 Double Half-Orientation Code Method

Gabor filter as an effective instrument to extract palmprint orientation, Gabor filter has the 2-D spectral characteristic of texture and 2-D variation characteristic of spatial position. Moreover, it is applicable to extract linear feature of palmprint image [1, 3]. This paper defines half-Gabor filter for palmprint half-orientation. The real component of Gabor filter has the general form as follows:

$$G(x, y, \theta, \mu, \delta) = \frac{1}{2\pi\delta^2} \exp\left\{\frac{x^2 + y^2}{2\delta^2}\right\} \cdot$$
$$\cos(2\pi\mu(x\cos\theta + y\sin\theta)) \tag{1}$$

Where μ is the radial frequency with radian as the unit; is the orientation with radian as the unit; is the standard deviation respectively along x-axis and y-axis; empirical parameters are $\mu = 0.0916$ and $\delta = 5.6179$ [3]. Jang et al. [5] put forward a half-Gabor filter (hereinafter referred to as HGF) so as to strengthen the ridge characteristics of the fingerprint. These characteristics are the dispersion and crossing masks of Gabor filter. Moreover, this paper defines the following continuous half-Gabor filters:

$$\overline{G}(x, y, \theta, \mu, \delta) = \begin{cases} G(x, y, \theta, \mu, \delta), & if(-x\sin\theta + y\cos\theta) \geq -T \\ 0 & else \end{cases} \tag{2}$$

$$\overline{G}(x, y, \theta, \mu, \delta) = \begin{cases} G(x, y, \theta, \mu, \delta), & if(-x\sin\theta + y\cos\theta) \leq T \\ 0 & else \end{cases} \tag{3}$$

Where, T is the size of the half-Gabor filter, and the Gabor filter is 35×35; T is set in the range of 0–17; $\overline{G}$ and $\overline{G}$ are respectively a set of double half-Gabor filters.

In the process of extracting double half-orientation codes, the image of the palmprint convolutes with six double half-Gabor filters orientated by $j\pi/6(j = 0, 1\ldots, 5)$. The directions of $\overrightarrow{G}$ and $\overline{G}$ is set as $j\pi/6(j = 0, 1\ldots, 5)$, respectively. The half-orientation code of the palmprint is calculated as below:

$$\overrightarrow{P}(x, y) = \arg\max_j \overrightarrow{G}_j * I(x, y) \tag{4}$$

$$\overrightarrow{P}(x, y) = \arg\max_j \overline{G}_j * I(x, y) \tag{5}$$

Where I is the palmprint image; "*" represents the discrete convolution; $\overrightarrow{P}$ and $\overline{P}$ are The orientation indexes of the maximum filter responses of $\overrightarrow{G}_j$ and $\overline{G}_j$ are $\overrightarrow{P}$ and $\overline{P}$, respectively $\left(\vec{P}, \overleftarrow{P}\right)$ is called as the double half-orientation code of the image of the palmprint, and expressed as $P = \left(\vec{P}, \overleftarrow{P}\right)$.

T directly defines the size of the half-Gabor filter indirectly. If we define the size of Gabor filter as and T is set in the range of 0~[N/2], the corresponding half-Gabor filter would have the size in the range of [N/2]~N. The half-Gabor filter is utilized to extract the half-orientation of the palmprint image, which is related to T. If T is smaller, the half-Gabor filter can extract the image of the half-orientation palmprint more accurately. When T is smaller, the palmprint image can only convolute with partial Gabor filter so as to extract the directional feature. Consequently, the extracted half-orientation is both sensitive and instable. Based on the comparison, the noise robustness and stability of the half-orientation which is extracted by the half-Gabor filter with large T is stronger. However, the half-orientation extracted thereby may be deviated from the correct half-orientation characteristic of palmprint. The optimal T should be a certain orientation between 0 and [N/2]. Additionally, T is relevant with the palmprint image resolution. In most cases, T is set empirically. Empirically, this paper set T as 2 (unless otherwise additionally specified).

3   Double Half-Orientation Nonlinear Matching

Hamming distance is generally utilized to determine the similarity between two images of palmprint in the orientation coding method [7]. For instance, Hamming distance is utilized in Fusion code, Palm code and Ordinal code methods in matching stage. The competitive code method put forward the angle distance to recognize the palmpring, and it is equal to the sum of the three bit-by-bit Hamming distance. Guo et al. [6] put forward the uniform Hamming distance measurement formula. If the corresponding bits are the same, the Hamming distance is calculated as 0. Otherwise, it is calculated as 1. In the case that the corresponding bits of the two sequences are not the same, they are two different bits. The final matching result is the sum of Hamming results of the binary codes. Therefore, Hamming distance is linear digitally. In order to enhance the discriminability, we propose a nonlinear angle matching score method so as to evaluate the double half-orientation similarity. In the matching stage, the matching score of the similar orientation codes is high. The matching score is enough small in the case that the orientation code difference is the maximum. This paper defines the nonlinear matching method as follows:

$$ori\_score(code\_dis) = \frac{1}{e^{k*code\_dis}} \tag{6}$$

$$code\_dis = \min((|O_d - O_t|, n_\theta - |O_d - O_t|) \tag{7}$$

Where O_d and O_t are two "unidirectional codes", and k is a parameter. When the two "unidirectional codes" are the same (code distance $code\_dis = 0$), the perfect matching score is 1. In other words, when the distance between the two "unidirectional codes" reaches the maximum $n_\theta/2$, ori_score should be less than ξ.

$$\frac{1}{e^{k*(n_\theta/2)}} < \xi, \text{ thus } k > \frac{2}{n_\theta} \ln \frac{1}{\xi} \tag{8}$$

Where ξ is empirically set as 0.01; when $n_\theta = 6$, k is equal to $k = 1.6$; when $n_\theta = 12$, $k = 1$is acceptable; Fig. 1(a) shows the nonlinear matching scores of code distances under $n_\theta = 12$ and $k = 1$ respectively. In comparison, Fig. 1(b) shows the bit-dependent change of Hamming distance.

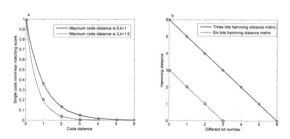

Fig. 1. (a) Nonlinear matching (b) Hanmming distance

In order to calculate the matching score of double half-orientation, we define two crossing matching scores on the basis of the code difference as follows:

$$p_1\_score(i,j) = ori\_score\left(code\_dis_{pp}\right) \\ + ori\_score\left(code\_dis_{ss}\right) \tag{9}$$

$$p_2\_score(i,j) = ori\_score\left(code\_dis_{ps}\right) \\ + ori\_score\left(code\_dis_{sp}\right) \tag{10}$$

Where O_p^i and O_s^i are represented as (O_p, O_s) of DHOC extracted from palmprint image i; $code\_dis_{\alpha\beta}$ is the code distance between the pixels of palmprint images i and j; $code\_dis_{\alpha\beta}$ is in the range of $\{0, 1, \dots, n_\theta/2\}$. The larger one of p_1_score and p_2_score is the final matching score of the two DHOCs.

$$p\_score(i,j) = \max(p_1\_score(i,j), \\ p_2\_score(i,j)) \tag{11}$$

Table 1 shows the corresponding DHOC based horizontal nonlinear matching scores. p_1_score and p_2_score are calculated according to ori_score obtained by DHOCs of the two pixels. The final matching score is calculated through normalizing the sum of the maximum values of p_1_score and p_2_score obtained by Formula (11). If the two DHOCs are the same, 1 is the perfect matching score. If the two DHOCs only have the same single sub-orientation codes, the ultimate matching score is greater than 0.5. Otherwise (the two DHOCs are absolutely not the same), the ultimate matching score is smaller than or equal to 0.2020. The ultimate matching score of the two images of the palmprint is calculated as follows:

$$matching\_score(A, B) = \frac{\sum\limits_{i=1}^{M} \sum\limits_{j=1}^{N} p\_score(i,j)}{2MN} \quad (12)$$

Table 1. Nonlinear matching score

DHOC distance	0	1	2	3
0	1	0.6011	0.5203	0.5042
1	0.6011	0.2020	0.1213	0.1052
2	0.5203	0.1213	0.0409	N/A
3	0.5042	0.1052	N/A	N/A

Where, M and N are the line number and the column number of the image of the palmprint; MN is the quantity of the pixels in the image; A and B are images of the palmprint. If the corresponding DHOCs of the palmprint image are the same, 1 is the perfect matching score.

4   Experimental Results Test

4.1   Palmprint Verification

The double half-orientation nonlinear matching method is adopted for the calculation of false matching score and true matching score. The distribution of true and false matching scores in PolyU database [5], IITD database and Red database is as shown in Fig. 2(a)–(c). The distribution of the matching score in other spectral databases is basically the same as that in Red database. According to such distribution, we can know that the true matching score is usually more than the false matching score.

Table. 2. Nonlinear matching

Err Rates	Comp Code	Ordi Code	Fusn Code	Palm Code	DHOC $n_\theta = 6$	DHOC $n_\theta = 12$
PolyU	3.56	4.82	4.11	3.86	2.56	2.46
Red	5.33	6.32	6.66	6.43	4.54	4.48
Green	5.64	7.48	8.04	8.61	5.06	4.71
Blue	5.48	7.28	7.48	8.72	5.01	4.68
NIR	4.86	6.01	7.41	7.54	3.92	3.88
IITD	34.91	38.85	40.39	49.41	32.24	31.96

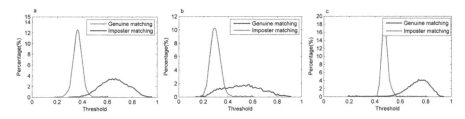

Fig. 2. (a) PolyU matching distribution (b) IITD matching distribution (c) Red matching distribution

Nonlinear classifier can effectively distinguish the true matching and the false matching of PolyU database and multispectral database. In comparison, the distribution in IITD database is different from that in PolyU database. This may be because of the obvious rotation and projection changes of palmprint images in IITD database. In most cases, the false matching score is still less than the true matching score, so this method is a simple and effective palmprint recognition algorithm.

4.2   Comparison Between Nonlinear Matching and Hamming Distance Measurement

Hamming distance measurement is utilized for the double half-orientation matching to evaluate the efficiency of the nonlinear matching score method. The recognition error rates obtained through "double half-orientation code Hamming measurement" method (TRAIN = 1) are shown in Table 3. Compared with the results in Table 2, DHOC difference between the nonlinear matching scheme and the Hamming distance scheme is minimally as 0.29 and maximally as 0.27 under $n_\theta = 6$; DHOC difference between the nonlinear matching scheme and the Hamming distance scheme is minimally as 0.29 and maximally as 0.90 under $n_\theta = 12$. In conclusion, the nonlinear matching method is applicable to the matching of double half-orientation code.

Table. 3. Hamming matching

Err rates	PolyU	Red	Green	Blue	NIR	IITD
DHOC	$n_\theta = 6$	2.60	4.60	5.14	3.96	31.97
DHOC	$n_\theta = 12$	2.80	4.81	5.26	4.26	32.86

5   Conclusion

The palmprint recognition method based on orientation code mainly includes two steps: (1) palmprint orientation feature extraction; and (2) palmprint orientation feature matching. It is required to apply a rational orientation code method and design an effective code matching algorithm, which are two crucial aspects of the coding method. In this paper, the half-Gabor filters are adopted for orientation feature extraction, namely: the double half-orientation code (DHOC) extraction method is adopted. The palmprint orientation features can be correctly and robustly presented by DHOC. Meantime, the nonlinear matching score calculates the DHOC distance. The nonlinear angle matching score method has been proved to be more effective compared with the traditional Hamming distance measurement in DHOC evaluation.

References

1. Kong, A., Zhang, D., Kamel, M.: A survey of palmprint recognition. Pattern Recognit. **42**, 1408–1418 (2009)
2. Jain, A.K., Feng, J.: Latent, palmprint, matching. Proc. IEEE Trans. Pattern Anal. Mach. Intell. **30**, 1032–1047 (2009)
3. Zhang, D., Kong, W.K., You, J., Wong, L.M.: Online palmprint identification. Proc. IEEE Trans. Pattern Anal. Mach. Intell. **25**, 1041–1050 (2003)
4. Kong, A.W.K., Zhang, D.: Competitive coding scheme for palmprint verification. In: Proceedings of the Seventeenth International Conference on Pattern Recognition (ICPR), pp. 520–523 (2004)
5. Jang, W., Park, D., Lee, D., Kim, S.: Fingerprint image enhancement based on a half-Gabor filter. In: Proceedings of the Advances in Biometrics, pp. 258–264. Springer, Berlin, Heidelberg (2005)
6. Guo, Z., Zuo, W., Zhang, L., Zhang, D.: A unified distance measurement for orientation coding in palmprint verification. Neurocomputing **73**, 944–950 (2010)
7. Zhang, D., Guo, Z., Lu, G., Zhang, L., Zuo, W.: An online system of multi-spectral palm-print verification. Proc. IEEE Trans. Instrum. Meas. **59**(2), 480–490 (2010)

Design of Digital Filters for the Contactless Conductivity Detection of Micro-fluidic Electrophoresis Chip Based on CIC and FIR

Hong-hua Liao[✉], Hao Fu, Bing-bing Zhou, Yi Lv, and Hai-lin Yuan

School of Information and Engineering, Hubei University for Nationalities,
No. 39 Xueyuan Road, Enshi, Hubei, China
esliaohonghua@qq.com

Abstract. To solve the noise problem of contactless conductivity detection (CCD) system in micro-fluidic electrophoresis chip, solution scheme of CIC and FIR digital cascade filter based on FPGA was presented. By selecting different filter, which includes CIC filter, FIR filter, and CIC and FIR digital cascade filter, microchip electrophoresis chromatographic signal with noise is compared and analyzed. In this paper, under the conditions of satisfying maximum distortionless output contactless conductivity detection chromatographic amplitude of micro-fluidic electrophoresis chip, the optimum parameters of CIC and FIR digital cascade filter were found, and frequency characteristic of the filter was analyzed. The results show CIC and FIR digital cascade filter can more effectively eliminate the noise of microchip electrophoresis chromatographic signal than CIC filter or FIR filter. CIC and FIR digital cascade filter can be meet real time demand for CCD of micro-fluidic electrophoresis chip.

Keywords: Micro-fluidic chip · Contactless conductivity detection (CCD) · FIR filter · Cascade filter · CIC filter

1 Introduction

Micro-fluidic chip electrophoresis technology has developed rapidly since the miniaturized total analysis system concept was first proposed by Manz project team [1]. Because of the advantages of universal, sensitivity and good reproducibility, and so on, a universal detection method of MCE avoids many problems caused by the contact between electrode and solution [2, 3]. Therefore, combined with needs of the subject, the CIC and FIR cascade digital filter based on FPGA is proposed.

2 Principle of CCD for Micro-fluidic Chip

Micro-fluidic chip contactless conductivity detector usually uses two electrode or four electrode structure [4]. The electrodes are integrated at the bottom of detection cell. The schematic of micro-fluidic chip structure contactless conductivity detector is shown in Fig. 1.

© Springer International Publishing AG 2018
F. Qiao et al. (eds.), *Recent Developments in Mechatronics and Intelligent Robotics*,
Advances in Intelligent Systems and Computing 690, DOI 10.1007/978-3-319-65978-7_7

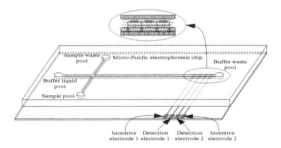

Fig. 1. The schematic of micro-fluidic chip structure contactless conductivity detector

In Fig. 1, Electrode and solution to be measured in the detection cell can be equivalent to a capacitor with an insulating layer as a medium. The two electrode can be equivalent to a capacitor formed by a buffer solution as a medium, and a resistance formed by the solution.

3   The Basic Principle of CIC Filter and FIR Filter

The CIC filter is the composition of one or more pairs of integrator and comb filters [5]. In order to meet the actual requirement, multilevel cascade design is usually used. CIC multi-level cascaded filter transfer function is:

$$H(z) = \left(\frac{1 - Z^{-MD}}{1 - Z^{-1}} \right)^N \tag{1}$$

In the formula: D is the decimation factor. M is the delay factor. N is the cascade number of the integral filter and comb filter.

The basic structure of FIR filter can be considered as thing that original data passes through a finite delay line. The data after filtering is obtained by using each delay line cumulative with weighted method. FIR filter is the composition of delay, multiplier and adder [6].

4   Analysis and Discussion

4.1   CCD Chromatographic Peak Signal Simulation of Micro-fluidic Chip

In order to compare the filtering performance of different filters, the Lorenz function and the Gauss function are used to simulate the chromatographic peak of contactless conductivity spectrum of micro-fluidic chip. The Lorenz function and the Gauss function can be expressed as:

$$L_i(x_i) = H_i / \left(1 + \frac{(x_i - \mu_i)^2}{\sigma_i^2} \right) \tag{2}$$

$$G_i(x_i) = H_i \exp\left[\frac{(x_i - \mu_i)^2}{2 * \sigma_i^2}\right] \tag{3}$$

In the formula: x_i is time required to complete the i spectral peak. σ_i is half peak height of the i spectral peak. μ_i is peak extreme point coordinate of the i spectral peaks. G is peak extreme of the i spectral peaks.

4.2 Filter Performance Analysis

CIC digital filters, FIR filters, and CIC and FIR cascaded filters with different series and different order are used to filter and analyze the analog signal of contactless conductance peak micro-fluidic chip of SNR = 1.9826 db. For the convenience of analysis, different series and different order CIC and FIR cascade filters are represented by (Nc, Nf). The Signal-to-noise Ratio curve of different series and different order filters is shown in Fig. 2.

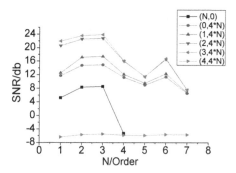

Fig. 2. The signal-to-noise Ratio curve of different series and different order filters

As can be seen from Fig. 2, when the CIC filter (Nf = 0) is used, the signal-to-noise Ratio of output signal increases with increase of cascaded series. When the series of CIC filter is increased to Nc = 4, the signal-to-noise Ratio of output signal is greatly reduced, and output signal is seriously distorted. When the FIR filter (Nc = 0) is used, the signal-to-noise Ratio of output signal reaches the maximum value at Nf = 12. When the CIC and FIR cascaded digital filters are in Nc = 1, 2, 3, with the increase of series, the change law of output signal-to-noise Ratio is similar to that of FIR filter.

4.3 Performance Analysis of CIC and FIR Cascaded Digital Filters

In order to better analyze the CIC and FIR cascade digital filter of different parameters, the analog signal of contactless conductance peak micro-fluidic chip of different signal-to-noise Ratio as input, the change curve of output signal-to-noise ratio is observed. The results are respectively shown in Fig. 3.

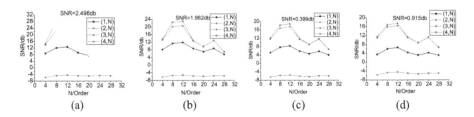

Fig. 3. The change curve of CIC and FIR cascaded digital filter output signal-to-noise ratio

It can be seen from Fig. 3 that when the CIC and FIR cascade digital filters are used in the Nc = 1, 2, 3, with the increase of series, the change law of output signal-to-noise ratio increases first, then decreases, and reaches the maximum at Nf = 12. When Nc = 4, the output signal-to-noise ratio changes little, and the output signal-to-noise ratio is small. Observe the output waveform under the condition (Nc = 4): output waveform has a serious distortion. Because of taking into account the logic resources of FPGA, under the premise that the output signal-to-noise ratio is similar, the 8 order FIR filter with less hardware resource is selected as much as possible.

When Nc = 3, Nf are 4, 8, 12, 16, 20, 24, 28, output waveform of CIC and FIR cascaded filter shown in Fig. 4(a). When Nf = 8, Nc are 1, 2, 3, 4, output waveform of CIC and FIR cascaded filter shown in Fig. 4(b).

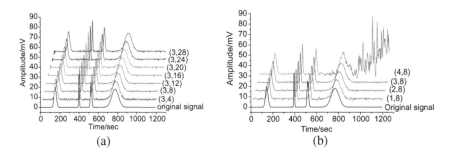

Fig. 4. Output waveform of CIC and FIR cascaded filter (figure a Nc is 3, figure b Nf is 8)

It can be seen from Fig. 4(a) that the CIC and FIR cascade digital filter can be used to filter the peak signal of contactless conductivity spectrum of microfluidic chip, and the overall characteristics of original signal can be well preserved. As can be seen from Fig. 4(b), when Nf = 8, CIC and FIR cascade digital filter Nc = 3, the filtering effect is better. When Nc = 4, filter output signal of CIC and FIR cascade digital filter has serious distortion. In a word, CIC and FIR cascaded filters of Nc = 3, Nf = 8 are suitable for contactless conductivity spectrum signal of microfluidic chips.

5   Testing and Verification

The actual signal that is the peak signal of contactless conductivity micro-fluidic chip with tail peak passes through the Nc = 3, Nf = 8 CIC and FIR cascaded filters. The comparative analysis diagram of actual signal and output signal is shown in Fig. 5.

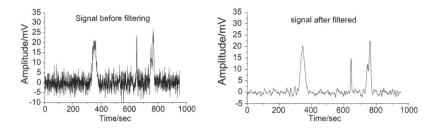

Fig. 5. Comparative analysis of CIC and FIR digital filter output signal and original signal

It can be seen from Fig. 5 that CIC and FIR cascade digital filter can be used to filter the peak signal of contactless conductivity micro-fluidic chip with tail peak. Under the condition of signal maximum nondistortion, the signal-to-noise ratio is increased from 0.9101 db to 16.1343 db.

The simulation results of CIC and FIR cascade filter IP core are shown in Fig. 6.

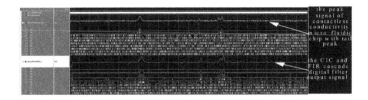

Fig. 6. Modelsim simulation results of CIC and FIR digital filter

It can be seen from Fig. 6 that test_org is the peak signal of contactless conductivity micro-fluidic chip with tail peak, out_1 is the CIC and FIR cascade digital filter output signal. The output waveform can keep the characteristic of original peak signal, and the noise is well suppressed.

6   Conclusion

In this paper, the detection principles of contactless conductivity detector and the characteristics of spectral peak signal are presented. Under the condition that spectral peak signal is not distorted, the effect of CIC filter, FIR filter and CIC and FIR cascade digital filter on the peak signal of contactless conductivity spectrum of micro-fluidic chip is analyzed from the perspective of lifting SNR. CIC and FIR cascade digital filter IP core of design based on ModelSim is verified. The results show that CIC and FIR cascade digital filters can be suitable for the digital filter of contactless conductivity detection system in micro-fluidic electrophoresis chip, and it can maintain basic characteristics of spectral peak signal of micro-fluidic electrophoresis chip.

Acknowledgements. This work is supported by National Natural Science Foundation of China (61463014, 61263030).

References

1. Zemann, A.J., Schnell, E., Volgger, D., Bonn, G.K.: Contactless conductivity detection for capillary electrophoresis. Anal. Chem. **70**(3), 563–567 (1998)
2. da Silva, J.A.F., do Lago, C.L.: An oscillometric detector for capillary electrophoresis. Anal. Chem. **70**(20), 4339–4343 (1998)
3. Arora, A., Mello, A.J.D., Manz, A.: Sub-microliter electrochemiluminescence detector-a model for small volume analysis systems. Anal. Chem. **34**(12), 393–395 (1997)
4. Klett, O., Nyholm, L.: Separation high voltage field driven on-chip amperometric detection in capillary electrophoresis. Anal. Chem. **75**(6), 1245–1250 (2003)
5. Guowen, L., Weiguo, Z.: Optimal design of high performance CIC filter. Comput. Simul. **33**(2), 234–238 (2016)
6. Stosic, B.P., Pavlovic, V.D.: On design of a novel class of selective CIC FIR filter functions with improved response. Int. J. Electron. Commun. **68**(8), 720–729 (2014)

An Improved Artificial Fish-Swarm Algorithm Using Cluster Analysis

Zhang Li-hua[✉], Dou Zhi-qian, and Sun Guo-long

School of Mechanical Engineering, Jiangsu University of Science and Technology,
Zhenjiang 212000, China
912142275@qq.com

Abstract. AFSA has been widely used as its super global search ability. However, AFSA still has the problem of falling into the local optimal value due to the randomness of the initial states of AFs, this paper introduces k-means clustering method into AFSA to ensure the randomness and the uniform distribution of the initial states of AFs by introducing distance factor. In the end this paper presents two types of testing functions are introduced to prove the improved method is better in convergence rate, accuracy and the effect of avoiding the local optimum.

Keywords: AFSA · Cluster · Optimization

1 Introduction

Artificial Fish-Swarm Algorithm (AFSA) is one of the best algorithms of optimization among the swarm intelligence algorithm proposed by the Chinese scholar Li Xiaolei [1] inspired by the fish movement behavior. The coordinates are called the state of each AF, the behavior includes prey, swarm, follow, move and evaluate [2].

Even with the advantages of simple, easy to implement [3] and the excellent performance of global search ability, the AFSA still have problems like parameters are defined by experience and slow convergence speed.

Most of the efforts to improve AFSA are concentrated on the optimization of the parameters [10]. In 2005, Wang Cuiru [4] improve AFSA with the method of adding a leaping behavior to obtain the global optimal value. In 2008 Wang Lianguo [5] proposed moving the coordinate of the AF directly to the sub-optimal position and gradually narrowing the individual AF visual scope and step size. In 2009 Liu Yanjun propose four adaptive AFSA methods to void the setting of the parameters to improving convergence speed and optimization accuracy [6]. However not too much attentions have been paid to the ways to optimize the initial values.

The AFSA algorithm start with a series of randomly distributed AFs, this always lead to slow convergence speed or even make the AFs falling into local optimum.

2010 Qu Liang dong [7] adopt Logistic mapping to optimize the initial state of AFs, 2009 He Dengxu optimize the initial values with a clustering method based on density

© Springer International Publishing AG 2018

F. Qiao et al. (eds.), *Recent Developments in Mechatronics and Intelligent Robotics*,
Advances in Intelligent Systems and Computing 690, DOI 10.1007/978-3-319-65978-7_8

and grid in solution space [8], these methods make the algorithm either lost its randomness of its initial states or will not quite efficient.

This paper presents an improved artificial fish swarm algorithm with a novel method
to optimize the initial values, which takes the randomness and the uniform distribution
of the initial values into account, and it turns out that the improved algorithm accelerates
the convergence process and avoid the local optimum to an extent.

2   Proposed Method

This paper presents a method based on improved K-means cluster to solve the problem.
Comparing the methods in other papers, this paper introduces distance factor into K-
means cluster method to control the randomness of the states of the AFs and to accelerate
the convergence speed, the progress are as follows:

Step 1. Generate n AFs with randomly distributed in the solution space;

Step 2. Generate a specific number n_s of AFs;

Step 3. Classify $n + n_s$ AFs according to K-means to form n clusters

Step 4. Calculate the distances of all the state of the cluster, forming the $n \times n$ matrix $d_{i,j}$

Step 5. Judge whether $\min(d_{i,j}) > \varepsilon \cdot dist$, if true then go to step 6, otherwise go to step 2; where $dist$ denote the distance when the clusters are evenly distributed, $dist = S_k \sqrt[k]{\dfrac{1}{n}}$, S_k denote $1 \times k$ matrix in where each of the element of the S_k is the maximum distance of its axis, k is the dimension of the solution space; ε denote the distance factor.

Step 6. Calculate food concentration of each AF and record the state of the AF with the best food concentration; evaluate each of the AF with the evaluation behavior; execute the behavior after the evaluation behavior, refresh the state of each AF; then refresh the state of all of the AF group.

Step 7. Judge whether satisfy the termination condition, if true then print the state of AF and its food concentration, otherwise jump to step 6.

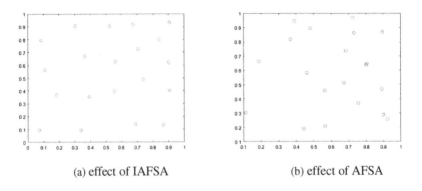

(a) effect of IAFSA   (b) effect of AFSA

Fig. 1.   (a) Effect of IAFSA, (b) effect of AFSA

Put the progress into Matlab, the effect of initialization of the AFs in the improved AFSA (IAFSA) and AFSA shows in Fig. 1(a) and (b) ($n = 20; \varepsilon = 0.8$).

Here we can clearly find out that the state of AFs in IAFSA is more evenly distributed in the solution space than AFSA.

In order to compare the performance of IAFSA and AFSA, here present two typical testing functions, the characters of these functions shows in Table 1:

Table 1. Parameters of the functions

Testing function	Seek max/min value	Value	Description of the figure of the function
$f_1(x, y)$	Max	2.118	Complex, multi-peak, peak value difficult to seek
$f_2(x, y)$	Max	210.4818	Complex, multi-peak, peak value difficult to seek

The tests are as follows:

$$f_1(x, y) = 1 + x \sin(4\pi x) + y \sin(4\pi y) + \frac{\sin(6\sqrt{x^2 + y^2})}{6\sqrt{x^2 + y^2} + 10^{-15}}, x, y \in [-1, 1]$$

In $f_1(x, y)$ the maximum value is 2.118, corresponding coordinates are (0.64, 0.64), (0.64, −0.64), (−0.64, −0.64), (−0.64, 0.64),

The figure of $f_1(x, y)$ shows in Fig. 2:

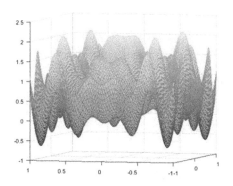

Fig. 2. The figure of $f_1(x, y)$

The size of the AF group AF_total = 20; visual = 0.1; step = 0.008; $\delta = 0.2$;

The testing results are as follows (Table 2):

Table 2. Testing results of $f_1(x, y)$

The algorithm	Coordinate	Max value	Iteration time(s)
AFSA	(0.6410, 0.6409)	2.1188	15.393722
IAFSA	(−0.6411, 0.6408)	2.1188	14.517551

The convergence effect graph of IAFSA and AFSA shows in Fig. 3(a) and (b).

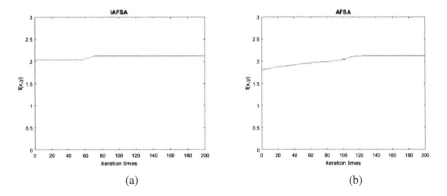

(a)   (b)

Fig. 3.

$$f_2(x, y) = \left(\sum_{i=1}^{5} i \cos((i + 1)x + i) \right) \left(\sum_{i=1}^{5} i \cos((i + 1)y + i) \right)$$

maximum value is 210.4818, and 9 maximum coordinates.
The figure of $f_2(x, y)$ shows in Fig. 4:

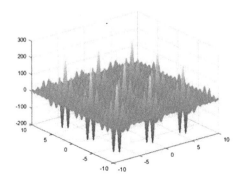

Fig. 4. The figure of $f_2(x, y)$

The size of the AF group AF_total = 20; visual = 10; step = 0.1; $\delta = 0.2$;
The testing data are as follows:

Algorithm	Coordinate	Maximum value	Iteration time(s)
AFSA	(5.4795, 6.6185)	55.8082	10.790126
IAFSA	(5.4821, −0.8002)	210.4808	11.824610

The convergence effect graph of IAFSA and AFSA shows in Fig. 5(a) and (b).

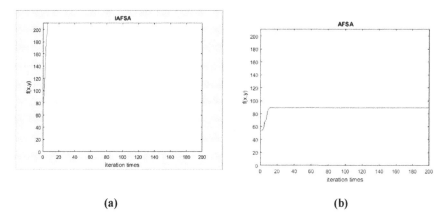

Fig. 5.

3 Conclusion

Prior work has documented that the IAFSA have better performance in convergence speed and closer to the optimal value than AFSA, which can be seen clearly in the testing function $f_1(x, y)$, furthermore, $f_2(x, y)$ shows in dealing with function like multi-extremes, IAFSA is easier to avoid the local optimum.

For comparison, this article does not add any other improvements into the algorithm, and in order to avoid the algorithm fall into the local optimum, the initial value can also be further optimized, which will be the future works of this paper.

Acknowledgments. This article was completed under the auspices of the BY2015065-09 fund, in which the author would like to expresses his sincere gratitude.

References

1. Li, X., Shao, Z.: An optimizing method based on autonomous animals: fish-swarm algorithm. Syst. Eng.-Theor. Practice **2002**(11), 32–38 (2002)
2. Neshat, M., Sepidnam, G., Sargolzaei, M., et al.: Artificial fish swarm algorithm: a survey of the state-of-the-art, hybridization, combinatorial and indicative applications. Artif. Intell. Rev. **42**(4), 965–997 (2014)
3. Shan, X., Jiang, M., Li, J.: The routing optimization based on improved artificial fish swarm algorithm. In: Intelligent Control and Automation, 2006. WCICA 2006. The Sixth World Congress on IEEE, pp. 3658–3662 (2006)
4. Wang, C.R., Zhou, C.L., Ma, J.W.: An improved artificial fish-swarm algorithm and its application in feed-forward neural networks. Mach. Learn. Cybern. **5**, 2890–2894 (2005)
5. Wang, L., Hong, Y., Zhao, F., Yu, D.: An improved artificial fish swarm algorithm. Comput. Eng. **34**(19), 192–194 (2008)
6. Liu, Y.J.: Improved artificial fish swarm algorithm based on adaptive visual and step length. Comput. Eng. Appl. **45**(25), 35–37 (2009)

7. Qu, L.-D.: Novel artificial fish-school algorithm based on chaos search. Comput. Eng. Appl. **46**, 40–42 (2010)
8. He, D., Qu, L.: Clustering analysis algorithm of AFSA. Appl. Res. Comput. **26**(10), 3666–3668 (2009)
9. Yin, Z., Zong, Z., Sun, H., et al.: A complexity-performance-balanced multiuser detector based on artificial fish swarm algorithm for DS-UWB systems in the AWGN and multipath environments. EURASIP J. Adv. Sig. Proces. **2012**(1), 1–13 (2012)
10. Zhang, Y., Li, Z., Feng, Z.: Improved artificial fish swarm algorithm based on dynamic parameter adjustment. J. Hunan Univ. (Natural Science Edition) **39**(5), 77–82 (2012)

The Heart Rate Monitoring Based On Adaptive Cancellation Method

Wang Wei, Xiong Ying, and Wang Xin[✉]

Hunan Provincial Key Laboratory of Intelligent Processing of Big Data on Transportation,
Changsha University of Science and Technology, Changsha, China
wangxin@csust.edu.cn

Abstract. For solving the problem that PPG signals were usually corrupted by motion artifact, a comprehensive approach for heart rate monitoring of real-time wearable devices was proposed in this paper. This method was based on the adaptive noise cancellation technique. Firstly, PPG signals and acceleration signals were pre processed through a band-pass filter. Acceleration signals and red PPG signals would be regarded as a set of noise reference signal. Next, an adaptive filtering algorithm was used to remove the motion artifacts. Finally, the heart rate could be extracted from the denoising PPG signals by using peak to peak value estimation method. The simulation results show that, compared with traditional LMS algorithm and FFT algorithm, the proposed method is more accurate, less error. And the method has the advantages of real-time, high efficiency and simplicity to monitor heart rate in different motion states.

Keywords: Photoplethysmography signal · Motion artifact · Adaptive filter · Acceleration signal · Heart rate

1 Introduction

Heart rate (HR) monitoring is a significant criterion for measuring health level. Estimating HR with PPG (photoplethysmography) signal is a useful method. PPG signals were used to monitor patients' HRs in the hospital, and it is also one of the primary signals with many portable wearable device. But, PPG signals can be easily influenced in the measurement process and it's difficult to make HR estimation [1, 2]. The PPG signal includes various of noise, such as the noise from subjects' motion or surroundings. Even though, linear filters have been used removing some artifacts from PPG signal, but some undesired noise cannot be removed. Most research in the field of motion artifact reduction focused on techniques which included adaptive filtering and wavelet transform, etc. Rasoul Yousefi [3] used adaptive cancellation technique to remove motion artifact. The noise reference signal was extracted from the corrupted PPG signals itself, so it was too sensible to reach the demands for real-time and efficiency of wearable device. Lee HW [4] performed artifact reduction with using Moving Average Filter (MAF), but it was difficult to remove motion artifact when there was a dramatic change in motion. Raghuram M, Zhang K [5, 6] considered wavelet transform according to non-stationarity of PPG signals. Wang Q [7, 8] performed motion artifacts reduction by using

© Springer International Publishing AG 2018
F. Qiao et al. (eds.), *Recent Developments in Mechatronics and Intelligent Robotics*,
Advances in Intelligent Systems and Computing 690, DOI 10.1007/978-3-319-65978-7_9

Empirical Mode Decomposition (EMD). Although, wavelet transform and EMD was effective to remove motion artifacts, but there was a common problem that how to choose the optimal threshold. Byung S. Kim, K. Venu Madhav [9] considered Independent Component Analysis (ICA). Motion artifacts can't be accurately removed in intense exercise because of complicated calculation and intricate process.

2   Analysis of PPG Signal

The collected PPG signals contain AC component and DC component. The DC component's form is baseline drift. It can be removed with a low-pass linear filter. The rest of AC component caused by abnormal heart pump blood cannot be filtered by digital filter directly, because frequency of AC component was overlapped with motion artifacts' [10]. The periodic variation of the PPG signal represents the heart rate information of the tester. The subject's voluntary or involuntary tremble caused motion artifacts which would be corrupted the cardiac synchronous pulsatile component of the arterial blood. Therefore, How to reduce the motion artifacts is one of the most challenging problems at present. Due to the limited memory capacity of wearable devices, the algorithm not only requires short computation times, but also requires high performance of noise reduction and low consumption of energy. With same experimental conditions and data sources, variable step size- LMS algorithm is better than ICA algorithm, wavelet transform and PCA algorithm in computation times and complexity obviously. So adaptive filtering can achieve less amount of computation and better real-time.

3   Adaptive Cancellation Method

The details of adaptive cancellation method are described as follows: (a) preprocessing, (b) adaptive filtering, (c) HR estimating. The flow chart of the proposed method is shown in Fig. 1. $PPG(i)$ is the corrupted PPG signal, $PPG(ref)$ is the red PPG signals, $ACC(x)$. $ACC(y)$ and $ACC(z)$ are the acceleration signals of three-axis accelerometer $ACC(m)$ is

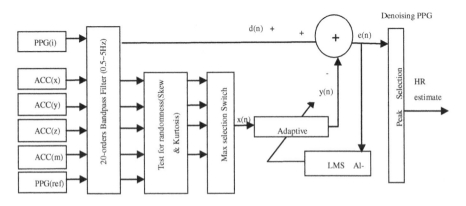

Fig. 1. Signal processing flow

the composite acceleration signal. Motion artifacts are difficult to be removed from the corrupted PPG signal $ACC(x)$, $ACC(y)$.$ACC(z)$, $PPG(ref)$ and $PPG(i)$ are extracted from the same motion signal source. $PPG(i)$ signal contains low frequency noise, high frequency noise and motion artifacts. So signal preprocessing is necessary. All signals are preprocessed by using a 20-order bandpass filter between 0.5–5 Hz frequency band. The low and high irrelevant frequency components of all signals will be filtered out.

3.1 Adaptive Filtering Algorithm and Improvement

The iterative formula of fixed step size LMS algorithm is,

$$y(n) = x(n)\omega(n) \tag{1}$$

$$e(n) = d(n) - y(n) \tag{2}$$

$$\omega(n+1) = \omega(n) + \mu(n)x(n) \tag{3}$$

$x(n)$ is the input signal. $y(n)$ is the output signal. $d(n)$ is the desired signal. $e(n)$ is the error signals. $\omega(n)$ is the estimate value of weights vector of filter. $\mu(n)$ is the step factor.

To ensure the convergence of the LMS algorithm, the range of step size is $0 < \mu < \dfrac{1}{\lambda_{max}}$, λ_{max} is the largest eigenvalue of the auto correlation matrix of the input signal. References [11] proposed a variable step size LMS algorithm (NVS-CLMS) based on Lorentz function, the step size $\mu(n)$ is the Lorentz function of $e(n)$:

$$\mu(n) = a \log\left[1 + (\frac{e(n)}{\sigma})^2\right] \tag{4}$$

In order to improve the convergence rate and tracking speed, we proposed a new nonlinear relation of the LMS filter algorithms (VLS-LMS) based on Lorentz function and sigmoid function. The algorithms eliminated the influence of irrelevant noise, and had good anti-interference ability. The formula of the improved step factor $\mu(n)$ is:

$$\mu(n) = \frac{a}{b^{m*e(n)*e(n)} + 1} + c \quad \left(\frac{a}{2} + c > 0, \ c > 0\right) \tag{5}$$

To make this algorithm converged, step factor should be satisfied at $0 < \mu < \dfrac{1}{\lambda_{max}}$.

$$\lim_{e(n)\to\infty} \mu(n) = c \quad (e(n) \to \infty) \tag{6}$$

$$\lim_{e(n)\to 0} \mu(n) = \frac{a}{2} + c \quad (e(n) \to 0) \tag{7}$$

In general, the range of $\mu(n)$ is $\dfrac{a}{2} + c < \mu(n) < c$, "C" are increased gradually, $\mu(n)$ is also increased. It meets the requirements of system design.

To measure the performance of VLS-LMS algorithm, we used adaptive noise canceling system for simulation. The simulation conditions are as follows: adaptive filter order is 2. The reference input signal $x(n)$ is the Gaussian noise with a zero-mean and the variance is 1. $V(n)$ is the Gaussian noise with zero-mean which irrelevant to $x(n)$. Making 500 times independent simulations with 1000 samples each time, we got the statistical average and learning curve. The optimal parameters of each algorithm were obtained through multiple experiments. The best parameters of NVS-CLMS are: $a = 0.006$, $\delta = 0.01$. The best parameters of LVS-LMS are: $a = -0.3$, $b = 8$, $c = 0.15$, $m = 10$.

It can be seen that LVS-LMS algorithm is superior to NVS-CLMS algorithm in convergence and tracking ability in Fig. 2.

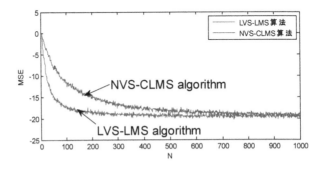

Fig. 2. The comparison MSE between LVS-LMS algorithm and NVS-CLMS

3.2 Generating the Reference Signal

In daily life, PPG signals of wearable device were easily disturbed by different motion states. These disturbances are caused by daily activities (such as walking, running, jumping, etc.). In addition, the identical motion may cause different motion artifacts. Reference signals have been used to reduce motion artifacts in many algorithms. Particular reference signals just work in particular motion states. For example, accelerometer signal and noise signal are irrelevant when fingers beat desktop and keep stationary of wrist. Therefore, accelerometer signal is not suitable to be reference signal in this state. So, five reference signals are considered for adaptive filtering method. They are accelerometer axis of $X(ACC_x)$ accelerometer axis of $Y(ACC_y)$ accelerometer axis of $Z(ACC_z)$ the magnitude of acceleration $ACC_m = ACC_x + ACC_y + ACC_z$ and the red PPG signal. The relationship between five reference signals and PPG signal are studied. Skew and kurtosis can be served as important features for detection of motion artifacts. A reference signal is generated from the five when kurtosis and skew are considerable.

3.3 Analysis and Results

The experiment data was gathered from the wearable device in independent research, and the real reference heart rate was collected by Mindray iMEC 8 monitor. The subjects were 12 healthy adults (age: 23–27). Test states were divided into two types: (1) micro-motion (typing, horizontal or vertical movement of wrist, etc.), (2) strong motion (walking, running, jumping, etc.). PPG signal would be corrupted by different motion noise in different motion conditions. Thus, it is necessary to reconstruction PPG signal for calculating heart rate in precisely. Figure 3 represents PPG signal which removed motion artifact through LMS algorithm.

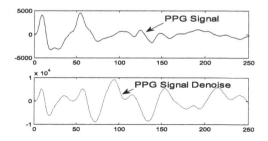

Fig. 3. a–b are the corrupted PPG signal and after noise cancellation in walking state

The peak-peak detection is used to calculating heart rate for denoising PPG signal. Figure 4 represent in different motion conditions, heart rate calculated by adaptive cancellation method and real-heart rate.

| (a)Sitting and arm swing | (b)irregular walking | (c)running |

Fig. 4. Heart rate in different motion states

We adopt mean absolute errors for analysis to better compare the performance of the proposed method with that of FFT algorithm and traditional LMS algorithm in different motion states.

The average error rate is defined as,

$$Error = \frac{1}{M} \sum_{i=1}^{M} \frac{|HR_{PPG} - HR_{REAL}|}{HR_{REAL}} \tag{8}$$

The mean absolute errors is defined as,

$$Error1 = \frac{1}{M} \sum_{i=1}^{M} |HR_{PPG} - HR_{REAL}| \tag{9}$$

M is total number of times, HR_{PPG}, HR_{REAL} are estimated heart rate and real- heart rate. Mean absolute errors unit is BPM. Table 1 shows that subjects' heart rate of mean absolute errors in different motion condition.

Table 1. The mean absolute errors of heart rate using different algorithm after de-noising

Motion	FFT algorithm	LMS algorithm	This paper
Resting	0.83%	0.54%	0.20%
Sitting and are swing	6.30%	4.65%	1.53%
Walking on treadmill	7.52%	4.03%	1.53%
Running on treadmill	12.67%	5.58%	2.76%
Static-motion in turn	10.55%	8.26%	2.45%

Table 1 illustrates the algorithm's mean absolute errors of measuring HR in different motion types. In stationary state, the error of all algorithms is tiny which less than 1.00%. For traditional FFT method it's as high as 12.67% in intensive state. For LMS, because the fixed LMS can not be adjusted in time, the mean absolute errors was 8.26% when motion type change suddenly. The mean absolute errors of adaptive cancellation method in this paper are controlled fewer than 3% in all motion types.

Due to the limited memory of wearable heart rate monitoring system, the algorithm must have less computation time. Therefore, the computational time of the algorithm is taken as the evaluating indicator. We apply the same simulation condition which simulating 20 times independently with 6000 samples each time. The corresponding results are listed in Table 2 which showing that both reference [12] and reference [13] use more average running time than the algorithm put forward in this paper.

Table 2. The comparison of running time

Method	Reference [12]	Reference [13]	This paper
Operation time T(s)	95.14	104.83	72.24

In summary, a new method for measuring heart rate is proposed in this paper. Its accuracy and effectiveness is higher than traditional method obviously.

4   Conclusion

In this paper, adaptive cancellation method of motion artifacts reduction is proposed. The adaptive filtering algorithm is improved by generating appropriate reference signal and a new LMS algorithm during the process of dynamic heart rate monitoring. Experiments have shown that proposed method is superior to ordinary LMS and FFT. The accuracy and effectiveness of monitoring heart rate is improved in continuous noninvasive monitoring. In future works, we will consider simpler methods for heart rate monitoring.

Acknowledgements. The work was supported by the science research plan of Education Department of Hunan Province: Sparse method for remote sensing image processing.

References

1. Shang, A.B., Kozikowski, R.T., Winslow, A.W., Weininger, S.: Development of a standardized method for motion testing in pulse oximeters. Anesth. Analg. **105**(6), S66–S77 (2007)
2. Han, H., Kim, J.: Artifacts in wearable photoplethysmographs during daily life motions and their reduction with least mean square based active noise cancellation method. Comput. Biol. Med. **42**(4), 387–393 (2012)
3. Yousefi, R., Nourani, M., Panahi, I.: Adaptive cancellation of motion artifact in wearable biosensors. In: 34th Annual International Conference of the IEEE Engineering in Medicine and Biology Society, San Diego, pp. 2004–2008. IEEE (2012)
4. Lee, H.W., Lee, J.W., Jung, W.G.: The periodic moving average filter for removing motion artifacts from PPG signals. Int. J. Control Autom. Syst. **5**(6), 701–706 (2007)
5. Raghuram, M., Madhav, K.V., Krishna, E.H.: Dual-tree complex wavelet transform for motion artifact reduction of PPG signals. In: 2012 IEEE International Symposium on Medical Measurements and Applications Proceedings (MeMeA), Budapest, pp. 1–4. IEEE (2012)
6. Zhang, K., Jiao, T., Fu, F.: Motion artifact cancellation in photoplethysmography using reconstruction of wavelet transform modulus maxima. Chin. J. Sci. Instr. **30**(3), 586–589 (2009)
7. Raghuram, M., Madhav, K.V., Krishna, E.H.: HHT based signal decomposition for reduction of motion artifacts in photoplethysmographic signals. In: 2012 IEEE International Instrumentation and Measurement Technology Conference (I2MTC), Graz, pp. 1730–1734. IEEE (2012)
8. Wang, Q., Yang, P., Zhang, Y.T.: Artifact reduction based on Empirical Mode Decomposition (EMD) in photoplethysmography for pulse rate detection. In: Engineering in Medicine and Biology Society (EMBC), 2010 Annual International Conference of the IEEE, Buenos Aires, pp. 959–962. IEEE (2010)
9. Kim, B.S., Yo, S.K.: Motion artifact reduction in photoplethysmography using independent component analysis. IEEE Trans. Biomed. Eng. **53**(3), 566–568 (2006)
10. Yushun, G., Baoming, W., Dandan, G., et al.: Adaptive elimination of motion artifact separation during oxygen saturation monitoring in ambulant environments. Space Med. Med. Eng. **25**(4), 266–270 (2012)
11. Leya, Z., Hua, X., Tianrui, W., et al.: Low computational complexity variable step-size CLMS algorithm based on Lorentzian function. Syst. Eng. Electron. **38**(5), 998–1003 (2016)
12. Young, A., Wentink, E., Wieringa, F.: Characterization and reduction of motion artifacts in photoplethysmographic signals from a wrist-worn device. In: 2015 37th Annual International Conference of the IEEE Engineering in Medicine and Biology Society (EMBC), Milan, pp. 6146–6149. IEEE (2015)
13. Raghuram, M., Madhav, K.V., Krishna, E.H.: A novel approach for motion artifact reduction in PPG signals based on AS-LMS adaptive filter. IEEE Trans. Instrum. Meas. **61**(5), 1445–1457 (2012)

An Intelligent Heating Device Using Central Heating Technique for Northern China Family

Shuhao Yu[(✉)], Fei Guo, Maosheng Fu, and Huali Xu

School of Electronic and Information Engineering, West Anhui University, Lu'an 237012, China
yush@wxc.edu.cn

Abstract. To overcome the defects of present central heating, we propose an intelligent heating device for northern China family. It provides accurate calculation for the heating capacity and heating costs without changing the large-scale building heating pipeline layout. The user can require the related information, set the relevant parameters and control the hot water valve by the provided APP. The proposed central heating system can automatically adjust the heating water valve based on the indoor temperature and has better application and extension value.

Keywords: Heating device · Central heating · Northern China family · APP

1 Introduction

At present, the heating system is used in central heating mode for most of the northern China. Because of the cold weather and heating for a long time, the central heating mode can help to reduce costs and save energy to some extent [1]. According to "2010 annual report on China building energy efficiency", in the northern part of China, every heating period, the total building energy consumption reaches 688.7 million tce (coal), Thereinto, the heating consumption of the total residential energy reaches 312.6 million tce (coal), so the total energy of residential heating consumption accounts for 45.3% of the building energy consumption. "Civil energy-saving regulations" puts forward clearly that the building which implements of central heating should be installed the heat metering devices and the room temperature control devices. However, with the larger heating coverage, Heat metering network monitoring management system has many problems in communications and constractions, as well as the high-accuracy of regulation at room temperature control accuracy. There are many deficiencies in central heating [2–5].

Firstly, it wastes energy seriously because the users' link facilities are relatively crude and simple. Meanwhile, the heating system will run all day no matter the weather is cold or hot. This will result in not only a lot of energy waste, but also environmental problems indirectly. Secondly, there exists an unfair phenomenon of heating charge. The heating system is used by the means of hot water cycle for northern China. This will lead to temperature difference for the different floor in user's home. Furthermore, the different geographical location can also cause the temperature difference. There exists the phenomenon that some users need to open window since overheating and some

© Springer International Publishing AG 2018
F. Qiao et al. (eds.), *Recent Developments in Mechatronics and Intelligent Robotics*,
Advances in Intelligent Systems and Computing 690, DOI 10.1007/978-3-319-65978-7_10

ones have troubles because of insufficient heating. So, the uneven heating will lead to the unfair phenomena of heating charge. There are three ways of charging: (1) According to the flow rate, this method only considers flow and ignores the quality of hot water. (2) According to the construction area, it is the most common method and also ignores the heating quality. (3) According to the number of radiators. This charging method is mainly to prevent the user to install additional radiators randomly and still ignores the heating quality. Thirdly, the heating quality is very poor. The reason as mentioned above is that the users are come from different geographical location and different floors [6–8].

In this work, we design an intelligent heating device based on Internet of Things (IoT). The proposed system can calculate the heating cost accurately, save energy, improve heating quality and need not to change the existing heating piping layout on a large scale.

2   The System Requirements of Proposed Device

In recent years, automation systems are applied to control lighting, cooling/heating, ventilation, and so forth. The role of heating system automation and control systems is a part of the whole problem. The majority of recent developments have followed the advances made in computer technology, telecommunications, and information technology. Although these systems has become increasingly popular, the necessity for smart tools and methods, to provide remote control and real time monitoring of heating system energy consumption. In many heating systems, the control and monitoring of the system are manually performed. Manual control leads to low efficiency in the case of incompetent supervision. Moreover, inadequate control of ambient and water temperature and water flow increases the complaints of users regarding poor or excessive heating, which causes a waste of energy. Consequently, the remote control-based energy management of building heating systems especially central ones is overlooked in spite of their extensive usages. Current heating systems' performance assessment tools are deficient in their ability to integrate and process building/system monitoring data to generate actionable information that can assist in achieving a higher level of heating system performance [7].

To overcome the defect of the existing heating mode, we propose an intelligent heating device based on the Internet of Things. It has the following advantages.

(1) Intelligent temperature control. It is different from the traditional rough hot water heating. The proposed heating system can gather indoor temperature through the real-time temperature sensor. The MCU can intelligent control the opening of the inlet valve by analysis the obtained data to maintain the indoor temperature at relatively stable state.

(2) Fair billing. The traditional charging mode is based on the housing area. It is unfair because the hot water heating energy dissipation is more serious and the water temperature of different users is significant difference. The proposed system takes a precise way to charge by the actual amount of heating. Therefore, it is more fair and reasonable.

(3) Remote control. Our intelligent heating system developed an APP which can provide users to inquire indoor temperature, flow, cost, heat supply and cost. Moreover, the user can control opening and closing of the water supply valve remotely.

(4) Energy saving and environmental protection. The traditional heating mode has a huge waste because it ignored the user is indoor or not. Heating has been run always. In our intelligent heating system, if the user is not in the house for a long time, he can control the heating water valve by supported APP. Thus, this can help to reduce energy waste and protect the environment.

3   The Design of Proposed Device

A. *Principle of heating calculation*

According to "Heat meter identification rules (JJG225 – 2001)" [9], the two heat calculation formulas are shown as Eqs. (1) and (2).

$$Q = CM\Delta t \tag{1}$$

$$Q = \int_0^t q_m * \Delta h * d_t \tag{2}$$

Where Q is the released heat (kJ), C denotes the specific heat capacity ($J\ kg^{-1} \cdot °C^{-1}$), M is the quality of heat carrier through the heat meter (kg), Δt is the temperature difference between inlet and outlet (°C), denotes the quality of the carrier through the heat meter ($kg \cdot s^{-1}$), denotes the enthalpy difference of pipeline inlet and outlet ($kJ \cdot kg^{-1}$) and t denotes the time (s) [10].

From the Eq. (1), we can find if we calculate the quantity of heat which dissipates through the user housing, it need to be known M and Δt. This two parameters can be measured by the flow sensor and the temperature sensor which included the inlet one and the outlet one. The specific heat capacity of water is different under different temperature and it gets minimum at 4 °C, then it will increase with decreasing temperature [11]. In the range of allowable error, we can use 4200 J/(kg · °C) for the specific heat capacity of water.

B. *System design*

The proposed intelligent heating device includes nine part as following: user's mobile terminal APP, controller module, display module, temperature measurement module, flow measurement module, WiFi module, IC card module, adjustable valve and the power supply module. The technical framework of the proposed intelligent heating system as shown in Fig. 1.

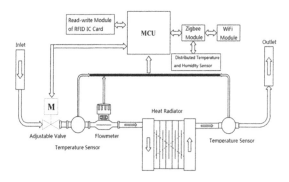

Fig. 1. The technical framework of intelligent heating device

The device can realize the information exchange with mobile terminal. The user may set the corresponding parameters through the mobile phone APP, control the hot water valve opening and closing to adjust the indoor heating condition. The system also can achieve the intelligent temperature regulation by the indoor temperature and automatically adjust the heating water valve opening. Indoor distributed temperature sensor data acquires by using ZigBee. The ZigBee node transfers the data of the indoor temperature and humidity to the coordinator, then the coordinator will transfer all the data to MCU (Microprogrammed Control Unit).

C. Controller selection

The power consumption is a very important index for the design of instrument. In computer system, we adopt W/MIPS (watt/million instructions per second) typically to measure processor power and performance. Our design use MSP430 single-chip which is 16-bit microcontroller. Its power consumption in the active mode is 250 uA/MIPS (traditional Mcs51 microcontroller, about 10–20 mA/MIPS). It has strong processing capability, fast running speed, low power consumption and it can meet the needs of the design.

D. Temperature sensor selection

In our design, it need two independent temperature measuring element for measuring the temperature of inlet and outlet. To simplify the design, we use the production of DS18B20 temperature sensor of Dallas company. The resolution of the DS18B20 is 12 bits, the accuracy is 0.5 °C and the detection range is −55 °C to +125 °C. It has the advantages of small volume, low cost, strong anti-interference ability and high precision. Its package shown as Fig. 2. Indoor temperature measurement uses DHT11 temperature and humidity sensor unit.

Fig. 2. Package diagram of DS18B20

E. *Flow monitoring module program*

To calculate the quantity of heat, we also need to measure the quality of heat carrier (water) through the users' home. For measuring the quality of liquid flowing through the pipeline, the required physical quantity as following: the cross-sectional area of the pipeline (S), heat carrier velocity (V). The quality of the heat carrier then flows through the pipe (M) can be calculated as Eq. (3).

$$M = \int \rho Sv \, d_t \tag{3}$$

Where ρ is the carrier density, its allowable range of water density is 1 g/cm^3. The cross-sectional area of the pipe can be measured.

F. *Communication module design*

The intelligent heating device uses WiFi as the wireless communication to exchange information with the mobile phone terminal. The heating device transfers all the kinds of information through the WiFi network to the mobile phone terminal APP. Then, the user can require the information and real-time control by APP. The indoor distributed temperature sensor data acquisition uses ZigBee wireless communication as previously mentioned.

G. *Other module design*

Power supply module: this device uses low power consumption design. It adopts the single-chip sleep mode to set the data update cycle as 10 min. When the microcontroller goes into the sleep mode, its working current is 0.1 uA and only the timer to maintain the state of work. When the timer interrupt is triggered, the SCM will turn into the normal operation mode and detect, calculate, update information, send data and then go into the sleep mode automatically. Therefore, our design uses battery power supply mode. The battery provides 3.7 V DC voltage by converting the ASM1117 – 3.3 chip from 3.7 V to 3.3 V.

The main interface of the APP shows the some important data such as: current room temperature, current hot flow, current amount of heating and account balance. By using the given APP, the users can inquire the quantity of heat and consumption record, set the temperature and heat flow rate. Thus, the accurate calculation of heat supply and heating cost is realized, the energy is saved, and the heating quality is improved.

4   Conclusion

To overcome the defects of present central heating, we propose an intelligent heating device for northern China family. The device is designed based on Zigbee and low-cost, low-power wireless sensors. It provides accurate calculation for the heating capacity and heating costs without changing the large-scale building heating pipeline layout. By the provided APP, the user can require the related information, set the relevant parameters and control the hot water valve. The proposed central heating system can automatically adjust the heating water valve based on the indoor temperature. The proposed intelligent heating device has better application and extension value.

Acknowledgements. The authors gratefully acknowledge the support of the Natural Science Foundation of the Higher Education Institutions of Anhui Province (No. KJ2015ZD44), the Universities Natural Science Foundation of Anhui Province (No. KJ2014A277) and the Universities Excellent Young Talents Foundation of Anhui Province (No. gxyqZD2016249).

References

1. Arabkoohsar, A., Farzaneh-Gord, M., Ghezelbash, R., Koury, R.N.N.: Energy consumption pattern modification in greenhouses by a hybrid solar-geothermal heating system. J. Braz. Soc. Mech. Sci. Eng. **39**(2), 631–643 (2017)
2. Chen, X., Zhang, Z., Shi, J., Yang, Z., Chen, J.: Research on the operating characteristics of floor heating system with residential EVI air source heat pump in China. In: International Refrigeration and Air Conditioning Conference (2016)
3. Fidchunov, A.L.: Influence of coking parameters on the nitrogen oxides in the coke-battery heating system. Coke Chem. **58**(8), 290–295 (2015)
4. Li, L., Zaheeruddin, M.: Pressure and temperature control analysis of a high-rise building hot water heating system: a simulation study. Energy Effic. **8**(4), 773–789 (2015)
5. Liu, L., Fu, L., Jiang, Y.: An on-off regulation method by predicting the valve on-time ratio in district heating system. Build. Simul. **8**(6), 665–672 (2015)
6. Poraj, J., Gamrat, S., Bodys, J., Smolka, J., Adamczyk, W.: Numerical study of air staging in a coke oven heating system. Clean Technol. Environ. Policy **18**(6), 1815–1825 (2016)
7. Şahin, V., İpek, O., Başoğul, Y., Gürel, B., Keçebaş, A.: Remote control-based energy management for energy savings in a central heating system. Environ. Prog. Sustain. Energy **36**, 600–609 (2016)
8. Shen, X., Liu, B.: Changes in the timing, length and heating degree days of the heating season in central heating zone of China. Sci. Rep. **6**, 33384 (2016)
9. Zhang, X.-R.: Study on the solar energy heat pump space heating system in the agricultural and pastoral areas in Inner Mongolia. In: Zhang, X., Dincer, I. (eds.) Energy Solutions to Combat Global Warming, pp. 29–45. Springer International Publishing, Cham (2017)
10. Stepanov, A.V., Egorova, G.N.: Simulation of heat transfer of heating-system and water pipelines under northern conditions. J. Eng. Phys. Thermophys. **89**(5), 1284–1288 (2016)
11. Yamankaradeniz, N.: Thermodynamic performance assessments of a district heating system with geothermal by using advanced exergy analysis. Renew. Energy **85**, 965–972 (2016)

Study on Estimation Method of Oil-Immersed Transformer Insulation

Ying Liu[(✉)], Xiang-jun Wang, and Xin-yi Jiang

College of Electric Engineering, Naval University of Engineering, Wuhan 430033, China
graceful618@163.com

Abstract. To investigate the aging effects of transformer oil and insulation paper on the extended Debye model parameters and determine oil change time, accelerated aging experiment at 130 °C and PDC test on oil-paper insulation samples were carried out in laboratory. Then corresponding extended Debye models corresponding to every sample were established, and the changing rules of those models' branch parameters with aging of transformer oil and insulation paper were analyzed respectively. It is suggested that the capacitance variation rate of intermediate time constant branch be taken as a threshold for changing transformer oil. It is demonstrated that: (1) transformer oil mainly influences the branch parameters with time constant less than 500 s, oil aging products will increase steering polarization of dipoles in oil and oil-paper interface polarization, the initial part amplitude of depolarizing current rises, the resistance of smaller time constant branch decreases while the capacitance increases; (2) insulation paper mainly influences branch parameters with time constant larger than 500 s, paper aging products only enhance the overall polarization in paper with aging time increasing, the end of depolarizing current curve rises, the branch resistance with larger time constant decreases while the capacitance increases; and (3) the capacitance variation rate of intermediate time constant branch of extended Debye model can be taken as a threshold for changing oil. When capacitance change rate rises over 0.36% and dielectric loss of transformer oil reaches 4.0%, oil should be changed.

Keywords: Oil-paper insulation · Extended Debye model · Depolarizing current · Oil-paper interface polarization · Time constant · Capacitance variation rate

1 Introduction

In high-voltage or super high-voltage power transmission network, oil-immersed transformers are the great majority at present. If the oil-immersed transformer breaks down, power of great area will be cut. Therefore, it is important to prevent transformer against fault.

Aging of oil-paper insulation structure in oil-immersed transformer will weaken its electrical and mechanical properties. Since the winding properties keep stably, life-span of transformer is determined by oil-paper insulation properties.

© Springer International Publishing AG 2018
F. Qiao et al. (eds.), *Recent Developments in Mechatronics and Intelligent Robotics*,
Advances in Intelligent Systems and Computing 690, DOI 10.1007/978-3-319-65978-7_11

Normal methods are adopted usually to measure the oil-paper, such as analysis of oil chromatography paradigmatic, local electro-discharge test, content analysis of aging product, oil-paper polymerization analysis. But those methods are complicated to lift up the core and affected by environment easily [1–3].

Oil-paper insulation samples are aged for 7, 18 and 29 days at 130 °C in this paper and compared with new oil-paper insulation sample. Those samples are polarized and depolarized. Relationship of oil-paper insulation status with the parameters in extend Debye model of are studied and explained. A threshold for changing transformer oil is got from experiment.

2　Transformer Oil-Paper Insulation Equivalent Model

2.1　Extend Debye Model

While the transformer oil-paper is charged with direct-current voltage, two kinds of current will flow in the conductor. One is caused by current carrier; another is caused by polarized dipole. Dipole will be polarized by electric field and return to random status while electric field is removed. The polarization and depolarization process can be described by a circuit, which is composed of an energy consumption resistance R in tandem with an accumulation energy capacitance C. As the transformer oil-paper is compound insulator, polarization and depolarization time of dipole varies in different position of oil-paper. The oil-paper insulation equivalent circuit should be composed of several resistances in tandem with capacitances (Fig. 1).

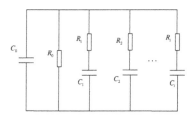

Fig. 1. Extended Debye model for oil-paper insulation

R_0 is insulation resistance, which represents the conductivity of oil-paper; C_0 is geometric capacitance; R_i and C_i represent different relaxation time.

Geometric capacitance C_0 almost does not vary during operation of transformer. It can be measured by capacitance gauge. Other parameters in this circuit can be identified from polarization and depolarization current curve, as listed below

$$R_0 = \frac{U_0}{i_p(t_p) - i_d(t_p)} \tag{1}$$

$$i_d = -\sum_{i=1}^{n} \left[A_i \cdot e^{-t/\tau_i} \right] \tag{2}$$

$$A_i = U_0 \cdot \frac{1 - e^{-t_p/\tau_i}}{R_i} \tag{3}$$

$$\tau_i = R_i \cdot C_i \tag{4}$$

The number of branches is three. The branches are arranged in order of time constant, that is $\tau_1 < \tau_2 < \tau_3$.

3   Preparation of Samples

Samples are prepared in terms of Fig. 2. Normal cellulose insulation paper and 25th transformer oil are selected. Samples are processed before test. Specific procedure is listed in Fig. 3. Polymerization of sample has come to terminal after 29 aging days.

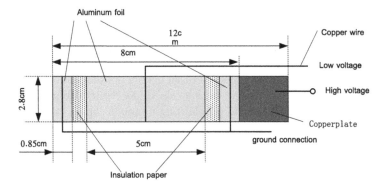

Fig. 2. Oil-paper insulation sample

Samples are separated into 4 groups, which are aged respectively for 0, 7, 18 and 29 days at 130 °C.

Before PDC (Polarization and Depolarization Current) test, aged transformer oil is separated from aged insulation paper. Samples are immersed to remove aged transformer oil. The effect on experiment result caused by immersion at 40 °C is neglected, because the aging environment is 130 °C.

Resolution ratio of PDC test measurement is 1 pico-ampere (pA). Direct-current voltage is 500 V. Test time is 10,000 s.

4   Result and Analysis

4.1   Effect on Parameters of Debye Model Caused by Aged Transformer Oil

4.1.1   Variation of Resistance R_i

From Fig. 4a, it can be found that insulation resistance R_0 and branch resistance R_1 decrease obviously with the aging day of transformer oil increasing. Resistance R_2 decreases slightly and R_3 almost does not vary. Transformer oil is oxidized to alcohol,

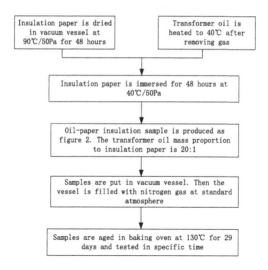

Fig. 3. Thermal aging flowchart of oil-paper insulation

acid, a little of low polymer hydrocarbon, carbon monoxide and carbon dioxide during aging process. Since those oxidization products are polarized, electric conductance of samples increases. The branch with middle time constant describes the polarized interface among insulation oil-paper. Resistance R_2 and R_3 do not vary because oil-paper are not aged.

4.1.2 Variation of Capacitance C_i

From Fig. 4b, it can be found that only capacitance C_1 increases obviously with the aging day of transformer oil increasing. Capacitance C_2 increases slightly in later stage. C_0 and C_3 almost do not vary. Since the structure is not changed during aging period, C_0 maintain constant. C_1 remarks intensity of polarized dipole in oil, and product of aging

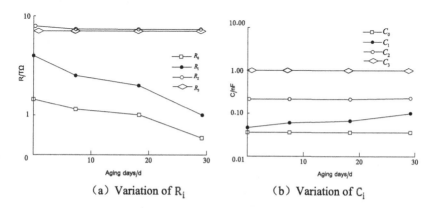

(a) Variation of R_i (b) Variation of C_i

Fig. 4. Variation of slip resistance and capacitance with transformer oil aging alone. a Variation of R_i, b variation of C_i

oil strengthens polarized intensity. C_2 describes intensity of polarized interface among insulation oil-paper. Only oil is aged to some extent, interface can be polarized distinctly. The explanations for C_3 is same as R_3 in 4.1.1.

4.2   Effect on Parameters of Debye Model Caused by Aged Insulation Paper

4.2.1   Variation of Resistance R_i

From Fig. 5a, it can be found that insulation resistance R_0 and branch resistance R_3 decrease obviously with the aging day of transformer oil increasing. R_1 and R_2 almost do not vary. Insulation paper is set off some chemical reaction, such as hydrolysis, pyrolysis, oxide degradation. Cellulose loop is broken to polarity product, such as moisture, low polymer acid, furfural and so on. So R_0 and R_3 decrease obviously. Different from aged transformer oil, aged product diffuses slowly and gathers at inner of insulation paper. The whole insulation oil-paper, not the interface among it, is polarized. So R_2 does not vary, which is a symbol of intensity of polarized interface among insulation oil-paper.

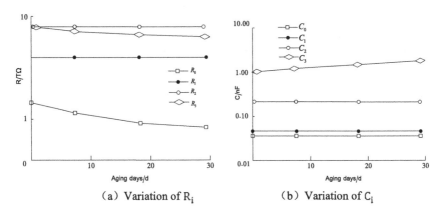

(a) Variation of R_i (b) Variation of C_i

Fig. 5. Variation of slip resistance and capacitance with insulation paper aging alone. a Variation of R_i, b variation of C_i

4.2.2   Variation of Capacitance C_i

From Fig. 5b, it can be found that only capacitance C_3 increases obviously, and C_0, C_1, C_2 almost do not vary. Since the structure is not changed during insulation paper aging period, C_0 maintain constant. The whole insulation oil-paper, not the interface among it, is polarized, which explains the behavior of C_0, C_1 and C_3.

5   Determination of Threshold for Oil Changing

5.1   Variation Ratio of Capacitance

From discussion above, aging polar product strengthens electric conductance of insulation oil-paper. In order to keep insulation, transformer oil needs to renew. But the aging ration varies for different environment and power load, a threshold for oil changing should be determined.

Common indexes of transformer oil are moisture content and dielectric loss. Moisture content fluctuates greatly and is unsuited for determining to change oil. General threshold of dielectric loss is 4.0%. The relationship between variation ratio of capacitance and dielectric loss is established in this paper. Variation ratio of capacitance with middle time constant is defined as below

$$\theta_i = \frac{C_2^{i+1} - C_2^i}{C_2^i \cdot \Delta T_i} \quad i = 0,\ 1,\ 2,\cdots \tag{5}$$

where θ_i is variation ratio of capacitance, F/d; ΔT_i is the interval days between ith and i + 1th PDC test; C_2^{i+1} and C_2^i are values of C_2 in ith and i + 1th PDC test respectively.

5.2   Experiment Result

Dielectric loss of three group of samples attain 4.0% in 14, 16, 12 days. Their variation ratio of capacitance approach each other closely, and their average value is 0.36%.

6   Conclusions

1. Aged transformer oil only affects the branch greatly in Debye model with minor time constant. The corresponding resistance decreases and capacitance increases. Aged insulation paper only affects the branch greatly with major time constant. The corresponding resistance also decreases and capacitance increases.
2. Oil aging products will increase steering polarization of dipoles in oil and oil-paper interface polarization. Paper aging products only enhance the overall polarization in paper with aging time increasing.
3. The capacitance variation rate of intermediate time constant branch of extended Debye model can be taken as a threshold for changing oil. When capacitance change rate rises over 0.36% and dielectric loss of transformer oil reaches 4.0%, oil should be changed.

References

1. Xiaomeng, H.U., Jianggui, D.E.N.G.: Typical case analysis and-countermeasures for 110 kV transformer anti-short circuit ability. High Volt. Appar. **45**(12), 133–136 (2013)
2. Yang, L., Liao, R., Sun, H., et al.: Investigation on properties and characteristics of oil-paper insulation in transformer during thermal degradation process. Trans. China Electrotech. Soc. **24**(8), 27–33 (2009)
3. Pahlavanpour, P., Eklund Martins, M.A.: Insulating paper ageing and furfural formation[C]. In: Proceedings Electrical Insulation Conference and Electrical Manufacturing and Coil Winding Technology Conference [S.l.], pp. 283–288. IEEE (2003)

Development of a Wireless Security Monitoring System for CCEL

Xue-jiao Liu, Xiang-guang Chen[✉], Zhi-ming Li, Ni Du, and Ling-tong Tang

School of Chemistry and Chemical Engineering,
Beijing Institute of Technology, Beijing 100081, China
xgc1@bit.edu.cn

Abstract. A wireless security monitoring system is developed for chemistry and chemical engineering laboratories (CCEL) (or dangerous goods warehouses). The field information of each laboratory node is received through the handheld terminal, which realizes the functions of remote wireless monitoring and controlling. A wireless sensor node for CCEL was designed in this paper to detect the information such as fire, gas leaks, and abnormal sound in the laboratory. Once the insecure information is detected, the corresponding execution devices can be controlled according to the node's decision algorithm. The detection and control information and the field image would be transmitted to the handheld terminal at the same time. It also can alarm quickly in order to minimize the risk of life and equipment loss in the laboratory. The experimental results show that the wireless security monitoring system developed in this paper can detect the on-line danger information in CCEL, so that the management personnel can monitor the safety status of the laboratory and take effective protection measures in time. Therefore, the system developed in this paper has applicative values.

Keywords: Wireless sensor network (WSN) · Chemistry and chemical engineering laboratories (CCEL) · WSN node · Security monitoring system

1 Introduction

At present, in the safety management measures of chemical and chemistry engineering laboratories (CCEL), many colleges and universities draw on the advanced management model and management system at home and abroad, have made great improvements [1]. A wireless security monitoring system for CCEL based on WSN is developed in this paper, which includes a number of monitoring nodes and handheld terminals. The indoor monitoring node can realize real-time monitoring of temperature, humidity, combustible and toxic gases concentration, smoke, flame and so on. And the node can transmit data, alarm information, field image to handheld terminal, so that to quickly alarm and timely take measures for the occurrence of fires, combustible gases or toxic gases leak.

In the security alarm system, the fire monitoring system plays an important role. According to the Chinese national standard "Fire automatic alarm system design specifications", the fire monitoring system generally consists of the fire detector, input and output modules, fire alarm controller and fire linkage control equipment and other

© Springer International Publishing AG 2018
F. Qiao et al. (eds.), *Recent Developments in Mechatronics and Intelligent Robotics*,
Advances in Intelligent Systems and Computing 690, DOI 10.1007/978-3-319-65978-7_12

components [2]. Currently, the teaching and research activities of CCEL in colleges and universities are becoming more and more frequent, and the number of research and comprehensive experiments has been greatly increased, and the safety of laboratory has become more and more prominent [3]. Traditional data monitoring and management occupies many system resources, the hardware and software are limited. The traditional databases can no longer meet the requirement of the majority intelligent devices and cannot be applied on these platforms [4]. Nowadays the management of the domestic laboratories in the research institute and universities has issues of poor real time, high cost and low precision. It is difficult to determine the quality of the environment of the laboratory [5]. In view of the present situation and existing problems of laboratory security management in colleges and universities, it is an important subject to do laboratory security management and prevent laboratory safety accidents [6]. The security management of CCEL has far-reaching influence on the formation of students' world view, outlook on life and values [7].

Reference [8] summaries from aspects of the laboratory safety education and training, safety and security facilities, safety management level and safety management institution and so on, and puts forward practical measures to strengthen laboratory safety. Reference [9] makes a brief comparison of CCEL between China and the United States in aspects of humanistic management, safety education, infrastructure construction, drug management, use management and institutional setting. As mentioned in reference [1], many colleges and universities learned from the advanced management model and management system at home and abroad in CCEL safety management measures and have made great improvements. In order to improve the reliability of the wireless security monitoring system, reference [10] is aiming at the diagnose of the wireless security monitoring system for CCEL.

2 Framework Design of Wireless Security Monitoring System

The system mainly includes several monitoring nodes and handheld terminal. The monitoring node can detect the indoor temperature, humidity, concentration of combustible and toxic gases, smoke, flame and so on, and also has image processing capability. It can achieve the flame identification and alarm using fusion technology of CCD and sensors and can remote transmit images wirelessly. The terminal uses color touch screen as a human-computer interaction interface, so that its operation is intuitive and fast. It can establish data wireless communication with the field sensor nodes. The information interaction mode between the handheld terminal and monitoring nodes is shown in Fig. 1.

At the same time, the monitoring nodes in different laboratories can communicate with each other. According to the speed of combustible gas or fire, the monitoring nodes in the non-disaster laboratory of the building will also send out different warnings through sound and light alarm to remind the building personnel or to guide the staff to evacuate the building.

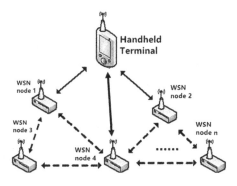

Fig. 1. The information interaction mode between the handheld terminal and the monitoring node

3   Hardware Design of the Wireless Security Monitoring System

Monitoring nodes need to acquire camera data, compress and decompress picture, so the nodes use STM32F103ZET6 processor with high frequency, 512 kB on-chip flash and 64 kB on-chip RAM. The processor is connected to the sensor, digital camera, voice module, LED alarm light, wireless communication module and color touch screen. The hardware block diagram of the monitoring node is shown in Fig. 2.

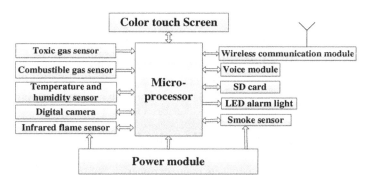

Fig. 2. The hardware block diagram of the monitoring node

Signals of sensors would be transmitted to single-chip microcomputer (SCM) and processed simultaneously. When the signal strength exceeds the threshold, the SCM control the voice module and LED alarm light module to provide an alarm. At the same time, sensor parameter values and field pictures in JPG would be transmitted to the handheld terminal through the wireless communication module.

The monitoring node can detect the indoor hazardous gas concentration, flame and other information, and remote transmit the field images wirelessly. When hazardous gases leak or smoke or flames appears, the person in charge of the laboratory safety can receive alarm information and field data at first, and also can view the field images on the terminal. Therefore, the managers can make further judgments and take disposal means at first time.

The handheld terminal uses STM32F103ZET6 as the core component, and it also includes external transmission module, voice module, power module, SD card and color touch screen. The hardware block diagram is shown in Fig. 3. In usual, the handheld terminal receives all sensor parameters through the wireless transmission module. When the indoor security monitoring node detects the danger information, the compressed field image is transmitted to the handheld terminal and display on the color touch screen after storage and decompression. The image resolution is 640 × 480 or 320 × 240, image compression ratio is adjusted to 64 files to adapt to different network environments and the actual needs.

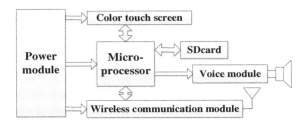

Fig. 3. The hardware block diagram of the handheld terminal

The terminal uses a color touch screen as a man–machine interaction interface, the operation is intuitive and rapid, and it can communicate with the sensor nodes through the wireless sensor network. When the field sensor detects combustible gas or fire, the corresponding node icon on the terminal display screen automatically turns red and displays the dangerous gas concentration and field image. At the same time, there will be voice prompting the operator to detect danger and prevent missing alarm information.

4   Software Design of Wireless Security Monitoring System

As the indoor monitoring node connects many peripherals. Running these peripherals, acquiring and storing, compressing and decompressing image all need to take up the processor storage resources, so need to design a memory manager program. The memory manager program establishes a memory management table, and count the situation of free and occupancy at any time to allocate memory to new tasks or free memory after completing old tasks. This allows the limited processor memory resources to be quickly allocated and used efficiently. SCM program uses C language and the software environment is keil uvision3 V3.90. Each module subroutine in the indoor monitoring node is completed by interrupt mode. Including gas sensor acquisition interruption, temperature and humidity sensor acquisition interrupt, camera capture interrupt, smoke particle sensor acquisition interrupt and timer interrupt. The main program flow chart is shown in Fig. 4.

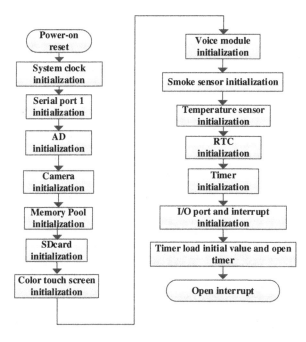

Fig. 4. Main program flow chart of monitoring node

The task of the handheld terminal is to receive the field sensor data and field image, and display the data and picture after decompression on the color touch screen. The main

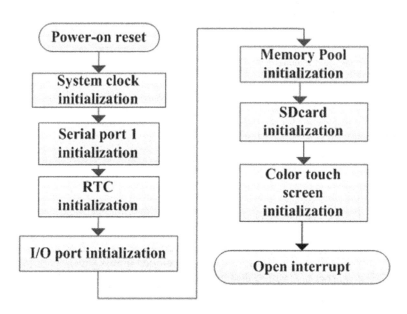

Fig. 5. Main program flow chart of the handheld terminal

program flow chart is shown in Fig. 5. The single-chip microcomputer control the ports
and receive the instructions through interrupt way. The wireless communication module
receives the subroutine.

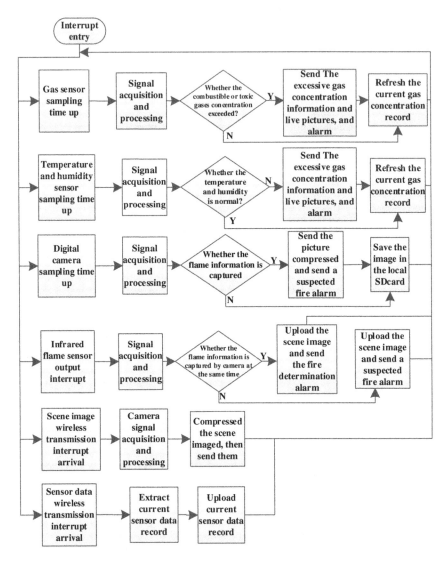

Fig. 6. The interrupt program flow chart of the monitoring node

The interrupt program flow chart is shown in Fig. 6.

The security monitoring system for CCEL developed in this paper can transmitted
data of several sensors and field image after decompression to the handheld terminal.
As the nodes often transmit large amounts of data, the wireless communication module

undertakes busy data transfer. It would have the phenomenon of power depletion or power down if it is put to use for a long time.

5   Test of the Indoor Security Monitoring Node

In order to carry out the testing experiments, two indoor security detection nodes are developed in this paper and placed in two laboratories respectively.

Indoor security monitoring node can real-time detect indoor temperature, humidity, combustible gas concentration and smoke. Under normal circumstances, all indoor security detection nodes displayed on LCD screen is coloured (such as pink). The temperature measurement accuracy and humidity measurement accuracy of Sensor SHT11 have been determined in the factory, respectively are ± 0.4 °C and 3%. The readings on the LCD screen of Sensor 4P-90 need to be adjusted to zero by adjusting the potentiometer in the air where is no combustible gas.

Programing to simulate the indoor smoke, so that the buzzer of the indoor security detection node would give out a sharp alarm and the smoke state on the LCD screen is displayed as "SS:Y". In Fig. 7, The letter N stand for there is no smoke, Y stand for there is smoke in the laboratory. Put some combustible gas near the combustible gas sensor with the lighter, the buzzer also give out a sharp alarm and the LCD screen shows the concentration change of the combustible gas (the alarm limit set to 2%). We can see the current combustible gas concentration is 3.2%, as shown in Fig. 7. When the lighter is removed, combustible gas concentration reading will slowly drop to 0%.

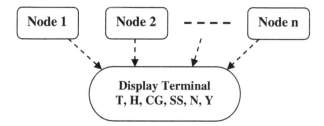

Fig. 7. The schematic diagram of indoor security monitoring nodes. *T* temperature, *H* humidity, *CG* combustible gas, *SS* smoke state

After the actual test of the indoor security detection node, the temperature and humidity detection, combustible gas concentration detection and smoke state detection are proved normal, response sensitively and alarm timely.

6   Conclusion

A fire, toxic gas leak and abnormal sound real-time monitoring system based on WSN for CCEL or dangerous goods warehouse is developed in this paper. The field data of combustible and toxic gas concentration, temperature and humidity, fire, smoke particle concentration and the image after compression can be transmitted to the handheld terminal. The handheld terminal would display the parameters after receiving data above, so that the safety responsible person can make a comprehensive judgment on the scene and take measures at first time.

Acknowledgements. Scientific research in this paper was supported by the Instrumentation Developed Project of Beijing Institute of Technology.

References

1. Qiu, Q., Luo, Z.K., Lv, W.Z., Wang, F.: Investigation on university safety management of chemistry laboratory in domestic and abroad. Guangzhou Chem. Ind. **38**(5), 272–274 (2010). (Chinese)
2. Bai, Z.W.: Application status and development trend of fire monitoring system in intelligent building. Secur. Sci. Technol. **2**, 24–26 (2003). (Chinese)
3. Bao, M.Q., Zhang, Y., Zhang, S.C.: Analysis of college chemistry laboratory safety problems and exploration of safety management measures. Exp. Technol. Manag. **29**(1), 188–191 (2012). (Chinese)
4. Nie, W., Xu, J., Liu, Y.: Design of automatic closed system about gas outburst monitor and its role of the reducing security cost. In: Proceedings of the 2nd International Conference Mine Hazards Prevention and Control, pp. 538–543. Atlantis Press, Paris (2010)
5. Wang, P., Wang, Z.: Design and implementation of open computer lab monitoring and management system. Comput. Mod. **11**, 125–128 (2007)
6. Liu, Y.Q., Bao, H.G., Song, S.J.: Research on university laboratory safety management. Exp. Sci. Technol. **10**(1), 173–175 (2012). (Chinese)
7. Qiu, N.W., An, X.C., Jia, J.W.: Challenges and solutions confronted with for laboratory safety management in colleges and universities during new period. Exp. Technol. Manag. **29**(1), 181–185 (2012). (Chinese)
8. Feng, J., Xiong, W., Qiu, H.D., Long, Y.X.: Exploration on strengthening the safety of chemistry and chemical engineering laboratory. Guangdong Chem. Eng. **43**(6), 209–210 (2016). (Chinese)
9. Qiu, H.D., Li, G., Wang, W., Long, Y.X.: Comparison on safety management of chemistry and chemical engineering laboratories between universities of China and United States. Exp. Technol. Manag. **31**(4), 203–209 (2014). (Chinese)
10. Li, Z.M., Chen, X.G., Wu, L., Wei, Y.F.: Fault diagnosis method for wireless security monitoring system in chemistry and chemical engineering laboratory. Trans. Beijing Inst. Technol. **35**(10), 1062–1066 (2015). (Chinese)

Research and Application of a Developing Approach for PIMS Software

Ni Du, Ling-tong Tang, Xue-jiao Liu, and Xiang-guang Chen(✉)

School of Chemistry and Chemical Engineering, Beijing Institute of Technology,
Beijing 100081, China
xgc1@bit.edu.cn

Abstract. Considering of the poor universality and high development costs in different productive process information management systems (PIMS), this paper presents a new developing approach for PIMS software. It applied the layered-design theory and Modular program development technology to design a general PIMS, which allows users alter the HMI by simple ways like Keyboard inputting and pictures inserting. Mean while, the approach proposed makes it possible to reselect monitoring data when field databases changed. Experimental results demonstrate that the approach proposed in this paper has a prominent personalized design, and it can be applied to different production process information management system.

Keywords: Process information management system (PIMS) · Layered-design · Modular program development · Personalized HMI design

1 Introduction

Supplies and energy in industrial production process fluctuating all the time cause many problems in monitoring and control system. To solve these problems, process information management system in industrial production comes into being.

Process information management system (PIMS) is a process oriented platform, for information integration and management, which is designed to cooperate with common control systems such as Distributed Control System (DCS), Programmable Logic Controller (PLC), Intelligent instruments and so on [1]. PIMS can realize many functions such as production data acquisition, real-time process view, trends display, alarm record, historical trend report generation, data storage, production statistics report generation under the network environment [2, 3].

PIMS has advantages of less investment and high efficiency by making full use of the enterprise existing computer local area network resources. Meanwhile, it can integrate all kinds of existing control system into an information platform to provide field data for more advanced management networks such as EPR, CRM, OA, MIS. In other words, PIMS can help the data and reports of management system more timely and accurate [4]. Therefore, since the concept of process information management system was put forward, a lot of enterprises have invested so much to develop all kinds of PIMS to improve their information and intelligence of production process [5, 6]. According

© Springer International Publishing AG 2018
F. Qiao et al. (eds.), *Recent Developments in Mechatronics and Intelligent Robotics*,
Advances in Intelligent Systems and Computing 690, DOI 10.1007/978-3-319-65978-7_13

to user's access ways, PIMS can be divided into C/S mode and B/S mode and C/S and B/S mode mixed mode [7–9].

Although PIMS has been widely applied in all kinds of process industries, many small and medium-sized enterprise (SMEs) has not developed their own information management system due to lacking of funds and understanding of information management. Without PIMS, enterprise is not able to ensure the quality of products or increase the profit. So it is significant to develop a general and simple PIMS to alleviate the financial press. On SMEs in improving the information management and efficiency of production.

2 Design

2.1 System Structure Design

PIMS is the bridge of fields and clients. Specially, PIMS collects and processes the field data, and then transfers them to the clients.

In Fig. 1, this system is divided into three layers, which contains fields layer, application layer and client layer. As is shown in Fig. 1, fields layer transfer the field production data through PLC, DCS and connect with database. Then the database system carries out on the field data's collection, analysis, processing and storage. Databases adopt distributed structure layout, form an database system with physical distribution but logic stored through the computer network. This layout can meet cases that data bases are stored in different computers or different locations. Data transmission interfaces employ OPC (OLE for Process Control) data transmission communication interface standard as the unified data transmission interface, and the field controllers such as PLC, DCS implement dynamic data transmission with data bases by installing the OPC server.

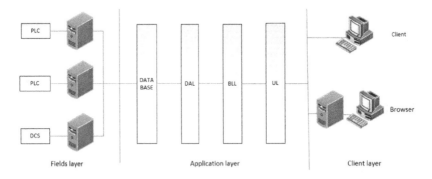

Fig. 1. Layered structure of PIMS

Application layer complies with the standards of the three-layer structure developing model to meet the demand of generalization. Three-layer structure developing model is a strictly hierarchical method, namely the data access layer (DAL) can only be access to business logic layer (BLL) and BLL can only be accessed by the user layer (UL). Specially, users send requests to BLL through UL, BLL completes the relevant business rules and logical relationship, and reads data from databases by accessing DAL. At last, BLL returns display data to the UL. So users can alter the HMI display and choose different database by themselves without worrying about affecting the business logic layer.

User layer applies a common access pattern-Combination of B/S and C/S. B/S is the abbreviation of Brower/Server, clients just need to install a Browser, then the Browser interacts with the database data by the Web Server. C/S is also called Client/Server, the Server usually adopts high performance PC, workstations, or minicomputer. In order to guarantee the system security, the main application functions are based on C/S model, the browser users can only be authorized to view field data and the production reports.

2.2 System Function Design and Implementation

In order to meet the demand of information management of SMEs and the character of both generality and convenience, we classify and integrate the basic functions of process information management system which is shown in Fig. 2.

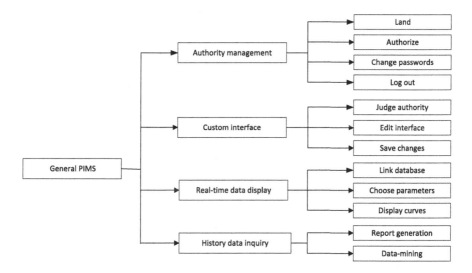

Fig. 2. Block diagram of the system functions

The system in Fig. 2 is divided into four function modules, there are authorization management, custom interface, real-time data display and the historical data inquiry respectively. In the module of authorization management, the program realizes the basic functions such as login, logout, authorization, change passwords. In the module of custom interface, as show in Fig. 3, the program can judge the user's authorization and alter the HMI display easily by keyboard inputting or mouse inserting. This module takes the first step to a general system. The real-time display module is used to implement the functions of database connection, selected parameters display and real-time data display. Figure 4 shows the core logic of the third modules. At first, the program judges if user has authorization to change the database, After the validation, user can connect another database and the program can identify the connected database column names automatically. Then user can choose field parameters to be monitored. It's worth mentioning that this system can only show two parameters at screen due to guarantee the interface clean and response speed. The last module is history data inquiry, it can complete report generation and data modification.

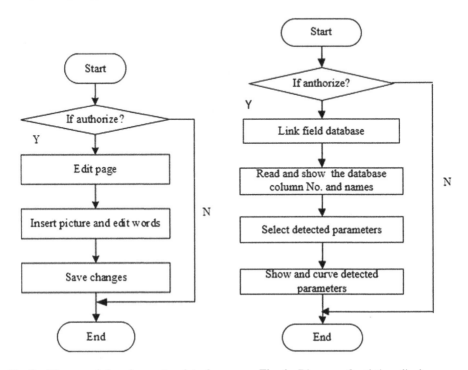

Fig. 3. Diagram of changing custom interface   **Fig. 4.** Diagram of real-time display

3   Application

This system has been applied to tire production process and chlortetracycline fermentation process, users can switch two systems easily by changing the login interface and data display interface without entering the bottom program.

3.1   Chlortetracycline Fermentation Process Information Management System

Chlortetracycline (CTC) is a kind of broad spectrum antibiotics, belongs to tetracycline class in medical treatment which has been widely used in animal husbandry and agriculture. CTC fermentation is multiple inputs and multiple outputs process of microbial breeding system, it has complex running mechanism and strong nonlinear, Fig. 4 illustrates CTC fermentation production process [10]. CTC fermentation PIMS can monitor the real-time production parameters such as temperature, PH, dissolved oxygen, and so

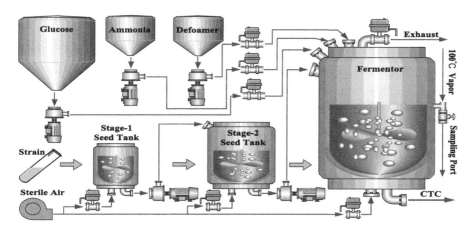

Fig. 5. CTC fermentation process flow diagram

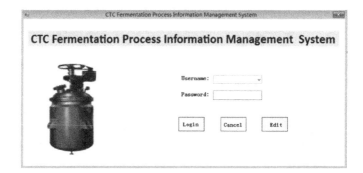

Fig. 6. Interface of CTC fermentation PIMS

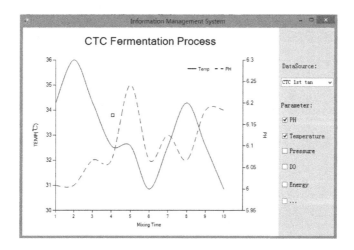

Fig. 7. Real-timedata display interface

on. It is easier to find problems such as bacterial infection and improve the production efficiency of CTC fermentation (Fig. 5).

CTC fermentation PIMS's interface and real-time data display is designed as shown in Figs. 6 and 7.

3.2 Tread Production Process Information Management System

As shown in Fig. 8, tread production process includes six-processes: mixing, squeezing, rolling, ply cutting, bead shaping, and sulfuring.

Fig. 8. Tread production process

By Changing the IP address and password of database in configuration files, the system can be easily switched CTC fermentation PIMS to tread production PIMS. The

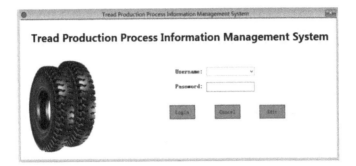

Fig. 9. Interface of tread production PIMS

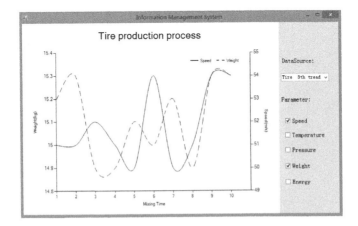

Fig. 10. Real-time data display of tread production PIMS

system will recognize column names of the new database and show the important parameters automatically. Then users can alter the interface and select the monitor parameters by themselves. The tread production PIMS's interface and real-time data display are designed as shown in Figs. 9 and 10.

4 Discussion

The experimental results show that the approach proposed in this paper is feasible to develop a general and simple PIMS. The system allows administrators alter the HMI by simple ways like Keyboard inputting and pictures inserting and the approach makes it possible to reselect monitoring data when field databases is changed. We believe this approach should be introduced to many SEMs to improve their efficiency for development management software.

5 Conclusions

This paper has presented a framework for general process information management system. By layered-design and modular-programming, this system can realize the basic functions of PIMS and meet the demands of production process. We discuss the user interface management module and real-time data display module emphatically. The first module realizes changing the HMI by keyboard inputting and pictures inserting, the second one makes it possible to connect other production field data bases and reselects new monitor parameters. Therefore, it is significant and valuable for SMEs to reduce their development cost and improve the developing level of process information management. In the next phase of research work, we will investigate how to add new functions such as parameter prediction modeling and data mining.

Acknowledgments. This scientific research work is financially supported by Beijing Research & Design Institute of Rubber Industry, China (No. 20151041012). The authors also thank Double Coin Holdings Ltd. for providing the dataset of industrial rubber mixing process.

References

1. Wang, Z.: Enterprise information and management change. Renmin Univ. China publ. **8**, 3–12 (2001). (Chinese)
2. Shao, H.: Development tread of automation and advanced control in process. Autom. Panor. **2**, 38–42 (2002). (Chinese)
3. Gao, C., PIMS, : Process information management. Syst. Electric Times **2**(4), 100–102 (2003). (Chinese)
4. Li, X., Wang, X.: Process information management system based on CIPS. J. Hun. Univ. (Nat. Sci.). **25**(1), 81–84 (1998). (Chinese)
5. Wang, B., Li, J., et al.: Production information management system application in fine chemical industry. Digit. Pet. Chem. **5**, 53–56 (2007). (Chinese)

6. Yi, X.: The design of process information management system and its application in chemical production. South China University Technology, Shanghai (2013). (Chinese)
7. Fan, Y., He, H.: The research of 2-tier and 3-tier structure based on the client/server architect. Appl. Res. Comput. **18**(12), 23–24 (2001). (Chinese)
8. Tan, G.: B/S structure of the software project practice [M]. Publishing House of Electronics Industry, Beijing (2004). (Chinese)
9. Yang, X., Zhou, C., et al.: Research on general aviation safety information management system based on B/S. J. Civil Aviat. Flight Univ. China **13**(3), 13–15 (2012). (Chinese)
10. Mayeli, P.-C., Cristina, C.-H., Perez-Carrillo, E., et al.: Fate of free amino nitrogen during liquefaction and yeast fermentation of maize and sorghums differing in endosperm text. Food Bio Prod. Process. **91**(1), 46–53 (2013)

Study of Prediction Methods for Contamination in the Chlortetracycline Fermentation Process

Ling-tong Tang[1], Jian-wen Yang[1], Xiang-guang Chen[1(✉)],
Min-pu Yao[2], Su-yi Huang[2], De-shou Ma[2],
Qing Yu[2], and Biao Zhou[2]

[1] School of Chemistry and Chemical Engineering,
Beijing Institute of Technology, Beijing 100081, China
xgcl@bit.edu.cn
[2] Pucheng Zhengda Fujian Biochemical Co. Ltd,
Pucheng 353400, Fujian, China

Abstract. During the chlortetracycline (CTC) fermentation process, if the fermenter is invaded by other harmful bacteria, the fermentation broth may be contaminated. Once this occurs, the broth must be discharged for preventing other schedules of fermentation production from contamination, otherwise it will waste more raw materials and bring great economical damage. In order to acquire some important comprehensive contamination feature information, an information fusion method is proposed in this paper based on Desert-Smarandache theory (DSmT), combining multiple process information that indirectly hints contamination. And experimental results based on field data show that the method can predict whether the process of CTC fermentation is contaminated, so if this method is applied into real fermentation production, the safety performance of production process will be improved.

Keywords: Chlortetracycline fermentation · Information fusion · Fermentation contamination prediction · DSmT

1 Introduction

Chlortetracycline (CTC),a sort of tetracyclic broad-spectrum antibiotic, is mainly used in medicine, agriculture, and animal husbandry. It has strong antibacterial effects on Gram-positive bacteria, chlamydia, spirochetes, rickettsia, negative bacteria, mycoplasma, and other bacteria [1, 2]. Meanwhile CTC is characterized by bacterial inhibition, high feed efficiency, promoting animal growth, less residue, lower production costs, mature technology, and so on. However, CTC is a kind of typical secondary metabolite whose microbial synthesis mechanism is very complicated, and it is difficult to accurately structure the models of the biochemical reaction mechanism [3–5]. Therefore, to obtain high-quality and high-yield of CTC, in addition to the careful selection of Streptomyces aureofaciens strains and researching more suitable medium for strains' growth and breeding, the optimization control of fermentation process is an effective method [6]. However, the production process of CTC fermentation is a

F. Qiao et al. (eds.), *Recent Developments in Mechatronics and Intelligent Robotics*,
Advances in Intelligent Systems and Computing 690, DOI 10.1007/978-3-319-65978-7_14

multi-input and multi-output microbial breeding growth system, and presents strong non-linear characteristic with complex running mechanism [7–9]. Making the fermentation process better through optimization control is still one of the main problems in the field of fermentation engineering [10].

Contamination refers to the invasion of other microorganisms that destroy the fermentation production, and it has been a big problem in the industrial fermentation production process. The number of rapid breeding phage or bacteria is soon significantly greater than the culture strains, and these harmful microbes release a lot of byproduct, badly inhibiting the growth and reproduction of the culture strains [11]. If a large fermenter is contaminated and leads to the discharge of fermentation broth, such a major production accident will waste a lot of production raw materials. This accident not only causes significant economic losses, but also pollutes the environment [12]. Even a small scale or part of the production stage occurs the bacteria contamination, the yield and quality of the metabolites will be decreased. The consequences caused by the fermentation process contamination are as follows: (1) The medium nutrients are converted to the products that do not need. (2) The living condition of the normal production bacteria is changed, and the normal fermentation production process cannot be established. (3) Produce the degrading enzyme decomposing the metabolites [13]. In this paper, we mainly study the prediction method of contamination in the CTC fermentation process.

2 Contamination Prediction Method

At present, chlortetracycline fermentation process is still very complicated. First, Streptomyces aureofaciens strains are removed from a sand spores tube and inoculated into the slant for culture, and then transferred to the sub-slant for culture. These two processes activate the strains in the incubator. After that, a small piece of slant spores is inserted into the shake flask to enlarge the culture. Finally, through the amplification culture of the primary and secondary seed pot, strains go to a large volume of fermenter for fermentation production. The schematic diagram of chlortetracycline fermenter is shown in Fig. 1.

Bacterial contamination involves the invasion of other microorganisms that hinder the fermentation culture during the fermentation process. Because many factors will cause contamination and some reasons have not yet been determined, it is difficult to avoid this kind of accidents. Therefore, establishing the soft-sensor model of the contamination characteristic variables of chlortetracycline fermentation process is one of the most important scientific problems.

2.1 Soft Sensor Model Solution of Contamination Prediction in the CTC Fermentation Process

Once contaminated, the physicochemical characteristics or biological characteristics of the fermentation broth are very different from those in the normal state. The emergence of this abnormal phenomenon occurs mainly in two stages of the fermentation process: seed culture stage and the production stage in the fermenter. Fermentation

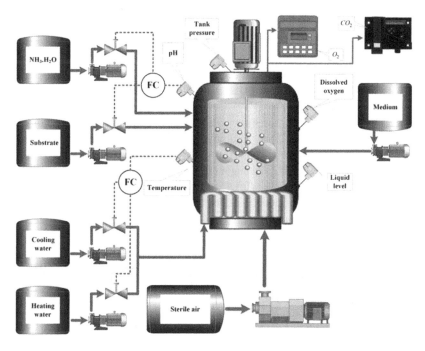

Fig. 1. The schematic diagram of chlortetracycline fermenter

contamination is hard to be seen with the naked eye, it is necessary to determine the specific situation with other detection methods and equipment. Conventional sterile detection methods mainly include plate culture method, broth culture method and microscopic examination method. The fermentation contamination detection with information fusion and reasoning is a new method based on DSmT theory, developing a soft sensor model in the fermentation process combined with physical methods, chemical methods, biological methods, detection instruments and data driven. The design scheme of contamination prediction system in the CTC fermentation process is shown in Fig. 2.

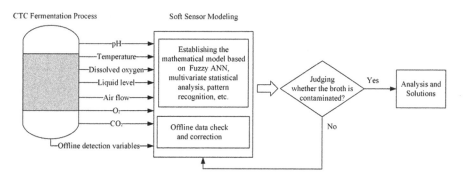

Fig. 2. Contamination prediction system in the fermentation process

2.2   Fusing Contamination Information in the CTC Fermentation Process

Building a contamination prediction system is an important part to realize the optimal control strategy of CTC fermentation process. The crux of the presented method is to fuse the various variables information got from online sensors and soft-sensors into a control decision. Here are some available variables: potential of hydrogen (pH), temperature (TE), fermentation time (FT), dissolved oxygen (DO), agitation rate (AR), ammonia accumulation (AA), supplying glucose accumulation (SGA), motor current (MC), liquid level (LL), air flow accumulation (AF), oxygen uptake rates (OUR), carbon dioxide excretion rate (CER), amino nitrogen concentration (AC), biological value (BV), viscosity (VS), and total sugar content (SC). As shown in Fig. 3, the entire information fusion process consists of five parts, including: (a) data preprocessing; (b) establishing the SRWNN model; (c) detecting variables; (d) fusing contamination information with DSmT; (e) making a final decision.

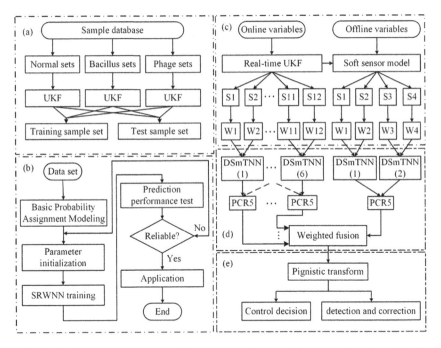

Fig. 3. Implementation process of fusing contamination information based on DSmT

3   Experiment Analysis

The whole CTC fermentation process consists of four main stages, namely, preparation of strains, amplification of primary seeds and secondary seeds, and fermentation process in the fermenter. In these phases, the possible links that introduce other bacteria into production process include air leak in the pipes, offline samples collection, tank-to-tank transfers, agitator malfunction, media sterilization procedures, and so on.

Some variable data has been obtained from the real-time contamination prediction experiment during the actual production process. The experimental data is divided into three types of situations to analyze and discuss. Figures 4, 5 and 6 have described these three different contamination conditions. The three curves in every figure respectively express the probability of the normal state, bacillus invasion and phage invasion. The next section, we will explain the distribution trends of the three curves in detail.

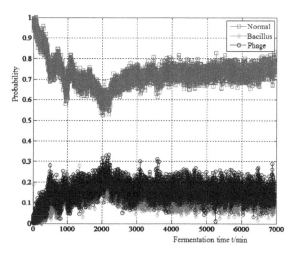

Fig. 4. Probability change trend of normal condition

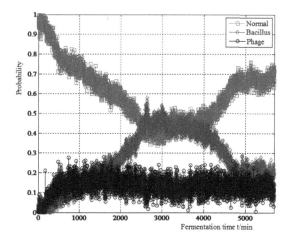

Fig. 5. Probability change trend of bacillus-invaded condition

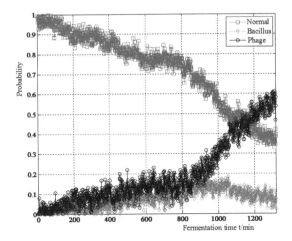

Fig. 6. Probability change trend of phage-invaded condition

In Fig. 4, the 'Normal' trajectory is above 60% although it approaches to 50% at about 2000 min, but never crosses the boundary, and the other two curves have been fluctuating around 15% all the time. Hence, the culture process can be regarded as a normal condition. In Fig. 5, the probability of 'Bacillus' curve exceeds the lowest threshold by about 40% from about 2500–4000 min, as the track of 'Normal' is below 50% and that of 'Phage' is about 10%. This condition means that the overgrowth of bacillus would harm the fermentation strains. Figure 6 describes a phage-infected batch. At about 1100 min, the probability of 'Phage' is over 40%, and the trajectory of 'Normal' is less than 50%. whatever bacillus or phage invades the fermentation process, it is necessary to take some remedial measures.

4   Conclusions

Once the CTC fermentation broth is contaminated, it will affect the quality and yield of the product, and even cause significant economic losses. There are many factors that lead to the emergence of bacterial contamination, but many companies cannot achieve real-time online detection because of the equipment prices and maintenance costs. The general methods to judge the contamination of the broth are microscopic examination and bacterial culture, however, their poor timeliness often causes delays in the prevention and control of contamination. In this paper, a prediction method with information fusion based on DSmT is proposed, which fuses the information of measured and unmeasurable variables and can accurately predict the trend of contamination in the CTC fermentation process in time.

Acknowledgments. The research work was supported by Pucheng Zhengda Fujian Biochemical Co. Ltd. We also thank Charoen Pokphand Group for providing the industrial datasets offed-batch CTC fermentation process.

References

1. Li, Y.Q., He, X.R.: Study on breeding chlortetracycline streptomyces by microwave mutagenesis and laser mutagenesis. Chin. J. Biotechnol. **14**(4), 445–448 (1998)
2. Fu, J.L.: Optimization and metabolic validation of chlortetracycline fermentation process based on parameter. East China Univ. Sci. Technol. (2003)
3. Pang, G.F., Cao, Y.Z., Zhang, J.J., et al.: Simultaneous determination of oxytetracycline, tetracycline, chlortetracycline and doxycycline in poultry meat by high performance liquid chromatography. J. Instrum. Anal. **24**(4), 61–63 (2005)
4. Wang, H.Z., Luo, Y., Xu, W.Q., et al.: Ecotoxic effects of tetracycline and chlortetracycline on aquatic organisms. J. Agro Environ. Sci. **27**(4), 1536–1539 (2008)
5. Liu, J.G., Zhao, Z.H., Zhang, B.K., et al.: Effects of a Chinese herb feed additive and chlortetracycline on the growth of growing piglets and their mechanism. J. Southwest Agric. Univ. **27**(6), 877–880 (2006)
6. Li, W., Yan, Z.: Determination of oxytetracycline, tetracycline and chlortetracycline residue in animal meat. Chin. J. Health Lab. Technol. **11**(6), 730–731 (2001)
7. Olsson, L., Nielson, J.: On-line and in situ monitoring of biomass in submerged cultivations. Trends Biotechnol. **15**(12), 517–522 (1997)
8. Wang, Y.J., Fan, Y.: Studies of on-line and in-situ measuring method for biomass concentration. Prog. Biochem. Biophys. **27**(4), 387–390 (2000)
9. Wang, B., Sun, Y.K., Huang, Y.H., et al.: Soft sensor method for lysine fermentation process based on adaptive FLSVM. Chin. J. Sci. Instrum. **32**(2), 469–474 (2011)
10. Wang, J.L., Feng, X.Y., Yu, T., et al.: Research progress of optimal control techniques in fermentation process. Chem. Ind. Eng. Prog. **27**(8), 1210–1214 (2008)
11. Camu, N., González, A., De Winter, T., et al.: Influence of turning and environmental contamination on the dynamics of populations of lactic acid and acetic acid bacteria involved in spontaneous cocoa bean heap fermentation in Ghana. Appl. Environ. Microbiol. **74**(1), 86–98 (2008)
12. Lv, D.M., Cai, L.P., Shi, W., et al.: Study on the problem of bacterial contamination. Coal Chem. Ind. **3**(2), 86–87 (2013)
13. Junker, B., Lester, M., Leporati, J., et al.: Sustainable reduction of bioreactor contamination in an industrial fermentation pilot plant. J. Biosci. Bioeng. **102**(4), 251–268 (2006)

Technical Study of an Improved PWM Controlling DC/DC Converter

Hui Li[(✉)]

Department of Mechanical Engineering, Hunan Institute of Science
and Technology, Yueyang, 414006, China
932101798@qq.com

Abstract. For the shortcoming of the conventional PWM DC/DC converter has huge switching loss for the main switches in the case of high switching frequency, the circuit system has the disadvantage of low efficiency. An improved PWM DC/ DC converter-PWM DC/DC converter for phase-shifted controlling is discussed, and the new circuit designs the auxiliary circuit for the conventional PWM DC/DC converter, and the purpose is to realize the soft switching of the main switches. Compared to conventional PWM DC/DC circuit, all the main switches work in a state of soft-switching for the new PWM DC/DC circuit. So the new PWM DC/DC circuit can eliminate the opening loss of the main switch and improve the efficiency of the system, and can also ensure the service life of the circuit. Finally the experimental results coincide with the theoretical analysis.

Keywords: DC/DC converter, phase-shifted controlling, ZVC · Simulation

1 Introduction

Now the PWM DC/DC converter is widely used in pulse width modulation (PWM) technology. When the control mode is adopted, the switching devices are turned on and off at high voltage and high current. As the switches are not an ideal device, the switches with opening loss and turning off losses, and there are referred to as witching loss. Under certain conditions, the switch loss in each switch cycle is constant, so the converter switching loss and switching frequency is proportional to the total. The switching frequency is higher, and the switching loss is also bigger, so the efficiency of the converter is correspondingly lower. On the other hand, the existence of the switching loss affects the improvement of the switching frequency, which limits the miniaturization and light weight of the converter. In addition, the traditional switching device for PWM controlling for hard switching mode, it has the problems of inductive turn off, capacitive switch and diode reverse recovery of the three defects, in addition to the big problems of turn-on and turn off loss. Thus these problems further hinder the switching device frequency increasing and the miniaturization of the circuit and light weight [1–3]. Aiming at the common PWM DC/DC converter of four switches with great loss in the hard switching mode, the paper studies the new full-bridge and zero-voltage switching PWM circuit. In the ordinary full-bridge PWM DC/DC converter by

© Springer International Publishing AG 2018
F. Qiao et al. (eds.), *Recent Developments in Mechatronics and Intelligent Robotics*,
Advances in Intelligent Systems and Computing 690, DOI 10.1007/978-3-319-65978-7_15

increasing the resonant inductor, resonant capacitor and diode, it can realize the soft-switching of four switches and reduce the turn-on and turn off losses, and also improve the efficiency of circuit system.

2   The Work Principle of Improved PWM DC/DC Converter

The conventional PWM DC/DC converter consists of full bridge inverter and an output full-wave rectifier circuit, as shown in Fig. 1. V_{in} is the DC input voltage, Q_1 and Q_3, Q_4 and Q_2 constitute two of the converter bridge arms, known as the leading bridge arm and lagging bridge arm. VD_{r1} and VD_{r2} are rectifier diodes, and L_f is the output filter inductor, and C_f is the filter capacitor, and R_L is the load. In Fig. 1, the four switches Q_1 to Q_4 are working in the hard switching state, and the switching loss is very large. In order to reduce the switching losses of four switches, adding a resonant inductor L_r, each switch in the resonant capacitor C_1 to C_4 parallel and anti parallel diode D_1 to D_4 in the main circuit shown in Fig. 1. Through creation of some external conditions, and by adding the components and the original transformer, the circuit can realize soft-witch for four switches, and the circuit working diagram as shown in Fig. 2. In the first half and the second half of the cycle, the each cycle of improved PWM DC/DC converter has seven working modes, and the working mode of the upper and lower cycle is opposite, and the seven operating modes and operating waveforms in the upper half cycle are shown in Figs. 3 and 4 [4–6].

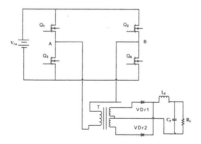

Fig. 1. The circuit diagram of traditional PWM DC/DC circuit

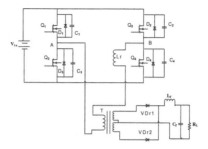

Fig. 2. The circuit diagram of improved PWM DC/DC circuit

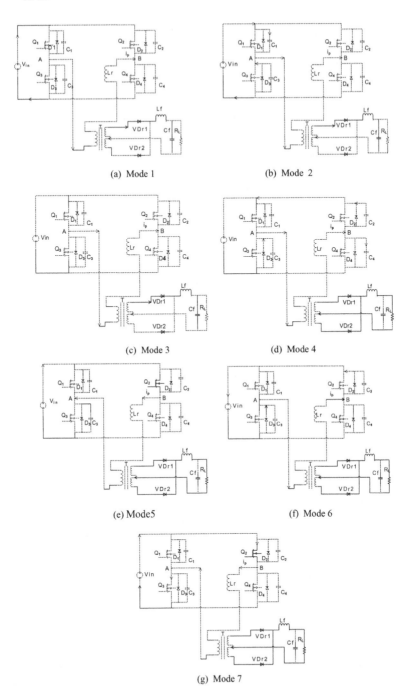

(a) Mode 1

(b) Mode 2

(c) Mode 3

(d) Mode 4

(e) Mode5

(f) Mode 6

(g) Mode 7

Fig. 3. Five work stages of the improved PWM DC/DC circuit

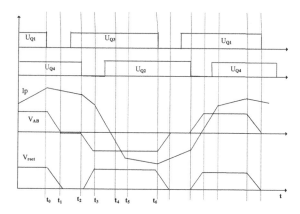

Fig. 4. The work waveform of the improved PWM DC/DC circuit

3 The Experiment Results and Analysis

For the sake of confirming the correctness of theoretical analysis of the improved PWM DC/DC converter, the paper gives the simulation results in the software of Saber, and the experimental results are shown in Figs. 5, 6, 7, 8, 9. Figure 5 is the trigger pulse voltage waveforms of Q_1 to Q_4, Q_1 and Q_3 is the advance of the bridge arm, than the lagging arm Q_2 and Q_4 ahead of the 1.5 μs. Figure 6 is the voltage and current waveforms of the leading bridge arm switch Q_3 and Q_1, Fig. 7 is the voltage and current waveforms of lagging arm of Q_4 and Q_2, and from Figs. 6 and 7 it can be seen that each switch is zero voltage switch when the voltage has dropped to zero. Figure 8 is the waveforms of the transformer primary voltage U_{ab} and the original current I_{ab}, and the simulation results are in good agreement with the theoretical analysis. Figure 9 is the input DC voltage waveform of the circuit, and the output voltage of the transformer after the transformer filter and filter capacitor rectifier DC voltage waveform.

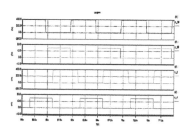

Fig. 5. The drive simulation waveforms of four switch tubes Q_1 to Q_4

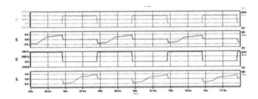

Fig. 6. The experimental results of voltage and current waveforms for Q_1 and Q_3

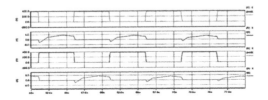

Fig. 7. The experimental results of voltage and current waveforms for Q_2 and Q_4

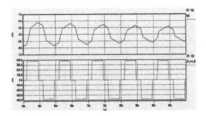

Fig. 8. The experimental results of transformer primary voltage U_{ab} and primary current I_{ab}

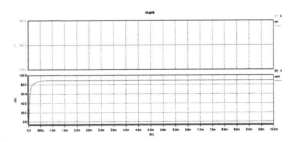

Fig. 9. The experimental results of input DC voltage and output DC voltage

4   Conclusion

In summary, in industrial and agricultural production the new PWM DC/DC PWM circuit replaces the traditional DC/DC PWM circuit, it can make the four tubes working in soft switching, and it can reduce switching loss, and enhance the efficiency of the converter, so it has very important application value in the country vigorously promote energy saving society.

References

1. Wang, Z., Liu, J.: Power electronics technology, pp. 101–103. Machinery Industry Press, Beijing (2011). (in Chinese)
2. Rong, J., Su, X., Wan, L., et al.: The technology research on phase-shifted and full-bridge zero voltage switching PWM circuit. Manag. Technol. Electr. Energy Effic. **4**, 16–21 (2015). (in Chinese)
3. Bodur, H., Bakan, A.F.: A new ZVT PWM DC-DC converter. IEEE Trans. Power Electr. **17**(1), 40–47 (2002)
4. Liang, Z.: Research on the phase-shifted and full-bridge DC/DC converter based on DSP, pp. 32–48. Zhejiang University, Zhejiang (2006)
5. Ruan, X., Yan, Y.: Soft-switching techniques for PWM full bridge converters. In: Proceedings of 31st Annual IEEE Power Electronics Specialists Conference, vol. 2, Piscataway (NJ), pp. 634–639 (2000)
6. Redl, R., Sokal, N.O., Balogh, L.: A novel soft-switching full-bridge DC/DC converter: analysis, design considerations, and experimental results at 1.5 kW, 100 kHz. IEEE Trans. Power Electr. **6**(3), 408–418 (1991)
7. Shen, Z., Li, S.: Study of phase-shifted and full-bridge ZVS-PWM inverter based on UC3879. Foreign Electronic Measurement Technology, pp. 25–29 (2003) (in Chinese)
8. Zhang, Z., Cai, X.: Principle and Design of Switching Power Supply. Publishing house of electronics industry, Beijing (2004)

Head-Related Transfer Function Individualization Based on Locally Linear Embedding

Xu Ming[1(✉)], Yan Binzhou[2], Guo Shuxia[2], and Gao Ying[2]

[1] Science and Technology on Avionics Integration Laboratory,
Shanghai 200233, China
ybz1123581321@sina.com
[2] Northwestern Polytechnical University, Xi'an 710072, China

Abstract. A head-related transfer function personalized algorithm based on Locally Linear Embedding is proposed for the precise localization of human beings with different physiological parameters. HRTF data was processed to reduce dimensionality by Locally Linear Embedding at first and linearly fitted in the low-dimensional space to extract the representative HRTF. Correlation analysis was used to select the physiological parameters which had a great influence on HRTF. The nonlinear mapping between physiological parameters and the representative HRTF was established by Artificial Neural Network, and then the individual HRTF can be calculated by a small number of body parameters. The experimental results show that the proposed algorithm can achieve personalized HRTF quickly and accurately, and solves the problem that personalized HRTF is difficult to measure.

Keywords: Virtual sound localization · HRTF · LLE · Neural network

1 Introduction

In the virtual sound theory, the process of the sound source to send the sound spread to the ears, through the head, trunk, shoulder and other human body parameters of the integrated filtering effect known as the Head-Related Transfer Function. If the use of headphones to reproduce the sound through their corresponding head related function processing, it can well reconstruction of the virtual sound field, thereby increasing environmental sense. However, the relationships between HRTF data and human physiological parameters are relatively close, different physiological parameters corresponding to the HRTF are different. At present, many research institutions have made personalized HRTF data measurement, but most of the work is time-consuming, so how to quickly and accurately get different listeners HRTF has become the focus of the study of virtual sound field.

Based on the above theory, a head-correlation function personalized method is proposed in combination with the measured data in the CIPIC database. The local linear embedding algorithm is used to reduce the HRTF data, and the HRTF is extracted from the low-dimensional space. The HRTF can be calculated by measuring a

F. Qiao et al. (eds.), *Recent Developments in Mechatronics and Intelligent Robotics*,
Advances in Intelligent Systems and Computing 690, DOI 10.1007/978-3-319-65978-7_16

small number of human parameters, and the characteristic HRTF linear interpolation and LLE weight can be calculated by using the neural network to construct the HRTF, and finally derived personalized HRTF data.

2 Personalized Method of HRTF

All of the following data analysis is based on CIPIC published database, the main block diagram of HRTF personalized algorithm shown in Fig. 1.

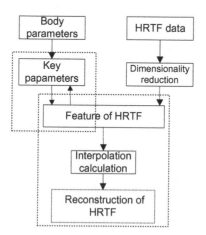

Fig. 1. Algorithm block diagram of individual HRTF

In Fig. 1, the arrow represents the flow of data in the algorithm. Two dashed box, one is the neural network to establish the human body parameters and characteristics of HRTF nonlinear mapping, and the other is the characteristics of HRTF reconstruction. The characteristic HRTF is obtained by local linear analysis of the dimensionality reduction data, and the nonlinear mapping relationship between the human parameters and the characteristic HRTF is established by the neural network algorithm to realize the personalized HRTF reconstruction.

2.1 LLE Dimensionality Reduction Processing

For the input sample $X = \{x_1, x_2, \ldots, x_n\}, x_n \in R$, of the high dimension D, k is the number of neighborhoods, the low-dimensional sample $Y = \{y_1, y_2, \ldots, y_n\}, y_n \in R$ of the output is the dimension d, the LLE algorithm can be described as the following three steps:

Calculate the k neighboring points of each sample. Calculate the Euclidean distance between all points and the sample points, and select the nearest point. Euclidean distance is calculated by Eq. (1).

$$\rho = \sqrt{(y_1 - x_1)^2 + (y_2 - x_2)^2 + \cdots + (y_n - x_n)^2} \tag{1}$$

Calculate the local reconstruction weight matrix of the sample points. Define a reconstruction error function, as shown in Eq. (2).

$$\min(\alpha) = \sum_{i=1}^{N} \left| X_i - \sum_{i=1}^{K} W_{ij} X_j \right|^2 \tag{2}$$

Where X_j is the jth adjacent point of X_i, W_{ij} is the weight when X_i is calculated for X_j, and W_{ij} satisfies two conditions: if X_j is the neighboring point or vector of sample X_i, then $\sum_{i=1}^{K} W_{ij} = 1$, if there is no adjacent point or vector, then $W_{ij} = 0$. In the calculation W_{ij}, first construct a size of $K \times K$ local covariance matrix Q, as shown in Eq. (3).

$$Q_{ijd} = (x_i - x_j)^T (x_i - x_d) \tag{3}$$

Combined with $\sum_{i=1}^{K} W_{ij} = 1$, using the Lagrangian multiplier method to calculate the local reconstruction matrix, as shown in Eq. (4).

$$W_{ij} = \sum_{d=1}^{k} Q_{ijd}^{-1} \Big/ \sum_{p=1}^{k} \sum_{q=1}^{k} Q_{ipq}^{-1} \tag{4}$$

In order to eliminate the possibility that Q is singular, it is subjected to regularization as shown in Eq. (5).

$$Q = Q + rE \tag{5}$$

Where E is the unit matrix, the size is the same as Q, and r is the regularization parameter.

Map all sample points into low-dimensional space. Assume that the condition satisfies Eq. (6).

$$\min(\beta) = \sum_{i=1}^{k} \left| y_i - \sum_{j=1}^{k} w_{ij} y_{ij} \right|^2 \tag{6}$$

Where β is the reconstruction error and y_i is the low-dimensional point corresponding to the high-dimensional point mapping. To maintain the structure of high-dimensional data, the following two conditions should be satisfied: $\sum_{i=1}^{N} y_i = 0, \frac{1}{N} \sum_{i=1}^{N} y_i y_i^T = E$, where E is the unit matrix of size, the weight w_{ij} is stored in the sparse matrix W of size N * N. The adjustment error function is Eq. (7).

$$\min(\beta) = \sum_{i=1}^{N} \sum_{j=1}^{N} M_{ij} y_i^T y_j \tag{7}$$

Where M is a symmetric matrix of size $N \times N$ and calculated using Eq. (8).

$$M = (E - W)^T (E - W) \tag{8}$$

There are two advantages to the use of LLE algorithm data: (1) LLE algorithm in reducing the low-dimensional data points to maintain high-dimensional data points between the manifold structure unchanged, to retain the perception factors. (2) Using the relatively large correlation of HRTF data in the adjacent direction, combined with the local linearity of LLE algorithm, it can make the error less in reconstruction.

In the use of LLE algorithm to reduce the dimension of the data, not only can reduce the HRTF data storage capacity, but also can make it in low-dimensional space for feature analysis. Here is a relatively simple method to extract the characteristics of HRTF; the idea is to point the data points in order to fit. Specific steps are as follows:

1. Straight fit of two points on the edge of low-dimensional manifold.
2. The next point with the previous line to fit, in the calculation of this section of the slope of the straight line and the slope of the previous slope of the error δ_i, it will be set with the allowable error δ comparison, if $\delta_i < \delta$ is satisfied, it can be assumed that there is still a linear relationship between the two linear endpoints. On the contrary, not a linear relationship will be the corresponding high-dimensional HRTF data as a characteristic HRTF.
3. Repeat (2) until all points are calculated.

The method of extracting HRTF is to use the local linear relation of LLE and the principle of HRTF correlation in adjacent direction. As shown in Fig. 2, the result of the selection of the characteristic HRTF on the basis of the data points, and the black dot represents the characteristic HRTF point obtained by the above method.

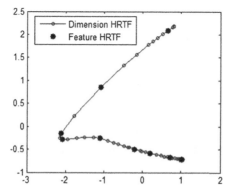

Fig. 2. Line fitting of one-dimension manifold

2.2   Selection of Key Human Body Parameters

Personalization of HRTF is done by measuring human parameters to determine HRTF data for these human parameters. However, in the CIPIC given human parameters data, there is no description of the parameters of HRTF data impact. Therefore, in the personalization of these physiological parameters of the correlation analysis, and thus select the key human parameters, the correlation analysis as shown in Eq. (9).

$$r = \left| \frac{\sum\limits_{i=1}^{N}(X_i - \overline{X})(Y_i - \overline{Y})}{\sqrt{\sum\limits_{i=1}^{N}(X_i - \overline{X})^2(Y_i - \overline{Y})^2}} \right| \tag{9}$$

Where X is the body parameter matrix; Y is the HRTF or body parameter matrix.

The correlation between the individual parameters is analyzed, and the correlation coefficient matrix as shown in Fig. 3. For convenience, here will be relatively small correlation coefficient directly zero, leaving only a large correlation coefficient. The meaning of parameters of the human is shown in the literature [9].

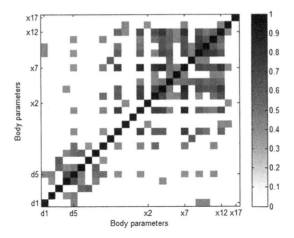

Fig. 3. Correlation analysis of physiological parameters

As can be seen from the Fig. 3, some physiological parameters between the relatively large, we can choose some of the better measured as key human parameters.

By the perception of the perception mechanism, the human in 1 k–8 kHz frequency of the sound more sensitive. Therefore, in the choice of parameters should follow two principles: (1) select the relatively large human body parameters associated with HRTF; (2) 1 k–8 kHz within the focus of the relevant analysis. Figure 4 is the human body parameters with the horizontal angle of $\theta = 0$, elevation angle $\varphi = 0$ HRTF data correlation analysis.

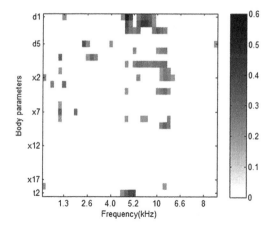

Fig. 4. Correlation analysis between physiological parameters and HRTF

The above two steps, as well as the measurement of the feasibility of the parameters, the final selection of key human parameters shown in Table 1.

Table 1. The key physiological parameters

Selection	d_1	d_5	d_6	d_7	x_3	x_6	x_{12}	t_2
Physical meaning	Cavum concha height	Pinna height	Pinna width	Intertragal incisure width	Head depth	Neck width	Shoulder width	Pinna flare angle

3 Experiment Analysis

HRTF spectrum difference and spectral distortion are used to qualitatively analyze and quantify the error. Spectral distortion SD is expressed by Eq. (11).

$$SD = \sqrt{\frac{1}{N}\sum_{i=1}^{N} 10\log\left(H_i - \hat{H}_i\right)^2}(dB) \qquad (11)$$

Which $\hat{H}$ is the predicted HRTF data; H is measured CIPIC database HRTF data.

The test experiment uses the CIPIC database, the HRTF data of Subject_018 and Subject_003 as the test group, and the remaining 33 sets of data as the training group, where Subject_003 is the artificial head model. Figures 5 and 6 show the comparison of the HRTF in some directions after the test group is personalized with the measured data spectrum in the database.

There is a certain error between the reconstructed HRTF data and the measured data, but it can be analyzed: (1) the personalized HRTF data obtained relative to the measured value, the peak value, the valley point position and so are only slightly

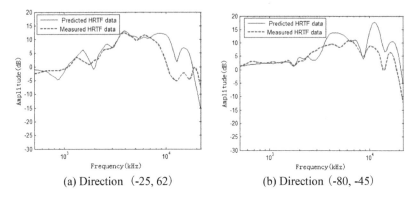

(a) Direction （-25, 62）   (b) Direction （-80, -45）

Fig. 5. Individualization spectrum of Subject_003

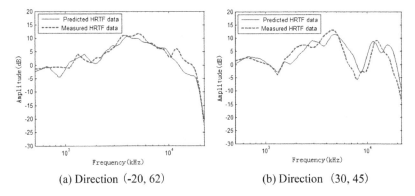

(a) Direction （-20, 62）   (b) Direction （30, 45）

Fig. 6. Individualization spectrum of Subject_018

biased, and the spectrum trend has not changed; (2) relative to other regions, in the human sensitive 1 k–8 kHz listening range spectrum distortion smaller. Therefore, these errors do not deviate greatly from the sound localization.

Table 2 shows the spectral distortion in some directions.

Table 2. SD of Individualization HRTF of different orientation (dB)

Number	Direction (horizontal angle, elevation angle)/°					
	$(-80, -45)$	$(-45, -17)$	$(-30, 0)$	$(0, 17)$	$(30, 45)$	$(45, 90)$
Subject_003	4.41	4.12	3.21	3.87	3.91	3.05
Subject_018	3.15	2.26	2.57	3.11	2.14	2.24

As can be seen from Table 2: (1) Subject_018 reconstruction of personalized HRTF data spectral distortion in most of the direction are less than Subject_003. (2) The two subjects in the $(-80, -45)$ direction to restore the spectral distortion is relatively large.

1 Is used in the CIPIC database Subject_003 experimental body is a dummy model, so the body structure and real people have a certain gap, not exactly meet the training results, resulting in personalized HRTF reconstruction when there are some deviations. $(-80, -45)$ direction is the direction of the horizontal and elevation angles measured in the CIPCI database. In the process of reconstructing the matrix by LLE, the neighborhood HRTF in this direction is the same as that of the same side, and its direction is the left and right sides, the reconstruction error is slightly larger than the other direction.

References

1. Xie, B.: Head-Related Transfer Function database and its analysis. Sci. China Ser. G **50**, 267 (2007). doi:10.1007/s11433-007-0018-x
2. Xu, S., Zeng, L., Li, Z., et al.: A pilot measurement of head-related transfer function blur in spatial localization. In: 2007 IEEE International Conference on Industrial Engineering and Engineering Management, Singapore, pp. 467–471 (2007)
3. Zotkin, D.N., Duraiswami, R., Davis, L.S., et al.: Virtual audio system customization using visual matching of ear parameters. In: 16th International Conference on Pattern Recognition (ICPR 2002), vol. 3, pp. 3–6 (2002)
4. Yin, T., Yu, G., He, Y., et al.: Simulation of individual head-related transfer functions using an improved head-neck-shoulder model. In: 21st International Congress on Sound and Vibration 2014: ICSV 21, vol. 3(6), pp. 1948–1953, Beijing, China, 13–17 July 2014
5. Ramos, O.A., Tommasini, F.C.: Magnitude modeling of HRTF using principal component analysis applied to complex value. Arch. Acoust. J. Pol. Acad. Sci. **39**(4), 477–482 (2014)
6. Xiang, S., Nie, F., Pan, C., et al.: Regression reformulations of LLE and LTSA with locally linear transformation. IEEE Trans. Syst. Man Cybern. Part B Cyber. **41**(5), 1250–1262 (2011). A publication of the IEEE Systems, Man, and Cybernetics Society
7. Zhang, T., Li, S., Wu, S., Tao, L.: Face recognition dimensionality reduction based on LLE and ISOMAP. In: Du, W. (ed.) Informatics and Management Science V, vol. 208, pp. 775–780. Springer, London (2013)
8. Jihua, Y., Shuxia, S., Yahui, C., et al.: A face recognition algorithm based on LLE-SIFT feature descriptors. In: 2015 10th International Conference on Computer Science Education (ICCSE 2015), pp. 729–734, Cambridge, UK, 22–24 July 2015
9. Algazi, V.R., Duda, R.O., Thompson, D.M., et al.: The CIPIC HRTF database. In: 2001 IEEE Workshop on Applications of Signal Processing to Audio and Acoustics, pp. 99–102 (2001)

A Research on Information Security of University Libraries in the Era of Big Data

Likun Zheng[✉], Yongxin Qu, Hui Zhang, Huiying Shi, and Xinglan Wang

Harbin University of Commerce, Harbin, China
glczlk@163.com

Abstract. In the era of big data, knowledge storage and access are growing rapidly. As centers of network information in universities, libraries are important supports for teaching and researching. Combined actual situation of university libraries with information security issues, the paper puts forward three main methods to ensure. They are reliable technologies, advanced equipment and effective managements. Teachers and students will be taught how to using and not being deceived by false information as well.

Keywords: The era of big data · University libraries · Information security

1 Introduction

On September 14, 2016, CNNIC (China Internet Network Information Center) published *the 38th China Internet Network Development State statistic Report.* It was reported that in June 2016, there were 710 million Internet users in China. About 21.32 million new Internet users added in the first half of the year. The growth rate is 3.1%. China's Internet penetration rate has reached 51.7%, increasing 1.3% compared to the end of 2015. It is 1.3% higher than the global and 3.1% higher than the Asian average rate. Messages sending and usages are both increasing greatly. As centers of network information of universities, university libraries are the important support for teaching and research. Information security problems in university libraries become more and more serious. Only to ensure securities of library information can guarantee teachers and students a good use of resources. However, the current information security managements in libraries have remained in the original basis in China. It must be further strengthened on reliable technologies, advanced equipment and efficient managements. In addition, teachers and students should retrieve information in vast amounts of information. False information usages should be avoided. Meanwhile, it is teaching contents that information literacy education in the libraries.

F. Qiao et al. (eds.), *Recent Developments in Mechatronics and Intelligent Robotics,*
Advances in Intelligent Systems and Computing 690, DOI 10.1007/978-3-319-65978-7_17

2   Information Security Problems that University Libraries are Facing with in the Era of Big Data

Now, utilization rates for paper resources have been declining year by year. At the same time, digital resources orders are increasing in university libraries. Information security problems on digital resources have become more and more important and urgent for Chinese university librarians. There are two threats on information security in digital literature resources: one is media information storages lost caused by unexpected damages, the other is network virus threats [1].

2.1   The Surrounding Environment and Equipment Configuration

Computer network systems need very strict external environmental conditions. If external environmental conditions fail to conform to relevant standards would bring permanent damages to computers. The key conditions include ground temperature and humidity in digital network control centers and environmental requirements at workstations in the libraries. Designs of computer rooms are unreasonable without antistatic, anti magnetic interference, dust and fire prevention, waterproof and lightening protection, leakage and theft prevention. All these above problems can lead to improper information storages directly, such as information losses caused by equipment with slow speed [2].

2.2   Virus Risks

With the growth of information and openness of the Internet in the era of big data, various computer network viruses are increasing rapidly. At the same time, viruses and hackers' attacks in kaleidoscopes are closely related to our information work. Viruses will appear in processes of mail access, network shopping payment and download literature resources behaviors, etc. In recent years, the outbreaks of CIH virus, iloveyou virus and nimda virus made many library information destroyed greatly.

2.3   Security Problems Teachers and Students are Facing with in Network Resources Usages in the Era of Big Data

In processes of network retrieval, teachers and students will encounter many unrelated software download bundles. These bundles have brought great inconvenience to computer users. In processes of searching for journals and papers published, once teachers and students encounter false journal information published by cheaters to defraud publication fee from them, then, they will probably waste not only time but also money.

3   Information Security Protection Measures for University Libraries in the Big Data Era

In the big data era, university libraries should ensure information security to guarantee the stability of digital resource storages well. It is necessary to assist teachers and students to use resources normally and support teaching activities and researches in the universities. Main protection measures can be seen in Fig. 1.

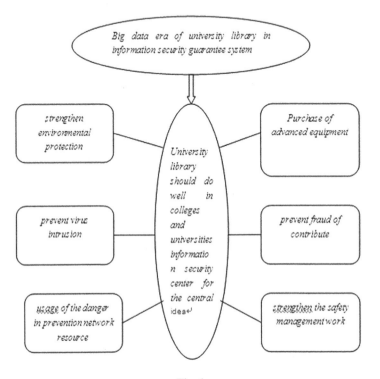

Fig. 1.

3.1   Strengthen Environmental Protections and Purchase Advanced Equipment

Computer network systems require very strict environmental conditions. If not been handled in time, heavy rain or thunders, even a sudden power-off can cause information resources losses. Therefore, emergency plans for safety should be made in order to tackle any problems at any time. Central air-conditioners should be installed in computer rooms to ensure them under proper temperature and humidity situations. Also, locations for the network systems should be at quiet places away from large equipment with static electricity. In addition, from an overall consideration, university libraries should purchase advanced equipment, operating systems and application software sets to fundamentally solve problems on loopholes and lack of stored information. Thus passive situations for network information security can be obtained fundamentally [3].

3.2 Prevent Virus Intrusion

In order to prevent virus invasion effectively, firstly, we should use firewall technologies as a first line of defense at present. Firewalls will control and monitor exchanges for information and access them to achieve effective managements of network security. They can help protect organization networks from external attacks. Firewall works are a priority for our virus intrusion protection efforts.

Secondly, we should use Invasion Detection System (IDS) alarms. IDS can identify authorized users and malicious invasion. One of common intrusion detection mechanisms is antivirus software usage. These certain software can be used in network systems and important computers to prevent virus intrusions and information theft [4].

Finally, network technology centers of universities should fit each computer with the latest anti-virus software visions. Teachers and students should be taught about specialized antivirus knowledge. Teach them how to use anti-virus software and how to kill viruses regularly. While using USB to transfer documents, it is critical to kill viruses first in order not to bring them to other users.

3.3 Teachers and Students Should Use Resources Correctly in the Network Resources in the Era of Big Data

Not only university libraries but also teachers and students should focus on information security issues. First of all, teachers and students should learn to check network resources carefully when download software and/or rule out bundled software useless. Only in this way can users use software fast and efficiently. In addition, a lot of cheaters made similar web sites as journals have by only changing their own email addresses and/or information so that some teachers and students ignored and have been cheated. Therefore, librarians should provide teachers and students with reliable journal submission emails, web sites, telephone numbers, submission information in databases, and so forth. Also, they should make relevant micro learning resource classes for teachers and students to learn and use.

3.4 Strengthen Managements

In order to reduce network information security risks, managers in charge of school libraries should do well as the following: make risk assessments regularly, strengthen managements of equipment, record equipment running processes, set up a special position to check operation records of equipment daily. At same time, store information regularly, strengthen personnel managements, make sound operating procedures, and so on. Staff in charge of network systems should regularly check out software system operations and records carefully. Once any abnormal reactions are found, they should report it immediately and discuss how to deal with it effectively. Managers should appoint someone to find security problems and solve them. The person should dig out risks of network information security, classify risk information retrieved and establish information database. What's more, he/she must constantly update and expend information in databases at the same time, and on this basis, establish related risk decision support systems [5].

4 Conclusion

In the big data era, it is very crucial to keep information safe in university libraries. According to the present problems, this paper offers four main aspects to handle them well. To strengthen environmental protections and purchase advanced equipment, to prevent virus invasion, to use network resources correctly in this era and to intensify managements. Thus we can guarantee teachers and students with safe scientific information centers and prevent information security crises.

Acknowledgement. This thesis is one of the stage achievements for the project (No.16GLD05) of the Heilongjiang Province Society Scientific Fund. Many people have contributed to it. This thesis will never be accomplished without all these invaluable contributions from my dear group members. It is my greatest pleasure and honor working with my team. I'm deeply grateful to them.

References

1. Yuan, M., Wu, S., Wang, W.: Compound library information security and strategy. Modern Intell. **24**, 19–20, 24 (2005)
2. Liu, J., Wang, C.C., Wang, W.: Introduction to digital library network Information security. Modern Intell. **12**, 87–88, 92 (2005)
3. Wang, H.: The library network information security idea. Modern Intell. **5**, 48–50 (2004)
4. Rong, X.: University library network information security issues and solutions. Modern Intell. **11**, 67–68 (2006)
5. Chai, W., Zhou, N.: The network information security and the integration of Web data mining technology research. J. Theory Pract. Intell. **1**, 97–101 (2009)

An Improved PID Algorithm Based on BP Neural Network of Ambient Temperature Controller

Yanfei Liu$^{(\boxtimes)}$, Jieling Wang, Jingjing Yang, and Qi Li

Xi'an Research Institute of High Technology, Xi'an, China
bbmcu@126.com

Abstract. In order to create an environment suitable for crop growth, this paper aims at the characteristics of crop growth environment, put forward an improved PID algorithm controller which is based on BP neural network. The controller use BP neural network to improve PID control algorithm, and use this PID algorithm to control the temperature of crop growth. The algorithm is used to simulate the control system by Matlab. The results show that the algorithm not only improves the fastness of step response, but also greatly reduces the overshoot.

Keywords: BP neural network · Temperature · PID algorithm · Crop growth environment

1 Introduction

There are some environmental problems in crop growth in the remote areas. This paper put forward an improved PID algorithm controller which is based on BP neural network to solve these problems, Crop growth environment is a complex object which contain non-linear, distributed parameter, time-varying, large delay and multivariate coupling. However, through the decoupling of multivariate, the crop growth environment can be reduced to a link with first-order large inertia and large delay. This paper merge BP neural network and PID control, it can complement each other, give full play to their advantages, and can achieve the best control effect.

2 The Design of Overall Control Program

There are many factors to control in the environment of crop growth, such as temperature, humidity, light, carbon dioxide concentration and moisture, etc. Because the temperature is one of the most important environmental factors for crops, but also the most difficult controlling of the control system, the more advanced control system theory and control algorithm is used in the temperature control system, therefore, the temperature control is taken as an example. The system structure of the control algorithm proposed in this paper is shown in Fig. 1.

© Springer International Publishing AG 2018
F. Qiao et al. (eds.), *Recent Developments in Mechatronics and Intelligent Robotics,*
Advances in Intelligent Systems and Computing 690, DOI 10.1007/978-3-319-65978-7_18

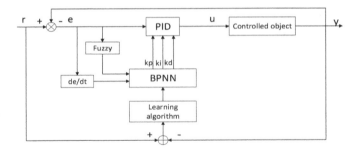

Fig. 1. System structure of control algorithm

In the figure, r is the target temperature value, y is the current temperature value, e is the error between them, and u is the control quantity. the parameters Kp, Ki and Kd are establish by BP neural network, and it can make a best adjustment effect.

3 BP Neural Network Control

The learning process of BP neural network consists of two stages: The first stage is the forward propagation process, the input signal processed by the hidden layer after through the input layer, and the actual output value of the neuron is calculated at the output layer; The second stage is the back propagation process of error, if the desired output value is not obtained at the output layer, the error between the actual output and the desired output is calculated by layer by layer, and the weight coefficient is adjusted according to the error, the structure of three-tier BP neural network shown in Fig. 2.

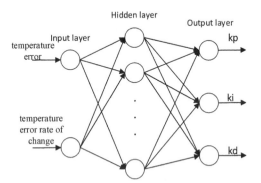

Fig. 2. Three-tier BP neural network structure

The input of the input layer is the temperature error and the temperature error rate of change, and the role of the hidden layer is to blur the input variables, that is, to solve the membership grade of the input variables. The Gaussian-type function is selected as the membership grade function, and then select Gauss function as the fuzzy inference function, and realize the function of fuzzy inference. The output layer realizes the function of defuzzification. we use the gravity method to blur.

(1) Gaussian-type function

$$f(x) = 1/\{1 + \exp[1 - 0.5(\frac{x - c}{\sigma})^2]\} \tag{1}$$

(2) Gauss function

$$f(x) = \exp[1 - 0.5(\frac{x - c}{\sigma})^2] \tag{2}$$

3) the gravity method to blur

$$v_0 = \frac{\int v u_v(v) dv}{\int u_v(v) dv} \tag{3}$$

4   The Improved PID Algorithm

The PID algorithm used in this paper is improved on the basis of the traditional incremental digital PID algorithm. Aim at this design, in order to reduce the error of measurement temperature, and make the measurement temperature as far as possible accurate, the traditional incremental PID algorithm were made the following improvements: Anti integral saturation and Flexible use of differential terms.

4.1   Anti Integral Saturation

The integral saturation refers to the output $U(k)$ of the PID controller cause the actuator to achieve the limit position (maximum or minimum). When calculate the output $U(k)$. If $U(k - 1) > U_{max}$, only accumulate the negative deviation; and if $U(k - 1) < U_{min}$, only accumulate the positive deviation. This algorithm can avoid the long for controller to stay in the saturation zone.

4.2   Flexible Use of Differential Terms

According to the experimental, it is found that the original differential terms is not obvious $K_d \times [E(k) - E(k - 1)]$ in the use of positional PID expression. Based on this reason, the differential term is changed to $K_d \times T(k - 1)/T(k)$, $T(k - 1)$ is the time to maintain the temperature constant, $T(k)$ is the time to maintain the current temperature unchanged. If $T(k)$ is larger than $T(k - 1)$, the temperature change rate becomes smaller, and the role of the differential term should be weakened, opposite, when $T(k)$ is smaller than $T(k - 1)$, the temperature change rate becomes larger, and the differential term should be strengthened. $T(k - 1)/T(k)$ is used to influence the value of the differential term, so that $K_d \times {}^{T(k-1)}\!/_{T(k)}$ can be used instead of the traditional differential term.

4.3 The Overall Structure of the Improved PID Control Algorithm

Based on the above analysis of the PID control algorithm and the research of the improved algorithm, the output expression of the PID control algorithm is as follows:

$$U(k) = K_p \times [E(k) - E(k-1)] + Ki \times E(k) + K_d \times {}^{T(k-1)}\!/_{T(k)} \qquad (4)$$

the whole PID control framework is shown in Fig. 3

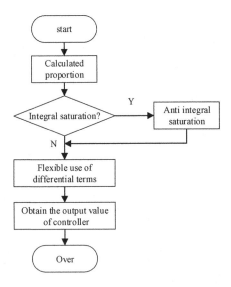

Fig. 3. Improved PID control algorithm framework

5 The PID Algorithm Based on BP Neural Network

Compared with the model training algorithm used in the past, the improved genetic algorithm is not easy to fall into local extremumin the search process, and can quickly find the optimal solution in the global scope, and the objective function is not required to be differentiable and Non concave, it also has the advantages of simple programming.

5.1 The Selection of Controlled Object

The object considered in this paper is temperature, and can be expressed by a first order inertia link with pure delay. The transfer function is:

$$G(s) = \frac{Ke^{-\tau_s}}{TS+1} \tag{5}$$

The K is a static gain; T is the time constant; τ is a pure lag time.

5.2 PID Parameter Adjustment

In order to obtain a better control effect, it is necessary to adjust the proportional, integral and derivative, and form a relationship that cooperate and restrict each other between the control value. This relationship is not a simple "linear combination", but to find the best combination from the nonlinear combinations. The neural network has the ability of any nonlinear expression, and can achieve the best combination of fuzzy PID control by learning the system performance. In this paper, the BP neural network is used to build the Kp, Ki and Kd of the PID control system. The whole control system is divided into 3 parts:

(1) First, the crop growth environment system is identified to obtain the current temperature value and the predetermined temperature value of the crop growth environment, and then provide a learning signal to the BP neural network.
(2) According to the learning signals provided identification system, Neural network adjust the three parameters proportion, integral and differential of PID controller.
(3) The improved incremental PID controller is used to control the controlled object directly, and the three parameters Kp, Ki and Kd can be adjusted by the neural network.

6 Experimental Test

We use Matlab as a computer-aided design tool, and write the corresponding program, the simulation results shown in Fig. 4.

From the above two figures we can see: the traditional incremental PID control overshoot is too large, too many shocks, the dynamic performance is inferior to the improved PID control system which based on BP neural network; The traditional

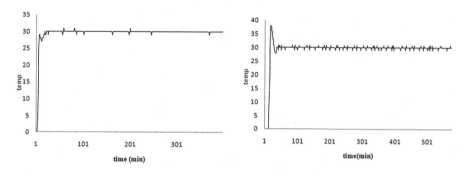

Fig. 4. Improved PID control system and traditional incremental PID controller

incremental PID control system pursues the rapidity of the ascending phase, thus leading to the shock of the late adjustment; It will inevitably affect the rising speed, if we pursue the small and less shock and other dynamic performance of the late adjustment, but the improved PID control system which based on BP neural network has a small step response overshoot, and the transition time is obviously shortened, and the performance is obviously superior to the traditional incremental PID control system.

7   Conclusion

After computer aided software design and simulation, it can be proved: Compared with the traditional incremental PID controller, the dynamic performance of the improved PID control system controller which based on BP neural network has been greatly improved, the controller completes the PID parameter tuning and control mode conversion of the BP neural network better, and achieve a simple method, the control effect is good.

References

1. Lai, Q.Y.D., Wings, N., Qi, C., Mu, L.N.: The greenhouse temperature control system which based on neural network PID control. J. Agric. Eng. **2**, 307–311 (2011)
2. Zhang, M., Zhang, H.: Neural network PID controller based on genetic algorithm optimization. J. Jilin Univ. **5**, 91–96 (2005)
3. Sen, O., Wang, J.: A new improved genetic algorithm and its application. J. Syst. Simul. **8**, 1066–1068, 1073 (2003)
4. Liu, Y., Zhai, H.L., Chai, T.: Nonlinear adaptive PID control based on neural networks and multiple models. J. Chem. Eng. **59**, 1671–1676 (2008). (in Chinese)
5. Li, G., Who, Z.: A nonlinear PID neural network algorithm based on controller. J. Cent. South Univ. (Natural Science Edition) **5**, 1865–1870 (2010)
6. Guo, Y., Yao, Z., Nan, W.: Nonlinear PID controller. J. North Cent. Univ. (NATURAL SCIENCE EDITION) **05**, 423–425 (2006)
7. Xiong, J.: Neural network self-tuning PID controller design and simulation. Northeastern University, p. 6 (2013)
8. Li, Z.X., Xie, X., Mao, W.: PID controller based on neural network parameters self-tuning. Ind. Instrum. Autom., 6–8 (1999)

Study on Security of Mobile Payment

Yining Jin[1](✉), Shi Wang[1], Yongxin Qu[1], Qingtian Guo[2], and Jinping Li[1]

[1] University of Commerce Harbin, Harbin, China
hcujyn@163.com
[2] Harbin Branch of Shanghai Pudong Development Bank, Harbin, China

Abstract. With the rapid development of mobile Internet and mobile terminal, mobile pay is being applied by more and more people. But mobile pay has become a high incidence of security incidents and various types of security events are often seen in various reports. The thesis analyzes the security threats of mobile payment and discusses how to improve its security methods to reduce the security risks to the lowest level.

Keywords: Mobile · Payment · Security · Protect

1 Introduction

With the rapid development of mobile Internet and mobile terminal, mobile pay is being applied by more and more people because it can complete the payment timely and easily and unlimited on time and place. In china Internet users reached 688 million and mobile Internet users reached 620 million by the end of 2015. Mobile payment users are about 400 million. But mobile pay has become a high incidence of security incidents and various types of security events are often seen in various reports as the people and the scale of transactions is growing. Therefore it is particularly important how to avoid risks during the mobile payment application.

2 Mobile Payment Scene Analysis

2.1 The Mode of Mobile Payment

Mobile payment refers to the two parties achieve commercial transactions for some goods or services through mobile devices with the mobile network. Currently, the mobile payment terminal mainly refers to intelligent phones.

Mobile payment is mainly used in the following ways.

(1) Short Message Pay. Short Message Pay is the first mobile payment application. It establishes a correspondence relationship of the user's mobile phone SIM card with the user's own bank card account. User who sends text messages complete the transaction payment requests under the guidance system of SMS. It can be operated simply and traded at any time.

© Springer International Publishing AG 2018
F. Qiao et al. (eds.), *Recent Developments in Mechatronics and Intelligent Robotics*,
Advances in Intelligent Systems and Computing 690, DOI 10.1007/978-3-319-65978-7_19

(2) Scan code pay. Scan code pay, which is mainly used by Alipay and WeiXin users, is an account-based system on the wireless payment scheme. It completes payment transactions of goods or services through users to scan business two-dimensional code or through business to scan users two-dimensional code.

(3) NFC short-range communication. NFC short-range communication is a short-range wireless connectivity technology. Users can use the "mobile wallet" to finish on-site consumption in cooperation merchant POS machines.

2.2 Mobile Payment Application Scene Analysis

Currently it is mainly the three methods of payment scenarios. The comparison of Mobile Payment Application is shown as follows Table 1.

Table 1. The comparison of Mobile Payment Application

Payment method	Short Message Pay	NFC short-range communication	Scan code pay
Representative form	China Union Pay	Apple Pay, Huawei Pay, Xiaomi Pay etc.	Alipay, WeiXin, Mobile QQ, etc.
Device	All phones	Prescribed equipment	All Intelligent phones
Download APP	No need	No need	Need
Payment terms offline	Not be achieved	China Union Pay POS machines with cloud lightning pay	Support business cooperation
Payment terms offline	Support business cooperation	14 APP such as meituan, qunar, etc.	Support the most APP
Support bank	All banks	19 Banks	WeiXin:80 banks Alipay:180 banks
Web environment	No network	No network	Scan code with network
Advantage	Simple operation, the support by wide businesses and financial	Safe, easy step, fast operation without network	Higher popularity, pay scenes richer, convenient payment
Disadvantaged	Utilization drop, susceptible virus	Prescribed equipment, system requirements, pay scene less	Third-party payment account, susceptible virus

The comparison of mobile payment displays their distinctive. Scan code pay is absolute dominance in current mobile pay market because of users' adhesion in WeiXin or Alipay. NFC short-range communication is a new form of third-party payment, currently it is difficult to change the public's mobile payment habits because of the limitation of intelligent phones and business POS devices requirements. It will be more favored by high-end users who focus on more security and privacy. Short Message Pay is being ignored more and more without adhesion scenarios.

3   Mobile Payment Security Threats

3.1   Higher Risk of Mobile Internet

It transfers the communication contents such as verification codes and text etc. through electromagnetic signals in mobile network. Compared with a wired Internet, the risk of a wireless Internet is higher as mobile Internet system is imperfect and wireless channel is open. Mobile network is more vulnerable by physical attack such as eavesdropping, spoofing or denial service.

3.2   The More Vulnerability of Mobile Phone

Firstly, mobile phone performance and operating environment is very limited. Resources can not be chose freely. Due to limitations of memory and processing equipment, it affected the work of data encryption, data integrity and reliability. Due to space limitations display, some transfer methods of the input data and the performance had to be given up. Secondly a relatively probability of stolen or loss mobile terminal is higher. The replacement cycle is shorter. Information security awareness of some people is not strong. It increases potentially the vulnerability of mobile payments without the necessary security settings or disposal. Otherwise security issues are growing exponentially because of the immature of mobile soft environment, open operating system, diversity platform.

3.3   Comprehensive Threats of Mobile Payment

Users often bound phones with bank accounts during the mobile payment. In order to profit more directly, black industrial chain about mobile payment is rapidly growing. The virus about mobile payment is changing more high-risk and intelligently. Annual security report released by the cheetah organization that 9.59 million computer virus were intercepted in 2015, which increased by 342.5% than 2014. Over 60% of them were related to payment. China has the most uses who infected virus in the world.

Defects such as pseudo base stations, fake wifi, application of counterfeit tampering, phishing, mobile application security vulnerabilities, android system vulnerabilities, etc. in users' phone are used to obtain sensitive information such as bank card numbers, ID numbers, phone numbers, verification codes, etc. Users may lost their property. So various mobile payment security threats of mobile payment are comprehensive.

3.4   The Weak of User Security Awareness

In mobile payment few people concern about mobile security. Many users only concern the features, performance and convenience and availability, etc. They lack of common sense precautions and safety skills are poor, so frequent security incidents happen frequently.

4   The Security Precautions of Mobile Payment

4.1   Improve Safety Awareness and Skills

Users need to raise awareness of their own defense and concern related to security incidents while they enjoy the convenience that new technologies bring They should learn actively and master a variety of security-related skills so security incidents can be avoided. The offense and defense process is one foot step ahead and upward spiral. The security protection can't be once and for all. It is rules and characteristics of security issues.

4.2   Scientific Management and Use

(1) Emphasis on setting a password. The password is an important barrier in the management and use of security. It should avoid that all passwords are all the same and simple. It is easy to be guessed if password is your birthday, telephone, ID card or continuous and repeat numbers. Passwords of power or screen unlock must be set in phones.

(2) Ensure the phone safety. Phones should be installed professional security software such as anti virus or malicious programs. APP should be downloaded from official site or professional download app. Phone, ID card and bank card shouldn't be put together. Don't tell others your privacy information such as check code, etc. easily. Don't save the account or password information in your phone's browser. Clean periodically the cache, forms and other information Cookies in phone. Phone should be formatted and ensured the information which can be restored when it need be replaced.

(3) Avoid the risk of mobile networks. It needs cautious to connect free wifi. If you need mobile pay, you should use GPRS or 3G, 4G as much as possible.

(4) Good habits for safe operation. Close small free password payment function in phones. The amount of credit and debit card bounded don't be too much and need to set daily trading limit. It is safer through the specialized APP of bank or third-party payment companies than through the related browser.

4.3   Effective Emergency Measures

It is an important part about security and protection to know how to deal with emergency measures rapidly, orderly and effective if the security accidents happen. When your phone is lost or stolen, First of all, you should call immediately the bank and third party payment vendor to freeze related businesses, then report the loss of phone numbers and re-submit. Last, you should login account and turn off the wireless payment service. If account funds are stolen, you should call the police. Once the phone has suddenly no signals and may be re-submitted by others. If it is not, you should transfer out balances of pay platform exclude cell phone signal problems or failures.

5   Conclusion

We need a dialectical attitude and awareness for mobile pay. We needn't be so worried about the existence of security risks that we dare not use it, or too optimistic to use it without protection. And as the technology update and progress, there will be new security threats, so we should be careful and keep learning and mastering safety knowledge and skills, use related preventive methods flexibly to control their own security risks actively and minimize possible losses.

Acknowledgement.   This thesis is a stage achievement of the Project (No. 16GLD05) of the Heilongjiang Province Society Scientific Fund. Many people have contributed to it. This thesis will never be accomplished without all the invaluable contributions selflessly from my group members. It is my greatest pleasure and honor working with my group. I'm deeply grateful to them.

References

1. Wang, L., Yao, X.: Research on the problem of mobile security. Information Security and Technology, January 2015
2. Ren, G., Liu, L.: The security strategy of mobile payment at the internet of things times. Automatic Machine, August 2011
3. National Institute of Standards and Technology. NICE. http://csrc.nist.gov/nice/. Accessed 11 May 2010
4. Information of http://baike.baidu.com/view/30156.htm
5. Jose, R.: Design lessons from deploying NFC mobile payments. In: LNCS, vol. 8276, pp. 86–93 (2013)

An Undetermined Coefficients Method for a Class of Ordinary Differential Equation with Initial Values

Li Gao[✉]

Liberal Arts Experimental Teaching Center, Neijiang Normal University,
Neijiang, Sichuan 641112, People's Republic of China
249435911@qq.com

Abstract. In this paper, based on the Taylor series, we present a method of undetermined coefficients to solve a class of ordinary differential equation with initial values. Theoretical analysis and examples show this method can achieve accuracy $O(h^{m+1})$ where m is the order of Taylor series of the right function $f(x)$. Furthermore, compared with traditional methods such as the finite difference method and the finite element method, this method can avoid solving complicated and large linear systems.

Keywords: Ordinary differential equation · Euler type · Undetermined coefficient

1 Introduction

For numerically solving ordinary differential equations (ODEs) and partial differential equations (PDEs) in various engineering problems, some typical and traditional methods have been presented so far, including the finite difference method (FDM), the boundary element method (BEM) and so on, too many results about their developments, accuracy, convergence, stability have arisen since they were presented. These can be partly seen in [1–11] and references therein.

In this paper, based on the Taylor series, we present a new numerical method for the ordinary differential equation

$$\sum_{k=0}^{n} a_{n-k+1} x^k y^{(k)} = x^n f(x), x \in (0,1),$$

$$y^{(k)}(0) = 0, k = 0, 1, \cdots, n-1,$$

(1)

where a_k, $k = 1, 2, \cdots, n + 1$ are all constants, $y(x)$ is the function to be determined. This method is quite different from such the existed methods as the FDM, FEM and BEM.

In the next sections, by some theoretical analyses and experimental demonstrations, we find this method has some evident characteristics:

- It can provide the unique numerical solution of (1);
- It can achieve accuracy $O(h^{m+1})$ when $f(x) \in C^{m+1}[0,1]$;

© Springer International Publishing AG 2018
F. Qiao et al. (eds.), *Recent Developments in Mechatronics and Intelligent Robotics*,
Advances in Intelligent Systems and Computing 690, DOI 10.1007/978-3-319-65978-7_20

- It does not lead to complicated and large linear systems.

The remainder of the paper is organized as follows. In Sect. 2, we describe our method in details, including the construction of this method, the existence, uniqueness, the accuracy of the numerical solution and the computational complexity of this method. In Sect. 3, some examples are given to check this presented method in Sect. 2. Finally, we draw some conclusions about our method, point out some shortages which are expected to be overcome.

2 The Method of Undetermined Coefficients

Supposing the equation

$$y^{(2)}(x) + y'(x) + y(x) = d(x), \ x = [0, 1] \tag{2}$$

has a unique solution $y(x) \in C^2[0, 1]$.

Case 1: If the function $d(x)$ is a given 2 order polynomial, namely,

$$d(x) = d_0 + d_1 x + d_2 x^2, \tag{3}$$

we can guess that $y(x)$ is also a 2 order polynomial of the form

$$y(x) = t_0 + t_1 x + t_2 x^2. \tag{4}$$

Substituting (4), (3) into (2), we get the equality

$$(t_0 + t_1 + 2t_2) + (t_1 + 2t_2)x + t_2 x^2 = d_0 + d_1 x + d_2 x^2$$

for all $x = [0, 1]$. Consequently, by comparing the coefficients, one immediately gets

$$t_2 = d_2, t_1 = d_1 - 2t_2, t_0 = d_0 - t_1 - 2t_2, \tag{5}$$

and thus the solution $y(x)$ is exactly found.

Case 2: When the $d(x)$ is not a polynomial, we can subdivide the domain $(0, 1)$ into $\cup_{i=1}^N \Omega_i$. On each sub-domain Ω_i, $(i = 1, 2, \cdots, N)$, we replace $d(x)$ with its 2 order Taylor series, and the remaining work is repeating the above course from (3) to (5) for seeking an approximate solution of (2) in Ω_i.

For convenience, we assume that $y(x) \in C^{m+1}(\bar{\Omega})$, $f(x) \in C^{m+1}(\bar{\Omega})$, $m \geq 1$. For $\bar{\Omega} = [0, 1]$, let $\Delta_x = \{0 = x_0 < x_1 < \cdots < x_M = 1\}$ be uniform partitions of Ω with mesh sizes $h_x = h$. Throughout this paper, we denote by $\tau_i, i = 1, 2, \ldots, M$ the midpoints of Δ_x, and by $\Omega_i = (x_{i-1}, x_i)$.

According to the conditions $y^{(k)}(0) = 0, k = 0, 1, \ldots, n - 1$, the solution $y(x)$ can be expressed as $y = x^n u(x)$ with $u(x)$ the new unknown function. Substituting this into (1), by some computations, we obtain the equivalent system

$$\sum_{k=0}^{n} b_{k+1} x^k u^{(k)} = f(x), x \in (0,1), \tag{6}$$

where

$$b_{k+1} = \sum_{t=1}^{n-k+1} a_t C_{n-t+1}^k n! / (t+k-1)!, k = 0,1,\ldots,n.$$

In every sub-interval Ω_i, let

$$\hat{f}(x) = \sum_{k=0}^{m} f^{(k)}(\tau_i)(x - \tau_i)^k / k! \tag{7}$$

after expanding all the polynomials $(x - \tau_i)^k$, $k = 0,1,\ldots,m$ and combining like terms, (7) reads

$$\hat{f}(x) = \sum_{k=0}^{m} g_{k+1} x^k, \tag{8}$$

where

$$g_{k+1} = \sum_{t=k}^{m} f^{(t)}(\tau_i) C_t^{t-k}(-\tau_i)^{t-k} / t!.$$

Then, for the right function $f(x)$ in (6), we can easily write its m order Taylor series on the point τ_i in sub-domain Ω_i $(i = 1,2,\ldots, M)$ as

$$f(x) = \hat{f}(x) + (x - \tau_i)^{m+1} f^{(m+1)}(\theta_i) / (m+1)! \tag{9}$$

With $\theta_i \in \bar{\Omega}_i$.

By using the idea similar to (4), taking approximate solution u_h as

$$u_h = \sum_{t=0}^{m} c_t x^t, \tag{10}$$

with $c_t,\ t = 0, 1,\ldots, m$ undetermined coefficients, we get

$$\sum_{k=0}^{n} b_{k+1} x^k u_h^{(k)} = \sum_{k=0}^{n} s_k x^k, \tag{11}$$

where

$$s_k = \begin{cases} \sum_{t=0}^{k} k!b_{t+1}c_k/(k-t)!, & \text{if } k = 0, 1, \ldots, n, \\ \sum_{t=0}^{n} k!b_{t+1}c_k/(k-t)!, & \text{if } k = n+1, \ldots, m. \end{cases} \tag{12}$$

Substituting (8)–(12) into (6), and dropping the remainder term

$$(x - \tau_i)^{m+1} f(\theta_i)^{(m+1)} \Big/ (m+1)!$$

in sub-domain Ω_i, we obtain the approximate system

$$\sum_{k=0}^{n} b_{k+1} x^k u_h^{(k)} = \hat{f}(x), x \in (0,1), \tag{13}$$

namely

$$\sum_{k=0}^{m} s_k x^k = \sum_{k=0}^{m} g_{k+1} x^k$$

in Ω_i. Let the corresponding coefficients be equal to each other in this equality, then the coefficients c_k, $k = 0,\ldots, m$ can be expressed as:

$$c_k = \begin{cases} g_{k+1} \Big/ \sum_{t=0}^{k} k!b_{t+1}/(k-t)!, & \text{if } k = 0, 1, \ldots, n, \\ g_{k+1} \Big/ \sum_{t=0}^{n} k!b_{t+1}/(k-t)!, & \text{if } k = n+1, \ldots, m. \end{cases} \tag{14}$$

From (14), we immediately have the following result about existence and uniqueness for approximate system (13).

Theorem 1. *Supposing $\sum_{t=0}^{k} k!b_{t+1}/(k-t)! \neq 0$ when $k \leq n$, $\sum_{t=0}^{n} k!b_{t+1}/(k-t)! \neq 0$ when $n+1 \leq k \leq m$, and $f(x) \in C^{m+1}(\bar{\Omega})$, then the approximate solution u_h defined by (10) can be uniquely solved by the Eq. (13).*

Now, we analyze the convergence of this method, we denote by

$$Lu = \sum_{k=0}^{n} b_{k+1} x^k u^{(k)}, u \in \Omega, \tag{15}$$

then the following theorem is true.

Theorem 2. *Assuming that Eq. (1) has unique solution $y(x)$, and $y(x), f(x) \in C^{m+1}(\bar{\Omega})$. Let $e \equiv y - y_h$ with $y_h = x^n u_h$. Then*

$$|e| \leq C_1 C_2 (h/2)^{m+1}, \tag{16}$$

where h is the step length, C_1 is a constant corresponding to the operator L^{-1}, and

$$C_2 = \max_{i=1}^{M} f(\theta_i)^{(m+1)} \Big/ (m+1)!, \theta_i \in \bar{\Omega}_i, i = 1, 2, \ldots, M.$$

Furthermore, this error satisfies

$$\lim_{|a_n+1| \to +\infty} |e| = 0. \tag{17}$$

Proof. In fact, by (6), (13), we can get

$$L(u - u_h) = f - \hat{f},$$

on each sub-domain $\bar{\Omega}_i$, $i = 1, 2, \ldots, M$.

Because Eq. (1) has unique solution $y(x) \in C^{m+1}(\bar{\Omega})$, by the relation $y(x) = x^n u(x)$, we know (1) is equivalent to (6), and the operator L is an invertible bounded linear operator on $C^{m+1}(\bar{\Omega}_i)$, $i = 1, 2, \ldots, M$, which shows that (16) is true.

Furthermore, by $b_{k+1} = \sum_{t=1}^{n-k+1} a_t C_{n-t+1}^{k} n! / (t+k-1)!, k = 0, 1, \ldots, n$, we know $\lim_{a_n+1 \to +\infty} b_1 = \infty$. Combining with (13), (6), we have $u - u_h = (f - \hat{f})/b_1 - \sum_{k=1}^{n} b_{k+1} x^k u^{(k)} / b_1$ in $\bar{\Omega}_i$, $i = 1, 2, \ldots, M$, which leads to (17), and the proof of Theorem 2 is completed.

3 Numerical Examples

In this section, we give some numerical examples to show the performance of our method. In these examples, we mainly check the result (16) and (17) in Theorem 2: the relation of accuracy with the order of Taylor series of the function $f(x)$ and a_{n+1}. We always take the step length $h = 0.1$, and in each sub-domain Ω_i, $i = 1, 2, \ldots, N$, we compute Taylor series of $f(x)$ in the center point τ_i of this sub-domain.

In the following tables, for convenience, the notation $x.y_1y_2 - p$ means $x.y_1y_2 \times 10^{-p}$. We test errors in the center and all endpoints of all sub-domains:

- E_n – the maximum absolute errors at the centers $\{\tau_i\}_{i=1}^{N}$;
- E_v – the maximum absolute errors at the endpoints $\{x_i\}_{i=0}^{N}$.

The tested equations have respectively the following information:

Example 1.

$$a_1 = a_2 = 1, n = 2, y = x^2 e^x;$$

Table 1. Results of Example 1

a_3	$m = 2$		$m = 3$		$m = 4$		$m = 5$	
	E_n	E_v	E_n	E_v	E_n	E_v	E_n	E_v
50	5.41−3	5.69−3	5.69−4	4.06−4	1.56−5	1.09−5	3.74e−6	5.82−6
100	2.00−3	1.58−3	6.52−5	2.58−5	8.66−6	1.38−5	1.25−6	1.14−6
500	1.10−4	8.07−5	4.56−6	7.38−6	2.95−7	2.37−8	6.40−11	1.18−8
1000	2.87−5	5.14−5	1.50−6	1.13−6	4.31−8	3.87−8	7.23−10	1.45−9
5000	1.18−6	2.68−5	7.26−8	1.66−7	3.91−10	1.61−9	1.19−11	2.47−11

Table 2. Results of Example 2

a_6	$m = 2$		$m = 3$		$m = 4$		$m = 5$	
	E_n	E_v	E_n	E_v	E_n	E_v	E_n	E_v
0	2.61−3	3.22−3	8.35−4	1.02−3	1.42−5	1.73−5	2.32−6	2.82−6
10	2.46−3	3.04−3	7.82−4	9.58−4	1.31−5	1.60−5	2.41−6	2.59−6
100	1.51−3	1.82−3	4.41−4	5.25−4	6.44−6	7.47−6	9.73−7	1.10−6
1000	1.37−8	2.25−4	7.35−5	1.05−4	1.87−6	2.48−6	4.74−7	6.12−7
5000	5.73−5	6.97−5	5.24−6	3.32−6	1.87−7	3.26−7	1.04−7	1.74−7

Example 2.

$$a_1 = a_2 = a_3 = a_4 = a_5 = 1, n = 5, y = x^5 \sin(0.2x + 1).$$

From Tables 1 and 2, we can clearly see that the results are in accordance with the theoretical analysis in Sect. 2: the method basically achieves accuracy of $O(h^{m+1})$, in the same time, just as we expected, the errors of u and u_h is inversely proportional to $|a_{n+1}|$.

4 Conclusions

In this paper, we introduced a new numerical method for solving a class of ordinary differential equation. By giving direct formulas of the undetermined coefficients, we showed this method can avoid solving complicated and large linear systems. Theoretically analysis and numerical experiments demonstrated this method can achieve accuracy $O(h^{m+1})$ when $f(x) \in C^{m+1}(\bar{\Omega})$.

References

1. Morton, K.W., Mayers, D.F.: Numerical Solution of Partial Differential Equations. Cambridge University Press, Cambridge (2005)
2. LeVeque, J.R.: Finite Difference Methods for Ordinary and Partial Differential Equations. Society for Industrial and Applied Mathematics, Philadelphia (2007)
3. Fairweathe, G., Karageorghis, A., Maack, J.: Compact optimal quadratic spline collocation methods for the Helmholtz equation. J. Comput. Phys. **230**, 2880–2895 (2011)
4. Christara, C.C.: Quadratic spline collocation methods for elliptic partial differential equations. BIT **34**, 33–61 (1994)
5. Abushama, A.A., Bialecki, B.: Modified nodal cubic spline collocation for Poisson's equation. SIAM J. Numer. Anal. **46**, 397–418 (2008)
6. Knabner, P., Angermann, L.: Numerical Methods for Elliptic and Parabolic Partial Differential Equations. Springer, New York (2003)
7. Cecka, C., Darve, E.: Fourier-based fast multipole method for the Helmholtz equation. SIAM J. Sci. Comput. **35**, A79–A103 (2013)
8. Hewett, D.P., Langdon, S., Melenk, J.M.: A high frequency hp boundary element method for scattering by convex polygons. SIAM J. Numer. Anal. **51**, 629–653 (2013)
9. Barnett, A.H., Betcke, T.: An exponentially convergent nonpolynomial finite element method for time-harmonic scattering from polygons. SIAM J. Sci. Comput. **32**, 1417–1441 (2010)
10. Chen, K.: Matrix Preconditioning Techniques and Applications. Cambridge University Press, Cambridge (2005)
11. Saad, Y.: Iterative Methods for Sparse Linear Systems. Society for Industrial and Applied Mathematics, Philadelphia (2003)

A Gaussian-Surface Fit to Oceanic Mesoscale Eddies

Song Li[1(✉)] and Liang Sun[2]

[1] Insitute of Marine Science and Technology, National University of Science and Technology,
Changsha, China
pineli@nudt.edu.cn
[2] School of Earth and Space Science, University of Science and Technology of China,
Hefei, China

Abstract. This study shows a Gaussian-surface fit method to obtain key parameters of oceanic mesoscale eddies from maps of sea level anomaly (MSLA). The mesoscale eddies are firstly divided into mononuclear eddies and multiple eddies through a simple splitting strategy and then fitted in a corresponding Gaussian model. Two examples are selected to show the fitting result for mononuclear eddies and multiple eddies, respectively. The result shows that the method works well and efficiently.

Keywords: Mesoscale eddies · Gaussian-surface fit · MSLA

1 Introduction

Mesoscale eddies exist almost everywhere in the global ocean. Their spatial and temporal scales range from tens to hundreds of kilometers and days to months respectively. Mesoscale eddies play an important role in heat, salt and biochemical properties transportation on a global scale. Technical improvements in remote sensing satellites in the past few decades, which provide long-term global altimeter data, has inspired growing interest in mesoscale eddies research.

The first step to investigate mesoscale eddies is to identify them accurately and efficiently. There have been several different automatic eddy identification algorithms and they are separated to three groups: (1) method based on physical parameter [1], (2) method based on flow geometry [2], (3) method based on sea level anomaly (SLA) [3]. Li, et al. has developed a hybrid method based on the SLA data to identify mononuclear eddies [4]. For multi-eddy structures, which usually form when eddies closely located and connected, Yi, et al. raised a Gaussian-surface-based method to identify them out of MSLA [5].

However, few studies about physical parameters of mesoscale eddies have been taken yet. The aim of the present paper is to use SLA maps to study important physical parameters of mesoscale eddies by the simulation with Gaussian model. The authors mainly focus on mononuclear eddies and multiple eddies with only two local extremes. This paper selects two samples, one of which is a mononuclear eddy and the other is a

F. Qiao et al. (eds.), *Recent Developments in Mechatronics and Intelligent Robotics,*
Advances in Intelligent Systems and Computing 690, DOI 10.1007/978-3-319-65978-7_21

multiple-structure eddy, to present the possibilities and goodness of fitting in Gaussian-surface model. And the results show that it is possible and helpful using Gaussian-surface model to simulate mesoscale eddies, though there are some problems remained to be solved.

The rest of the paper is organized as follows. First we introduce the altimeter data and vortex identification information which has been used in this work and the general procedures for simulating a specific mesoscale eddy. Then we show the fitting result of two selected examples. Finally, some discussion is offered about this study, where some limitations for current work and some suggestion for the forward work are contained.

2   Data and Method

2.1   Data

The SLA data is obtained from the merged and gridded satellite product distributed by AVISO. The data is daily available with a 0.25° × 0.25° resolution in the global scale. In this study, the SLA data used were taken on Jan 1st, 1993 (Referencing Fig. 1).

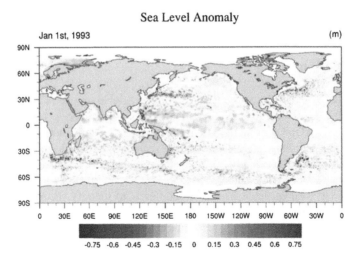

Fig. 1. Global distribution of SLA on January 1st, 1993. As the figure shown, there are a large number of vortexes existing every day. Blue means cold vortex while red means warm vortex. (Color figure online)

Besides, we have used vortex identification information provided by Li [4], which is generated from SLA data as well.

2.2   Method

An identification product was used as the data source, and it provided daily mononuclear information, including the extreme values, the amplitude, the location, etc. Due to different formations for different eddies, it is necessary to classify eddies before the simulation in Gaussian-shape model.

Firstly, a simple splitting strategy was adopted to identify the mononuclear eddies [4], which are defined to contain only one local extreme point. A local extreme point is defined as the point whose SLA value is the maximum or the minimum compared to its surroundings. For those failed in the identification or identified as multiple eddies with more than one local extreme points, the number of adjoining grids was accounted. According to the proportion of the total grids adjoining its local extreme neighbor, they were recognized as mononuclear eddies or multiple eddies. Finally a Gaussian-shape fitting was taken to obtain the main parameters of eddies according to their classification.

For mononuclear eddies, the specific model formulation is defined as

$$h_i(x, y) = b_i + G_i(x, y) \tag{1}$$

$$G_i(x, y) = b + a \cdot \exp\left\{ -\frac{1}{2}\left[\frac{((x - x_0) \cdot \cos\theta + (y - y_0) \cdot \sin\theta)^2}{s_1^2} + \frac{((y - y_0) \cdot \cos\theta - (x - x_0) \cdot \sin\theta)^2}{s_2^2} \right] \right\} \tag{2}$$

where x, y represent the horizontal and vertical distance from the grid point to the eddy center accordingly. x_0, y_0 denote the offset of the eddy extreme point to the eddy center. θ represents the angle rotation of axes. s_1, s_2 represent the eddy scale along the rotated "horizontal" and "vertical" axis respectively. b_i and a denote the basal SLA height and eddy amplitude respectively.

For multiple eddies, which include more than one local extreme, the formulation is defined as

$$h_i(x, y) = b_i + \sum G_i(x, y) \tag{3}$$

where $\sum$ includes all adjoining eddies described in the mononuclear formation. However, as there are too many parameters for multiple eddies, we only simulate those multiple eddies which include two local extremes.

3 Result

Two cases were selected to represent the Gaussian-shape fit result for mononuclear eddies and multiple eddies respectively.

3.1 Mononuclear Eddy Fitting

Figure 2 shows the result of a mononuclear eddy Gaussian-shape, which was selected from thousands of mononuclear eddies on January 1st, 1993. The main fit parameters are listed in Table 1. The fitting parameters are all fallen within a reasonable range: the amplitude is 0.1577 m, which is a typical eddy amplitude value; the eddy scales s_1, s_2 are 80.34 and 50.33 km respectively, which are also reasonable scale values for meso-scale eddies [6]. Besides, the fitting correlation coefficient R^2 equals 0.9662, highly closed to 1, which means an excellent fitting.

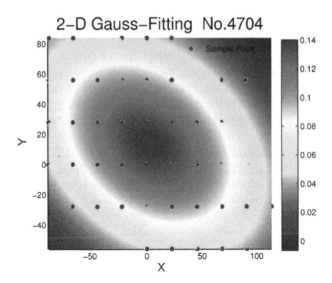

Fig. 2. A two dimensional Guassian-shape fit to a specific mononuclear eddy (its extreme number is 4704, on January 1st, 1993). The deep blue points represents real sample points. (Color figure online)

Table 1. Main Gaussian-fit parameters for a specific mononuclear eddy. R^2 represents correlation coefficient for the fitting.

a	b	s_1	s_2	R^2
0.1577	−0.01758	83.34	50.33	0.9662

3.2 Multiple Eddy Fitting

Figure 3 presents the result of a multiple eddy Gaussian-shape fit, which was selected from the Western Pacific region on January 1st, 1993. It was combined with two components, each of which contains one local extreme. And the two components make a portion to 22.50 and 13.24% adjoining each other, respectively. The main fit parameters are listed on Table 2. The fitting amplitude $a_1 = 0.2635$ m and $a_1 = 0.2736$ m are both typical for mesoscale eddies while the scale parameters s_{i1} and s_{i2} ($i = 1, 2$) are also reasonable values. The correlation coefficient $R^2 = 0.9769$, highly closed to 1, means it is a good fitting.

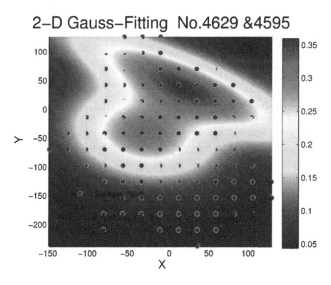

Fig. 3. A two dimensional Guassian-fit to a specific multiple eddy (their extreme number are 4629 and 4595, on January 1st, 1993).

Table 2. Main Gaussian-fit parameters for a specific multiple eddy. a_i, s_{i1}, s_{i2} ($i = 1, 2$) represent the corresponding parameters of each component, respectively.

a_1	a_2	b	s_{11}	s_{12}	s_{21}	s_{22}	R^2
0.2635	0.2736	0.0437	67.55	124.2	51.55	33.59	0.9769

4 Discussion

The main purpose of this article is to obtain the main characteristic parameters of mesoscale eddies in global oceans by simulate them with Gaussian model, including amplitude and scales. We tried to fit mesoscale eddies in Gaussian model and there were different fitting behaviors between mononuclear eddies and multiple eddies. For mononuclear eddies, we found that the most typical amplitudes are between 0.05–0.2 m, especially concentrated on 0.1 m; the most typical radius are ranged from 50 to 150 km, especially concentrated on 100 km. For multiple eddies, we fit them successfully with Gaussian model in two-extremes condition. It illustrates eddies' dynamical combining process well. This is one of the few corresponding efforts to fit multiple eddies compared to other research. It is significant to obtain major parameters of multiple eddies because it helps us to obtain a deeper and clearer understanding of eddies.

However, several limitations of this research should be noticed. Firstly, not all eddies are fitted successfully, especially for multiple eddies. In spite of this, we just simulate multiple eddies with two local extremes, though there are many real eddies contain more than two local extremes. One most important reason is that there are too many parameter to fit for more than two-extremes conditions and we fail to simulate it well. Thus, future studies may focus on solving the fit problem for multiple eddies with more than two local extremes and promote fitting success rate for all kinds of eddies.

References

1. Okubo, A.: Horizontal dispersion of floatable particles in the vicinity of velocity singularities such as convergences ☆. Deep Sea Res. Oceanogr. Abstr. **17**(3), 445–454 (1970)
2. Dong, C., Nencioli, F., Liu, Y., Mcwilliams, J.C.: An automated approach to detect oceanic eddies from satellite remotely sensed sea surface temperature data. IEEE Geosci. Remote Sens. Lett. **8**(6), 1055–1059 (2011)
3. Chelton, D.B., Schlax, M.G., Samelson, R.M., De Szoeke, R.A.: Global observations of large oceanic eddies. Geophys. Res. Lett. **34**(15), 87–101 (2007)
4. Li, Q.Y., Sun, L., Liu, S.S., Xian, T., Yan, Y.F.: A new mononuclear eddy identification method with simple splitting strategies. Remote Sens. Lett. **5**(1), 65–72 (2014)
5. Yi, J., Liu, Z., Du, Y., Wu, D., Zhou, C., Wei, H., et al.: A Gaussian-surface-based approach to identifying oceanic multi-eddy structures from satellite altimeter datasets. In: 22nd International Conference IEEE on Geoinformatics, pp. 1–5 (2014)
6. Wang, Z., Li, Q., Sun, L., Li, S., Yang, Y., Liu, S.: The most typical shape of oceanic mesoscale eddies from global satellite sea level observations. Front. Earth Sci. **9**(2), 202–208 (2015)
7. Yi, J., Du, Y., Zhou, C., Liang, F., Yuan, M.: Automatic identification of oceanic multieddy structures from satellite altimeter datasets. IEEE J. Sel. Top. Appl. Earth Obs. Remote Sens. **8**(4), 1–9 (2015)

Recurrent Neural Networks for Solving Real-Time Linear System of Equations

Xuanjiao Lv[1(✉)], Zhiguo Tan[2], and Zhi Yang[3]

[1] Nanfang College of Sun Yat-Sen University, Guangzhou 510970, China
lxj_258369@163.com
[2] School of Automation Science and Engineering,
South China University of Technology, Guangzhou 510641, China
tanzhiguo136@163.com
[3] School of Data and Computer Science,
Sun Yat-Sen University, Guangzhou 510006, China
issyz@mail.sysu.edu.cn

Abstract. In this paper, a new recurrent neural network (RNN) model is proposed and investigated for solving real-time linear system of equations. The proposed model has an advantage over the existing RNNs, specifically, the gradient-based neural network, Zhang neural network in terms of convergence performance. In addition, theoretical analysis is given, and illustrative example further demonstrates the effectiveness and efficiency of the presented model for the real-time solution of linear equations.

Keywords: Recurrent neural network · Global convergence performance · Linear system of equations

1 Introduction

Considered to be a basic problem that often arises in mathematics and engineering fields, linear system of equations solving finds many applications in robotic vision systems [1], robot inverse kinematics [2]. This problem could be mathematically described as $Ax(t) = b$, our goal is to obtain the unknown column vector $x(t) \in R^n$ for the given constant coefficient matrix $A \in R^{n \times n}$ and the constant vector $b \in R^n$. Generally, numerical algorithms are adopted to solve such a problem [3]. However, it may take much time for them to solve large-scale online linear system of equations due to their serial-processing nature and low computational efficiency. Recently, the recurrent neural network approach, which is of parallel and distributed storage nature, is viewed as one of the important and powerful ways for real-time problems solving involved simultaneous linear system of equations. Many neural network models have been proposed, investigated and exploited by researches to speed up the processing due to in depth study [1, 3–5]. In [1], to solve linear system of equations in real-time, Wang proposed a neural network, which is based on the gradient descent algorithm and depicted in an explicit neural dynamic equation. In [4], robustness of such a neural network was analysed for solving real-time linear system of equations. In [3], Zhang et al. presented a neural

© Springer International Publishing AG 2018
F. Qiao et al. (eds.), *Recent Developments in Mechatronics and Intelligent Robotics*,
Advances in Intelligent Systems and Computing 690, DOI 10.1007/978-3-319-65978-7_22

network (ZNN), which is based on a new descent manner for online linear system of equations solving. Recently, in order to solve online linear system of equations, a novel neural network was presented and analysed in [5].

2 Proposed Recurrent Neural Network

For the purpose of solving the linear system of equations $Ax(t) = b$ in parallel and in real time, Wang [1] proposed the gradient-based neural network as follows:

$$\dot{x}(t) = -\eta A^T Ax(t) + \eta A^T b \tag{1}$$

where A^T denotes the transpose of matrix A, $\eta > 0 \in R$ is termed the convergence parameter, which is designed for adjusting the convergence time of the neural network. Obviously, (1) is an explicit dynamic differential equation (i.e. $\dot{x}(t) = \ldots$).

Recently, Zhang et al. [3, 5] proposed an implicit dynamic system for solving the constant (or say static) and time-varying linear system of equations. For constant linear system of equations solving, the proposed implicit dynamic system (i.e. $A\dot{x}(t) = \ldots$) could be written as:

$$A\dot{x}(t) = -\eta Ax(t) + \eta b \tag{2}$$

In this paper, we propose the following recurrent neural network model for solving constant linear system of equations:

$$A\dot{x}(t) = -\eta(AA^T + A^TA + I)(Ax(t) - b) \tag{3}$$

where I is the $n \times n$ identity matrix. In addition, for the convenience of comparison, we give the model presented in [5] as following:

$$A\dot{x}(t) = -\eta AA^T(Ax(t) - b) - \eta Ax(t) + \eta b \tag{4}$$

Clearly, models (2), (3), (4) are implicit dynamic systems, while model (1) is explicit dynamic system. Implicit systems are in accord with actual systems well, and seem to possess superior performances in describing dynamic systems in comparison to explicit systems (1) [3]. Besides, it is worth noting that, implicit dynamic systems could be converted into explicit dynamic systems by some simple manipulations. Furthermore, compared with the neural network based gradient descent algorithm (1), and the ZNN model (2), and model (4), our model (3) could achieve higher global exponential convergence performance. For a better understanding on this point, in the following section of this paper, we pay attention to the exponential convergence of the implicit dynamic system (3) for real-time linear system of equations solving.

3   Global Exponential Convergence of the Presented Model

Now, we give the theoretical results about global exponential convergence of our model (3).

Theorem 1. For the linear system of equations $Ax(t) = b$, suppose $A \in R^{n \times n}$ is a nonsingular matrix, the neural state $x(t) \in R^n$ of the recurrent neural network (3), starting from any initial state $x(0) \in R^n$, will converge to the theoretical value $x^* = A^{-1}b$ in a global and exponential manner. In addition, the exponential convergence rate is at least $\eta(2\alpha + 1)$ with α denoting the minimum eigenvalue of AA^T.

Proof. We first define the error function $\zeta(t) \in R^n$: $\zeta(t) = Ax(t) - b$

Evidently, $\zeta(t)$ denotes the solution error by using the RNN model (3). The error-function derivative $\dot{\zeta}(t)$ could be written as: $\dot{\zeta}(t) = A\dot{x}(t)$. In the model (3), by replacing $Ax(t) - b$ and $A\dot{x}(t)$ with $\zeta(t)$ and $\dot{\zeta}(t)$, respectively, we could get: $\dot{\zeta}(t) = -\eta(AA^T + A^TA + I)\zeta(t)$. Let $\|\zeta(t)\|_2$ represent the Euclid norm of vector $\zeta(t)$. Now, we choose the Lyapunov function $\nu(t) = \|\zeta(t)\|_2^2/2 \geq 0$. The derivative of $\nu(t)$ with respect to time is:

$$\frac{dv}{dt} = \left(\frac{\partial v}{\partial \zeta}\right)^T \frac{d\zeta}{dt} = \zeta^T \dot{\zeta} = -\eta\zeta^T(AA^T + A^TA + I)\zeta(t) \tag{5}$$

Considering that matrix A is nonsingular, and AA^T and A^TA are symmetrical, there exists two orthogonal matrixes P and Q such that

$$P^{-1}AA^TP = P^TAA^TP = diag(\lambda_1, \lambda_2, \ldots, \lambda_n)$$

$$Q^{-1}A^TAQ = Q^TA^TAQ = diag(\lambda_1, \lambda_2, \ldots, \lambda_n)$$

where $\lambda_j > 0, j = 1, 2, \ldots n$ is the eigenvalue of AA^T. Therefore,

$$AA^T = Pdiag(\lambda_1, \lambda_2, \ldots, \lambda_n)P^T \tag{6}$$

$$A^TA = Qdiag(\lambda_1, \lambda_2, \ldots, \lambda_n)Q^T \tag{7}$$

Substituting (6) and (7) into (5), we obtain

$$\frac{dv}{dt} = -\eta\zeta^T(Pdiag(\lambda_1, \lambda_2, \ldots, \lambda_n)P^T + Qdiag(\lambda_1, \lambda_2, \ldots, \lambda_n)Q^T + I)\zeta \tag{8}$$

We further have

$$\frac{dv}{dt} \leq -\eta(2\alpha + 1)\|\zeta(t)\|_2^2 \leq 0 \tag{9}$$

where $\alpha = \min (\lambda_1, \lambda_2, \ldots, \lambda_n)$ is the minimum eigenvalue of AA^T. We conclude $\dot{v}$ is negative definite. Therefore, by Lyapunov stability theory [6, 7], the equilibrium point $\zeta(t) = 0$ is globally and asymptotically stable. Moreover, it follows from (9) that $\dot{v}(t) \leq -2\eta(2\alpha + 1)v(t)$, $v(t) \leq v(0) \exp(-2\eta(2\alpha + 1)t)$, where $v(0) = \|\zeta(0)\|_2^2/2$. In addition, by defining $M: = \|A\|_F\|x(0) - x^*\|_2$, where $\|A\|_F$ denotes the Frobenius norm of the matrix A, we have $\|\zeta(t)\|_2^2 = \|A(x(0) - x^*)\|_2^2 \leq M^2$, which results in

$$\|\zeta(t)\|_2 \leq M \exp(-\eta(2\alpha + 1)t), t \geq 0 \tag{10}$$

Clearly, the exponential convergence rate is at least $\eta(2\alpha + 1)$. Thus, we complete the proof. The global exponential rate $\eta(2\alpha + 1)$ of the presented model (3) is larger than $\eta\alpha$ of the neural network model based on the gradient descent algorithm (1), and η of the ZNN model (2), and $\eta(\alpha + 1)$ of the model (4) [5], which can theoretically ensure the presented implicit system model (3) could achieve better convergence performance.

4   Simulation Results

For the purpose of demonstrating better global exponential convergence of our model (3) in comparison with the gradient-based neural network model (1), the ZNN model (2), and model (4), we could consider the same example as in [1, 5]. The minimal eigenvalue α is 0.2345 for this special matrix A. Without losing generality, zero initial state is set for the simulation. With design parameter $\eta = 1000$, computer simulation results are shown in Figs. 1, 2. For model (2), the convergence time to achieve a tiny solution error $\exp(-7)M$ is about 7 ms, which is shown in Fig. 1(a). With the same design parameter $\eta = 1000$, as illustrated in Fig. 1(b), the convergence time to achieve a tiny solution error $\exp(-7)M$ by the model (4) is about 5.7 ms. As seen from Fig. 2(a), for our model (3), the convergence time to achieve a tiny solution error $\exp(-7)M$ is only roughly 0.3 ms. Moreover, Fig. 2(b) illustrates the convergence performances of the gradient-based

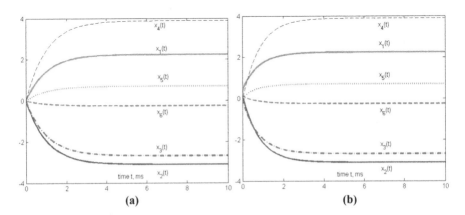

Fig. 1. (a) Online solution to $Ax(t) = b$ synthesized by model (2) with $\eta = 1000$. (b) Online solution to $Ax(t) = b$ synthesized by model (4) with $\eta = 1000$

neural network model (1), the ZNN model (2), the model (4), and our model (3) for real-time linear system of equations solving. The blue solid curve in Fig. 2(b) shows the solution error $\|Ax(t) - b\|_2$ generated by the presented model (3). Clearly, this blue solid curve trends to 0 with much less time than other three curves which are produced by the model (1), model (2) and model (4).

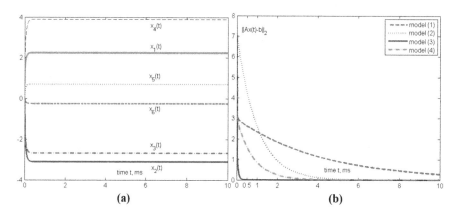

Fig. 2. (a) Online solution to $Ax(t) = b$ synthesized by model (3) with $\eta = 1000$. (b) Convergence of solution error $\|Ax(t) - b\|_2$ synthesized by model (1), model (2), model (3), and model (4) for online linear equations solving with $\eta = 1000$

It is worth mention here that $\eta(2\alpha + 1)$ is theoretically the minimum convergent rate of our model (3), the actual convergent rate is larger than $\eta(2\alpha + 1)$ for this special matrix A and initial state $x(0)$. Therefore, simulation results agree well with the aforementioned theoretical result, and validate the advantage of our model (3) over gradient-based neural network model (1), ZNN (2) and RNN model (4) in terms of convergence rate.

5   Conclusion

In this paper, we proposed a RNN model, which is described by an implicit dynamic equation. This model is able to generate online solutions to large-scale simultaneous linear system of equations. In addition, we have investigated and analyzed the globally exponentially convergence performance of such a model. The theoretical convergence rate is higher than the gradient neural network model (1), the ZNN model (2), and the neural network model (4) for real-time linear system of equations solving. Computer simulation results agree well with the theoretical analysis.

References

1. Wang, J.: Electronic realization of recurrent neural work for solving simultaneous linear equations. Electron. Lett. **28**(5), 493–495 (1992)

2. Sturges, R.H.J.: Anolog matrix inversion (robot kinematics). IEEE J. Robot. Autom. **4**(2), 157–162 (1988)
3. Yi, C., Zhang, Y.: Analogue recurrent neural network for linear algebraic equation solving. Electron. Lett. **44**(18), 1078–1079 (2008)
4. Chen, K.: Robustness analysis of Wang neural network for online linear equation solving. Electron. Lett. **48**(22), 1391–1392 (2012)
5. Chen, K.: Implicit dynamic system for online simultaneous linear equations solving. Electron. Lett. **49**(2), 101–102 (2013)
6. Zhang, Y., Ge, S.S.: Design and analysis of a general recurrent neural network model for time-varying matrix inversion. IEEE Trans. Neural Netw. **16**(6), 1477–1490 (2005)
7. Zhang, Y., Wang, J.: Global exponential stability of recurrent neural networks for synthesizing linear feedback control systems via pole assignment. IEEE Trans. Neural Netw. **13**(3), 633–644 (2002)

Analysis of Relationships Between Amount of Physical Activity of Patients in Rehabilitation and Their ADL Scores Using Multidimensional PCA

Akio Ishida[1(✉)], Keito Kawakami[2], Daisuke Furushima[3], Naoki Yamamoto[4], and Jun Murakami[2]

[1] Faculty of Liberal Studies, Kumamoto College, National Institute of Technology, Kumamoto, Japan
ishida@kumamoto-nct.ac.jp
[2] Advanced Course of Electronics and Information Systems Engineering, Kumamoto College, National Institute of Technology, Kumamoto, Japan
[3] Division of Health Sciences, Graduate School of Medicine, Osaka University, Osaka, Japan
[4] Department of Human-Oriented Information Systems Engineering, Kumamoto College, National Institute of Technology, Kumamoto, Japan

Abstract. This paper investigates relationships between the ADL evaluations of convalescent stroke patients in rehabilitation and their physical activity amounts measured by an accelerometer. A previous study performed a correlation analysis between the activity amount and two FIM items (i.e., motor FIM and cognitive FIM), which is one of the ADL evaluations, and showed the existence of a significant correlation between both of them. In this paper, we performed a correlation analysis and also did a multidimensional PCA (MPCA) of data matrix and tensors, respectively, constructed from more detailed FIM items than above. As the results of the correlation analysis, we showed that the correlation coefficients of the activity amount of walking when compared the motor and cognitive FIM were relatively large, and that the similar tendency to the result of the previous study was obtained. By considering the detailed FIM items, we noticed that correlation coefficients of the locomotion subscale of FIM compared with the walking calories and the number of walking steps were the largest. Furthermore, from the results of performing MPCA, we found several pairs of principal component scores with higher correlation coefficients. The above-mentioned results suggest that the use of accelerometer is considered to be effective in grasping patients' FIM scores.

Keywords: Stroke patient · ADL · Physical activity amount · Accelerometer · MPCA

1 Introduction

It is predicted that the ratio of elderly people will reach one third of the population in Japan by cause of the falling birthrate and the aging of the baby boomer generation [1, 2]. This prediction leads to concerns about the increase in patients who need

© Springer International Publishing AG 2018
F. Qiao et al. (eds.), *Recent Developments in Mechatronics and Intelligent Robotics*,
Advances in Intelligent Systems and Computing 690, DOI 10.1007/978-3-319-65978-7_23

rehabilitation, the shortage of rehabilitation facilities, and the lack of personnel worked for those facilities. From the above, it is deduced that the role of in-home rehabilitation will increase in importance in the near future. We have studied on this problem, and in this paper describe about a way of evaluation of rehabilitation data acquired from in-home cared elderly people.

The functional independence measure (FIM) is a widely used index in the rehabilitation field which is exploited to evaluate the activities daily living (ADL) of the person in rehabilitation [3]. We plan to use the accelerometer to measure rehabilitation data from that person. In former research, the significant correlation relationship acknowledged between physical activity amount and FIM score of convalescent stroke patients [4]. In this paper, we report more minute analysis than the former research about the relationship by taking up more detailed items of FIM. For this purpose, we use multidimensional principal component analysis (MPCA) [5, 6], which is the extension of PCA [7] to multidimensional data and enables detailed analysis.

Generally speaking, it requires manpower of experts to measure FIM score from rehabilitation patients. Since the accelerometer measures the physical activity amounts of them automatically, we consider that a family member can easily measure the data to be used to rehabilitate the patients at their home if it is possible to estimate the FIM score from those data. From this reason, our research can be considered useful to solve the problem mentioned in the beginning of this chapter.

2   Data for Analysis

In this study, we deal with the FIM data and that of accelerometer, both of which were measured from 14 rehabilitation inpatients in a hospital in order to analyze the relationships between them. FIM is a scale which is used to measure the degree of independence of patients' daily life by even a non-expert. This scale is divided into two subscales, that is, motor FIM and cognitive FIM. To explain further, motor FIM has 4 subscales and cognitive FIM does 2 subscales, and these total 6 subscales are subdivided into 18 items as shown in Table 1 [8]. Each of those items is evaluated on a scale from 1 to 7, and it is referred to as the FIM score. The FIM score is usually evaluated by hospital staffs periodically.

Table 1. Evaluation items of FIM

FIM scale	FIM subscale (6)	FIM item (18)
Motor FIM	Self-care	Eating, grooming, etc.
	Sphincter control	Bladder management, etc.
	Transfers	Bed/chair, toilet, etc.
	Locomotion	Walk/wheelchair, stairs
Cognitive FIM	Communication	Comprehension, expression
	Social cognition	Social interaction, etc.

The accelerometer is a wearable device that used to measure and record the physical activity, such as step counts, moving distance, and consumed calories. This time, we use a high accuracy accelerometer "Active style Pro HJA-750C" (OMRON Corp., Tokyo Japan) [9], which can acquire many kinds of physical activity data. In these data, walking calories, living activities calories, walking exercise, living activities exercise, and step counts are chosen to analyze. For reference, the rehabilitation patients wore the accelerometer in the daytime for the measurement 387 ± 111 min including the time of rehabilitation.

3   Correlation Analysis

3.1   Construction of Data Matrix

As the first step to analyze the relationships between physical activity amount and FIM score, the data matrix is constructed from convalescent stroke patients as mentioned in the Sect. 2. The size of this matrix is 32×14 whose column vectors consist of 14 patients' individual data as illustrated in Fig. 1. The way of selecting row vectors is particular, that is, those are consist of 18 FIM items, 6 FIM subscales, 2 FIM scales, and 5 acquisition data from the accelerometer.

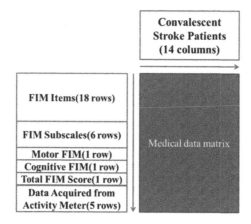

Fig. 1.   Image of medical data matrix.

3.2   Analyzing Result of Data Matrix

In the previous study [4], Pearson's correlation coefficient was exploited to compare physical activity amount with FIM score of convalescent stroke patients, and the value of correlation was shown as follows: 0.59 for the motor FIM and 0.62 for the cognitive FIM in the relationships between walking time and these FIM items, respectively. Similarly with this study, we calculated the correlation coefficients between column vectors of the matrix illustrated in Fig. 1 and each item of physical activity amount acquired from the accelerometer.

Figure 2 shows the resultant correlation coefficient values extracted from those of all combination of the row vectors, that is, relationships between 8 FIM items and

5 items from the accelerometer, as rows and columns, respectively. It is known that from the figure, at first, motor and cognitive FIM scales are correlated with the items concerned with walking, such as walking calories and step counts. This result is similar to that of ref. [4]. In the next place, as the result of more detailed analysis, which deals with FIM subscales, we see that the correlation values are the highest (0.78) in relationships between locomotion FIM subscale and walking calories, and the same subscale and step counts.

Item	Walking Calories	Living Activities Calories	Walking Exercise	Living Activities Exercise	Step Counts
Self-Care	0.65	0.57	0.4	0.42	0.61
Sphincter Control	0.49	0.44	0.31	0.36	0.44
Transfers	0.69	0.66	0.51	0.6	0.58
Locomotion	0.78	0.62	0.59	0.44	0.78
Communication	0.46	0.39	0.3	0.35	0.41
Social Cognition	0.53	0.39	0.38	0.38	0.48
Motor FIM	0.69	0.61	0.47	0.48	0.64
Cognitive FIM	0.51	0.4	0.36	0.38	0.47

Fig. 2. Correlation coefficient of each data.

4 Analysis Using MPCA

4.1 PCA and MPCA

Principal component analysis (PCA) is one of the multivariate analysis methods [7]. Since this method calculates uncorrelated synthetic variables called "principal components" from the original multivariate data, we could extract the characteristics of that data. In PCA, the original data is represented by a matrix, for example, like with samples and variables in rows and columns, respectively.

MPCA is thought as an extension of PCA to tensor data, which is defined in the next section, as mentioned in Chap. 1, and this method can calculate principal components of that data [5, 6].

4.2 Definition of Tensor

In this paper, the word "tensor" means a multi-dimensional array, which can be thought of as an extension of matrices to higher-dimensional arrays, that is, N-dimensional arrays are called "N th-order tensor". Figure 3 illustrates a third-order tensor $\mathcal{A}$ with a size of $3 \times 3 \times 3$, and three arrows in the figure indicate each mode (or way) of the tensor. A third-order tensor is also called a three-mode or three-way tensor [10].

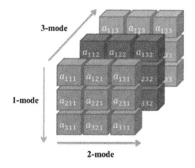

Fig. 3. Example of a third-order tensor.

Now, in this figure, the (i, j, k)-th element of $\mathcal{A}$ is denoted as a_{ijk} for $i, j, k = 1, 2, 3$, where each subscript expresses the mode of dimensionality, that is, from i to k corresponds 1-mode to 3-mode, respectively. As just described, categorical data composed of three modes can be stored in a third-order tensor.

4.3 Construction of Data Tensor and General Flow of Analysis

For the sake of analyzing multivariate medical data effectively, the data structures illustrated in Fig. 4 is used in this study. Below, these two structures are described. The tensor $\mathcal{A}$ in the left side of the figure, whose size is $6 \times 4 \times 10$, is for analyzing the characteristics of patients' ADL function. In this tensor, the 1-mode denotes FIM subscales with six items, the 2-mode does the number of evaluations from 1 to 4 times, and 3-mode does the number of ten patients. Next, the $5 \times 5 \times 10$ tensor $\mathcal{B}$ in the right side of the figure is constructed in order to extract the features of physical activity amount of the patients. Each mode of this tensor from 1 to 3 represents five measuring items of accelerometer, the number of evaluations from 1 to 5, and ten patients who are the same as the tensor $\mathcal{A}$, respectively.

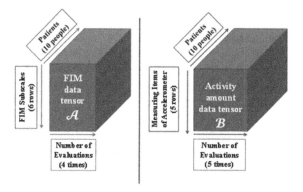

Fig. 4. Structure of medical data.

The general flow of the analysis is as follows:

(Step 1) MPCA is applied to both tensors $\mathcal{A}$ and $\mathcal{B}$.
(Step 2) The principal component scores of the 3-mode for both tensors are calculated by using 1 and 2-mode principal components obtained from Step 1.
(Step 3) The correlation coefficients between the principal component scores of tensors $\mathcal{A}$ and $\mathcal{B}$ are calculated

4.4   Calculation Procedure of MPCA

In this section, the calculation procedure to apply MPCA to the tensor $\mathcal{A}$ is described below. Additionally, the principal components and scores of the tensor $\mathcal{B}$, which is constructed from activity amount data, also can be calculated in the same way as for $\mathcal{A}$.

4.4.1   Standardization of Tensor

To reduce the influence on the calculation results caused by the difference of units of the 1 and 2-mode variables of $\mathcal{A}$, the 3-mode data is necessary to be standardized. The (i, j, k)-th elements f_{ijk} of a standardized tensor $\mathcal{F}$ can be obtained by the following equation:

$$f_{ijk} = \frac{a_{ijk} - \bar{a}_{ij}}{s_{ij}}, (i = 1, \cdots, 6; j = 1, \cdots, 4; k = 1, \cdots, 10), \tag{1}$$

where $\bar{a}_{ij}$ and s_{ij} in Eq. (1) denote the average and the standard deviation, respectively, and can be computed by

$$\bar{a}_{ij} = \frac{1}{10} \sum_{k=1}^{10} a_{ijk}, s_{ij} = \sqrt{\frac{1}{10} \sum_{k=1}^{10} \left(a_{ijk} - \bar{a}_{ij}\right)^2}, (i = 1, \cdots, 6; j = 1, \cdots, 4). \tag{2}$$

4.4.2   HOSVD Calculation

The higher-order singular value decomposition (HOSVD) is a generalization of the singular value decomposition (SVD), which is a method of matrix decomposition, and can be applied to decompose a tensor [11]. The principal components for each mode of the data tensor, which are explained in the Sects. 4.2 and 4.3, can be obtained by applying HOSVD.

The tensor $\mathcal{F}$ obtained in the Sect. 4.4.1 is decomposed into a product of a core tensor S and matrices $U^{(1)}$, $U^{(2)}$, and $U^{(3)}$ by HOSVD as follows:

$$\mathcal{F} = S \times_1 U^{(1)} \times_2 U^{(2)} \times_3 U^{(3)}, \tag{3}$$

where $\times_n$ denotes a "n-mode product" which is a product operation between a tensor and a matrix. The matrices $U^{(1)}$ and $U^{(2)}$ in Eq. (3) have coefficients of 1 and 2-mode principal components (i.e., FIM subscales and the number of evaluations) in the column of the matrix, respectively.

4.4.3   Calculation of Principal Component Scores

By using $U^{(1)}$ and $U^{(2)}$ obtained from the Sect. 4.4.2, principal component scores of 3-mode (i.e., patients) can be calculated by next equation:

$$\mathcal{Z} = \mathcal{F} \times_1 U^{(1)^\mathrm{T}} \times_2 U^{(2)^\mathrm{T}}, \tag{4}$$

where "T" denotes a transpose of a matrix. In Eq. (4), $\mathcal{Z}$ represents a third-order tensor arranged in order of the score, in which 1 and 2-mode represent the number of principal component, and 3-modes does the patients. That is, for example, each element of the vector z_{ijk}, $(i, j\text{:fixed})$ of $\mathcal{Z}$ is the principal component score of the k-th patient $(k = 1, \cdots, 10)$, which is referred to as $FIM(i, j)$. $FIM(i, j)$ is a combination of the i-th principal component of 1-mode and the j-th one of 2-mode.

The contribution rate of k-th patient's score $(k = 1, \cdots, 10)$ $cont_{ij}^{(FIM)}$ related to $FIM(i, j)$ is computed by using the variance of the vector z_{ijk}, $(i, j\text{:fixed})$ as follows:

$$cont_{ij}^{(FIM)} = \frac{\mathrm{var}(z_{ijk})}{\sum_{ij} \mathrm{var}(z_{ijk})}, (i = 1, \cdots, 6; j = 1, \cdots, 4), \tag{5}$$

where a notation var($\bullet$) means to compute the variance.

4.5   Analyzing Result of Data Tensor

In this section, the analyzing results by MPCA are described from four viewpoints in order.

4.5.1   Principal Component Coefficients

Figures 5 and 6 show the principal component coefficients of the data tensors $\mathcal{A}$ and $\mathcal{B}$, where the orders of both tensors are 3, as described in Sect. 4.3. The vertical axis of Fig. 5 is the principal component coefficient value of the FIM items and the number of evaluations, which are obtained from the tensor $\mathcal{A}$ by MPCA as $U^{(1)}$ and $U^{(2)}$, respectively. Likewise, Fig. 6 shows the principal component values of the tensor $\mathcal{B}$, whose meaning are same as those of Fig. 5. The horizontal axis of those figures is the element of each principal component in common, which means the FIM items for the upper graph and the number of evaluation for the lower graph.

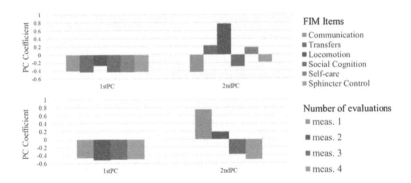

Fig. 5. PC coefficient of FIM data tensor $\mathcal{A}$.

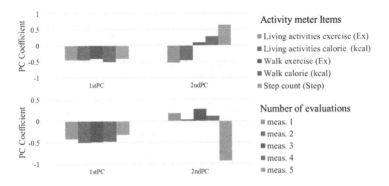

Fig. 6. PC coefficient of Activity amount data tensor $\mathcal{B}$.

From those figures, we see the following:

(1) Each first principal component of tensor $\mathcal{A}$ shows the general characteristics for the items of the horizontal axis mentioned above (see the left side of Fig. 5).

(2) The second principal component concerning the FIM items which obtained from tensor $\mathcal{A}$ shows mainly the locomotion FIM subscale, and on the other hand, that concerning the number of evaluations does the early part of the number of evaluation (see the right side of Fig. 5).

(3) As for the tensor $\mathcal{B}$, the principal components express the general characteristics about each items similarly with (1) (see the left side of Fig. 6).

(4) The second principal component of the measuring items of the accelerometer, which was obtained from the tensor $\mathcal{B}$, shows the items relating to walking, such as walking calories and step counts (see the upper right of Fig. 6). Another second principal component of $\mathcal{B}$, which is that of number of evaluation, shows mostly the last time of evaluation (see the lower right of Fig. 6).

4.5.2   Principal Component Scores

In the next place, the principal component scores are calculated for important components by Eq. (4) in Sect. 4.4. Hereafter, we use the notation $FIM(i,j)$ to express the principal component which is a combination of the i-th column vector of $U^{(1)}$ and the j-th column vector of $U^{(2)}$, where those matrices are obtained from the tensor $\mathcal{A}$. Similarly, the notation $ACT(i,j)$ expresses the same combination above in concerning with the tensor $\mathcal{B}$.

Here, we take up $FIM(1, 1)$ component as an example of interpretation of the results. Since all the elements of the first column vectors of $U^{(1)}$ and $U^{(2)}$ are negative as seen from Figs. 5 and 6, we see that all the elements of the matrix, which is constructed by multiplying each element of the vectors mutually, have positive value as shown in Table 2. This fact leads that the value, by which the original data matrix of each patient in Eq. (4) is multiplied, is always positive. Therefore, we know that the patients having high value of the principal component score possess the characteristics of both of combined components.

Table 2. Example of positive coefficients of matrix by taking product each PC coefficient.

	1st PC in number of evaluation items			
	meas1 (−)	meas2 (−)	meas3 (−)	meas4 (−)
1st PC in FIM items				
Communication (−)	+	+	+	+
Transfers (−)	+	+	+	+
Locomotion (−)	+	+	+	+
Sphincter control (−)	+	+	+	+
Self-care (−)	+	+	+	+
Sphincter control (−)	+	+	+	+

By such a way of thinking, we enumerate below some features of $FIM(1, 1)$ and $FIM(2, 1)$, and as well as $ACT(1, 1)$ and $ACT(2, 1)$.

(1) $FIM(1, 1)$ and $ACT(1, 1)$ components express the general characteristics relating to the FIM item and the measuring items of the accelerometer, respectively, in the whole period of evaluation.

(2) $FIM(2, 1)$ component represents the general characteristics about locomotion in the whole period of evaluation. On the other hand, $ACT(2, 1)$ is thought as the general characteristics relevant to walking in the same period.

4.5.3   Correlation Among Principal Components

Figure 7 shows the Spearman's correlation relations among the principal component scores of four components described in B. For example, we can see that the correlation value between those scores of $FIM(1, 1)$ and $ACT(2, 1)$ is 0.62 from the figure. Below, the main points which we can see from the figure about the principal component score are summarized.

(1) The correlation value between $FIM(1, 1)$ and $ACT(1, 1)$ is 0.66, and this fact shows the existence of a significant correlation relationship between the general characteristics of FIM and the measuring items of the accelerometer. It is because that this value can be thought of as a middle-level correlation (the same hereinafter).

(2) Regarding to $ACT(2, 1)$, the correlation relationships between it and $FIM(1, 1)$ and between it and $FIM(2, 1)$ are significant, since the correlation values of those relationships are 0.62, as described as an example, and 0.67, respectively. It is inferred from this that the walking characteristics measured from the accelerometer is correlated to the general characteristics and also the locomotion subscale of FIM.

	$FIM(1,1)$ Score	$FIM(1,2)$ Score	$FIM(2,1)$ Score	$FIM(2,2)$ Score
$ACT(1,1)$ Score	0.66	0.48	0.02	-0.53
$ACT(1,2)$ Score	-0.2	0.18	0.15	-0.42
$ACT(2,1)$ Score	0.62	-0.18	0.67	-0.01
$ACT(2,2)$ Score	0.3	0.39	0.68	-0.41

Fig. 7. Correlation coefficient of each PC score.

As a sequel to above summary, the following inference can be deduced: There is a possibility of grasping the improvement of FIM score of rehabilitation patients by analyzing the principal component scores $ACT(1, 1)$ and $ACT(2, 1)$.

4.5.4 Detailed Analysis of Relationships Between Principal Components

Among combinations of principal components, here, we pay attention to $FIM(2, 1)$ and $ACT(2, 1)$ for the reason mentioned above. As additional remarks, since both of those components are related to walking, the correlation between them shows mid-high value.

Figure 8 plots their principal component scores, where $ACT(2, 1)$ is horizontal coordinate and $FIM(2, 1)$ is vertical one, and the points in the figure represent each patient. The meaning of the magnitude of those points, which is seen by analyzing the principal component coefficients in the same way as Table 2, is noted in the figure. We can show the following findings obtained from the figure:

(1) The lower the $FIM(2, 1)$ component, the higher the value of the locomotion FIM subscale. Similarly, the lower the $ACT(2, 1)$ component, the higher the items relating to walking of the activity amount.

(2) The FIM scores relating to walking are not improved for the patients who are plotted on the first quadrant. In contrast, the patients on the third quadrant are improved in it.

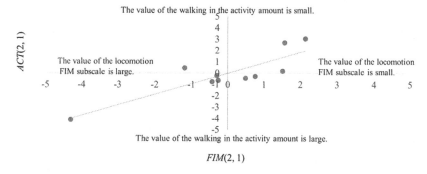

Fig. 8. Scatter plot of PC scores.

5   Conclusions

In this paper, we applied correlation analysis and MPCA to FIM data and those of accelerometer of convalescent stroke patients for the purpose of analyzing the relationships between those two kinds of data. At first, in correlation analysis, by conducting more detailed analysis than the former research, we see that there is a strong correlation between transfer and locomotion of FIM subscales and walking calories and step counts of the items of accelerometer. Secondly, as a result of applying MPCA to both data from the FIM evaluation and the accelerometer measurement, we find some combinations of the principal component scores of those data with high correlation value by analyzing the correlation between them. From the above results, the following can be inferred: It is possible to estimate the improvement of the FIM scores from the measurement values of the accelerometer to some extent.

As the future work, we would like to analyze by increasing the sample data obtained from rehabilitation patients, and to use other multivariate method extended to multidimensional data.

Acknowledgements.   The authors thank Kumamoto Rehabilitation Hospital for providing the rehabilitation data of its inpatients.

References

1. Nagasawa, S.: Long-term care insurance act and home care. Jpn. Med. Assoc. J. **58**(1–2), 23–26 (2015)
2. Tokudome, S., Hashimoto, S., Igata, A.: Life expectancy and healthy life expectancy of Japan: the fastest graying society in the world. BMC Res. Notes **9**, 482 (2016)
3. Smith, P.M., Illig, S.B., Fiedler, R.C., Hamilton, B.B., Ottenbacher, K.J.: Intermodal agreement of follow-up telephone functional assessment using the functional independence measure in patients with stroke. Arch. Phys. Med. Rehabil. **77**, 431–435 (1996)

4. Sawamura, Y., Ito, Y., Minakata, N., Kawagoe, A., Terui, Y., Shiotani, T., Satake, M.: Relationships between physical activity amount and physical function in convalescent stroke Patients (Kaifukuki Nousocchuu Kanja no Shintai Katsudouryou to Shintai Kinou tono Kankei). In: The 48th Congress of Japanese Society of Physical Therapy, vol. 40, suppl. no. 2, B-P-18 (2013). (in Japanese)
5. Inoue, K., Hara, K., Urahama, K.: Matrix principal component analysis for image compression and recognition. In: Proceedings of the 1st Joint Workshop on Machine Perception and Robotics (MPR), pp. 115–120 (2005)
6. Ishida, A., Aibara, U., Murakami, J., Yamamoto, N., Saito, S., Izumi, T., Kano, N.: Analysis of rehabilitation data by multi-dimensional principal component analysis method using the Statistical Software R. Adv. Mater. Res. **823**, 650–656 (2013)
7. Madsen, R.E., Hansen, L.K., Winther, O.: Singular value decomposition and principal component analysis. Neural Netw. **1**, 1–5 (2004)
8. Hetherington, H., Earlam, R.J., Kirk, C.J.C.: The disability status of injured patients measured by the functional independence measure (FIM) and their use of rehabilitation services. Injury **26**, 97–101 (1995)
9. Product Information, FUKUDA COLIN Co., Ltd. http://www.fukuda.co.jp/colin/products/basic/231.html/. Accessed 18 Feb 2017
10. Cichocki, A., Zdunek, R., Phan, A.H., Amari, S.: Nonnegative matrix and tensor factorizations: applications to exploratory multi-way data analysis and blind source separation, Chap. 1. John, New York (2009)
11. Lathauwer, L.D., Moor, B.D., Vandewalle, J.: A multilinear singular value decomposition. SIAM J. Matrix Anal. Appl. **21**, 1253–1278 (2000)

A Feature Selection Algorithm for Big Data Based on Genetic Algorithm

Bo Tian[(⊠)] and Weizhi Xiong

Big Data Institute, Tongren University, Tongren, Guizhou, China
tianbomail@163.com

Abstract. Features selection is an important task since it has significant impact on the data mining performance. This paper present an algorithm to perform feature selection based on the adaptive genetic algorithm. First, the method to compute the crossover probability and mutation probability were proposed. Therefore, the subset feature selection operation can be seen as a process of evolution, and realized adaptive feature subsets selection and optimization. Experimental results demonstrate that the proposed algorithm achieves notable classification accuracy improvements and reduced the total computing time compare to the conventional algorithm.

Keywords: Big data · Feature selection · Genetic algorithm

1 Introduction

With the development of the information and computer technologies, big data mining has been widely employed in various industries, such as finance and stock market, traffic, tourism, health records, social security, science data, and so forth. As everyone knows, the big data containing three problems that is velocity, variety and volume problems, and big data comprises various types of data [1]. These problems are the main challenges for big data mining. Therefore, the traditional data mine method is lead to extreme time consuming and complexity [2]. It cannot meet the demand of big data process and time requirement obviously.

In data mining algorithms, the feature selection model is a key issue. It select the most important feature to improve the performance of classifier, which is employed to predict classes for new samples. There are some studies have been made according to this problem. As an efficient algorithm, Heuristics has been widely applied in feature selection [3], which most is top down supervised learning. Additionally, the heuristics require full set of data in training processing. It is not suitable to dynamic stream processing environment [4], and many feature selection algorithm is only developed for some special application areas. As a general rule, the convergence speed of traditional algorithm is still slow and may not converge to a global minimum [5]. Therefore, it necessary to design a high performance algorithm for feature selection.

© Springer International Publishing AG 2018
F. Qiao et al. (eds.), *Recent Developments in Mechatronics and Intelligent Robotics*,
Advances in Intelligent Systems and Computing 690, DOI 10.1007/978-3-319-65978-7_24

2   Feature Selection Algorithm Based on Genetic Algorithm

Genetic algorithm is an efficient heuristic method which is widely employed to get global solution within solution space. Therefore, to solve the feature selection problem, an adaptive genetic algorithm based was proposed. And the genetic operators such as selection, crossover and mutation were designed in following section.

2.1   Initial Population

Initial population contains n features as $F = (f_1, f_2, \cdots\cdots, f_n)$. The initial population which referred as $G = Ln$. L is represent the number of bits to the corresponding feature. In order to improve calculative efficiency and accelerate the convergence, we set its value to 5.

The fitness value of individual was calculated according to the following formula.

$$P = \frac{kr_{cf}}{\sqrt{k + k(k+1)r_{ff}}} \tag{1}$$

Where k is constant. The r_{cf} and r_{ff} represent the forward and backward time delay respectively. In calculating process, if $P > \varepsilon$, then the algorithm is terminated. Where the ε is threshold.

2.2   Selection

In order to avoiding premature convergence sometimes occurs, the adaptive selection according to the following probability.

$$Q = \alpha P + \frac{(e - e^{k/k\max})}{e + e^{k/k\max}} \tag{2}$$

Where Q is the fitness value of next generation. The selection operation are produced according to the probability Q.

2.3   Crossover

The adaptive crossover probability can be computed as follows:

$$P_c = \begin{cases} P_{c2}, f_{avg} > f', P_{c2} \le 1 \\ P_{c1}(f_{\max} - f')/(f_{\max} - f_{avg}), f_{avg} \le f', P_{c1} \le 1; \end{cases} \tag{3}$$

Where f' is the individual which fitness is bigger than another in crossover process. f represented the fitness of individual will be crossover. $P_{c1}, P_{c2}, P_{m1}, P_{m2}$ were parameters and $f_{\max}, f_{avg}$ is the max fitness and average fitness of last generation, respectively.

2.4 Mutation

The purpose of mutation is to introduce a slight perturbation to increase the diversity of trial individuals after crossover, preventing trial individuals from clustering and causing premature convergence of solution. The probability of mutation is calculated as follows:

$$P_m = \begin{cases} P_{m2}, f_{avg} > f, P_{m2} \leq 1 \\ P_{m1}(f\text{max} - f)/(f\text{max} - f_{avg}), f_{avg} \leq f, P_{m1} \leq 1; \end{cases} \tag{4}$$

The steps involved the proposed algorithm are listed below:

(Step 1) Initialize the population. Crossover probability and the probability of mutation were computed by using (3) and (4) respectively.

(Step 2) The fitness of individual were computed by using (1). And the algorithm is whether return determined by termination condition.

(Step 3) The adaptive selection was completed according to the (2), and the next generation was obtained.

(Step 4) The crossover probability was computed by using (3), A random λ was generation for individual as pair. If $P_c > \lambda$, then start crossover operation.

(Step 5) The mutation probability was computed by using (4), and a random η in the range [0, 1] was generation. When $P_m > \eta$, the mutation is started.

(Step 6) The next generation is obtained. If the termination condition is satisfied, then the optimal feature subset is return, else goto (step 2).

3 Experimental Results

In this section, we evaluate the performance of our proposed feature selection algorithm through computer simulation. The simulation is performed by the matlab. To validate the performance of proposed algorithm, the BIF and C-F which are two kinds of represented algorithms in feature selection are compared [6]. The common dataset is obtained from UCI machine learning repository [7]. The dataset is given as Table 1.

Table 1. The datasets using in experiments.

Dataset	The number of features	Sample size	The number of classes
Anneal	898	38	6
Mfeat-factors	2000	216	10
Mushroom	8124	22	2
Spectrometer	531	100	4
Winc	13	178	3

In order to ensuring the fairness of experiment, Each algorithm would select the same number of features. As the classical leaning algorithm, the decision tree is used in test. And the experimental environment is the Weka. The leaning algorithm run three times and took the average. The result is show as Table 2.

Table 2. Size of feature subset for BIF algorithm

Datasets	Algorithm	The number of selected features	Percentage
Anneal	BIF	14	36.8
	C-F	9	23.6
	Proposed algorithm	7	18.4
Mfeat-factors	BIF	114	52.7
	C-F	173	80.1
	Proposed algorithm	126	58.3
Mushroom	BIF	15	68.1
	C-F	9	40.9
	Proposed algorithm	7	31.8
Spectrometer	BIF	51	51.0
	C-F	45	45.0
	Proposed algorithm	42	42.0
Winc	BIF	8	61.5
	C-F	7	53.8
	Proposed algorithm	4	30.7

The results shown in Tables 1 and 2 revealed that our proposed algorithm could select the smallest feature subset, because it can take out of the sample which has been identified. Furthermore, the classification accuracy of each feature subset was shown as following.

Experiment results are shown in Table 3 for each algorithm. From the simulation results, we can observe that the proposed algorithm can has better classification accuracy over the competing algorithm by using fewer features.

Table 3. Classification accuracy of each feature subset

Dataset	Accuracy		
	BIF	C-F	Proposed algorithm
Anneal	89.24	85.67	89.56
Mfeat-factors	90.43	88.72	92.33
Mushroom	84.39	76.84	86.94
Spectrometer	80.17	74.21	84.26
Winc	74.96	83.31	86.99

The accuracy of classification are shown in Figs. 1 and 2 for each algorithm. From the result, we can conclude that the proposed algorithm have better accuracy when comparing with other algorithm.

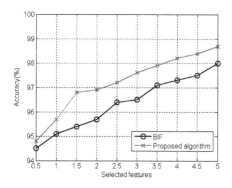
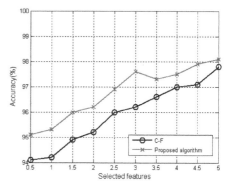

Fig. 1. The accuracy of classification for Anneal   **Fig. 2.** The accuracy of classification for Winc

4   Conclusions

A feature selection algorithm based on the adaptive genetic was proposed in this paper. The method to compute the crossover probability and mutation probability were designed according to adaptive genetic. The algorithm realized adaptive feature subsets selection and optimization. Experimental results show that the proposed algorithm achieves notable classification accuracy improvements, and reduced the total computing time, comparisons with the conventional scheme.

Acknowledgements.   This study was supported by the Science and Technology Cooperation Project of Guizhou province under Grant No. Qiankehe LH [2015] 7251. Science Foundation of Tongren University under Grant No. trxyDH1503.

References

1. Qiu, M., Ming, Z., Li, J.: Phase-change memory optimization for green cloud with genetic algorithm. IEEE Trans. Comput. **64**, 3528–3540 (2015)
2. Chiang, C.L.: Improved genetic algorithm for power economic dispatch of units with valve-point effects and fuels. IEEE Trans. Power Syst. **20**, 1690–1698 (2005)
3. Ronowicz, J., Thommes, M., Kleinebudde, P.: A data mining approach to optimize pellets manufacturing process based on a decision tree algorithm. Eur. J. Pharm. Sci. **73**, 44–51 (2015)
4. Yang, H., Fong, S.: Countering the concept-drift problems in big data by an incrementally optimized stream mining model. J. Syst. Softw. **102**, 158–165 (2015)
5. Ying, X.J., Xin, X.W.: Several feature selection algorithms based on the discernibility of a feature subset and support vector machines. Chin. J. Comput. **37**, 1705–1710 (2014)
6. Pinheiro, R.H.W., Cavalcanti, G.D.C.: A global-ranking local feature selection method for text categorization. Expert Syst. Appl. **39**, 2851–2857 (2012)
7. Information on http://www.ics.uci.edu/~mlearn/MLRepository.html

Bat Algorithm for Flexible Flow Shop Scheduling with Variable Processing Time

Han Zhonghua[1,2(✉)], Zhu Boqiu[1], Lin Hao[3], and Gong Wei[1]

[1] Faculty of Information and Control Engineering, Shenyang Jianzhu University,
Shenyang, China
hanzhonghua@sia.cn
[2] Department of Digital Factory, Shenyang Institute of Automation, CAS, Shenyang, China
[3] School of Foreign Languages, Shenyang Jianzhu University, Shenyang, China

Abstract. In order to solve the flexible flow shop scheduling problem with variable processing time (FFSP-VPT), it is analyzed, the paper analyzes the selection mode of processing time, jobs detection mode as well as rework mode, sets up FFSP-VPT mathematical model and puts forward a two-step encode mode for FFSP-VPT. By means of simulation experiment, the results are compared to mature genetic algorithm and differential evolution algorithm so that effectiveness of bat algorithm will be verified to solve practical production of FFSP-VPT.

Keywords: Flexible flow shop · Bat algorithm · Variable processing times · Hamming distance · Adaptive position update

1 Introduction

In the research of classical production scheduling, processing time is usually considered as an uncontrollable "rigid" parameter. But in the actual production processes, processing time of the same job in the same machine is always "flexible". According to customer's requirements, processing time is controllable to some extent. The concept of controllable processing time promotes theoretical study to be closer to practical production. But when one stage with variable processing time is added, n processing time variables will also be introduced so that solution space will be increased n dimensions and then this greatly increases complexity of problem-solution.

In 1980, Vickson was one of the earliest scholars who studied production scheduling with variable processing time [1]. Inspired by Vickson's research, more scholars paid more attentions to the production scheduling problem with variable processing time. This paper will focus on how to find the optimal solution under condition that processing time affects the rework rate.

In 2010, Professor Yang proposed bat algorithm [2]. There has been many studies on bat algorithm in many fields, such as production scheduling [3, 6]. Until now, bat algorithm has not been used for the FFSP-VPT, so this paper will give full play to the characteristics of the bat algorithm to solve the FFSP-VPT.

© Springer International Publishing AG 2018
F. Qiao et al. (eds.), *Recent Developments in Mechatronics and Intelligent Robotics*,
Advances in Intelligent Systems and Computing 690, DOI 10.1007/978-3-319-65978-7_25

2 Problems Description

Problems in this paper can be described as: n jobs follow the same processing sequence at m stages to be processed. At least one of those stages is a stage with variable processing time. There are buffers between stages. Stages with variable processing time will be based on the actual machine or the manual work to set a number of options for the processing time and to set the rework rate corresponding to each processing time. And rework rate increases as processing time decreases.

Before jobs enter stages, processing time will be selected in advance and rework rate will be determined at the same time. After the completion of the stage with variable processing time, jobs will immediately enter quality inspection stage to detect quality of each job. If the job is qualified, it will be moved to next stage, or it will return to previous stages. So, the process for the job which returns to previous stages is called rework. Optimization algorithm is used to determine processing sequence of jobs in the first stage and the processing mode of the stage with variable processing time. After above steps, start time and completion time of jobs in each process will be determined so as to more effectively reach requirements of optimization target.

2.1 Notations

n: Number of jobs to be scheduled;

J_i: Job i, $i \in \{1, 2, \cdots, n\}$;

m: Number of stages;

OP_j: Stage j, $j \in \{1, 2, \cdots, m\}$;

OPV_j: Stage j' with variable processing time, $j' \in \{1, 2, \cdots, m\}$, $OPV_j \in \{OP_1, OP_2, \cdots, OP_m\}$;

$S_{i,j}$:   Start time of job i at stage j;

$C_{i,j}$:   Completion time of job i at stage j;

M_j: Number of machines in stage j;

$WS_{j,w}$:   The machine w at stage j;

$T_{i,j}$:   Processing time of job i at stage j;

$Nt_{j'}$:   Number of the selective processing time at stage j with variable processing time;

$Tv_{i,j',b}$:   Job i selected the bth processing time at stage j with variable processing time, $i \in \{1, 2, \cdots, n\}$, $b \in \{1, 2, \cdots, Nt_{j'}\}$;

$Pq_{j',b}$:   Qualified rate of job j' after selected the bth processing time at stage j with variable processing times, $b \in \{1, 2, \cdots, Nt_{j'}\}$;

$Prw_{j',b}$:   Rework rate of job j' after selected the bth processing time at stage j with variable processing times, $b \in \{1, 2, \cdots, Nt_{j'}\}$;

Nrw: Number of stages when jobs rework;

$OPR_{j''}$:   Rework stage j'', $j'' \in \{1, 2, \cdots, Nrw\}$, $OPR_{j''} \in \{OP_1, OP_2, \cdots, OP_m\}$.

2.2 Mathematical Model

2.2.1 Problem Assumptions

In order to more effectively solve FFSP-VPT, three problem assumption variables are proposed.

$$A_{i,j,w} = \begin{cases} 1 & \text{job } i \text{ is processed in machine } w \text{ at stage } j \\ 0 & \text{otherwise} \end{cases} \tag{1}$$

$$Z_{i,j',b} = \begin{cases} 1 & \begin{array}{l} \text{job } i \text{ is selected the bth processing time} \\ \text{at stage } j \text{ with variable processing time} \end{array} \\ 0 & \text{otherwise} \end{cases} \tag{2}$$

$$Rw_{i,j',b} = \begin{cases} 1 & Prw_{j',b} \geq r \text{ and } (1) \\ 0 & Prw_{j',b} < r \text{ and } (1) \end{cases} \tag{3}$$

In (3), $Rw_{i,j',b}$ refers to the assumed variable rework sign. Rework rate is compared with the random number to generate $Rw_{i,j',b}$ which determines whether jobs will be reworked or not.

2.2.2 Constraints

$$C_{i,j} = (S_{i,j} + T_{i,j}), i \in \{1, 2, \cdots, n\}, j \in \{1, 2, \cdots, m\} \tag{4}$$

$$S_{i,j} \geq C_{i,j-1}, i \in \{1, 2, \cdots, n\}, j \in \{1, 2, \cdots, m\} \tag{5}$$

Generally basic constraints of FFSP can also be applied to FFSP-VPT. Constraint (4) illustrates the relationship among the start time, the processing time and the completion time of job i at stage j. Constraint (5) ensures that the next stage starts after this stage completes.

2.2.3 Mathematical Model of FFSP-VPT

The heart of FFSP-VPT mathematical model is to establish the relationship between the qualified rate and the rework rate. According to put in the concept of variable processing time based on FFSP mathematical model, the FFSP-VPT mathematical model is established.

$$\sum_{b=1}^{Nt_{j'}} Z_{i,j',b} = 1, i \in \{1, 2, \cdots, n\} \tag{6}$$

$$\sum_{b=1}^{Nt_{j'}} (Z_{i,j',b} \cdot Tv_{j',b}) = T_{i,j}, i \in \{1, 2, \cdots, n\}, \quad OPV_{j'} = OP_j \tag{7}$$

Constraint (6) shows each job can only select one fixed processing time at a stage with variable processing time. Constraint (7) shows if stage j is a stage with variable processing time, the processing time of job i at stage j must be one of the selective processing time.

$$\sum_{i=1}^{n} \sum_{w=1}^{M_j} A_{i,j,w} = n, j \in \{1, 2, \cdots, m\} \tag{8}$$

$$\sum_{j=1}^{m} \sum_{w=1}^{M_j} A_{i,j,w} = m, i \in \{1, 2, \cdots, n\} \tag{9}$$

$$\sum_{i=1}^{n} \left(\left(\sum_{w=1}^{M_{j''}} A_{i,j'',b} \right) \cdot Rw_{i,j',b} \right) \leq n, j'' \in \{1, 2, \cdots, NrW\} \tag{10}$$

$$\sum_{j''=1}^{Nrw} \sum_{w=1}^{M_{j''}} A_{i,j'',w} = Nrw, i \in \{1, 2, \cdots, n\} \tag{11}$$

Constraint (8) shows that the number of jobs processed at a stage is equal to the number of jobs if there is no rework. Constraint (9) shows the stage number of the jobs processed is equal to the number of stages if there is no job rework. Constraint (10) shows the number of reworked jobs at the same stage is less than or equal to the number of jobs, and the number of jobs processed at a stage is $[n, 2n]$ if there is rework. Constraint (11) shows the rework stage number of a job is equal to the number of specified rework stages if there is rework. The stage number of jobs to be processed is $n + Nrw$.

$$Prw_{j',b} + Pq_{j',b} = 1 \tag{12}$$

$$\exists b, \, b' \in \left\{ 1, 2, \cdots, Nt_j \right\}, \, Tv_{i,j',b} \leq Tv_{i,j',b'}, \, Tv_{i,j',b} \leq Tv_{i,j',b'}, \, Pq_{j',b} \text{ and } Pq_{j',b'}$$

meet the following relationship:

$$Pq_{j',b} \leq Pq_{j',b'} \tag{13}$$

$$Prw_{j',b} \geq Prw_{j',b'} \tag{14}$$

The constraint in (12) refers to the relationship between the qualified rate and the rework rate. The constraints in (13) and (14) show the qualified rate after selecting the *bth* processing time is proportional to the processing time at stage j' with variable processing time. The rework rate after selecting the *bth* processing time is inversely proportional to the processing time at stage j' with variable processing time. That is, the shorter the processing time, the lower the qualified rate, the higher the rework rate.

$$C_{i,max} = \begin{cases} C_{i,m} & Rw_{i,j',b} = 0 \\ C_{i,Nrw} & Rw_{i,j',b} = 1 \end{cases} \tag{15}$$

In (15), $C_{i,max}$ refers to the completion time of job i. If there isn't rework, $C_{i,max}$ is the completion time of the last stage, otherwise, $C_{i,max}$ is the completion time of the last stage with variable processing time.

2.2.4 Optimization Objective

$$\min C_{max} \tag{16}$$

$$C_{max} = \max\{C_{1,max}, C_{2,max}, \cdots, C_{n,max}\} \tag{17}$$

Functions (16) and (17) show that the optimization objective of this paper is to minimize the maximal completion time of all jobs at the last stage (makespan).

3   Encode and Decode of Bat Algorithm

3.1   Encode and Decode

The solution to the problem can be divided into two parts. The previous part is to determine processing sequence of jobs in the first stage. The latter part refers to selected the processing time number, that is, to determine the processing mode at stage with variable processing time.

Jobs sorting encoding method is applied to encode. Individual is $\{g_1, g_2, \ldots, g_i, \ldots, g_{2n}\}$. Genes from 1 to n stand for processing sequences of jobs in first stage. Genes from $n + 1$ to $2n$ stand for the selection of processing time number at the stage with variable processing time. The range of genes from $n + 1$ to $2n$ is $(1, b + 1)$. Value of gene g_i is $\lfloor g_i \rfloor$. $\lfloor \rfloor$ indicates downward rounding. $\lfloor g_i \rfloor$ represents selection of processing time number of job i.

Firstly, the individual is decoded into the processing sequence of jobs in the first stage. Then machines are successively arranged according to the decoded processing sequence. The first Available Machine First Rule (FAMFR) will be used to arrange machine on the basis of the assigned processing sequence. In the following processes with invariable processing times, according to the workpiece in the process of the completion of the time from small to large to arrange the processing order. At successive stages with invariable processing time, jobs will be ranked from large to small to arrange the processing sequence of this stage according to completion time at last stage. When completion time of jobs is the same, Most Work Remaining Rule(MWKR)will be used to arrange processing sequence of jobs. And FAMFR rule can also be used to arrange machine during assigning processing machine. During stage with variable processing time, genes from n + 1 to 2n will be transformed into the corresponding variable processing time number and the processing time can be determined by the number. Moreover, First In Fist Out, FIFO rule is used to determine start and completion sequence at each stage till all jobs complete all stages to output the makespan.

3.2 Establishing Initial Population

The initial population should be established to ensure that initial solution is dispersed in solution space to maximum extent. Therefore, BA uses random initialization to generate individuals as initial population, and calculates fitness value of all individuals.

4 Experimental Results

This experiment is based on Matlab, the operating environment is Win10x64, Intel (R) Core (TM) i7-5500U@2.40 GHZ, Matlab2012a, and computer memory is 8 GB.

In the processing, pattern printing stage in painting shop is a typical stage with variable processing time. According to requirements, users can select different capacity types to set scheduling plan. For example, high-production mode has a short processing time, a low qualified rate and high rework rate; medium-production mode has a medium processing time, a medium qualified rate and a medium rework rate; low-production mode has a long processing time with a high qualified rate and low rework rate. Therefore, pattern printing stage is a typical stage with variable processing time. Towards the application case of scheduling problem in painting stage, GA, DE and BA are used as global optimization tools to solve the FFSP-VPT. Among them, OP_4 is set as the stage with variable processing time. In the BA, the factor of updating loudness is set as 0.9, and the factor of updating pulse emission is set as 0.9. NP is set as 30, and set Gen_{max} is set as 1000 (Table 1).

Table 1. Relationship between processing times and reworking rate

Numbers of processing time p	Processing time T	Rework rate of jobs Prw
1	5	50%
2	10	40%
3	20	20%
4	30	10%

Red-marked job in Fig. 2 are reworked jobs.

As is in Fig. 1, the optimal solution appears in the eighth experiment of BA when each algorithm respectively runs ten times in experiment and optimal solution is 1518. The mean of results is 1540.4 after GA algorithm runs ten times. The mean of results is 1523.7 after BA algorithm runs ten times and the mean of results is 1530.8 after DE algorithm runs ten times.

Based on the mean of results, BA mean has a greater advantage than other algorithms. As is shown by Fig. 2, the range is 15 and the variance is 15.512 after GA runs 10 times; the range is 14 and the variance is 15.9 after BA runs 10 times; and the range is 14 and the variance is 24.52 after DE runs 10 times.

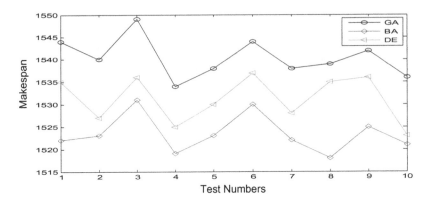

Fig. 1. The results got by GA, DE and BA in 10 runs

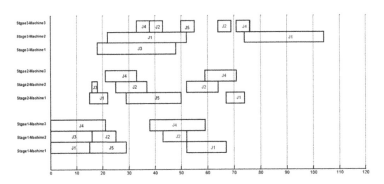

Fig. 2. Gantt of FFSP-VPT

5   Conclusion

In this paper, a specific form of FFSP with variable processing time which is called
FFSP-VPT is investigated. A two-stage coding mode is proposed which is designed for
FFSP-VPT. BA is used as the global optimizing tool. Bat individuals are decoded as the
first stage processing sequence and the processing mode. Through scheduling application cases test in painting shop, the test results are compared to BA and DE. Test results
show that BA is extremely effective to solve FFSP-VPT in practical production. Multiple
experiments are designed on painting shop instances, and GA and DE are used for
comparison purpose. The results demonstrate the effectiveness of BA algorithm in
solving the FFSP-VPT in the actual production

Acknowledgements. The work was supported by Natural Science Foundation of Liaoning
Province of China (201602608) and Sciences and Discipline Content Education Project
(XKHY2-61).

References

1. Vickson, R.G.: Two single machine sequencing problems involving controllable job processing times. AIIE Trans. **12**(3), 258–262 (1980)
2. Yang, X.-S.: A new metaheuristic bat-inspired algorithm. In: González, J.R., Pelta, D.A., Cruz, C., Terrazas, G., Krasnogor, N. (eds.) Nature Inspired Cooperative Strategies for Optimization, vol. 284, pp. 65–74. Springer, Berlin (2010)
3. Gao, M.L., Shen, J., et al.: A novel visual tracking method using bat algorithm. Neurocomputing **177**, 612–619 (2016)
4. Adarsh, B.R., Raghunathan, T., et al.: Economic dispatch using chaotic bat algorithm. Energy **96**, 666–675 (2016)
5. Liu, C.P., Ye, C.M.: Bat algorithm with the characteristics of Levy flights. CAAI Trans. Intell. Syst. **8**(3), 240–246 (2013)
6. Dao, T.K., Pan, T.S., Nguyen, T.T., Pan, J.S.: Parallel bat algorithm for optimizing makespan in job shop scheduling problems. J. Intell. Manuf. (2015). doi:10.1007/s10845-015-1121-x

Recognition of Facial Expressions and Its Applications Based on the GABOR Multi-directional Feature Integration

Siyu Wang, Li Zheng, Hua Ban, and Shuhua Liu[✉]

Northeast Normal University, Changchun, China
liush129@nenu.edu.cn

Abstract. In order to overcome problems including weak Gabor global feature recognition ability, long extraction time, and too high data dimensions, a facial feature extraction algorithm based on the Gabor multi-directional feature integration and block histogram is adopted to conduct accurate classification of extracted expression features using SVM. First, facial expression images are extracted visually by the Nao robot. After pre-processing of images, Gabor transform is conducted, and the expression features of five scales and in eight directions are extracted. Next, features of the same scale but in different directions are integrated through LBP filtering. The integrated image is further divided into sub-blocks of the same size. The histogram distribution within every sub-block area is calculated to jointly realize image representation. Second, the characterization will be classified. Third, through the voice module of the Nao robot, the recognized expressions are spoken out. The algorithm proposed in this paper can achieve a high expression recognition rate in the Japanese Female Facial Expression (JAFFE) Database.

Keywords: Facial expression recognition · Gabor features · Feature integration · Block histogram · Nao

1 Introduction

Facial expression can show emotions, feelings and intentions of humans. In this paper, the focus is on studying textural feature extraction methods based on pixel information. Expression features are extracted using Gabor multi-directional feature integration and block histogram. The method can not only effectively reduce the number of data dimensions and the calculation difficulty, but also maintain critical information of feature extraction. To be specific, this paper first explores defects of the traditional Gabor features, including weak global features and redundant feature data. Second, from a brand-new perspective, Gabor features are studied and improved, and Gabor features of the same scale but in different directions are integrated based on the local binary system model. The principal component analysis (PCA) is conducted to further reduce the feature size. Third, the local binary pattern (LBP) is used to achieve filtering of integrated feature images. Then, the histogram distribution is calculated to obtain features describing texture images. The method not only maintains advantages of Gabor filtering in representing changes of local

F. Qiao et al. (eds.), *Recent Developments in Mechatronics and Intelligent Robotics*,
Advances in Intelligent Systems and Computing 690, DOI 10.1007/978-3-319-65978-7_26

image texture, but also compensates defects of weak global feature capability, efficiently reduces the number of dimensions of feature data, and achieves favorable classification results. Finally, the proposed expression recognition method is applied to the Nao robot.

2   Research Background

The facial expression recognition system mainly consists of three parts, namely facial image pre-processing, feature extraction and classification of facial expressions. The focus is on studying feature extraction of facial expressions. Then, Support Vector Machine (SVM) is employed to conduct classification of expressions. There are mainly two approaches to extracting facial features. The first approach is based on geometric features of shape changes. The second approach is based on texture characteristics of pixels. The extraction approach based on geometric features can be further divided into the active shape model (ASM) [1] and the active appearance model (AAM) [2]; while the extraction approach based on texture features into LBP [3] and Gabor wavelet transform [4]. Recently, a method based on Gabor wavelet transform has been widely applied to feature extraction of facial expressions. It can successfully recognize detail changes of textures of different scales and in different directions without being significantly influenced by illumination changes. The feature extraction of different scales and in different directions based on Gabor wavelet transform can enrich feature details, and increase the feature dimensions and the data redundancy, making it necessary to conduct data dimension reduction of Gabor characteristics [5].

3   Feature Extraction of Facial Expressions

The facial expression extraction framework studied in this paper is shown in Fig. 1. First, Gabor features of five scales and in eight directions are used to extract facial expression images. Next, Gabor features of the same scale but in different directions are integrated. Third, Gabor features after integration undergo LBP filtering to obtain the feature integration graph. Fourth, the feature integration graph is blocked to obtain the histogram sequence.

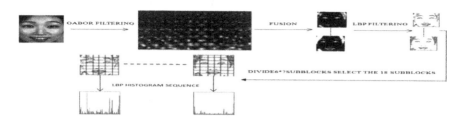

Fig. 1. Facial expression feature extraction framework

3.1 Feature Extraction Based on Gabor Wavelet Transform

The two-dimensional Gabor kernel function is defined as below:

$$\varphi_{\mu,v}(Z) = \left(\left\| K_{\mu,v}^2 \right\| / \sigma^2 \right) e^{-\frac{\left\| K_{\mu,v} \right\|^2 \|Z\|}{2\sigma^2}} \left(e^{iK_{\mu,v}Z} - e^{-\sigma^2/2} \right) \tag{1}$$

In Eq. (1), parameters, μ and v, stand for the direction and scale of the Gabor filter, respectively. $Z = (x, y)$ is the image coordinate of the given position. $K_{\mu,v}$ is the central frequency of the filter, which decides the direction and scale of the Gabor kernel. $\left\| K_{\mu,v} \right\|^2 / \sigma^2$ is defined under the framework of the two-dimensional Gabor filter, and is

used to compensate the energy spectrum confirmed by the frequency. $e^{-\frac{\left\| K_{\mu,v} \right\|^2 \|Z\|}{2\sigma^2}}$ is the Gaussian function. $e^{iK_{\mu,v}Z}$ is an oscillation function, of which the solid part is a cosine function and the imaginary part is the sine function. $e^{-\sigma^2/2}$ is the DC component removing the filter. In this way, the reliance of the two-dimensional Gabor wavelet transform on the image's absolute luminance can be avoided. Where, σ is the radius of Gaussian function, which can adjust the value of the two-dimensional Gabor wavelet. When σ is large enough, the filter can ignore the influence of direct components, and is not sensitive to global illumination. The advantage is it that it allows description of the spatial frequency structure and maintenance of the spatial relationship. $K_{\mu,v}$ in certain direction and of certain scale can be expressed as:

$$K_{\mu,v} = K_v e^{i\varphi_\mu} \tag{2}$$

Where, $K_v = K_{max}/f^v, v \in \{0, 1, \cdots, 4\}; \varphi_\mu = \dfrac{\pi\mu}{8}, \mu \in \{0, 1, \cdots 7\}.$ Set $K_{max} = \pi/2, f = \sqrt{2}, \sigma = 2\pi$. The two dimensional Gabor of the facial expression image conducts convolution through the Gabor function and the facial expression images. Different μ and v are selected to obtain response of texture characteristics in different directions and of different scales (frequencies). Concerning a given point, the convolution between images and Gabor filter, $\varphi_{\mu,v}(Z)$, is defined as below:

$$O_{\mu,v}(Z) = I(Z) \otimes \varphi_{\mu,v}(Z) \tag{3}$$

In Eq. (3), $O_{\mu,v}(Z)$ represents description of Gabor characteristics of Image $I(Z)$ the μ scale and in the v direction; $\otimes$ is the convolution operator. Gabor feature image after convolution is shown in Fig. 2.

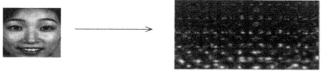

a. Expression image   b. Gabor characteristics of the image

Fig. 2. Gabor feature image after convolution. a. Expression image, b Gabor characteristics of the image

3.2   Gabor Multi-directional Feature Integration

Equation (3) is adopted for convolution operation. Results of Gabor transform contain not only phase information featuring periodical changes, but also relatively smooth and stable amplitude information. The amplitude information reflects the image energy spectrum. Since every expression image is transformed into 40 pieces of Gabor feature images, Gabor features are about 40 times of the original image features. Obviously, the high-dimension feature vector should not be directly used for classification. Therefore, in this paper, Gabor features of the same scale but in different directions are integrated so as to better reduce the redundancy of the original Gabor feature data. Some facial expression recognition algorithm based on Gabor features just applies Gabor amplitude information, but the phase information is not used. The major cause is that Gabor phase information shows periodical changes along with changes of the spatial position. Changes of amplitude are relatively smooth and stable. Amplitude can reflect the energy spectrum of images. Therefore, this paper adopts amplitude characteristics of Gabor transform. Gabor amplitude image features of the same scale but in different scales are defined as a channel. Every channel stands for the sum of Gabor amplitude images of the same scale but in eight different directions. The integration rule of every channel is defined is below:

1) Convert Gabor features of the same scale but in eight different directions on the facial image pixel points into the binary coding mode

$$P_{\mu,v}(Z) = \begin{cases} 1, (f(Z) \le G(z, \mu, v)) \\ 0, (\text{Others}) \end{cases} \tag{4}$$

Where, $G(Z,\mu,v)$ is the feature of Gabor amplitude of the pixel, $Z(x, y)$, in eight directions; $I(Z)$ is the original image. When the Gabor amplitude feature is larger than the original image grayscale, it suggests features are easier to be extracted in the direction. Therefore, if the binary system position is "1", Eq. (4) will keep it; if it is "0", Eq. (4) will abandon it.

2) Integrate the binary coding rules in eight directions into a decimal coding rule. See Eq. (5) below:

$$T_v(Z) = \sum_{\mu=0}^{7} P_{\mu,v}(Z) \cdot 2^{\mu} \tag{5}$$

Where, every $T_v(Z)$ stands for a local direction, and all local directions constitute the set of all features of the scale. Every facial expression is converted into multi-directional integrated feature image of five scales. Figure 3 stands for the feature integration graph to which the facial expression image in Fig. 2 is corresponding to, $v = 0, \cdots, 4$.

v = 0   v = 1   v = 2   v = 3   v = 4

Fig. 3. Feature integration graph of facial images

3.3 LBP Filtering

LBP filtering is an operator which can effectively describe texture images. It is a method to extract texture features based on the pixel information, Being small in the calculation amount and constant in rotation and illumination, LBP has found wide applications in the field of texture classification and facial image analysis. Therefore, LBP filtering of integrated feature image can highlight the signal texture. The LBP algorithm is generally defined as a "3 × 3" window. The gray value of the window center is adopted as the threshold value, which is compared with the adjacent pixel value. The binarization processing is conducted, which is to set points which are larger than the threshold value to be "1" and which are smaller than the threshold value to be "0". According to differences of pixel positions, the LBP value of the window is obtained via weighted summation. The process can be defined as below:

$$\begin{cases} H_v(Z) = \sum_{n=0}^{7} K\left(T_v^p(Z) - T_v^\omega(Z)\right) \cdot 2^n \\ K\left(T_v^p(Z) - T_v^\omega(Z)\right) = \begin{cases} 1, \left(T_v^p(Z) \geq T_v^\omega(Z)\right) \\ 0, (Others) \end{cases} \end{cases} \tag{6}$$

Where, $T_v^p(Z)$ and $T_v^\omega(Z)$ stand for the central pixel and the domain pixel in a feature image of the scale. Figure 4 represents the LBP filtering results of Fig. 3.

v = 0   v = 1   v = 2   v = 3   v = 4

Fig. 4. Facial images after LBP filtering

3.4 Block Histogram Sequence Characteristics

Every integrated feature image is divided into 18 sub-blocks not overlapped with each other. The histogram distribution of every sub-block is calculated to obtain the histogram sequence. Then, the histogram distribution of all sub-blocks forms a histogram sequence, thus finishing the feature representation of facial images.

4   Experiment

4.1   Experimental Environment

The algorithm proposed in this paper is experimented through the Nao robot. The robot collects facial images through the visual system, and recognizes human expressions based on the above steps. Next, the recognized results are expressed through voice. The development environments of this paper are Visual Studio 2010 and Open CV. The image databases adopted include the Japanese Female Facial Expression (JAFFE) Database and the self-built databases. Among various expressions of ten persons, one image is chosen as a test sample. The remaining data serve as the training samples. In order to train the video collection module, the expression image database built in this paper is adopted. In total, there are 70 pieces of expression images covering seven different expressions, including happiness, hate, fear, distress, wonder, anger and neutrality. The pixel of every image is 256*256. Figure 5 shows sevens self-built expression databases.

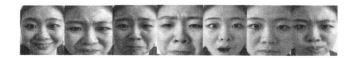

Fig. 5. Self-built expression databases

The recognition system consists of four modules, namely the loading module, the video collection module, the single image collection module and the exit system module. They are all shown in Fig. 6

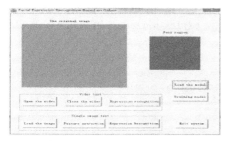

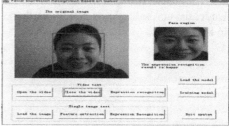

Fig. 6. Facial expression recognition system   **Fig. 7.** Video test module

First, the training samples and the test samples are all uploaded to the system via the loading module. The purpose of training samples is to build module parameters. The training samples can be trained by clicking on the "training model", and the model system is built. Second, the video collection model extracts images from the

Nao visual system. Third, the images are pre-processed. The approach based on the Gabor multi-directional feature integration stated above is used for feature extraction of images after pre-processing. Then, the SVM with the kernel function as RBF is employed for classification. Figure 7 shows the video test module. The Nao robot collects facial images through videos. Figure 8 shows the single image test module. The module loads images through the PC end. All these images are images of the jpg format. Figure 8 (a) is the image in JAFFE while Fig. 8 (b) is the image photographed by the author. The feature extraction and expression recognition module is the same to the video collection module.

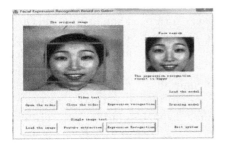   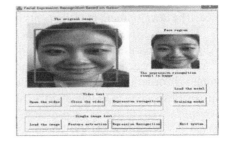

(a)Images from database   (b)Images from author

Fig. 8.   Single image test interface. (a)Images from database, (b)Images from author

4.2   Experimental Results

Click on the button of "Start the video" on the video test module. Start the Nao camera to collect facial images to be tested. The button of "expression recognition" in the right is used to finish facial expression recognition. Below the face region are expression results recognized. When the Nao Robot is adopted for video collection, the light ray in the environment cannot be directly cast on the human face area nor can it be too dark. If it is too dark, the Nao camera cannot locate the human face area. Under the normal lighting condition, the average expression recognition rate is shown in Fig. 9.

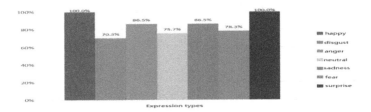

Fig. 9.   Expression recognition results

5   Conclusions

(1) A facial expression recognition algorithm based on Gabor multi-directional feature integration and block histogram sequence is proposed. SVM is used for accurate classification to realize recognition of seven expressions.

(2) A facial expression recognition system based on video test via the Nao robot is developed. The Nao visual module is employed for collection of expression mages, feature extraction and recognition. Finally, the Nao voice module is adopted to obtain the expression recognition results.

(3) Since differences of major feature areas in different directions are relatively significant.18 sub-blocks with significant texture changes are chosen from these sub-blocks, which include eyebrows, forehead wrinkles, eye corners, decrees and mouth corners. The histogram distribution of these parts is calculated to obtain the histogram sequence as expression characteristics to improve recognition effects.

Acknowledgments. This work is supported by Education Department of Jilin Province, under the Grant#([2016]503).

References

1. Cootes, F., Taylor, C.J., Lanitis, A.: Multiresolution search using active shape models, Procedings of 12th International Conference on Pattern Recognition,Los Alamitos,CA: IEEE CS Press,l: pp. 610–612 9–13 October 1994
2. Fadi, D., AbdelMalik, M., Bogdan, R.: Facial expression Recognition using tracked facial actions: classifier performance analysis. Eng. Appl. Artif. Intell. **26**(1), 467–477 (2013)
3. Zhou, S.R., Yin, J.P.: LBP texture feature based on Haar characteristics. J. Softw. **24**(8), 1909–1926 (2013)
4. Guo, Z., Zhang, L., Zhang, D.: A completed modeling of local binary pattern operator for texture classification. IEEE Trans. Image Process. **19**(6), 1657–1663 (2010)
5. Li, K., Yin, J.P., Li, Y., et al.: Local statistical analysis of Gabor coefficient and adaptive feature extraction. J. Comput. Res. Dev. **49**(4), 777–784 (2012)

A Novel Wind Power Prediction Technique Based on Radial Basis Function Neural Network

Yaqing Zhu[✉], Shihe Chen, Jia Luo, and Yuechao Wang

Electric Power Research Institute of Guangdong Grid Company, Guangdong 510080, China
zhuyaqing@163.com

Abstract. To ensure the stability of power system and wind farm operation, it is important for power system dispatch in to forecast wind power outputs exactly. The historical data are acquired from an operating wind farm. According to a well-developed Radial Basis Function (RBF) neural network, a wind power predictive model is established, using the historical data such as wind speed, environmental temperature, wind power and so on. Comparing with the actual power output of the wind, the forecasting results show that the proposed method can predict a comparatively accurate and lead to stable results. The proposed power prediction method can be used to make more reasonable dispatching plans.

Keywords: Wind power · Wind farm · Artificial intelligence method · RBF neural network

1 Introduction

The global energy crisis and climate warming have led to the rapid development of new energy sources. Wind energy, as a new energy source, has received great attention and affirmation from all over the world. However, wind power prediction has always existed as a problem. The prediction of wind power can improve the safety and reliability of power system. At present, the commonly used wind power prediction method is divided into the following three categories: time series analysis based on historical data, support vector machine method and neural network method [1].

Due to the high capacity intermittent wind power generation, uncertainty increases in the power grid and thus more spinning units are needed. Prediction of wind power can make considerable reduction of the spinning capacity, as well as the corresponding extra charge [2, 3]. The research shows that the application of short-term power prediction technology to the output power of wind farm is of great significance to improve the safe and economical operation of power grid.

In [4], time series method is used to forecast the output of wind power, and establish the autoregressive moving average model with the prediction error of 5%. In [5], the support vector machine (SVM) method is analyzed by physical and statistical methods, the results show that the SVM has a lot of application space in wind power prediction. In [6], the present situation about wind power prediction are introduced, and a model is optimized by neural network, and the simulation results show that the accuracy of the predicted data is higher.

© Springer International Publishing AG 2018
F. Qiao et al. (eds.), *Recent Developments in Mechatronics and Intelligent Robotics*,
Advances in Intelligent Systems and Computing 690, DOI 10.1007/978-3-319-65978-7_27

RBF network is put forward in 1980s, which is famous for the characters of stronger extrapolating capability, rapid calculation rate and non-linear mapping function [7]. In the paper, a novel wind power prediction method is proposed based on RBF algorithm. And the wind power of a wind farm in China is predicted and the effectiveness of the method is verified.

2 RBF Neural Network

RBF network belongs to feed-forward neural network. It is used for function approximation, control, etc. The network generally consists of three layers. The input layer only serves to transmit the signal. The hidden layer is to adjust the parameters of the activation function, in which most of the activation function using nonlinear optimization strategy to adjust the parameters, so the Gaussian function is selected. The third layer is the output layer, which provides a network response to the input layer's activation pattern and serves as an adder unit. Figure 1 shows the Radial Basis Function (RBF) network model.

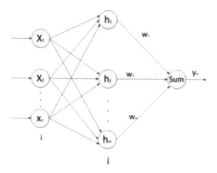

Fig. 1. The model of the network

In Fig. 1 $X_{input} = [x_1, x_2, \cdots, x_n]^T$ is selected as the input, $Hidden = [h_1, h_2, \cdots, h_m]^T$ is selected as radial basis vector, and the Gauss function is selected as h_j, where

$$h_j = \exp\left(-\frac{\left\|x - c_j\right\|^2}{2b_j^2}\right), j = 1, 2 \cdots, m \tag{1}$$

Besides, $Center_j = [c_{j1}, c_{j2}, \cdots, c_{ji}, \cdots, c_{jn}]^T$ is the basis function centers of the jth nodes, $B = [b_1, b_2, \cdots, b_m]^T$ is the basis function radius parameters of the jth nodes, $Weight = [w_1, w_2, \cdots, w_m]^T$ is the output weight.
So the identifying output is

$$y_{model}(k) = w_1 h_1 + w_2 h_1 + \cdots + w_m h_m \tag{2}$$

and defining the objective function as

$$J = \frac{1}{2}(y(k) - y_m(k))^2 \tag{3}$$

$y(k)$, $y_{model}(k)$ are respectively the set point and model output. Therefore, the gradient descent method is used to correct the weight, center and radius parameters, which are shown

$$\begin{cases} \Delta\omega_j(k) = -\eta\dfrac{\partial J}{\partial\omega_j} = \eta(y(k) - y_{model}(k))h_j \\ \omega_j(k) = \omega_j(k-1) + \Delta w_j(k) - \alpha(\omega_j(k-2) - \omega_j(k-1)) \end{cases} \tag{4}$$

$$\begin{cases} \Delta b_j(k) = \eta(y(k) - y_{model}(k))w_j h_j \dfrac{\left\|X - Center_j\right\|^2}{b_j^3} \\ b_j(k) = b_j(k-1) + \Delta b_j(k) - \alpha(b_j(k-2) - b_j(k-1)) \end{cases} \tag{5}$$

$$\begin{cases} \Delta c_{ji}(k) = \eta(y(k) - y_{model}(k))w_j \dfrac{x_j - c_{ji}}{b_j^2} \\ c_{ji}(k) = c_{ji}(k-1) + \Delta c_{ji}(k) - \alpha(c_{ji}(k-2) - c_{ji}(k-1)) \end{cases} \tag{6}$$

where the learning rate is η and the momentum factor is α. $\eta \in (0, 1)$ $\alpha \in (0, 1)$, $\dfrac{\partial y(k)}{\partial\Delta u(k)}$ is Jacobian information, which can be obtained by identification according to the neural network. Suppose the first input of the network is $\Delta u(k)$, there are

$$\frac{\partial y(k)}{\partial\Delta u(k)} \approx \frac{\partial y_m(k)}{\partial\Delta u(k)} = \sum_{j=1}^{m} \frac{c_{ji} - x_j}{b_j^2} \tag{7}$$

3   The Wind Power Prediction Model

A wind farm is selected as study object. The model input is selected strictly. Large numbers of parameters can be applied as inputs: wind speed V_i, environmental temperature T_i and wind power P_i. After data preprocessing, the wind speed V_i and environmental temperature T_i is selected as the input of wind power prediction model, constructing characteristic vector $E = [V_i, T_i]$ and normalizing

$$E' = \left[\frac{V_i}{\sqrt{V_i^2 + T_i^2}}, \frac{T_i}{\sqrt{V_i^2 + T_i^2}} \right] \tag{8}$$

The wind power prediction model is shown in Fig. 2:

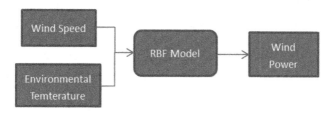

Fig. 2. The wind power prediction model

4　Simulation

The paper adopts the data of wind turbines from wind farm, the acquisition cycle of the data is 10 min. To guarantee the diversity of the training sample, 1900 groups of experimental data are selected randomly as samples from the wind farm, then randomly selecting 100 groups from the rest data as test samples for model validation.

From Fig. 3, the predicted wind power basically agrees well with actual value.

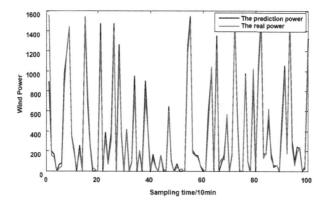

Fig. 3. The prediction of wind power

The prediction error curve of wind power is shown in Fig. 4. The average value of prediction error is 0.02 W, the standard variance is 52 W, which show the RBF algorithm has higher prediction accuracy.

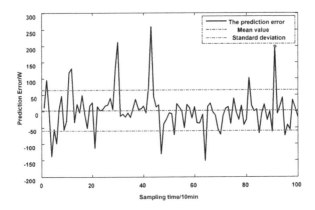

Fig. 4. The prediction error of wind power

5   Conclusion

A novel model of wind power prediction is established using a fully developed RBF network based on historical data of operating wind power plant selected wind speed, ambient temperature, wind force as feature vector. Comparing the predicted results with the actual wind power output, it is clear that the proposed method leads to acceptable and stable forecast results. The proposed power prediction method can be used to make a more reasonable dispatch plan.

References

1. Yuan, X.M.: Framework of problems in large scale wind integration. J. Electr. Power Sci. Technol. **27**(1), 16–18 (2012)
2. Yang, X., Xiao, Y., Chen, S.: Wind speed and generated power forecasting in wind farm. Proc. CSEE **25**, 1–5 (2005)
3. Sánchez, I.: Short-term prediction of wind energy production. Int. J. Forecast. **22**(1), 43–46 (2006)
4. Sideratos, G., Hatziargyriou, N.D.: An advanced statistical method for wind power forecasting. IEEE Trans. Power Syst. **22**(1), 258–265 (2007)
5. Milligan, M., Schwartz, M., Wan, Y.: Statistical wind power forecasting models: results for U.S. wind farms. Paper presented at WINDPOWER 2003 Austin, Texas, May 2003
6. Barbounis, T.G., Theocharis, J.B.: Locally recurrent neural networks for long-term wind speed and power prediction. Neuro Comput. **69**, 466–496 (2006)
7. Han, H.G., Qiao, J.F.: An efficient self-organizing RBF neural network for water quality predicting. Neural Netw. **24**, 717–725 (2011)

A Quantum Particle Swarm-Inspired Algorithm for Dynamic Vehicle Routing Problem

Bo Li[1,2(✉)], Guo Chen[1], and Ning Tao[2]

[1] Institute of Information, Dalian Maritime University, Dalian 116023, China
[2] Institute of Software, Dalian Jiaotong University, Dalian 116045, China
`daliannt@126.com`

Abstract. In order to solve the scheduling of dynamic vehicle routing problem, this paper establishes the simulation model to minimize the cost and stability value and maximize the loading rate. Then a quantum particle swarm-inspired algorithm is proposed. At first, it introduces the method based on the DCSC (double chains structure coding) including vehicle allocation chain and goods chain. Finally, the proposed method is applied to a dynamic simulation and the result of comparing with other classical algorithms verifies its effectiveness.

Keywords: Quantum particle swarm-inspired algorithm · Double chains structure coding · Virtual routing problem

1 Introductions

It is assumed all the relevant information is known before the route scheduling in the classical research for VRP, and the information will not be changed with time. However, there are many dynamic factors in real life, such as customer requirement, transportation demand, and the subjective idea of the dispatcher, these problems are called as DVRP (dynamic vehicle routing problem). The vast researchers inland and outland have used many kinds of algorithms to solve the DVRP and have obtained some achievement in the model and algorithm. Literature [1] studied stochastic demand VRP using neural dynamic planning method to solve the DVRP of single vehicle, and the model was applied to the postal service express delivery, production distribution and production scheduling and other fields. Literature [2] studied the VRP with real-time information and established the constraint model to solve the problem using the branch and bound method. Literature [3] studied the DVRP with travel time and dynamic requirement, but the result was lack of a clear explanation for the algorithm of dynamic time requirement. Literature [4] studied the DVRP with ant colony system, but the result was lack of a clear explanation for the algorithm of dynamic time requirement.

Considering the goods allocation and vehicle optimization in scheduling and rescheduling of DVRP, the novel quantum particle swarm-inspired algorithm is proposed here.

© Springer International Publishing AG 2018
F. Qiao et al. (eds.), *Recent Developments in Mechatronics and Intelligent Robotics*,
Advances in Intelligent Systems and Computing 690, DOI 10.1007/978-3-319-65978-7_28

2 Mathematical Model

2.1 Model Description

The definition of DVRP is that: A transportation network system is composed of 1 yard and m customers and they are expressed as 0, 1, 2,..., m; the vehicles should serve n ($n \leq m$) customers after they left the yard. The customers may be divided into static ones and dynamic ones according to the different demand. When the demand of the customers is changed, the plan may be modified to reschedule [5, 6].

2.2 Objective Function

The meaning of symbols in model can be described as follows: 0 is the yard, 1, 2, 3, ..., N are the customer numbers; 1, 2, 3, ..., K are the vehicle numbers; i, j ($i, j \in \{1, 2, ..., N\}$) are the distribution depot point and customer point respectively; the vehicle's maximum load is Q; the total capacity of vehicle k from the customer point i is $q_{ik}(t)$ ($q_{ik}(t) < Q$); q_i is the point of the demand i; and the common cost of vehicle is F_k; the running time for the vehicle from i to j is t_{ij}; the distance between demand i to demand j is d_{ij}; the cost from demand i to demand j is c_{ij}; the transport volume of vehicle k from customer i to customer j is ω_{ijk}; the business hours for vehicle k in customer i is s_{ki}.

The new demand of the customers may be inserted into the existing route if the loading capacity is enough, otherwise, one or several vehicles would be provided. When the demand of the initial demand increase so much to exceed the load limit of vehicle, then last demand will be look as new ones until the vehicle quantity restriction are met. Assuming that W is the sum of which haven't been served in static stage and the later ones in dynamic stage; here M is the number of the customers who have become new distribution yards, and the numbers are $N + 1, N + 2, ..., N + M$. Then the code of the initial yard has become into $N + M + 1$ increasing L vehicles. The vehicle code will become $W + i$, where i represents the customer location and the rest of the loading capacity in static stage is represented as $Q - q_{ik}(t)$.

Equation 1 is the mathematical model of DVRP in moment t:

$$\min Z = \sum_{k=1}^{K} \sum_{i=1}^{N+M+1} \sum_{j=1}^{N+M+1} c_{ij}x_{ijk} + \sum_{k=K+1}^{K+L} \sum_{i=1}^{N+M+1} \sum_{j=1}^{N+M+1} c_{ij}x_{ijk} + \sum_{k=K+1}^{K+L} F_k \sum_{j=1}^{N} x_{(N+M+1)jk}; \quad (1)$$

It is the objective function describing the transportation cost, the cost of new customers in dynamic stage and the cost of new vehicles in dynamic stage in Eq. 1.

The maximum transport time weight of the vehicle is set to 5 and the delivery tardiness weight is 2 according to Ishibuchi [7, 8], then the efficiency is designed as follow:

$$Efficiency = 5 \cdot (\max(F_n) - \min(b_n)) + 2 \cdot \sum \psi_n(F_n - DL_n) \quad (2)$$

The meaning of the symbol in Eq. 2 is as follows:

F_n is the delivery completion time in the customer n;

b_n is the time of reaching the customer n;
DL_n is the time window of the customer n;

$$\psi_n = \begin{cases} 1, & (F_n - DL_n) > 0 \\ 0, & \text{else} \end{cases} \tag{3}$$

3 Hybrid Quantum Particle Swarm Algorithm

In this paper, two kinds of sub-problems of vehicle selecting and goods selecting are included. In order to resolve the above problems, the quantum particle swarm-inspired algorithm is proposed to solve the DVRP.

3.1 The Quantum Particle Swarm-Inspired Algorithm

There are several advantages of the proposed quantum particle swarm-inspired algorithm such as: less parameter involved in the iteration, faster computing speed and stronger ability of global searching. If the dimension of the population is considered as D, the ith particle is $P_i = (p_{i1}, p_{i2}, \ldots, p_{iD})$, then the speed of the particle is $V_i = (v_{i1}, v_{i2}, \ldots, v_{iD})$ and the global optimal particle is $gbest = (g_{i1}, g_{i2}, \ldots, g_{iD})$, on the other hand, the local optimal particle can be described as $lbest_i = (l_{i1}, l_{i2}, \ldots, l_{iD})$, the renewal equation of the particle is shown as follows:

$$V_i(t + 1) = \omega(t) \times V_i(t) + c_1 \times r_1 \times (lbest_i(t) - P_i(t)) + c_2 \times r_2 \times (gbest(t) - P_i(t)) \tag{4}$$

$$P_i(t + 1) = P_i(t) + V_i(t + 1) \tag{5}$$

In Eq. 4, t presents the iterative algebraic, c_1 and c_2 present the acceleration factors, r_1 and r_2 are the random digits existing between $(0, 1)$ uniformly; $\omega(t)$ is the iteration weight, in which $\omega(t) = \omega_{\max} - \dfrac{(\omega_{\max} - \omega_{\min}) \cdot t}{M_1}$, where $\omega_{\max} = 1.4$ and $\omega_{\min} = 0.5$, M_1 here represents the maximum iteration algebraic of the population.

If r_1 and r_2 are the random numbers in some interval, it hasn't the ability to search each demand point. In this paper, he value in the $(t + 1)^{th}$ generation is described as follows:

$$r(t + 1) = \mu \times r(t) \times (1 - r(t)) \tag{6}$$

In which, $r(t) \in (0, 1)$, μ represents the control variable and $\mu = 4$.

3.2   The Steps of the Improved Algorithm

In this paper, the flow chart of the improved algorithm is in Fig. 1:

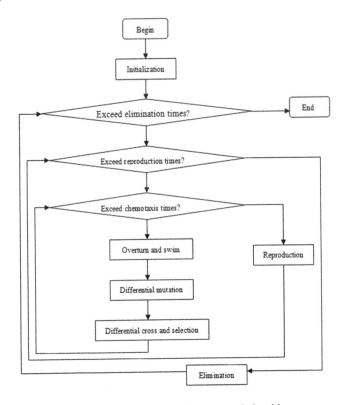

Fig. 1. The flow chart of the proposed algorithm

The specific description of the algorithm is as following:

Step 1: Select the appropriate particle location;
Evaluate the initial fitness value according to the objective function, then save the best position and the best individual fitness of each particle, in the meanwhile, the initially global best position and the best fitness value are saved;

Step 2: The sub phase particle swarm is established and the ith sub phase particles are optimized in order to insure the local optimization at different speeds in the opposite directions. Besides, the speed of the main phase particle swarm is updated through the global optimal point based on the current all sub phases;

Step 3: Judge whether i is exceed the limit of the maximum particle number or not, if yes, make $i = 0$ and go to Step 5, otherwise make $i = i + 1$, and go to Step 3;

Step 4: If the error of the fitness exceeds the preset limitation, the algorithm will be terminated, otherwise, return to Step 2 and continue the algorithm.

4   Simulation Experiment

In this paper the simulation experiment includes 1 distribution depot, 4 demands with dynamic demand and 24 customers with static demand, and the distribution area is square of 50×50 (unit: km) in order to verify the effectiveness of the proposed algorithm. The numbers of dynamic customers are a, c, f, h and that of static ones are 1, 2, ..., 24. If the demand of each customer is less than 2 m^3, the maximum cargo volume of all the vehicles is 8 m^3, and the maximum travel distance for the vehicles to complete the task continuously is 150 km. Meanwhile, the coordinates of the distribution depot is set as (20 km, 20 km). The number of initial vehicles is known as $m = \left(\sum_{i=1}^{n} q_i / \alpha Q \right) + 1 = 5$, in which q_i is the demand of customer i^{th}, Q is the maximum cargo volume, α is the complexity of the cargo compartment and the value is 0.8. When the demand of dynamic customers appears at moment 1, it should be rescheduled to the vehicles using the strategy and method which are proposed in this paper.

The five vehicles are at the point 5, point 24, point 20, point 12 and point 13 at moment 1 according to the initial scheme, that is, the key points set of dynamic factor is {5, 12, 20, 24}, and the cargo volume set of the vehicles is {1.2, 3.6, 0.8, 2.8}.

5   Conclusion

In this paper, a novel quantum particle swarm-inspired algorithm is established according to the different constraint conditions, and the mathematical model of DVRP is designed to describe the dynamic vehicle routing problem. The effectiveness of the proposed method to solve DVRP is verified through solving the simulation example and the comparison with several existing algorithms.

Acknowledgements. This work is financially supported by the National Natural Science Foundation of China (Nos. 51579024, 61374114), Dr scientific research fund of Liaoning Province (No. 201601244), Liaoning Provincial Social Science Planning Foundation of China (No. L16BGL008), China. Postdoctoral Science Foundation Funded Project (No. 2017M611231), Dalian Social Science Planning Foundation of China (No. 2016dlskyb104).

References

1. Teodorovic, D., Radivojevic, G.: A fuzzy logic approach to dynamic dial-a-ride problem. Fuzzy Set Syst. **116**(16), 23–33 (2000)
2. Zhang, J., Guo, Y., Li, J.: Research of vehicle routing problem under condition of fuzzy demand. J. Syst. Eng. **19**(1), 74–78 (2004)
3. Fu, Z., Eglese, R., Lyo, L.: Corrigendum: a new tabu search algorithm for the open vehicle routing problem. J. Oper. Res. Soc. **57**(8), 1008–1018 (2006)
4. Montemanni, R., Gambardella, L.M., Rizzoli, A.E.: Ant colony system for a dynamic vehicle routing problem. J. Comb. Optim. **10**(4), 327–343 (2005)

5. Khebbache-Hadji, S., Prins, C., Yalaoui, A., et al.: Heuristics and memetic algorithm for the two-dimensional loading capacitated vehicle routing problem with time windows. Cent. Eur. J. Oper. Res. **21**(2), 307–336 (2013)
6. Ning, T.: Study of application of hybrid quantum algorithm in vehicle routing problem. Dalian Maritime University, Dalian (2013)
7. Ishibuchi, H., Murata, T.: A multi-objective genetic local search algorithm and its application to flow shop scheduling. IEEE Trans. Syst. Man Cybern. Part C **28**(3), 392–403 (1998)
8. Fattahi, P., Fallahi, A.: Dynamic scheduling in flexible job shop systems by considering simultaneously efficiency and stability. CIRP J. Manuf. Sci. Technol. **2**(2), 114–123 (2010)

Design of Distributed Pipeline Leakage Monitoring System Based on ZigBee

Zhonghu Li[✉], Jinming Wang, Junhong Yan, Bo Ma, and Luling Wang

School of Information Engineering, Inner Mongolia University of Science and Technology,
Baotou 014010, China
lizhonghu@imust.edu.cn

Abstract. With the rapid increase of the retain quantity and service time, pipeline leak detection and leak location problem have gotten extensive attention. In order to monitor the industrial pipeline, this paper develop a distribute pipeline leakage signal acquisition and processing system based on Zig Bee. In this system, STM32 microcontroller is the core and Wireless sensor is the network node. LabVIEW is used in PC as the monitor. The data transmit by Zig Bee wireless network between the upper and lower computer. Cross-correlation operation is used as the Leak location algorithm. Correlation analysis of the four sensor output signals is executed. This method eliminate the measurement error caused by the change of sound wave propagation speed, and further improved the positioning accuracy.

Keywords: ZigBee · Distributed system · Pipeline leak detection · Leak location · Signal acquisition and processing

1 Introduction

Pipeline transportation is a new and economical way of transportation. This method is playing an increasingly important role in the national economy. But with the rapid increase of the retain quantity of pipeline and the increase of service time, and due to corrosion, aging, natural disasters and man-made damage and other factors, pipeline leakage accidents occur from time to time. These accidents will not only cause waste of resources and environmental pollution, but also may cause casualties and property losses and other major accidents. The safe operation of pipeline transportation system has been widely concerned by the industry. This paper research on distribute pipeline leak monitoring system, and realize the accurate positioning and real-time monitoring of pipeline leak leakage condition. The system is not only beneficial to the people's livelihood, but also has the sustainable development of the social and historical significance.

2 System Overall Structure and Principle

Compare to the wired communication technology, wireless communication technology has many incomparable advantages such as easy and flexible use, good expansibility and convenient installation etc. It can solve the network connection problem. In view

© Springer International Publishing AG 2018
F. Qiao et al. (eds.), *Recent Developments in Mechatronics and Intelligent Robotics,*
Advances in Intelligent Systems and Computing 690, DOI 10.1007/978-3-319-65978-7_29

of the outstanding advantages, it has been widely used in many industry fields. This technology is also the research focus in the automatic testing field. According to different communication protocols and wireless modules, the distribute test system mainly include the following: distributed test system based on Bluetooth, distributed test system based on ZigBee network and distributed test system based on WLAN etc. [1, 2]. Based on the specific requirements of pipeline leakage monitoring, this system adopt distributed structure. And ZigBee wireless communication system is used for data transmission.

The system includes four wireless sensor network nodes, Zigbee wireless communication network and upper monitoring center. Wireless sensor network node is an independent system composed by acoustic emission sensor, signal conditioning circuit, MCU, GPS module and Zigbee module. It can complete data acquisition, data initial processing and data exchange with the upper monitoring center. Zigbee wireless communication network is used to realize the communication between the sensor network node and the upper monitoring center. The wireless network is realized by configuring corresponding Zigbee wireless module and monitoring center. The upper monitoring center mainly consists of PC and Zigbee center nodes. It is the nerve center of the whole system. It sends control instructions to the sensor network nodes. It also collects and processes pipeline leakage data and monitors the network status of the whole system. The PC and Zigbee wireless modules are connected by RS232.

The pipeline leakage signal processing determines leakage condition and leak location. According to the characteristics of acoustic signals, some methods have been developed. Such as regional positioning method based on signal amplitude attenuation measurement, time difference measurement positioning method based on the waveform cross-correlation and positioning method based on waveform interference [3, 4]. This system adopts time difference measurement and waveform cross-correlation.

The traditional pipeline leakage point location system is a cross-correlation operation system. Two sensors, which are installed on the pipeline, provide the test signals. The experiment show that the leakage signal propagation time of the two sensors is different. Leakage location can be calculated by propagation velocity and the distance between the two sensors. However, the propagation velocity is influenced by many factors. Such as pipe material, diameter, wall thickness and internal stress. It will lead to some errors in traditional positioning methods. In order to improve the positioning accuracy, this system set four sensors in the pipeline. The system structure is shown in Fig. 1.

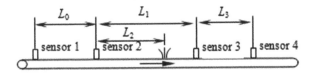

Fig. 1. Schematic diagram of four sensor arrangement

It is assumed that the acoustic propagation velocity in the pipeline is v. The time from the leakage point to the sensor 1 to sensor 4 is t_1, t_2, t_3 and t_4. There relations are shown as follows:

$$L_2 = \frac{2L_1 - L_0 + L_3 + \dfrac{L_0 + L_3}{t_1 + t_4 - t_2 - t_3} \times (t_1 + t_2 - t_3 - t_4)}{4}$$

$$= \frac{2L_1 - L_0 + L_3 + \dfrac{L_0 + L_3}{(t_1 - t_3) + (t_4 - t_2)} \times [(t_1 - t_3) - (t_4 - t_2)]}{4}$$

(1)

L_2 is the distance from leakage point to sensor 2. By L_2, we can calculate the leakage point location. According to the formula (1), the positioning result is independent of v. It will eliminates the measurement error of velocity, and improved the precision of leak location.

3 Wireless Sensor Network Node Design

The nodes of wireless sensor network include sensors, signal conditioning circuit, MCU system, GPS module and ZigBee wireless transmission module etc. The network nodes mainly complete signal acquisition, signal storage, signal transmission, system timing synchronization and other functions. The hardware structure of wireless sensor network node is shown in Fig. 2. The software system is written by Keil uVision 5 software. And the software includes initialization program, A/D conversion program, real-time storage program, wireless transmission program and time synchronization program.

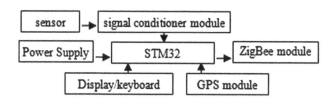

Fig. 2. Hardware structure of wireless sensor network node

3.1 Sensor Selection

The water leakage point is inconsistent. The intensity and frequency of the leakage acoustic signal are different in the whole process. From beginning to the end, the leakage sound signal has an average period. There are different period of the leakage in the process. The frequency characteristics of each period are also different [5, 6]. According to the existing laboratory pipeline system, PXI-6143 data acquisition card is used to collect the signal of the leakage of the pipeline system. And then MATLAB software is used to do Fourier analysis. It can get the leakage signal frequency bandwidth. The leakage signal is mainly concentrated between 1.25 kHz to 3 kHz. Combined with the sensitivity parameters, this design uses a SR10 acceleration sensor of Beijing soundwel Technology Co. Ltd. The sensor is low frequency sensor. And the sensor's frequency range is 1 Hz–15 kHz.

3.2 Signal Conditioning Circuit

The signal from acoustic emission sensor is generally small. It is amplified before they can be collected. In order to further improve the accuracy of signal acquisition, we should try to make the collected signal to fill the full range of the A/D converter. The peak to peak value of the output signal is about 200–400 mV. So it should be enlarged about 10 times. The amplifier is designed with low cost and low power consumption instrumentation amplifier AD620. The common mode rejection ratio of AD620 is 100 dB. The peak to peak value noise is 0.28 μV (0.1–10 Hz). And bandwidth is 120 kHz (G = 100).

3.3 MCU System

Network node processing chip is STM32F103RBT6 MCU. This system mainly uses on-chip A/D, on-chip direct memory access controller and on-chip real-time clock controller. MCU internal ADC is a 12 bit successive approximation analog-to-digital converter. It has many advantages, such as high precision, fast speed, convenient use and development, etc. On-chip direct memory access controller is used to provide high speed data transmission, which does not require CPU intervention. It can move the data quickly, reduce the burden of CPU and improve the efficiency of data carrying [7]. Real-time clock is an independent timer. And it can provide the function of the clock calendar. Modify the value of the counter can reset the current time and date of system. In order to ensure the synchronization of data acquisition of four network nodes, clock calibration is required. Network nodes and upper monitor will send time order to network nodes at a specified time interval. After receiving the instruction, the lower computer read the GPS time information from the serial port. Then it refreshes the real-time clock to ensure the consistency of the four network nodes.

3.4 GPS Module

In order to improve the precision of leak location, the clock synchronization of four network nodes must be guaranteed. This problem can be solved by receiving GPS time [8]. MCU serial port 2 interrupt receive and process GPS module message. The MCU conducts time on the STM32 internal real-time clock. The least theoretical error of GPS calibration time is 30 ns. GPS module adopt the GPS NEO-6M of ALIENTEK company.

3.5 ZigBee Wireless Transmission Module

ZigBee communication is used between upper and lower computer. This system uses ZAuZH_T ZigBee chip of Zhuo Wan technology. The module has a ZigBee protocol its self. The routing node and the main node can be automatically networked. Because it does not involve the study of ZigBee protocol, it is convenient to configuration. When the sampling time set by upper monitor arrive, lower computer started to sample. When the sampling points arrived to 4096, lower computer stop sampling. Then the sampled data is transmitted to ZigBee routing node through the serial port and sent to the main node subsequently. Finally the data is sent to the upper monitor.

4 Design of the Computer-Monitored System

PC monitoring system is programmed by LabVIEW. It mainly achieves data acquisition, filtering, denoising, cross correlation, data storage and other functions. The program include the main program, data acquisition, filtering, wavelet Denoising, cross-correlation arithmetic, database storage and query subroutine etc.

This system is programmed for serial port by using NI-VISA advanced application programming interface. The serial port protocol analysis is realized. Signal filtering function is achieved by using sub VI in the LabVIEW filter, where elliptic filter in IIR filter is used. The signal denoising function needs to call the MATLAB script node in LabVIEW firstly. Then call the MATLAB function of wavelet analysis in the script node. Based on the different properties of the wavelet coefficients, filter is achieved by using corresponding rules. This process can deal with the nonlinear caused by the noisy signal, such as the selection, extraction and cutting [9, 10]. Cross correlation operation call "cross-correlation operation" subroutine in LabVIEW. The location of pipeline leakage is calculated through the cross-correlation calculation results. This system carries out the multiple mutual correlations to the signals from four sensors. It will improve the precision of leak location. The ActiveX data object ADO provided by Microsoft corporation is used to link the database file. Data storage and query functions are realized by the database toolkit in LabVIEW.

5 Conclusion

This paper has developed a distributed pipeline leakage monitoring system based on Zig Bee wireless network. This design set four sensors in the pipeline, which eliminated the positioning error caused by the change of sound wave propagation speed. The introduction of Zig Bee wireless network makes the application of the system more flexible. Especially for the industrial site and wiring difficulties site, the implementation of the project will be more convenient.

Acknowledgments. Project researches on leakage signal mechanism and method of weak signal detection and dispose in water-supply pipelines (61362023) supported by National Natural Science Foundation of China.

References

1. Xuan, Z.: Wireless Distributed Test System Based on WLAN. North University of China, Taiyuan (2014)
2. Cheng, Y., Luo, J., Dai, S.: Distributed wireless temperature and humidity measurement system based on Zig Bee network. Electron. Meas. Technol. **32**(12), 144–146 (2009)
3. Miller, R.K., et al.: A reference for the development of acoustic emission pipeline leak detection techniques. NDT&E **32**(1), 1 (1999)
4. Jiao, J., He, C., Wu, B., et al.: Application of wavelet transform on modal acoustic emission source lo cation in thin plates with one sensor. Int. J. Press. Vessels Piping **81**(5), 427 (2004)

5. Lu, W., Wen, Y.M.: Leakage noise and its propagation in water pipeline. Tech. Acoust. **26**(5), 871 (2007)
6. Liu, Y.: Leak detection of water supply pipeline based on the sound wave theory. Nondestruct. Test. **30**(6), 355 (2008)
7. Meng, B.: STM32self-Study Notes. Beijing University of Aeronautics and Astronautics Press, Beijing (2012)
8. Beijing Soundwel Technology Corporation: The operating instructions of Wireless sound emitter, 8 (2013)
9. Zhou, P., Xu, G., Ma, X.: Proficient in LabVIEW Signal Processing. Tsinghua University Press, Beijing (2013)
10. Poorani, D., Ganapathy, K., Vaidehi, V.: Sensor Based Decision Making Inference System for Remote Health Monitoring. IEEE (2012)

Feature Extraction of Electrical Equipment Identification Based on Gray Level Co-occurrence Matrix

Qinghai Ou[1(✉)], Huadong Yu[1], Lingkang Zeng[1], Zhu Liu[1],
Xiao Liao[1], Yubo Wang[2,3], Guohua Liu[2,3], Yongling Lu[4],
Chengbo Hu[4], NanXia Zhang[5], and Qian Tang[5]

[1] State Grid Information and Telecommunication Group Co., Ltd.,
Beijing 102211, China
453682740@qq.com
[2] State Grid Key Laboratory of Power Industrial Chip Design
and Analysis Technology,
Beijing Smart-Chip Microelectronics Technology Co., Ltd.,
Beijing 100192, China
[3] Beijing Engineering Research Center of High-Reliability IC with Power
Industrial Grade, Beijing Smart-Chip Microelectronics Technology Co., Ltd.,
Beijing 100192, China
[4] State Grid Jiangsu Electric Power Company Research Institute,
Nanjing 211103, China
[5] North China Electric Power University, Beijing 102206, China

Abstract. With the development of the electricity industry in our country, the kinds and the quantities of electrical equipment are increasing quickly. When the electrical equipment is checked and evaluated, the recognition of equipment is very important. In this paper, based on gray level co-occurrence matrix to extract the texture of equipment, we proposed a way to identify the electrical equipment. First it uses the gray level co-occurrence matrix texture matching recognition of electrical equipment, and then adopts the method of fuzzy logic according to the result of the match on the classified recognition of electric equipment. By the experiments, the correctness and feasibility of identification method to the electrical equipment are proved.

Keywords: Electrical equipment · Feature extraction · Fuzzy probability · Image identification

1 Introduction

Nowadays, the operation of the power system is gradually realized multimedia and information [1], more and more substations have taken remote monitoring system instead of manual mode, which greatly improved the degree of automation of power plants and also reduced the demand of human resources [2].

Recognition technology based on color feature and recognition technology based on texture feature [3]. We use the recognition method based on the texture feature to

© Springer International Publishing AG 2018
F. Qiao et al. (eds.), *Recent Developments in Mechatronics and Intelligent Robotics*,
Advances in Intelligent Systems and Computing 690, DOI 10.1007/978-3-319-65978-7_30

analyze the characteristics of the structure in the image or to analyze the distribution of the color intensity in the image [4].

Fourier series method, texture spectrum method and so on.

2 Extraction of Image Features of Equipment

2.1 The Definition of Gray Level Co-occurrence Matrix

Set $f(x, y)$ as the gray value corresponding to the point of (x, y) in the image I ($N \times N$). That is if:

$$(m - 1) \times (N/k) \leq f(x, y) < m \times (N/k)$$

Then $f(x, y) = m$, among them, the pixel pair (i, j) of the gray level co-occurrence matrix is shown in Fig. 1.

Fig. 1. The pixels pairs of the gray level co-occurrence matrix

After the distance d and direction Θ are confirmed, the value of $G_{d, \Theta}$ (m, n) is incremented by 1 when the pixel pair (i, j) with direction Θ and distance d satisfies $i = m, j = n$.

$$G_{d,\theta} = \begin{bmatrix} G_\theta(0,0) & G_\theta(0,1) & \cdots & G_\theta(0,n) & \cdots & G_\theta(0,k-1) \\ G_\theta(1,0) & G_\theta(1,1) & \cdots & G_\theta(1,n) & \cdots & G_\theta(1,k-1) \\ \cdots & \cdots & \cdots & \cdots & \cdots & \cdots \\ G_\theta(m,0) & G_\theta(m,1) & \cdots & G_\theta(m,n) & \cdots & G_\theta(m,k-1) \\ \cdots & \cdots & \cdots & \cdots & \cdots & \cdots \\ G_\theta(k-1,0) & G_\theta(k-1,1) & \cdots & G_\theta(k-1,n) & \cdots & G_\theta(k-1,k-1) \end{bmatrix}$$

The matrix represents the number of times each combination (m, n) appears (in other words, $G_{d,\Theta}$ represents the number of gray pixels (i, j) in the original image).

2.2 Feature Extraction from Gray Level Co-occurrence Matrix

By Gray Level Co-occurrence Matrix, we can extract 14 texture features which exists redundancy.

(1) Angle second-order moment $E(\theta)$

Angle second-order moment E(θ) is also called Energy, and its formula is shown in Eq. (1):

$$E(\theta) = \sum_{m=1}^{k}\sum_{n=1}^{k}(P_\theta(m,n))^2 \tag{1}$$

(2) Moment of Inertia I(θ)

Moment of Inertia I(θ) is also called Contrast, and its formula is shown in Eq. (2):

$$I(\theta) = \sum_{m=0}^{k-1}\sum_{n=0}^{k-1}(m-n)^2 \times p_\theta(m,n) \tag{2}$$

(3) Entropy H(θ)

It represents the nonuniformity and complexity of image textures. Formula of H(θ) is shown in Eq. (3)

$$H(\theta) = -\sum_{m=0}^{k-1}\sum_{n=0}^{k-1}p_\theta(m,n)\lg p_\theta(m,n) \tag{3}$$

2.3 Texture Extraction with Gray Covariance Matrix

2.3.1 Improved Parameter K of Gray Level Co-occurrence Matrices

Selecting five transformers of the real shot pictures (transfm1–3), as shown in Fig. 2.

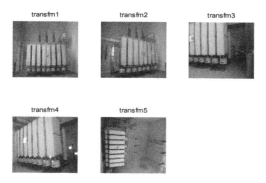

Fig. 2. The transformer pictures used as model

In the test, this paper sets four values of k, namely k = 16, 32, 64, 128. The following Fig. 3 shows the gray when the level quantization of image transfm1 set to K (k = 16, 64).

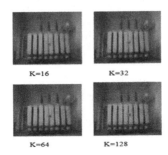

Fig. 3. Images under different k values

3 Classification of Texture Feature Based on Fuzzy Probability

3.1 Extracting Texture Features with Improved GLCM Method

The characteristic values of the transformer are extracted by the improved gray level co-occurrence matrix as shown in Table 1:

Table 1. The eight eigenvalues of transformer

k = 32	transfm1	transfm2	transfm3	transfm4	transfm5
mean(E)	0.133983	0.120534	0.13482	0.141422	0.131184
sqrt(E)	0.018469	0.018215	0.016713	0.016276	0.016272
mean(H)	2.474236	2.598716	2.443464	2.37036	2.484515
sqrt(H)	0.197542	0.191141	0.171333	0.145951	0.166928
mean(I)	0.422299	0.446666	0.363299	0.32598	0.367979
sqrt(I)	0.181114	0.169374	0.143121	0.11058	0.140351
mean(C)	0.467053	0.421123	0.417938	0.510414	0.460453
sqrt(C)	0.024994	0.018017	0.014974	0.017462	0.017354

3.2 Design Fuzzy Classifier

3.2.1 Defines the Value of the Variable

The possible range of the characteristic values of the GLCM of the transformer and the mutual inductor which has been previously obtained is now shown in Table 2.

Table 2. The classification interval of the eight eigenvalues of GLCM

transfm	min	max	Distance d = max − min	Radial r = d/2	Central point a
mean(E)	0.1205	0.1414	0.0209	0.010	0.130
sqrt(E)	0.0162	0.0185	0.0023	0.001	0.017
mean(H)	2.3703	2.5987	0.2284	0.11	2.48
sqrt(H)	0.1459	0.1975	0.0516	0.02	0.17
mean(I)	0.326	0.4467	0.1207	0.060	0.386
sqrt(I)	0.1105	0.1811	0.0706	0.035	0.14
mean(C)	0.4179	0.5104	0.0925	0.046	0.464
sqrt(C)	0.0149	0.025	0.0101	0.005	0.019
ct	min	max	Distance d = max − min	Radial r = d/2	Central point a
mean(E)	0.1047	0.2476	0.1429	0.071	0.176
sqrt(E)	0.0100	0.0129	0.0029	0.001	0.011
mean(H)	1.9097	2.7558	0.8461	0.423	2.332
sqrt(H)	0.0862	0.1271	0.0409	0.020	0.106
mean(I)	0.2308	0.5198	0.289	0.14	0.37
sqrt(I)	0.0646	0.1471	0.0825	0.041	0.105
mean(C)	0.3236	0.6435	0.3199	0.159	0.483
sqrt(C)	0.0092	0.0187	0.0095	0.004	0.013

3.2.2 Define the Membership Function of the Variable Value

The probability of belonging to a certain region is set according to the distance between the center and the center of each region, and the initial setting beyond the interval set as 0, as shown in Eq. (4).

$$\mu(x) = \begin{cases} 1 - \frac{|x-a|}{r} & \min \leq x \leq \max \\ 0 & x < \min, x > \max \end{cases} \tag{4}$$

3.3 Fuzzy Classification Rules

Among the eight categories μ_{ME}, μ_{SM}, μ_{MH}, μ_{SH}, μ_{MI}, μ_{SI}, μ_{MC}, μ_{SC}.

3.4 Model Validation

To test the reliability of the classification, this article found another three grid equipment picture as a test. The original image shown in Fig. 4:

In this paper, the improved gray covariance matrix method combined with the new design fuzzy approximation classification criteria for the grid transformer and mutual inductor identification can accurately distinguish whether a picture belongs to the transformer or mutual inductor.

Fig. 4. Three picture of unknown grid equipment to be identify

Acknowledgements. This work is supported by the Science and Technology Research Project of State Grid Corporation of China (526816160024).

References

1. Liu, H.-B., Hu, B., Wang, X.-Y.: Thinking about "much starker choices-and graver consequences-in" distribution network development. J. China Power **48**(1), 21–24 (2015)
2. Zhao, J.-J., Liu, H.: Remote monitoring system in the application of the computer room management. J. Shijiazhuang Inst. **9**(3), 89–93 (2007)
3. Yang, J.-H., Liu, J., Jian, Z., et al.: Combination of watershed and automatic seed region growing segmentation algorithm. Chin. J. Image Graph **15**(1), 63–68 (2011)
4. Stricker, M., Orengo, M.: Similarity of color images. Storage Retr. Image Video Databases III **2420**, 381–392 (1995)

Practical Research of Bidding Mode for the Whole Substation of Substation Equipment

Wang Jian[1(✉)], Yang Zhenwei[1], and Yi Wei[2]

[1] State Grid Jiu Jiang Electric Power Company, Jiujiang, China
sun270636982@163.com
[2] State Grid Huang Gang Electric Power Supply Company, Huanggang, China

Abstract. With the constant improvement of the infrastructure investment of the power system and intelligent degree, the procurement mode of material of power grid construction was difficult to satisfy different practical requirements among them, as an important material in power grid construction, a substation equipment has the characteristic of large quantity and high quality requirement, so procurement mode transition is imminent. Based on the transformer project in the taping 110 kV substation as an example in this paper, the whole business process of bidding mode for the whole substation was combed. Each stage of working effect of bidding mode for the whole substation of substation equipment was studied. The economic benefit of bidding mode for the whole substation of substation equipment was analyzed. The practice shows that the procurement efficiency and quality improvement of substation equipment is boosted for the bidding mode for the whole substation [1, 2].

Keywords: Substation equipment · Bidding mode for the whole substation · Integration · Power grid construction

1 Introduction

With the deepening application for the general design and general equipment of the new substation of State Grid Corporation in recent years, there was the miscellaneous planning in submitting and auditing. The conformable degree of bidding was not high. The construction period of equipment was uneven in conventional centralized batch purchasing which had caused adverse effects on actual demand of power grid construction to a certain extent. According to the needs of power grid construction and development, to improve the level of equipment integration and technical services, and innovate procurement mode, The State Grid Corporation has put forward the bidding mode for the whole substation of the primary equipment and secondary equipment of substation, since, 2015. According to the working principles of technology as a whole, combination and cost control, the bidding mode for the whole substation of substation equipment was researched and practiced. The State Grid Jiu Jing Electric Power Company regard it as the chance, the material in the Tai Ping of 110 kV substation was brought into the bidding mode for the whole substation. Currently, a experimental work makes great progress and significant achievements.

© Springer International Publishing AG 2018
F. Qiao et al. (eds.), *Recent Developments in Mechatronics and Intelligent Robotics*,
Advances in Intelligent Systems and Computing 690, DOI 10.1007/978-3-319-65978-7_31

2   The Implementation of Bidding Mode for the Whole Substation

According to the demand of bidding mode for the whole substation in design and whole-process control of construction, this project involving in twenty eight kinds of equipment such as the primary equipment, secondary equipment, communication equipment and building steel structure which the primary equipment and secondary equipment were adopted by the bidding mode for the whole substation.

A total integrator integrates thirty-eight kinds of materials, involving in 11 suppliers. A secondary integrator integrates fourteen kinds of materials, involving in four suppliers. The primary and secondary key equipment of the substation were produced by two integrators. After the contract signing, a integrated supplier had submitted the technical documents of bidding at the prescribed time.

The information of technical file was confirmed by construction unit, material company and integrated with company. According to bid documents of equipment, the material company checks integrity of bidding technical document and compiles confirmation of bid technical documents. The construction unit was to participate entirely, check again, finally ensure the integrity of the bid technical documents. Integrators make clear concrete implementing manager and design leader of the project according to responsibility.

Integrators rechecked the bid technical documents, confirmed the specific configuration scheme, technical parameters and interface mode, studied out delivery way and time node of equipment, filled in confirmation of technical data of equipment which strictly controled outsourcing equipment, checked the bid technical documents of outsourcing equipment supplier and technical clarify, confirmed the equipment configuration scheme, technical parameter, components selection and quantity whether it meets universal equipment four unification, later expansion of substation and device interfaces between different manufacturers.

The construction unit organized material, operating maintenance, scheduling control, marketing and other relevant departments, established combined consultation decision mechanism, held design connection in time, confirmed electric and civil interfaces whether it met the standard of universal equipment and the demand of antibugging measures. The arrival of goods schedule was established according to the project node and connecting with the construction progress. Integrators organized related equipment manufacturers to send professional technical personnel to the site, and to carry out the commissioning work after equipment delivery and inspection of merchandise received. The construction unit was to develop synchronously dynamic evaluation of standardization construction of integrators.

3   The Effective Analysis of Bidding Mode for the Whole Substation

3.1   The Reduction of Plan Preliminary Hearing

The same scale project in normal mode, the materials plan was needed by headquarters transmission and transformation equipment, and information equipment and provincial power companies bidding etc. three batch declarations involving in three times preliminary hearing. Each batch concentrated preliminary hearing requires 1 day at least, a set of process need 3 days shortest. Experts are organized to review 3 times. Experts were business mainstay of all departments, who were difficult to put together. The material plan in the taping only need one batch declarations of bidding mode for the whole substation, organize once plan concentrated preliminary hearing. Plan review time is shortened to 1 day, audit efficiency is greatly improved.

3.2 Raising the Accuracy of Planning Submission

The original material plan was easy to omit plan submission by three batch submission, according to the same scale project plan submission model. The material plan in the taping centralized at bidding mode for the whole substation which to unitily submit and check. The accuracy of plan submission is greatly increased, reducing the probability loss acceptance of the bid and improving the efficiency of the whole purchase. The delivery time of all kinds of material is at the same time which make change not unified delivery time of material, influence later inspection of merchandise received.

3.3 More Favourable Contract Implementation

The amount of supplier is decreased. The efficiency of contract signing and responsibility is increased. In the conventional batch, the primary equipment and secondary equipment of 110 kV substation had fifteen suppliers, and sign fifteen agreement. The primary and secondary equipment, only one contract is needed to sign after bidding mode for the whole substation. The workload of contract signing and running is greatly reduced, enhanced efficiency, and promoted accuracy of link of contract management. The link of contract implementation, tens of nanometers suppliers repeatedly cohesion. The drawing was lag and delivery equipment was not timely among many suppliers and so forth. The responsibility was more difficult. The primary and secondary equipment, just two high quality supplier are needed to coordinate after bidding mode for the whole substation. The workload is reduced in drawing confirmation, each link is more affluent.

3.4 Acceptance and Settlement are More Convenient

On account of Integration of high quality suppliers, the acceptance is more effective and the contract settlement is convenient. In batch bidding mode, the acceptance is more difficult to perform, the arrival of goods time is difficult to unity, the receipt of goods period is longer than the cycle, many suppliers is needed to participate in the acceptance, the acceptance time can not be unified arrangement, and even some suppliers refuse to participate in the acceptance, the problems of the goods make often argue back and forth, the responsibility is not divided clear. The primary and secondary equipment only need cooperate two suppliers in the bidding mode for the whole substation. The integrated manufacturer coordinate all the equipment of acceptance of the bid. The acceptance of arrival of goods is to ensure a smooth implementation. A amount of documents and invoice are greatly reduced. The work strength of contract settlement personnel is reduced, and the settlement accuracy is improved.

3.5   Compressing Level, Reducing Management Costs

A dozen suppliers were informed to attend discussion holding a design liaison meeting in the past. There were often some suppliers do not meet the requirement, not in time, and so on. Now only two representatives of integrated supplier for the primary and secondary equipment to be present. The efficiency of the overall supply and installation of the goods is to improve. The problems of equipment quality in the process of debugging of construction, which tender need to inform each manufacturer to process in batch bidding, now only need to correspond one to two integrators. The original "one-to-many" is turned into "one-to-two" mode, which significantly improves the communication efficiency and reduces the coordination management costs.

3.6   The Quality Improvement of After-Sales Service

The requirements for supplier were relatively low in batch bidding mode, which caused suppliers to present small, scattered, many characteristics and low integration. Some smaller suppliers did not meet the requirement in after-sales service. The suppliers often fudged in various reasons and delay in service when the product had problems which brought about delayed engineering construction to some extent and disguised increased the costs of engineering construction. The bidding integrator is usually the strongly comprehensive strength and the quality after-sales service supplier through the bidding mode for the whole substation. A integrated supplier can provide integrated services, improve the degree of adaptability of a manufacturer, and shorten the cycle of engineering construction.

4   Economic Benefit Analysis

With the same scale substation under the conventional bidding, the equipment cost is about 7.72 million yuan (The primary equipment is about 5.73 million yuan. The secondary and communication equipment are about 1.99 million yuan.). The settlement amount of Tai Ping of 110 kV substation is about 6.75 million yuan (The primary equipment is about 4.91 million yuan. The secondary and communication equipment are about 1.84 million yuan.) adopting bidding mode for the whole substation. The reduction ratio of cost for the bidding mode for the whole substation than conventional bidding mode as shown in Fig. 1.

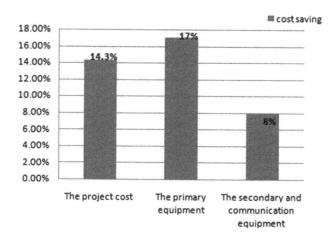

Fig. 1. Cost savings of bidding mode for the whole substation

5   Conclusion

Under the background of the construction of intelligent electric grid, the business model which the full station equipment of the substation supplied by the bidding supplier will be a tendency. The implementation of a bidding mode for the whole substation can effectively improved the efficiency of material purchase of the substation, promote the process of intensive management of company material, and accelerate the development of power grid construction. Based on the transformer project in the taping 110 kV substation as an example in this paper, the whole business process of the bidding mode for the whole substation of substation equipment was carded, the effects and economic benefits of a business model of the full station equipment of the substation supplied by a bidding supplier was studied. Practice shows that the bidding mode for the whole substation reduces the workload of substation equipment procurement, improves the efficiency and benefit of bidding.

References

1. Bi, Z., Wang, L., Lian, L.: System research of material supply chain adapting to the development of smart grid. Supply Chain Manag. **35**(1), 149–151 (2016)
2. Zhuang, Y.L.: Discussion and summary of bidding model of modular substation. Chinese E-commerce, No. 9, p. 91 (2014)

Automatic Age Estimation Based on Vocal Cues and Deep Neural Network

Guo Mei and Xiao Min[✉]

Software and Communication Engineering School, Xiangnan University, Hunan, China
rlhrdj@163.com

Abstract. The estimation of speaker age using vocal features is studied in this paper. Firstly, a large number of utterances from various speakers are collected for analysis. Secondly, the vocal features including prosodic and spectral features are extracted and compared among different age groups. The spectral energy ratios are proposed to effectively classify speakers from different age groups. Thirdly, artificial neural network is used to model the age features. Age model is learned for segment level classifiers and the probabilities of age distribution is used to generate effective features. Finally, age regression is implemented based on deep neural network. Experimental results show that the proposed model is effective, and the age features are robust to speaker variance.

Keywords: Feature analysis · Recursive neural network · Age estimation

1 Introduction

Human-Computer Interaction (HCI) is a fast developing research field in computer science and electrical engineering. Advanced signal processing and machine learning technologies have innovated many interesting applications in natural interaction with machines [1–4]. How to recognize the user's identity and status is an unsolved question.

Recent studies in age recognition show that vocal features can reflect the age difference of a person. Schoetz [5] from Lund University studied perception cues related to age, including both prosodic features and phonetic features. In their work, the individual variance is a major challenge. Brueckl et al. [6], studied the acoustic features from female subjects. In their experiment, the acoustic features seemed to be more related to the perceptual age than the actual age. The subjects amount is not enough for a complete age model. The generalization ability of the age estimation algorithm need to be further investigated.

Ramig et al. [7] focused on the pitch features and its relation with aging. According to their work, the pitch frequency was not closely related to the age of male adult subjects. However, their subjects group was not large enough, and the conclusion might not be suitable for all persons.

In this paper, we apply the deep neural network (DNN) to the age estimation problem. Deep learning provides us a powerful way for data representation. It is suitable for improving the current system since the feature extraction is very difficult for age estimation. The main contributions of this paper are two-fold: (i) we propose a set of

© Springer International Publishing AG 2018
F. Qiao et al. (eds.), *Recent Developments in Mechatronics and Intelligent Robotics*,
Advances in Intelligent Systems and Computing 690, DOI 10.1007/978-3-319-65978-7_32

effective acoustic features that may be used to estimate speaker age; (ii) an extension of Deep Neural Network application to the parallel linguistic information extraction.

The rest of the paper is organized as follows: Sect. 2 gives the general description of the database and a detailed analysis of feature changes is also provided; Sect. 3 describes the basic Gaussian mixture model method; Sect. 4 describes the age estimation algorithm used in our system; Sect. 5 provides the experimental results, and finally, conclusion is given in Sect. 6.

2   Age Data and Feature Analysis

Previous database is not designed for age estimation. There are very few available speech data can be used for the task of age modelling and testing. Therefore, we build a local age corpus including child, man, female, and elderly people. The speech corpus includes both English and Chinese sentences. A total of 128 subjects are participated in the recording where they were required to read the given materials.

Their age distribution is shown in Fig. 1.

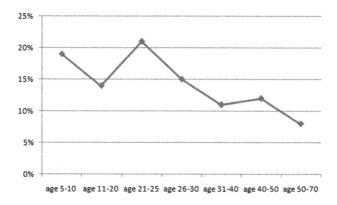

Fig. 1. A depiction of subjects distribution in the age database

Although the changes in acoustic features caused by aging are not as obvious as that caused by phonetic changes, it is possible to establish an aging model based on speech features. The major change in people's voice happens during adolescence, but the subtle changes last for the entire life. The dynamics of the vocal model also changes, which causes the differences in pitch frequencies. One obvious change related to age is the speaking rate, as shown in Fig. 2.

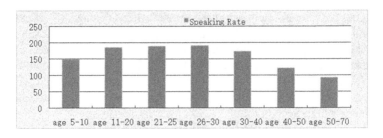

Fig. 2. Aging trends for syllables per minute

3   Gaussian Mixture Model

Gaussian Mixture Model (GMM) is a widely used algorithm that has strong data modelling ability. Theoretically it can model any probability distribution. The setting of Gaussian mixture number is important, and it can be empirically set in different applications.

The definition of Gaussian Mixture Model is:

$$p(\mathbf{X}_t|\lambda) = \sum_{i=1}^{M} a_i b_i(\mathbf{X}_t) \tag{1}$$

where $\mathbf{X}_t$ is a D-dimension random vector, $b_i(\mathbf{X}_t)$ is the i-th member of Gaussian distribution, t is the index of utterance, a_i is the mixture weight, and M is the number of Gaussian mixture members. Each member is a D-dimension variable follows the Gaussian distribution with the mean $\mathbf{U}_i$ and the covariance $\mathbf{\Sigma}_i$:

$$b_i(\mathbf{X}_t) = \frac{1}{(2\pi)^{\frac{D}{2}} |\mathbf{\Sigma}_i|^{\frac{1}{2}}} \tag{2}$$

$$\exp\left\{-\frac{1}{2}(\mathbf{X}_t - \mathbf{U}_i)^T \mathbf{\Sigma}_i^{-1}(\mathbf{X}_t - \mathbf{U}_i)\right\} \tag{3}$$

Note that:

$$\Sigma_{i=0}^{M} a_i = 1 \tag{4}$$

Expectation-Maximization (EM) algorithm is then used for the estimation of GMM parameters.

4   Deep Neural Network Based Age Estimation

Speaker age estimation is very challenging because it is unclear what features are effective for age analysis. The deep neural network is adopted in this paper to extract high level of age features from speech data.

Deep learning is an emerging field, and it has outperformed many of the previous state-of-the-art machine learning algorithms in computer vision and speech recognition. A strong attraction of DNN is that it can represent the high level feature without much variance from the low level data. Our work differs from the previous work in that we extend the application of DNN to a new area of parallel linguistic information analysis and combine the age acoustic features with the deep neural network architecture.

DNN is different from traditional multilayer neural network because it contains many hidden layers. DNN belongs to the family of directed graphical models and it models the posterior probability $p_{y|x}(y = s|\mathbf{x})$, where s stands for a class, $\mathbf{x}$ stands for an input vector (the observation) and y stands for the output of the neural network.

The proposed age estimation algorithm consists of the following steps:

Step One: Build an age group probability distribution at segment level from DNN.
Step Two: Extract static features at utterance level.
Step Three: Age estimation using Support Vector Regression (SVR).

The low level acoustic features are denoted as $\mathbf{f}_{lld}^i$ include pitch frequency p^i, energy ratio $\mathbf{e}^i$, and spectral features $\mathbf{s}^i$, where i stands for the segment index. These features are segment-wise, and each segment data $\mathbf{d}_w$ corresponds to one fixed-dimension feature vector $\mathbf{f}_{lld}^i$. w denotes the length of the time window. The segment data $\mathbf{d}_w$ is first converted into a sequence of short-time frames by Hanning window, and short-time Fourier analysis is used to extract energy and spectral features. The pitch frequency is estimated by wavelet analysis from each frame data. The feature vector is then formed by these extracted features with proper functional.

With these low level data, we are able to train a DNN to model the probability distribution $p(\mathbf{f}_{lld}^i|\lambda_j)$ of each age group, where λ_j is the age group model, and j is the group index.

5   Experimental Results

In order to verify the effectiveness of the proposed method, we select seven age groups to train and evaluate our age model. A total of 128 subjects are covered in the dataset, and the training to testing ration is 3 to 1. The dataset is randomly divided into training subset and testing subset. The recognition rates are averaged through out the subjects.

The output probabilities of DNN are shown in Fig. 3. The results are achieved from one utterance example from age group 30–40.

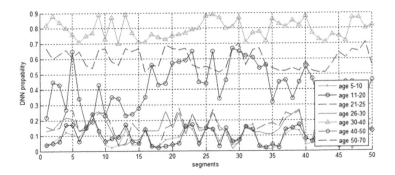

Fig. 3. DNN output probabilities of segment level classification

In order to verify the effectiveness of deep learning, the Gaussian Mixture Model (GMM) is used for the same feature representation described in Sect. 4. Gaussian mixture mode is used to model the low level feature distribution and its output likelihood is used as a high level feature. The mixture number is set to 6, and the Expectation-maximization algorithm is used to automatically estimate the mean vector, the covariance matrix and the weight.

The error of age estimation is defined by the difference between the machine estimated age and the subject's actual age. $e_k = abs(\widehat{a}_k - a_k)$, where e denotes the error, $\widehat{a}_k$ is the machine estimation, a is the actual age and k denotes the index of testing samples.

The accuracies of DNN based estimation and GMM based estimation are shown in Table 1. We can see that the DNN based method constantly improves the estimation results. The maximum age estimation error for DNN based method is 9.2 years, and the maximum error for GMM based method is 14.1 years. The averaged error for DNN is 5.9, and the averaged error for GMM is 9.0.

Table 1. Estimation errors of the machine estimation of speaker age (years)

Age group	DNN based method	GMM based method
5–10	8.1	14.1
11–20	4.3	10.1
21–25	7.8	8.0
26–30	5.5	6.7
31–40	2.3	5.8
40–50	9.2	13.1
50–70	4.3	5.5

The improvements may be brought by the desirable property of DNN. It represents the real-world data in a more abstracted and more generalized feature space than the traditional methods. Although GMM can model any distribution given enough training data, in real applications this kind of data requirement cannot be satisfied. GMM highly relies on the balanced training data and it has a relatively low ability of generalization.

6   Conclusion

In this paper we address the problem of automatic estimation of speaker age. Several acoustic features are proposed, and deep neural network is used to represent the high level features based on segment level features. Experimental results show that the proposed age estimation system achieves satisfactory results. In future work, culture and language differences will be considered in our age estimation algorithm.

References

1. Hu, M., Zheng, Y., Ren, F., Jiang, H.: Age estimation and gender classification of facial images based on Local Directional Pattern. In: Proceedings of IEEE International Conference on Cloud Computing and Intelligence Systems, Shenzhen, China, pp. 103–107 (2014)
2. Kalantari, S., Dean, D., Ghaemmaghami, H.: Cross database training of audio-visual hidden Markov models for phone recognition. Math. Probl. Eng. 6(2), 2141–2146 (2015)
3. Wang, K., An, N., Li, B.N.: Speech emotion recognition using fourier parameters. IEEE Trans. Affect. Comput. 6(1), 69–75 (2015)
4. Wang, F., Sahli, H., Gao, J.: Relevance units machine based dimensional and continuous speech emotion prediction. Multimed. Tools Appl. 74(22), 9983–10000 (2015)
5. Schoetz, S.: Perception, analysis and synthesis of speaker age. Ph.D. dissertation, Department of Computer Science, Lund University, Lund (2006)
6. Brueckl, M., Sendlmeier, W.: Aging female voices: an acoustic and perceptive analysis. In: Proceedings of Voice Quality: Functions, Analysis and Synthesis, ISCA Tutorial and Research Workshop, Geneva, 27–29 August, pp. 163–168 (2003)
7. Ramig, L.A., Ringel, R.L.: Effects of physiological aging on selected acoustic characteristics of voice. J. Speech Hear. Res. 26(1), 22–30 (1983)

Variable Forgetting Factor Based Least Square Algorithm for Intelligent Radar

Bin Wang[✉]

Northeastern University at Qinhuangdao, Qinhuangdao, China
wangbinneu@qq.com

Abstract. With the wide use of electromagnetic spectrum, the electromagnetic environment becomes more and more complicated. Complex electromagnetic environment will pose a severe challenge to radar system. Modern intelligent radar should provide feedback from receiver to transmitter, and transmit waveform according to working environment. Adaptive algorithm is the core problem of radar feedback. In this paper, after analysis of feedback in intelligent radar and adaptive filtering, we propose a novel variable forgetting factor based least square algorithm based on feedback system. In simulations, we compare the performances of vector estimation error, signal recovery, and equalization. Simulation results demonstrate that performances of the proposed variable forgetting factor based least square algorithm are better than traditional least square algorithm. Finally, the whole paper is summarized.

Keywords: Intelligent radar · Least square algorithm · Variable forgetting factor

1 Introduction

Radar uses electromagnetic wave to locate objects. However, with the continuous developments of science and technology, electromagnetic spectrum resources are increasingly tense. Traditional radar can only transmit single waveform, and it is difficult to adapt to the complex and changeable environment. More and more high requirements for radar intelligence are put forward. The core problem is adaptive algorithm of feedback system. More and more people put the research focus on adaptive algorithm.

Simon Haykin proposes the concept of cognitive radar. In cognitive radar, the radar can study the environment continuously through prior knowledge with the working condition and transmit waveforms adaptively to the environment [1]. In [2], the author proposes concept of intelligent adaptive radar. How to build an intelligent radar system imitating bats is discussed. In [3], for the three-phase distribution system, the authors put forward one implementation form of distribution static compensator. Based on control concept, a variable forgetting factor based recursive least square is proposed. In [4], the authors research on RLS algorithm and propose a variable forgetting factor algorithm splitting the difference. In [5], the authors focus on the application of a modified recursive least squares algorithm. The proposed method uses a variable forgetting factor to estimate frequency components. In [6], in interference suppression applications, a low-complexity variable forgetting factor mechanism is proposed which

© Springer International Publishing AG 2018
F. Qiao et al. (eds.), *Recent Developments in Mechatronics and Intelligent Robotics*,
Advances in Intelligent Systems and Computing 690, DOI 10.1007/978-3-319-65978-7_33

achieves superior performance. In [7], for system identification, the authors propose a variable forgetting factor based RLS method, which is corrupted by a noise-like signal. In [8], the authors focus on crosstalk cancellation problem and propose a new method which is more suitable in near-field crosstalk cancellation task. In [9], the authors propose an extension of TWLS algorithm to a generic reference frequency, which improve the performance of accurate estimation.

In this paper, we explain the feedback principle of intelligent radar. Then based on adaptive filtering model, we propose a novel variable forgetting factor based least square algorithm. In simulations, we will compare three types of performances: vector estimation error, signal recovery and equalization.

2　Feedback in Intelligent Radar and Adaptive Filtering

Modern intelligent radar should provide feedback from receiver to transmitter, and it forms a closed-loop system, which contains environment, control, feed-back channel and scene analyzer. Intelligent radar has the ability to study the environment and the transmitted waveform will be adaptive. The core problem of feedback mechanism is adaptive algorithm, which is embodied in the design of adaptive filter. Figure 1 is a basic signal-processing cycle.

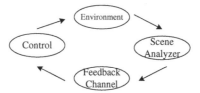

Fig. 1. Basic signal-processing cycle

Assume $s(n)$ is the true value of the input signal, $v(n)$ represents the noise of the interfering signal, the output $y(n)$ is the estimation value $\hat{s}(n)$ of $s(n)$.

$$E\left[e^2(n)\right] = E[(s - \hat{s})] = \min \tag{1}$$

It can realize that extracting a needed signal containing the desired signal from an input signal which is mixed with the noise. The greatest advantage of an adaptive filter is that the value of $h(n)$ can be automatically changed to achieve the related constraints.

$$y(n) = \sum_{m=0}^{L-1} h(m)x(n - m) = \sum_{j=1}^{L} h_j x_j \tag{2}$$

It can be seen that the output $y(n)$ is composed of sum of the L past moments, which is sum of the weights passed after a weighting process. This weighting factor is expressed by h_j or w_i. The expected output information signal is expressed by d. So

$$y_j = \sum_{i=1} w_i x_{i,j} \qquad (3)$$

In the process of radar system signal processing we generally use the adaptive filter, it has become an indispensable part of radar feedback system. According to the difference of the algorithm decision, the traditional algorithm can be divided into two main categories. One is based on least mean square criterion, the other is based on recursive least square criterion.

3   The Changes of Forgetting Factor

In recursive least square algorithm, the system performance is different with the changes of the forgetting factor. So forgetting factor is important to the whole radar system. How to reasonably determine the forgetting factor in different working environment seems particularly important. In other words, the forgetting factor should be adaptive to the environment.

Deploy

$$\varepsilon(n) = \sum_{i=1}^{n} \lambda^{n-i} e^2(i) \qquad (4)$$

We can obtain

$$\varepsilon(n) = e^2(n) + \lambda e^2(n-1) + \lambda^2 e^2(n-2) + \ldots + \lambda^{n-1} e^2 \qquad (5)$$

We can see that the latest degree of error square in front of the coefficient is 1, while the expression of the other error square corresponding to the power is forgotten factor λ corresponding power. λ is a number less than 1. If λ is small, we can see that when energy of the signal to be processed is closer to the nearest square of the error, and the effect of the system on the previous error is also forgotten and tracking channel information transmission capacity is even stronger. However, we all know that the expected signal in the recursive least square algorithm determines error, and if λ is a small value, then the expected signal on the error signal will be very large. So the output signal is very close to the expected signal, that is to say if the expected signal itself is wrong, then the output signal will also be wrong, and there is no role of the filter. To overcome the shortcoming, it is usually required that $0.95 \leq \lambda \leq 0.0955$.

We consider that when making error becoming smaller, value of the forgetting factor λ is close to 1, so that the error of the parameter is correspondingly smaller; otherwise, when the error is relatively large, the value of λ becomes smaller. So we can set the forgetting factor to a minimum value λ_{min}, which will make the system's tracking ability becomes stronger. By analyzing the forgetting factor λ, we can draw the following conclusions: As the error can affect the size of the λ value, we can analyze the error to find a suitable error point, such as the error mean square value $E[e(n)]$. When the error mean square is less than this error point, the pattern corresponding to the value of the forgetting factor is a curve that is gradually close to 1, and when the mean square error is greater than the error point, the pattern corresponding to the value of the forgetting factor is gradually close to λ_{min}. Here introduces a correction function to improve the forgetting factor to improve performance.

The expression of the correction function is as follows

$$\lambda = \frac{1 - \lambda_{min}}{\pi} arcctg\{e(n) - E[e(n)]\} + \lambda_{min} \tag{6}$$

4   Simulations

In order to test the advantages of the improved variable forgetting factor algorithm in the previous theoretical analysis, we make related simulations. Sine wave is adopted, with white Gaussian noise and SNR 10 dB. Figure 2 is curve of vector estimation error. It shows that the number of iterations required to achieve the convergence of the variable forgetting factor based least square algorithm is much less than that of the traditional least square algorithm. At the same time, value of the variable forgetting factor based least square algorithm is more stable. Therefore, the proposed algorithm has a strong engineering practical value.

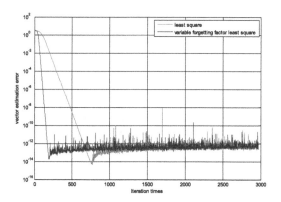

Fig. 2. Curve of vector estimation error

Following we will compare signal recovery. The simulation parameters are as follows: the signal used is the sine wave signal, and the signal to be processed is added noise. The noise type is Gaussian white, and SNR is 5 dB. The corresponding forgetting factor is 0.95, the minimum value of forgetting factor is 0.75. The result is shown in Fig. 3. It shows that variable forgetting factor based least square algorithm has better overall performance than traditional least square algorithm, so the improved algorithm has more application value than traditional algorithm.

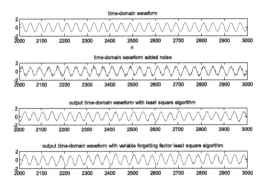

Fig. 3. Curve of signal recovery

Finally we simulate equalization. The signal used is sine wave signal, and the signal to be processed is added noise. The noise type is Gaussian white, and SNR is 5 dB. The corresponding forgetting factor is 0.95. The result is shown in Fig. 4. It shows that the signal is out of order before equalization. However, after equalization of variable forgetting factor based least square, the obtained signal approaches the expected signal.

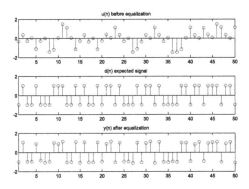

Fig. 4. Curve of equalization

5   Conclusions

With the increasing complexity of radar working environment, more and more requirements for the intelligence of radar are proposed. In this paper, after explaining the feedback principle of intelligent radar, we propose a novel variable forgetting factor based least square algorithm based on adaptive filtering model. Simulation results contain vector estimation error, signal recovery, equalization, and they demonstrate that performances of the proposed algorithm are better than traditional least square algorithm.

Acknowledgement. This work was supported by the National Natural Science Foundation of China (No. 61403067).

References

1. Haykin, S.: Cognitive radar: a way of the future. IEEE Signal Process. Mag. **23**(1), 30–40 (2006)
2. Guerci, J.R.: Next generation intelligent radar. In: Proceedings of IEEE Radar Conference, pp. 7–10 (2007)
3. Badoni, M., Singh, A., Singh, B.: Variable forgetting factor recursive least square control algorithm for DSTATCOM. IEEE Trans. Power Deliv. **30**(5), 2353–2361 (2015)
4. Paleologu, C., Benesty, J., Ciochină, S.: A practical variable forgetting factor recursive least-squares algorithm. In: Proceedings of 11th International Symposium on Electronics and Telecommunications, pp. 1–4 (2014)
5. Beza, M., Bongiorno, M.: Application of recursive least squares algorithm with variable forgetting factor for frequency component estimation in a generic input signal. IEEE Trans. Ind. Appl. **50**(2), 1168–1176 (2014)
6. Cai, Y., de Lamare, R.C.: Low-complexity variable forgetting factor mechanism for recursive least-squares algorithms in interference suppression applications. IET Commun. **7**(11), 1070–1080 (2013)
7. Paleologu, C., Benesty, J., Ciochina, S.: A robust variable forgetting factor recursive least-squares algorithm for system identification. IEEE Signal Process. Lett. **15**, 597–600 (2008)
8. Guang, P., Fu, Z., Xie, L. and Zhao, W.: Study on near-field crosstalk cancellation based on least square algorithm. In: Proceedings of Asia-Pacific Signal and Information Processing Association Annual Summit and Conference, pp. 1–5 (2016)
9. Belega, D., Fontanelli, D., Petri, D.: Dynamic phasor and frequency measurements by an improved Taylor weighted least squares algorithm. IEEE Trans. Instrum. Meas. **64**(8), 2165–2178 (2015)

Research on Logistics Cost Accounting of Iron and Steel Enterprises in the Environment of Big Data

Yulian Qiu and Huan Luo[✉]

School of Management, Wuhan University of Science and Technology, Wuhan, China
1149836749@qq.com

Abstract. Under the big data environment, it is necessary for the enterprises to develop from the traditional accounting model to the cloud accounting model. This paper aims at researching on the existing problems in logistics cost accounting of iron and steel enterprises, through the comparison of existing accounting methods and then puts forward the application of ABC in logistics cost accounting in the iron and steel enterprises under the environment of big data and cloud accounting.

Keywords: Iron and steel enterprise · Logistics cost accounting · ABC · Big data

1 Introduction

A enterprise produces a large number of data every day and conforming to the development of the times, big data and cloud computing arise at the historic moment. It is conducive to the enterprise to make the correct management decisions and promote the development of enterprises. Cloud accounting is the product of the development of accounting computerization and provides a new platform for accounting [1]. At present, the accounting system is only applicable to deal with the monetized and normalized data, so the logistics cost is difficult to be fully recognized by the original system. Big data and cloud accounting can excavate the deep-seated logistics data to help enterprises to get more comprehensive logistics cost information.

Iron and steel enterprises in China fall into the predicament of the severe overcapacity [2] and the proportion of logistics cost has been very high. It has become a hindrance to reduce cost and increase profits. Only by obtaining accurate logistics cost data, can we control the logistics cost effectively and improve the management level of logistics operation, so it is necessary to find out the most suitable method for steel enterprises.

2 Present Situations and Problems of Logistics Cost Accounting in Iron and Steel Enterprises

2.1 Present Situation of Logistics Cost Accounting

The production process of iron and steel enterprises is complex, but logistics cost is high without relatively accurate accounting method. Logistics cost mix in the relevant

© Springer International Publishing AG 2018
F. Qiao et al. (eds.), *Recent Developments in Mechatronics and Intelligent Robotics*,
Advances in Intelligent Systems and Computing 690, DOI 10.1007/978-3-319-65978-7_34

accounting subject [3], so the data is rough and fuzzy. It is difficult to achieve the purpose of effective control of logistics cost. The current "Accounting standards for enterprises" does not regulate how the logistics cost should be accounted [4]. Most of the iron and steel enterprises use the traditional cost method and use a single standard to evaluate the indirect logistics cost to obtain inaccurate logistics cost data.

2.2 Present Problems of Logistics Cost Accounting

(1) Lack Scientific Accounting Method
   Iron and steel enterprises have no related logistics cost accounting system. China introduced "The composition and calculation of logistics cost" in 2006. But for the small scope of content and practic, logistics process is very complex. The practicability of the rule is poor, so the majority of enterprises according to their own understanding to account logistics cost. The accounting method of iron and steel enterprise is not correct and with no standard specification, so it is especially difficult for enterprises to get complete logistics data.

(2) Distort Cost Information
   By using the traditional cost method, only a small part of the logistics cost has been accounted, which is just the tip of the logistics cost iceberg. Cost of capital occupation and other cost is not included in the logistics cost accounting. This method prevents enterprise managers to obtain accurate and comprehensive logistics cost data and the accuracy will also affect the accuracy of related products cost, thus leading to wrong pricing decisions.

(3) Responsibility Cost is Not Accurate
   Logistics activities need cooperation and coordination with other departments, and the cost of these activities should be distributed between them. The traditional cost method is only to get the total logistics information, such as wage, depreciation, rent and so on, and is not according to the departments to establish responsibility cost center. So iron and steel enterprises can not get the logistics costs of the departments. This method is not conducive to evaluate the performance of the responsibility of various departments, and is not conducive to the implementation of the responsibility accounting.

3 Logistics Cost Accounting Method of Iron and Steel Enterprises

3.1 Production Characteristics of Iron and Steel Enterprises

Compared with other enterprises, the production process of iron and steel enterprises is more complicated. The logistics cost includes direct cost and indirect cost [5]. The logistics cost is included in the link of supply, production and marketing. Supply and sales logistics cost is mostly direct cost which is easy to be collected. The indirect cost of production is high, and the distribution of indirect cost largely determines the reliability of logistics information. Taking a steel enterprise as an example, the production process is shown in Fig. 1.

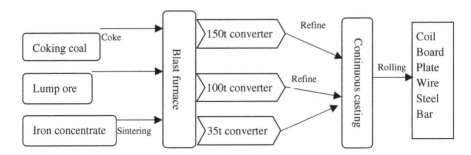

Fig.1 Production process of iron and steel enterprises

3.2 The Methods of Logistics Cost Accounting

(1) Statistical method

   Statistical method of accounting refers to that the enterprise extract the logistics cost from the accounting subjects, and then in accordance with the determination of a good logistics accounting object to re classification, distribution and summary.

(2) Traditional cost method

   The traditional cost method refers to that the distribution of indirect cost in logistics activities is determined according to the standard, which is mainly related to the quantity of production, such as direct working hours, machine hours and so on.

(3) Activity based costing

   ABC refers to the logistics activities as the center, logistics as the object of product cost accounting, then according to their resource drivers to confirm resources cost and according to the cost driver to distribute cost in cost pool. ABC can track the whole process of logistics activities, and analyze the effect of the non value-added operations to optimize the logistics process.

ABC distributes the indirect logistics cost based on a number of cost drivers, such as labor hours, machine hours, transportation mileage, storage volume and other variable. It pays more attention to the reasons for the consumption of resources. As shown in Fig. 2.

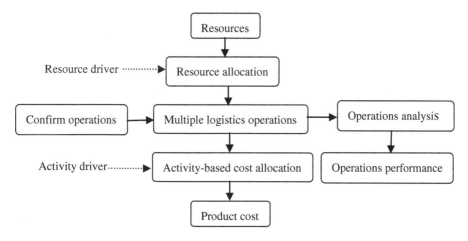

Fig. 2 The process of ABC for logistics cost accounting

3.3   Comparison of Traditional Cost Method and Activity Based Costing

(1)   Comparison of Traditional Cost Method and Activity Based Costing in Logistics Cost Accounting
The traditional cost method allocates indirect logistics cost according to a single variable.ABC is based on more logistics operations to allocate indirect cost, and pays more attention to the causes and methods of cost formation. The result of allocation is more accurate.

(2)   The Advantages of Activity Based Costing
① More Accurate Information
As iron and steel enterprises, the products are various, production processes are complex, but cost which has nothing to do with the product output is relatively large, such as system maintenance costs, loading and unloading costs and so on. The traditional cost method ignores the actual matching of product and cost in the distribution of indirect logistics cost, which is not conducive to control the cost of logistics activities. However, the activity-based costing method estimates the indirect cost of logistics by multiple allocation benchmarks, which can obtain relatively accurate information. The product cost information will be more accurate according to the logistics data, which will help the managers to make the right pricing strategy.
② Optimize Logistics Process
The traditional cost method gets relevant data only by the extraction and summary of accounting data. ABC further tracks and explores the way and causes of cost and analyzes whether all of the logistics operations will bring benefits to the enterprise. At the same time, it improves the correlation between the product and its actual cost and eliminates the no value-added operations to improve the logistics efficiency of the iron and steel enterprise [6].

③ More Suitable for Iron and Steel Enterprises

The indirect logistics cost of iron and steel enterprises is large and it is need to find the right allocation scheme to apportion. The traditional accounting method is generally applicable to the enterprise which has a relatively small proportion of indirect costs and a relatively simple product structure.ABC is suitable for the enterprise which has large proportion of indirect cost, complex production process and diverse types of products [7].

ABC can make up the requirement of the logistics data accuracy and it is adapted to the iron and steel enterprises.

④ Get More Accurate Product Cost

Logistics cost will eventually be collected into related products, as part of the cost of products and impact the product decisions.ABC and the traditional cost method have a great difference in the logistics cost calculation result. The reason of this result is that the distribution method of the indirect cost is different. The traditional cost method may overestimate the logistics cost of high yield product but simple procedure and underestimate the logistics cost of low yield product but complicated procedure. It seriously distorts the real situation of product logistics costs, but also affects the accuracy of product costs, thereby affects pricing and corporate profit.ABC method is more objective and more accurate to reflect the cost of the product.

3.4   The Selection of Logistics Accounting Method in Iron and Steel Enterprises

The use of statistical method is simple and flexible, but the disadvantage is that the logistics cost data is not accurate and rough. The logistics cost data is just the tip of the iceberg. The accounting method is the most commonly method for most enterprises, but the indirect cost allocation standard is single, especially iron and steel enterprises with more indirect logistics cost. The traditional cost method influence the accuracy of logistics cost, product pricing decision, logistics optimization and control failure.

4   The Implementation of Activity-Based Costing Under Big Data and Cloud Accounting Environment

The production and marketing link of iron and steel enterprises produce a large number of logistics data every day. The effective processing of these data with the best accounting method can get more accurate logistics cost information. Activity based costing with big data and cloud accounting environment in iron and steel enterprises provides useful information for management decision by analysing, processing and then obtain the logistics cost data. This method includes data collection, data processing, data analysis and decision making. As shown in Fig. 3.

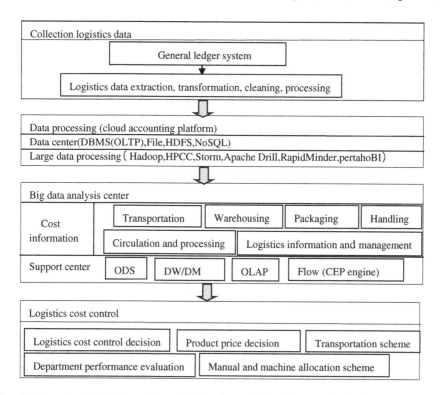

Fig. 3 Flow chart of ABC combine with big data and cloud accounting in the iron and steel enterprises logistics

5   Conclusions

With big data and cloud accounting era, it is very convenient for iron and steel enterprises to collect logistics information and select required data and then tap, process, analyze and export these data. It can make the cost more timely, accurate and effective under control. Can also be based on the output of the logistics cost of department as a benchmark for performance evaluation. The enterprises can also base on the output of the logistics cost of department as a benchmark for performance evaluation. Big data makes the iron and steel enterprises and suppliers and customers more closely. Enterprises adjust the purchasing strategy and sales strategy for selecting useful information, and then make their own profit but also improve customer satisfaction.

References

1. Jiaqian, C.: Cloud accounting "in the application of small and medium sized enterprises and influence—taking Shaoxing area as an example. Financial Times **7**, 125–126 (2016)
2. Wenjun, L.: On China's steel overcapacity problems and countermeasures of technology and enterprise **9**, 152 (2014)

3. Ruiqing, S.: Measurement and accounting of logistics cost. J. Shanghai Lixin University Commer. (2), 3–8 (2009)
4. Chun, Q.Y., Cui, Z.: The problems of enterprise logistics cost accounting of in 2013 **24**, 69
5. Aifang, W.: Logistics cost accounting of iron and steel enterprises metallurgical accounting **7**, 17–18 (2011)
6. Jufeng, Y., Rui, Y.: Operation cost calculation of medical logistics distribution center under the background of new. GSP J. Yuncheng Univ. **1**, 74–76 (2016)
7. Lili, C.: Discussion on the difference between traditional costing and activity-based costing accounting. entrepreneur world **1**, 25 (2012)

Selection of Part's Feature Processing Method Based on Genetic Algorithm

Yanwei Xu[1,2(✉)], Anbang Pan[2], and Tancheng Xie[2]

[1] School of Mechanical Engineering, Tianjin University, Tianjin 300072, China
xuyanweiluoyang@163.com
[2] School of Mechatronics Engineering, Henan University of Science and Technology, Luoyang 471023, China

Abstract. Genetic algorithm (GA) was applied to the decision-making of process and the optimal or near-operation choices for the production requirements were obtained. Firstly, according to technologic knowledge and components' designing requirement, every process method is expressed as a gene chromosome. Secondly, a new sort of fitness function was studied and defined. The conception named nearness and the main deviations were defined, and corresponding arithmetic was studied. Then, it introduced every step of process method decision on GA by examples. When Genetic algorithm (GA) was applied to the decision-making of process, it can overcome errors of decision-making from common process, thereby optimize process planning and improve intelligence of decision-making.

Keywords: Genetic algorithm · Process link · Optimization · Decision-making

1 Introduction

Many researchers pay attention to genetic algorithm at present. As a global optimization algorithm based on natural selection and heredity, its main characteristics are independent of gradient information and particularly suitable for machining complex problems and optimization of nonlinear problem which the traditional search method is hard to solve. The basic characteristics of GA [1, 2] is very suitable to make up for the shortcomings of traditional decision-making methods. This paper studies the application of genetic algorithms in the selection of feature process method.

2 GA for Process Decision Making

2.1 Gene Coding

According to the process knowledge and requirements, each process is represented as a chromosome in the genetic space. For example, the machining method of auxiliary hole can be divided into 20 kinds, and each machining method is coded. At the same time, in order to simplify the genetic operation, a mapping module for integer to binary coding

F. Qiao et al. (eds.), *Recent Developments in Mechatronics and Intelligent Robotics*, Advances in Intelligent Systems and Computing 690, DOI 10.1007/978-3-319-65978-7_35

and an inverse mapping module are developed. Because the machining method of hole is generally not more than 20, so the chromosomes take integer in [1, 2].The drill-hinge can be compiled into 1, for example, the drilling-boring can be compiled into 2 and so on. As the selection system of machining method for hole has a total of 17 coding of machining method encoding requires at least 5 bits of binary in the form of Table 1.

Table 1. Encoding instance of machining methods of holes

Machining methods	Gene coding	
	Decimal	Binary
Drill, hinge	1	00001
Drilling, rough reaming, fine hinge	3	00101
Drilling, expansion, hinge	6	00110
Rough boring, semi-fine boring, fine boring, fine boring	12	01100

2.2 Determination of Fitness Function

The fitness function is used to measure the adaptability of each gene to the environment and the GA is used to obtain the optimal solution which is also the most adaptive individual. Thence, how to design a good fitness function in the application of genetic algorithm is the most critical step. We carried out the appropriate information machining for machining parameters and machining conditions. For example, the machining conditions of low carbon steel blanks can be encoded as 2. This code may be the same as the value with the coding of the machining method, but they represent different meanings, as shown in the Table 2.

Table 2. Encoding of roughness

Ra	25	12.5	6.3	3.2	1.6	0.8	0.4	0.2	......
Coding	1	2	3	4	5	6	7	8	......

According to the machining manual and the expert experience, we find a close degree by comparing the technical requirements and machining conditions of the features to the final quality of each machining method can achieve. And it will play a role as a fitness function. First, we make the following definition.

Definition 1: The complete set of machining conditions (or machining requirements) of the feature is defined as the set of machining factors.

$$U = \{X|X = I_t, R_a, M_c, ...\} \tag{1}$$

here: It——Dimensional accuracy level

Ra——Surface roughness requirements

Mc——Other machining conditions, including blank type and feature size and other factors.

The elements of the machining factor set can be modified and expanded according to the actual requirements of the process designer.

Definition 2: Each item in set U is a set of numbers. It is defined as the machining condition set X and c_{max} is the maximum deviation of machining condition set X. In this,

$$\begin{cases} X = \{x | x \in [x_{min}, x_{max}]\} \\ c_{max} = x_{max} - x_{min} \end{cases} \tag{2}$$

Definition 3: The quality of the parts that can be achieved in each machining method must be a subset of the set of machining conditions. The subset of X corresponding to the machining method encoded as n is the n adaptive set.

$$X^{(n)} = \{x | x \in [x_1, x_2]\} \tag{3}$$

From the above three definitions we can see the following relationship:

$$\begin{cases} X^{(n)} \subseteq X \in U \\ x_{min} \leq x_1, x_2 \leq x_{max} \end{cases} \tag{4}$$

On the basis of the above definition, we can consider two kinds of close to the algorithm. The machining requirements and machining conditions which user input is integrated into the input set u after being encoded by the system.

$$U = \{x | x = I_t, R_a, M_c, \dots\} \tag{5}$$

The entry x in the input set corresponds to the term X of the set of machining factors, and the degree of closeness of x for X is

$$f(x, X) = 1 - \frac{\left| \frac{1}{2}(x_1 + x_2) - x \right|}{c_{max}} \tag{6}$$

$$f(x, X) = 1 - \frac{\min\{|x_1 - x|, |x_2 - x|\}}{c_{max}} \tag{7}$$

In Eq. (6), we take the input value of the user and compare the value of x. So that we can get an absolute value, then we use the maximum possible deviation minus the absolute value. Eventually we get a close degree. The closer the value of x0 to X is, the greater the close degree is.

Though the Eq. (6) is in line with the process of considering the problem and it is relatively simple to deal with. However, when the value range is too large or a number of machining methods have overlapping values, the results obtained by this equation are not satisfactory. n contrast, Eq. (7) is essentially a boundary method. It only accepts x that is closest to the input value x0 and ignores other values. After a number of data comparison, Eq. (7) can be satisfied with the results. Therefore, the final choice is Eq. (7). The fitness of the input set u for the machining factor set U is

$$f(u, U) = \sum_{i=1}^{m} k_i f(u(i), U(i)) \tag{8}$$

here, m——the number of elements of set u

u(i) ——i-th element of u

U(i) ——the elements of U that u(i) corresponding to

ki ——the weight of the corresponding machining factor.

For normalization, this value is generally met:

$$\sum k_i = 1 \cdot k_i > 0 \tag{9}$$

The process designer can select the coefficient according to the specific situation. In the following example, for the convenience of discussion, ki are equal. The blank of the hole to be machined is alloy steel, the machining accuracy is IT7 and the roughness requirement is 6.3. After processing, we constitute the input set

$$u = \{7, 6.3, 2\} \tag{10}$$

The calculation is shown in Table 2. This table is intended to illustrate the algorithm, so a number of machining factors are omitted. According to Table 2 and Eq. 2, when X is equal to it (i.e., on behalf of the machining accuracy requirements), the adaptive subset of the machining method is expressed as

$$\left\{ \begin{array}{l} X = \{x | x \in [0, 13]\} \\ X^{(1)} = \{x | x \in [11, 13]\} \end{array} \right. \tag{11}$$

The close degree of accuracy requirements which the user input is calculated as follows

$$\left. \begin{array}{l} x_{\min} = 0, x_{\max} = 13 \\ x_1 = 11, x_2 = 13 \\ x_0 = 7 \end{array} \right\} \Rightarrow f(7, I_t) = 0.6923 \tag{12}$$

We can deduce the rest from this, the calculation process for the fitness of the machining method coding 1 is following

$$\left. \begin{array}{l} f(7, I_t) = 0.6923 \\ f(6.3, R_a) = 0.748 \\ f(2, M_c) = 0.8333 \end{array} \right\} \Rightarrow f(u, U) = \frac{1}{3}(0.6923 + 0.748 + 0.8333) = 0.7579 \tag{13}$$

2.3 Determining the Rules to Stop Running

Because the number of alternative machining methods for different features is different, the maximum number of algebra m performed by the genetic algorithm is determined by the specific situation. Of course, the user can also, according to the situation, when the group's best fitness rate of increase (ΔE) is lower than a certain limit, terminate the operation. The genetic operation needs to met

$$\begin{cases} M < M_{\mathrm{max}} \\ \Delta E < \Delta E_{\mathrm{max}} \end{cases} \tag{14}$$

In the application of hole machining, we selected M = 50, $\triangle$E = 0.01.

3   Development Examples

Assuming that the blanks to be machined are non-ferrous metals, the machining accuracy is IT7 and the roughness requirement is 6.3. After the generation of genetic operation, the system eventually get the best fitness machining methods for the "rough boring, semi-fine boring (or precision expansion), fine boring, King Kong boring". And draw the relationship between algebra and fitness changes in the decision-making process, as shown in Fig. 1.

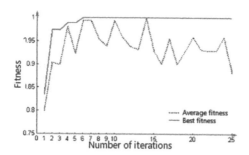

Fig. 1. Fitness variety of hole process method decision

4   Concluding

The GA is used to select the machining method table. The fitness function of the parting feature is designed. The concept of maximum deviation value and close degree is put forward for the first time, and the corresponding algorithm is put forward. This enriches the original process decision theory. Genetic algorithms can overcome the decision errors caused by errors in some conventional process decisions. Thus it optimizes the design process and improves intelligentialize of decision making.

Acknowledgements. This work was financially supported by the National Natural Science Foundation of China (51305127), the youth backbone teachers training program (2016GGJS-057) and scientific research key project fund of the Education Department Henan Province of China (14A460018).

References

1. Wang, X., Cao, L.: Genetic Algorithm: Theory, Application and Algorithm Implementation. Xi'an Jiaotong University Press, Xi'an (2002)
2. Li, Q., Liu, J.: Intelligent technology in CAPP developing tools. Mod. Manuf. Eng. **6**, 39–40 (2003)

Research and Realization of Criminal Investigation Instrument Based on CIS

Guofu Wang, Si Hu[✉], Jincai Ye, Xiaohong Wang, and Jianqiang Lu

School of Information and Communication, Guilin University of Electronic Technology, Guilin, China
husi_5819@126.com

Abstract. For the shortcomings of complex equipment, cumbersome operation and high cost when using CCD to detect criminal investigation material, an instrument of criminal investigation based on CIS (Contact Image Sensor) is designed. First, because of strong anti-interference and high sampling accuracy, the CIS is used to sample image information. Then, use Gabor algorithm for image filtering to meet the sampling image resolution of 1200 DPI. Finally, extract the physical evidence-related features Information to finish material detection of the criminal investigation (fingerprint matching, counterfeit identifying and covered character recognizing etc.).

Keywords: Criminal investigation instrument · CIS · Gabor algorithm

1 Introduction

Because of its internal integrated LED lamp, cylindrical lens and CMOS photosensitive unit, the contact image sensor (CIS) has a better image acquisition performance (less environment impact and faster image sampling), simpler operation and lower price comparing with the photoelectric coupling sensor (CCD). However, the function of current CIS image acquisition is simple, mainly for file fax, copy, etc., which cannot meet the demand of criminal investigation. In order to realize the material detetion needs on special occasions (such as crime scene, bank, library, etc.), it is necessary to study the advanced image processing technology, improve the image acquisition performance of CIS, optimize the image processing algorithm, and fundamentally solve the material detection difficult [1].

Criminal investigation instrument uses CIS to release the spectra of different bands, such as red (R), green (G), blue (B), infrared (IR) and ultraviolet (UV) five spectra, and then convert the reflected light signal into image information which can turn into material detection image after processed by the aid of image processing technology. The Gabor filter algorithm is used to obtain the image data related to the physical evidence and the clear image of the material evidence which can be used to re-veal the truth of the case is obtained through the method of distinguishing material. The instrument can realize fingerprint identification, Counterfeit/false passport identification, character altered or covered recognition and other functions.

© Springer International Publishing AG 2018
F. Qiao et al. (eds.), *Recent Developments in Mechatronics and Intelligent Robotics*,
Advances in Intelligent Systems and Computing 690, DOI 10.1007/978-3-319-65978-7_36

2   System Working Principle

The criminal investigation instrument is mainly consists of the image acquisition part and the image processing part. The specific working flow chart is shown below.

As can be seen from Fig. 1, the image acquisition part is responsible for sorting and sampling five different bands of image information, and then transmits the information to FPGA for image processing part which will turn the information into image processor and extract physical evidence [2]. The image processor mainly uses Gabor filter algorithm to extract image feature, to achieve fingerprint recognition, covered handwriting recognition and other functions.

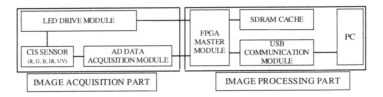

Fig. 1. System block diagram

3   Gabor Filter Algorithm

Gabor filter algorithm is used to extract the fingerprint, and according to the look-up table method, the extracted fingerprint and the fingerprint in the library are convoluted to find the most similar fingerprint [3]. Gabor algorithm is divided into four steps: (1) normalize data; (2) estimate direction; (3) estimate frequency; (4) filter.

3.1   Normalize Data

The normalized processing makes the sampled image reach a predefined mean and variance and enhances the overall contrast of the image. The normalized equation is as follows:

$$N(i,j) = \begin{cases} M_0 + \sqrt{\dfrac{V_0(I(i,j) - M)^2}{V}} \\ M_0 - \sqrt{\dfrac{V_0(I(i,j) - M)^2}{V}} \end{cases} \tag{1}$$

Among the variables above: M_0 represents the average of expectation value. V_0 is the variance of expectation. M represents the average of estimated value of grayscale $I(i, j)$. V represents the variance of estimated value of grayscale $I(i, j)$. $N(i, j)$ is the grayscale value of Normalized pixel (i, j). The output after processing is shown in Fig. 2.

Fig. 2. Normalized results, $M_0 = 90$ and $V_0 = 100$

As we can see in Fig. 2, after normalized, the result reduced the influence of the blurred image due to the unevenness of the light, adjusted the contrast of the image, improved the clarity of the fingerprint, provided an accurate fingerprint information.

3.2 Direction Estimation

The direction of fingerprints represents the characteristics of fingerprints at that point, and determines the performance of direction selecting in Gabor filter algorithm. It is a very important feature information in fingerprint extraction.

First, a 512×512 image is divided into 32 blocks with a size of 16×16 pixels. Then, the sober operator is used to calculate the gradient value of each pixel. The equations are as follows:

$$
\begin{cases}
\begin{aligned}
\partial_x(u,v) = {}& 2 \times N(u, v+1) + N(u-1, v+1) \\
& + N(u+1, v+1) - 2N(u, v-1) \\
& - N(u-1, v-1) - N(u+1, v-1) \\
\partial_y(u,v) = {}& 2 \times N(u-1, v) + N(u-1, v+1) \\
& + N(u-1, v-1) - 2N(u+1, v) \\
& - N(u+1, v+1) - N(u+1, v-1)
\end{aligned}
\end{cases}
\tag{2}
$$

u and v is the coordinates of the pixel, $u = 1, 2, \ldots, W-2$; $v = 1, 2, \ldots, H-2$, $W = H = 512$ pixel. And calculate the direction of each pixel. The calculation is expressed in (3).

$$
\begin{cases}
V_x(i,j) = \displaystyle\sum_{u=i-\frac{W}{2}}^{i+\frac{W}{2}} \sum_{v=j-\frac{W}{2}}^{j+\frac{W}{2}} 2\partial_x(u,v)\partial_y(u,v) \\[2em]
V_y(i,j) = \displaystyle\sum_{u=i-\frac{W}{2}}^{i+\frac{W}{2}} \sum_{v=j-\frac{W}{2}}^{j+\frac{W}{2}} [\partial_x(u,v) - \partial_y(u,v)] \\[2em]
\theta(i,j) = \dfrac{1}{2} \tan^{-1} \left| \dfrac{V_x(i,j)}{V_y(i,j)} \right|
\end{cases}
\tag{3}
$$

$\theta(i, j)$ represents the local direction of the box centered on pixel point (i, j), that is, the direction of the fingerprint lines, marked with O.

The experimental results are shown in Fig. 3. The Fourier spectrum perpendicular to the W × W window is taken as the dominant direction, and the direction of the ridge is estimated according to the direction [4].

ORIGINAL IMAGE   DIRECTION IMAGE

Fig. 3. Contrast pattern

3.3 Frequency Estimation

The reciprocal of the number of pixels between two adjacent peaks or valleys is called the frequency estimation [5]. The frequency estimation determines the performance of frequency selection in the Gabor filter algorithm. First, the algorithm starts with the center of each 16 × 16 pixel square to create a 32 × 16 rectangle along the direction of the box and make the rectangle's long side perpendicular to the direction of the ridge. Then, along the direction of the line, sum the gray value of 32 column pixels of the box to get 32 signal data S[k]. After that, fit the 32 data of S[k]. Equation (4) shows the method to calculate S[k] and Fig. 6 shows the fitting result of S[k].

$$
S[k] = \frac{1}{w} \sum_{n=0}^{w-1} N(u, v) \ \ k = 0, 1, \dots, w - 1.
$$

$$
u = i + \left| \frac{w}{2} - n \right| \sin \theta(i,j) + \left| k - \frac{1}{2} \right| \cos \theta(i,j) \tag{4}
$$

$$
v = j + \left| n - \frac{w}{2} \right| \cos \theta(i,j) + \left| k - \frac{1}{2} \right| \sin \theta(i,j)
$$

As can be seen from the Fig. 4, Fig. 4-1 shows the direction and frequency to be calculated of the window in the extracted fingerprint. Figure 4-2 shows the calculated sinusoidal curve of the gray value in the window, that is, the ridge frequency.

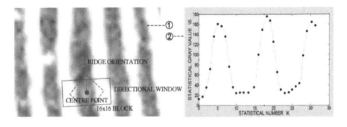

Fig. 4. Frequency estimation

3.4 Filter

Gabor filter uses the ridge direction value and the frequency value obtained in steps 2, 3 to eliminate the noise interference in the process of fingerprint extraction [6]. After that, clear and accurate fingerprint information can be extracted. The equations of Gabor filter is as follows:

$$h(x, y, \varphi, f) = \exp\left|-\frac{1}{2}\left|\frac{x^2}{\sigma^2} + \frac{y^2}{\sigma^2}\right|\right|\cos(2\Pi f_x)$$
$$x = x\cos\varphi + y\sin\varphi$$
$$y = -x\sin\varphi + y\cos\varphi$$

(5)

φ is the direction of Gabor filter, f is the frequency of sine plane wave δ_x and δ_y are the spatial constants of the Gaussian envelope along the x and y axes.

As shown in Fig. 5, the extracted fingerprint is clear.

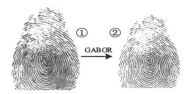

Fig. 5. Gabor filtering effect, (1) fingerprint samples, (2) filtering results

4 Experimental Verification

After specific experimental verifications, the criminal investigation material detector is capable of matching fingerprint, identifying modified or covered writing and other functions.

4.1 Verification of Fingerprint Matching Function

According to the look-up table, use the extracted fingerprints by Gabor algorithm to convolute with the fingerprints in the library respectively to find the similar fingerprints. The implementation process is shown below:

Figure 6-1 is the sampled image information of CIS. Figure 6-2 is the extracted fingerprints of the picture by image processing method. Figure 6-3 is the most similar fingerprints found in the fingerprint library. Figure 6-4 shows the matching similarity.

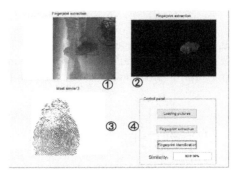

Fig. 6. Extracted fingerprints

4.2   Verification of Verified or Covered Character Recognizing Function

At the scene of the crime, there is often some evidence of altered documents, in which the handwriting is often the most important criminal information. The criminal investigation material instrument can effectively identify the covered handwriting information.

As we can see from Fig. 7, using the infrared light of 924 nm band can clearly get the covered text. And then the covered text can be extracted directly by using the image processing algorithm.

Fig. 7. Cover handwriting recognition

5   Conclusion

The criminal investigation instrument mainly carries on the high precision image sampling to the document through the CIS, and uses the image processing algorithm to carry on the feature extraction as well as matching discrimination to the sampled image. Thus it realizes the material examination detection. It can be seen from the experiment that the instrument can effectively complete the work of extracting and matching fingerprint, identifying counterfeit currency, getting covered text and other functions. So it is very suitable for the application of criminal investigation and inspection work and it also has the advantage of simple operation, a very high reliability and stability, and a very promising market application prospects.

References

1. Seo, M.W., Kawahito, S., Kagawa, K., et al.: A 0.27e-rms read noise 220-μV/e-conversion gain reset-gate-less CMOS image sensor with 0.11-μm CIS process. IEEE. Electron. Device. Lett. **36**(12), 1344–1347 (2015)
2. Ni, H., Liu, F., Li, Z.: Design of high-speed automated fingerprint identification system based on SoPC. In: International Conference on Consumer Electronics, Communications and Networks, pp. 258–261 (2013)
3. Lee, C.J., Yang, T.N., Chen, C.J., et al.: Fingerprint identification using local Gabor filters. In: International Conference on Networked Computing and Advanced Information Management, pp. 626–631. (2010)
4. Chavan, S., Mundada, P., Pal, D.: Fingerprint authentication using Gabor filter based matching algorithm. In: International Conference on Technologies for Sustainable Development, pp. 1–6. IEEE (2015)
5. Garg, B., Chaudhary, A., Mendiratta, K., et al. Fingerprint recognition using Gabor Filter. International Conference on Computing for Sustainable Global Development, pp. 953–958. IEEE (2014)
6. Chen, L., Cui, B., Fu, T., et al.: The study of counterfeit classification method based on image features. International Conference on Information Science and Control Engineering. IEEE, pp. 512–516. IEEE (2015)

Cilin-Based Semantic Similarity Calculation of Chinese Words

Xianchuan Wang[1,2(✉)], Xianchao Wang[3], and Zongtian Liu[1]

[1] School of Computer Engineering and Science, Shanghai University, Shanghai 200444, China
xch_wang@shu.edu.cn
[2] School of Computer Science, Fuyang Normal University, Fuyang 236037, China
[3] School of Mathematics and Statistics, Fuyang Normal University, Fuyang 236037, China

Abstract. Word similarity plays an important role in text classification, information retrieval, information extraction, machine translation etc. We treat extended Cilin as knowledge base, the shortest path of the two words in Cilin is the main factor, and the number of nodes in branch layer and the distance of the words in the layer are the correction factor. The results indicate the values of similarity we got are consistent with the value of Miller.

Keywords: Cilin · Chinese word · Semantic similarity

1 Introduction

Word similarity plays an important role in the field of natural language processing. It is widely used in text classification, information retrieval, information extraction, machine translation and so on. The semantic similarity of words can be seen as a coincidence degree between words and the corresponding concepts in the objective world, which is denoted by semantic distance. The less the semantic distance is, the greater the coincidence degree is, the greater the corresponding semantic similarity will be. Otherwise, the less the semantic similarity is.

There are the two main methods to calculate word semantic similarity. One is corpus-based statistics [1] and another is semantic knowledge-based calculation [2–6]. The former counts the features of each word, and then the similarity between these features as the similarity of the two words, which count probability distribution of words. The results are objective and can indicate the similarities and differences in the syntactic, semantic and semantic of words, but it depend on a large number of corpuses, the computational methods are more complex and the computational complexity is larger. The latter usually use semantic knowledge base such as Cilin, Wordnet and Hownet etc. it calculate the similarity by using the hierarchy of the knowledge base. The results are simple, intuitive, and easy to understand, can indicate the similarity and differences of semantic. It do not depend on corpus, but need knowledge base, the results are subjective.

In this paper, firstly, we introduced the Cilin and the coding method of it (Sect. 2), then we used the Cilin as the knowledge base to calculate the semantic similarity between the two words. The shortest path of the two words in Cilin is the main factor, the number

© Springer International Publishing AG 2018
F. Qiao et al. (eds.), *Recent Developments in Mechatronics and Intelligent Robotics*,
Advances in Intelligent Systems and Computing 690, DOI 10.1007/978-3-319-65978-7_37

of nodes in branch layer and the distance of the words in the layer are the correction factor (Sect. 3), and finally, we compared our method with the other three methods with some words in Miller (Sect. 4).

2   Cilin

We calculated the similarity of words by Cilin that was extended by Information Retrieval Lab of Harbin Institute of Technology [7]. They deleted the words whose appearance frequency is less than 3. The number of the words in Cilin is extended from 39.099 to 77.343. The Cilin [8] organizes the words with tree structure and divides the words into five levels. The structure of the extended Cilin is shown in Fig. 1. There are 12 big classes, 97 medium classes 1400 small classes in extend Cilin. The description fineness of the extended Cilin is superior to that of the Cilin.

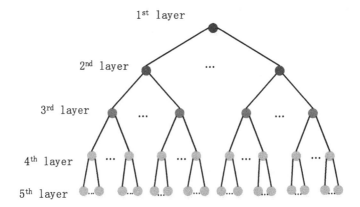

Fig. 1.   The layers of the extended Cilin

The extend Cilin respectively represents big classes, medium classes and small classes with capital letters, small letters and two decimal integers. And it respectively encodes the classes in the fourth and fifth level with capital letters and two decimal integers. The coding have five levels, which can denote the only word in extend Cilin.

In the fifth level in extended Cilin, the words in the same line are related, some words in the same word are synonyms and there is only one word in some line. In order to distinguish these three cases. We represent them with different markers. The coding in extend Cilin is shown in Table 1.

Table 1.   The coding of the extended Cilin

Bit	1	2	3	4	5	6	7	8
Marker	D	a	0	7	B	1	1	#, =, @
Meaning	Bigclass	Medium class	Small class	Word group		Atomic word group		
Layer	1st layer	2nd layer	3rd layer	4th layer		5th layer		

In Table 1, the coded bits are arranged from left to right. # denotes the words in the same line are related. = denotes the words in the same line are synonyms. @ denotes there is only one word in the line and there is on synonym or related word with it in Cilin.

3   The Word Similarity

Each big class is a tree structure with five level in extended Cilin. Therefor, the similarity between the words can be gotten by semantic distance in the tree structure. The semantic distance is mainly affected by the following three factors.

- The shortest path of the two words in cilin: l

The longer the shortest path of the two words in Cilin is, the less the similarity of the two words will be, the shorter the path is, the greater the similarity of the two words will be. The shortest path length is determined by the nearest common parent node of the two words in Cilin. The maximum depth of the extended Cilin is 5, therefore, the shortest path of the two words can be 2, 4, 6 and 8.

- The number of nodes in branch: n

The number of nodes in branch indicates the number of direct children of the nearest common parent node in extend Cilin, which indicates the density of the nearest common parent node. The greater n is, the more the direct children of the nearest common parent node will be, and greater the density will be, the less the semantic distance will be, and the greater the similarity between the two words will be. Otherwise, the less the similarity between the two words will be.

- The distance of the two words in branch: k

Words are classified and arranged according to a certain semantic order in extended Cilin. The greater the distance of the two words in branch is, the less the similarity of the two words will be. Otherwise, the greater the similarity of the two words will be.

If the two words do not belong to the same big class, we think the similarity of them is 0.1. In the condition that the words belong to the same big class, we treat the shortest path of the two words as the main factor, the number of nodes in branch layer and the distance of the words in the layer are the correction factor. And then, we got the semantic similarity of the two word with the formula (1).

The influence layer where the branch is on the similarity is different and the layer is given a different weight value. If the coding of the two words are the same and includes #, then we think the similarity of them is 0.5.

$$
\mathrm{Sim}(W_1, W_2) = \begin{cases} 1 & (W_1 = W_2 \text{ and } " = ") \\ m\left(1 - \dfrac{l}{20} + \dfrac{l}{20} * \dfrac{n-k+1}{n}\right) & (W_1 \neq W_2) \end{cases} \tag{1}
$$

Where m denotes the weight value of layer, which can be the following values: m = 0.65 for the second layer, m = 0.8 for the third layer, m = 0.9 for the fourth layer and m = 0.96 for the fifth layer [4].

Example 1:

Ib10B01 = injure, wound

Ib10B06 = casualty

In example, "Ib10B01=" and "Ib10B06=" have the difference in the 5th layer, then m = 0.96 and l = 2, there are 11 branch in the node "B" and n = 11. The distance between "01" and "06" is 5 and k = 5. Therefore, the similarity of the words is 0.9251, which is shown in the formula (2).

$$\mathrm{sim}(\mathrm{Ib10B01=, Ib10B06=}) = 0.96 \times \left(1 - \frac{2}{20} + \frac{2}{20} \times \frac{11 - 5 + 1}{11}\right) = 0.9251 \qquad (2)$$

4   Experiment and Discussion

We choose parts of Miller words that were processed by Zhu [6] to calculate the semantic similarity of words. The method we proposed in the text is compared with that of Liu [3], Tian [4] and Zhu [6]. Liuqun proposed Hownet-based method, Zhu Xinhua proposed Cilin-based and Hownet-based method. Tian Jiule proposed Cilin-based method. The results are shown in Table 2.

Table 2. The result about similarity of the words in Miller

Word1	Word2	Tian [4]	Liu [5]	Zhu [6]	Our method	Miller value
Car	Automobile	0.2119	1.0	0.9108	0.8880	0.98
Tourism	Travel	1.0	1.0	1.0	1.0	0.96
Secoast	Seaside	0.9577	1.0	0.9697	0.96	0.925
Noon	Midday	1.0	1.0	1.0	1.0	0.855
Food	Fruit	0.3091	0.1263	0.2822	0.5811	0.77
Tool	Instrument	0.1717	1.0	0.7653	0.6298	0.7375
Guys	Brother	0.6307	0.8000	0.5631	0.6393	0.415
Tourism	Car	0.1000	0.0741	0.1032	0.1000	0.29
Seaside	Forest	0.5495	0.1116	0.2807	0.6256	0.105

The values of the similarity between car and automobile, tool and instrument are relatively less with the method of Tian Jiule. The reason is that k is less and n is greater of the two groups words in Cilin. He think the similarity value is linearly negatively correlated with the ratio of them. Although we used the Cilin, the value of similarity is mainly determined by the shortest path and adjusted by the value of weight and adjustment factor, which improve the precision of the similarity. Compared with the method of Tian Junle, the value of similarity we got is more close to the value of Miller.

Compared with the method of Liu Qun, the results are not very different. However, there are some difference such as the similarity between seaside and forest, food and fruit. The reason is that they use different knowledge bases that have different semantic structures and have different focuses.

Compared with the method of Zhu Xinhua, although he combined Cilin with Hownet, the results of the two methods are not very different. In Table 2, the results of food and fruit by the mentioned methods are less than the value of Miller. The reason is that the nearest common parent nodes of them is in the 1st layer in extended Cilin. Compared with the three methods, the value of similarity we got is more close to the value of Miller.

5   Conclusion

The similarity calculation of words has been widely used in Chinese information processing such as text classification, information retrieval, information extraction, machine translation and automatic question and answer. We treat extended Cilin as knowledge base, the shortest path of the two words in Cilin is the main factor, and the number of nodes in branch layer and the distance of the words in the layer are the correction factor. We compared our method with the other three methods. The results indicate the values of similarity we got are consistent with the value of Miller.

Acknowledgements. The authors would like to thank the reviewers for their useful comments and suggestions for this paper. This work is supported by National Natural Science Foundation of China (61672006, 61273328, and 61305053).

Conflict of Interest. The authors claim that no conflict of interest exists in the submission of this manuscript, and the manuscript is approved by all co-authors for publication. None of the material in the paper has been published or is under consideration for publication elsewhere.

References

1. Zhan, Z., Liang, L., Yang, X.: Word similarity measurement based on Baidubaike. Comput. Sci. **40**, 199–202 (2013)
2. Wu, S., Wu, Y.: Chinese and English word similarity measure based on Chinese WordNet. J. Zhengzhou Univ. (Nat. Sci. Ed.) **42**, 66–69 (2010)
3. Qun, L., Sujian, L.: Vocabulary semantic similarity calculation based on HowNet. In: Proceedings of the Third Chinese Vocabulary Semantics, Taipei (2002)
4. Tian, J., Zhao, W.: Words similarity algorithm based on Tongyici Cilin semantic. J. Jilin Univ. (Inf. Sci. Ed.) **6**, 602–608 (2010)
5. Liu, Q., Li, S.: Words semantic similarity computing Basedon Hownet. J. Chin. Inf. Process. **7**, 59–76 (2002)
6. Zhu, X., Ma, R., Sun, L., Chen, H.: WordSemantic similarity computation based on Hownet and Cilin. J. Chin. Inf. Process. 29–36 (2016)
7. http://www.ltp-cloud.com/download
8. Jiaju, M.: Cilin. Commercial Press, Shanghai (1984)

Design and Implementation of Marine Engine Room Monitoring and Alarm System

Huibing Gan[✉], Bo Lv, and Guangsong Lu

College of Marine Engineering, Dalian Maritime University, Dalian, Liaoning, China
ghbzq@163.com

Abstract. The safety and reliability of the ship's monitoring and alarming system is of vital importance to the reliability and economy of the safe operation of the ship, improve the efficiency of the crew operation and management, and ensure the safety of the ship and all the personnel on board. The function of the engine room monitoring and alarm system was analyzed, the overall structure was designed, the hardware was selected and the software program was designed. A set of monitoring and alarming system was designed, which was based on PLC technology, OPC technology, WPF technology and database technology. The system is applied to the project of the land automation engine room. The results show that the system is stable and reliable, and the interface is friendly. It provides a certain reference to the research and design of engine room monitoring and alarm system.

Keywords: Engine room · Alarm and monitoring · PLC · WPF technology

1 Introduction

With the development of computer technology, the shipping industry is developing towards the direction of intelligent ship. The primary task of the intelligent ship is to make the operation on board easier and more uniform. The current emergence of more smart ship technology is mainly integrated bridge system [1–3]. Ship engine room monitoring and alarm system as an integrated ship bridge system is an important part of the realization of intelligent ship. The function of the ship cabin monitoring and alarm system is to realize the real-time monitoring of the running state and parameters of the machinery and electronic equipment in the cabin. When the system operates, if the system monitors the equipment operation faults or parameter exceeds the normal value of a certain range, the system will automatically according to the specific circumstances of the corresponding sound and light alarm, and indicate where the alarm is triggered, and print the alarm record. The marine engineers do not have to inspect the cabin at all times, as long as the monitoring room equipped with monitoring and control system will be able to understand the operation states of the mechanical and electrical equipment, in a timely manner to assist the duty officer to manage the ship's cabin equipment to improve the marine engineers' management efficiency, reduce work intensity, improve the working conditions of managers.

© Springer International Publishing AG 2018
F. Qiao et al. (eds.), *Recent Developments in Mechatronics and Intelligent Robotics*,
Advances in Intelligent Systems and Computing 690, DOI 10.1007/978-3-319-65978-7_38

2   The Overall Planning and Design System

The Marine engine room monitoring and alarm system is an important system of ship automation. The system mainly includes two aspects: software and hardware. The hardware of main control module of this system adopts SIEMENS S7 series of products as the core controller, through the acquisition of two sets of HP workstation display and processing cabin in the hard wired data and data communication, to achieve the key parameters of the signal on the electrical and mechanical equipment monitoring. Through the communication between PLC and upper computer to realize the real-time display, analysis and processing of the monitoring data of the key mechanical and electrical equipment in the engine room.

The system adopts the idea of hierarchical design, which is divided into three layers: executive layer, PLC control layer and upper computer monitoring layer. The system structure design is shown in Fig. 1.

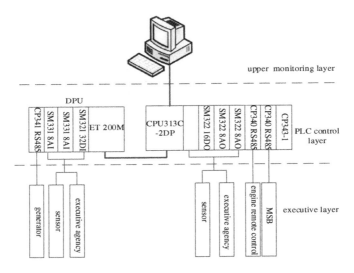

Fig. 1. System structure design

3   PLC Program Design of Monitoring and Alarm System

Lower computer program using STEP 7 programming software, STEP 7 include: trapezoidal table, statement list and function module of the three basic programming languages, and these can be converted to each other. This system adopts the idea of modular design, the function module of the system is divided into: switch hard wired data acquisition, analog connection data acquisition, communication data acquisition, communication console operation instructions, console output, alarm processing, instrument engineer safety alarm module. In STEP 7, the function of the corresponding module is edited, the function and the function block can be called in the main program OB1 to realize the function of the monitoring and alarm system. System program call planning as shown in Fig. 2.

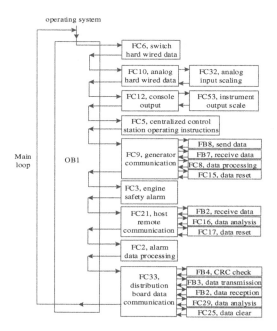

Fig. 2. PLC program call planning

Monitoring and alarm of equipment operating parameters and the abnormal state monitoring, equipment operating parameters or equipment failure, the system will automatically indicate the specific parameters of the device and send out the corresponding sound and light alarm. Sound and light alarm is the most basic function of the monitoring and alarm system. When the equipment is broken, the system can send out alarm sound by the centralized control console, the driver's control console buzzer and the alarm lamp post in the engine room. The alarm processing flow is referred to paper [1].

4   Upper Computer Software Design of Monitoring and Alarming System

The upper computer program of the ship engine room monitoring and alarm system is designed and developed by using the Visual C# language of Visual Studio 2013 platform with WPF (Windows Presentation Foundation) technology [4]. According to the characteristics of the operation of the monitoring and alarm system software, as well as the user's operating habits, the system software is divided into four areas, namely, the top functional area, alarm packet side column, navigation area and view display area as shown in Fig. 3. A desktop application takes a form as a unit, and an application corresponds to a main form that defines the visual boundaries of the application. The main form of the monitoring and alarm system is the framework of the whole system, the contents of which are filled by custom view, list view, trend view, Mimic system diagram. Scaling of visual elements of the content in the container in the view area.

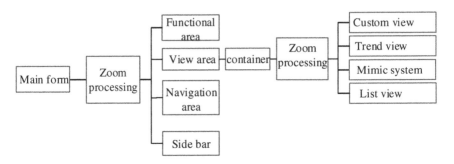

Fig. 3. The interface function division

Upper computer software using OPC technology [5–8], the use of server or client mode, the realization of communication with the hardware device. OPC server support interface access types are: custom interface and automation interface. This system client application uses C# language to realize.

The system software interface structure is shown in Fig. 4. After the system starts, the system enters the main interface, the user can access the table view, Mimic system diagram, trend view and custom view in the main interface. And then through the above four types of views to access a specific view.

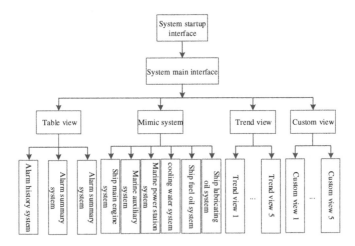

Fig. 4. Interface structure diagram

Figure 5 is shown the system main interface. The list view includes: alarm history, alarm suppression, active alarm points and all monitoring points. The Mimic interface is designed to monitor the parameters in the engine room more friendly, including main engine, generators, boiler and steam, lubricate oil system, fuel oil system and miscellaneous and etc.

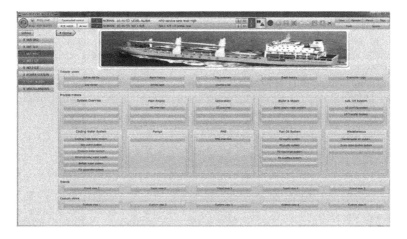

Fig. 5. System main interface

5   Conclusion

Through the development of the ship engine room monitoring and alarm system, all the functions of the design have been basically realized. The system can monitor the running state and parameters of the important equipment in the engine room. When the system is running, if a certain range monitoring system to monitor equipment operation failure or equipment operating parameters deviate from the normal set value, the system will automatically send out the specific situation of the corresponding sound and light alarm, display the alarm place, and alarm records. Timely and effective measures to manage. The research results have been successfully applied to the automatic engine room project and have good expected results.

Acknowledgment.   The study was supported by "Liaoning Provincial Natural Science Foundation of China" (No. 201602071), also supported by "the Fundamental Research Funds for the Central Universities" (No. 3132016316). The authors also gratefully acknowledge the helpful comments and suggestions of the reviewers, which have improved the presentation.

References

1. Gan, H., Ren, G., Zhang, J.: A novel marine engine room monitoring and alarm system integrated simulation. In: Proceedings of 2011 International Conference on Electronic and Mechanical Engineering and Information Technology, Harbin (2011)
2. Kongsberg K-Chief 600 Alarm and Monitoring System Operator Manul, Kongsberg (2013)
3. Pristrom, S., Yang, Z., Wang, J., et al.: A novel flexible model for piracy and robbery assessment of merchant ship operations. Reliab. Eng. Syst. Saf. **155**, 196–211 (2016)

4. Wang, J.G., Ji, Q., Jiang, X.Y.: The combined development of ogre and WPF based on its applications in the field of medical education software. Adv. Mater. Res. **1549**(403), 2296–2299 (2012)
5. Florea, G.G., Ocheană, L.A.: Towards total integration based on OPC standards. IFAC Proc. Vol. **45**(6), 1844–1849 (2012)
6. Li, D., Li, H., Zhang, K.: Development and realization of real-time data exchange between OPC client and multiple remote servers. Appl.Mech. Mater. **385–386**(1), 1655–1658 (2013)
7. Rosado-Muñoz, A., Mjahad, A., Muñoz-Marí, J.: Web monitoring system and gateway for serial communication PLC. IFAC Proc. Vol. **45**(4), 296–301 (2012)
8. Younis, M.B.: On the analysis of PLC programs: software quality and code dynamics. Intell. Control. Autom. **06**(1), 55–63 (2015)

Defining Limits of Resistance to Off-Line Password Guessing Attack and Denial-of-Service Attack in Multi-server Authentication Schemes

ChangYu Zhu[⊠] and Hong Wang

Xihua University, ChengDu 610039, China
zcy@xhu.edu.cn

Abstract. The offline password guessing attack and denial-of-service attack are two important security properties in multi-server authentication schemes. Recently, a large number of schemes have been published to against the two attacks. However, we find out that it is very difficult to against the two attacks at the same time. In this paper, the limits are pointed out using six recent authentication schemes as example. Then we analyze the reason of this case and give two conclusions to solve this problem. This conclusions can help us establish a more accurate provably secure model, so as to design a provably secure scheme in multi-server environment.

Keywords: Authentication scheme · Offline password guessing attack · Denial-of-service attack

1 Introduction

In recent years, many multi-server authentication schemes have been proposed based on the general multiple factors–password, smart card and biometric [1–6]. Password is susceptible to password guessing due to its low entropy. Secret information in smart card can be obtained by stealing smart card. Biometric can be easily copied, for example, fingerprint can be got by touching and iris can be obtained by requiring a camera [7]. Therefore, it is essential to resist the offline password guessing attack in multi-server authentication schemes even if the smart card and biometric are leaked. This is also called Multi-factor (assuming there are n factors) security which implies the protocol is still secure when $n - 1$ of n factors are lost.

The denial-of-service attack in multi-server authentication scheme mainly exploits the inefficient login phase or inefficient password update phase to increase the communication and computational cost of server. Although a wrong password in login phase or password update phase does not allow users to log the server, it is a serious drawback due to user's mistake causing denial-of-service attack. Moreover, once onetime mistake occurred in the password update phase, the user can no longer login the server using this same smart card. Therefore, multi-server authentication schemes should take in account of quickly detection mechanism to avoid the denial-of-service attack.

© Springer International Publishing AG 2018
F. Qiao et al. (eds.), *Recent Developments in Mechatronics and Intelligent Robotics*,
Advances in Intelligent Systems and Computing 690, DOI 10.1007/978-3-319-65978-7_39

Through the above discussion, we can see that the two attacks are important security attributes. Some of multi-server authentication schemes were claimed to resistance to the offline password guessing attack, some others pointed out these schemes cannot against denial-of-service attack and improved them. However, there are some papers pointed out those improved schemes cannot against offline password guessing attack and improved them. Again and again, a large number of papers about the two attacks have been published. Therefore, in order to terminate this conflict and establish a provably secure model, we point out it is very difficult to against the two attacks at the same time. In this paper, the limits of resistance to the two attacks have been pointed out by six recent examples. Due to space limitations, we give brief review on two examples and compared results on other examples. Finally, we demonstrate the cause and give two conclusions to solve this problem.

2 Brief Review and Analysis of Two Recent Schemes

2.1 He-Wang's Scheme

2.1.1 Brief Review

In He-Wang's scheme, smart card stores θ_i and z_i, where θ_i is the auxiliary string of biometric characteristic, which is used to restore biometric characteristic extractor, z_i is secret parameter stored in smart card. The user U_i has identity UID_i, password PW_i and biometric R_i.

In login phase of He-Wang's scheme, U_i inputs UID_i, PW_i and R_i into smart card. Smart card computes $\sigma_i = \text{Rep}(R_i, \theta_i)$, $X = xP$, $k_i = Z_i \oplus \text{H}(PW_i \| \sigma_i)$, $K_1 = xP_{pub}$, $CID_i = UID_i \oplus \text{H}(K_1)$, $h_1 = \text{H}(UID_i \| SID_j \| k_i \| X \| K_1)$, then U_i transmits $\{CID_i, X, h_1\}$ to the server. Upon receiving message, sever generates his own message and sends it combine with receiving message to trustworthy registration center RC. RC verify server at first, after the verification passed, he calculates $K_4 = kX = K_1$, $UID_i = CID_i \oplus \text{H}(K_4)$, $k_i = \text{H}(UID_i \| k)$, and checks whether $h_1 = \text{H}(UID_i \| SID_j \| k_i \| X \| K_4)$ holds or not. If it holds, RC verify server successful, and UID_i, PW_i and B_i are correct. Otherwise, login request is rejected.

In the password update phase, the user inputs old password PW_i and new password PW_i^{new}, then computes new secret value $z_i^{new} = k_i \oplus \text{H}(PW_i^{new} \| \sigma_i)$. Finally, smart card replaces old secret value z_i with z_i^{new}.

2.1.2 Security Analysis

(1) Denial-of-service attack

In login phase, if a user or an adversary inputs wrong password, the smart card sends this wrong message to sever. Then RC expends high computation cost due to using ECC point multiplication operation. Therefore He-Wang's scheme fails to the denial-of-service attack.

In password update phase, a user or an adversary inputs wrong old password PW_i' and PW_i^{new}. Then smart card computes new wrong secret value z_i^{new}, and replaces old z_i with z_i^{new}. Because there is no detection mechanism, the adversary or user will cause the denial-of-service attack. At the same time, the user can no longer login the server using this same smart card.

(2) Offline password guessing attack

If an adversary has obtained smart card, biometric R_i and message transmitted by public channel. The adversary can guess a candidate password PW_i' and computer secret value k_i'. However, he cannot verify whether PW_i' is correct or not. Therefore, He-Wang's scheme can resist offline password guessing attack.

2.2   Odelu et al.'s Scheme

2.2.1   Brief Review

In Odelu et al.'s scheme, the smart card store θ_i, z_i and s_i, where θ_i is the auxiliary string of biometric characteristic, z_i is secret parameter, s_i is password detection value. z_i and s_i are stored in smart card. The user U_i has identity UID_i, password PW_i and biometric R_i.

In the login phase of Odelu et al.'s scheme, U_i inputs UID_i, PW_i and R_i into smart card. Smart card computes the secret parameter $k_i = z_i \oplus H(PW_i\|\sigma_i)$ and the verification value $s_i = H(k_i\|UID_i\|H(PW_i\|\sigma_i))$. Because s_i is stored in the smart card, smart card checks whether UID_i, PW_i and R_i are correct by verifying the correctness of s_i.

In the password update phase, user inputs old password PW_i and computes $k_i = z_i \oplus H(PW_i\|\sigma_i)$, and checks whether $s_i = H(k_i\|UID_i\|H(PW_i\|\sigma_i))$ holds or not. If not, smart card rejects the request of password update. Otherwise, the user inputs new password PW_i^{new}, then computes new secret value $z_i^{new} = k_i \oplus H(PW_i^{new}\|\sigma_i)$, $s_i^{new} = H(k_i\|UID_i\|H(PW_i^{new}\|\sigma_i))$. Finally, smart card replaces old secret value z_i and s_i with z_i^{new} and s_i^{new}.

2.2.2   Security Analysis

(1) Denial-of-service attack

In the login phase, if a user or an adversary inputs wrong password, the smart card will quickly detect this wrong and reject the login request. Therefore, Odelu et al.'s scheme can resist denial-of-service attack caused by wrong password in login phase.

In password update phase, a user or an adversary inputs wrong old password PW_i', the smart card will detect this wrong and reject the password update request. This means that the password is correct once the password update successfully. Therefore, this scheme avoids to denial-of-service attack caused by the wrong password update.

(2) Offline password guessing attack

If an adversary has obtained smart card, biometric B_i and message transmitted by public channel. Then he can get $\sigma_i = Rep(R_i, \theta_i)$, z_i and s_i. The adversary computes $k_i' = z_i \oplus H(PW_i'\|\sigma_i)$, $s_i' = H(k'\|UID'_i\|H(PW_i'\|\sigma_i))$ by using guessing identity UID'_i and password PW_i'. Then he checks whether $s_i = s_i'$ are hold. If it is hold, the adversary finds the correct identity and password. Otherwise he repeats this steps until the correct identity and password are guessed. Therefore, we demonstrate that Odelu et al.'s scheme cannot against the offline password guessing attack.

3   Cryptanalysis

The existing multi-server authentication schemes are divided into two categories, local authentication like Odelu et al.'s scheme [2] and remote authentication like He-Wang's scheme [1]. The former category has password detection message in smart card. Before login request has been transmitted, smart card can detect whether password is correct. If it is wrong, login request will be rejected and login message cannot be transmitted. The latter category verifies the correctness of password in server or RC. If it is correct, server or RC will return message to smart card according to the scheme, otherwise, login request will be rejected.

Theorem 1. Multi-server authentication schemes using local authentication method can resist to the denial-of-service attack caused by wrong password. However, they are vulnerable to the offline password guessing attack unless the identity of the user used in login authentication has been participate in password detection and is enough random with high entropy.

Proof. Because login request can be detected before transmitted, it is obvious that multi-server authentication schemes using local authentication method can resist to the denial-of-service attack caused by wrong password. For offline password guessing attack, assuming detection function is $y = f(ID, PW, v_1, v_2, \ldots, v_{n-1}, d_1, d_2, \ldots, d_m)$, where ID is identity, PW is password, $v_1, v_2, \ldots, v_{n-1}$ are other factors in n factors authentication scheme in multi-server, $d_1, d_2, \ldots, d_m$ are secret data participated in password detection, which are stored in smart card or can be computed by existing data, and y is detection value stored in smart card. When ID equals to 0, it means that identity does not participate in password detection. In offline password guessing attack, once the adversary A obtains smart card and n-1 factors $v_1, v_2, \ldots, v_{n-1}$, A can get y and $d_1, d_2, \ldots, d_m$ via extracting from smart card. Then only ID and PW are unknown value. When ID equals to 0, the only unknown value is PW, hence A can guess password easily. When ID does not equal to 0, there exist two cases.

Case A. ID has weak strength with low entropy. The adversary A can obtain user's ID and PW according to the following procedure.

Step 1: A guesses a candidate ID' and PW', and computes

$$y' = f(ID', PW', v_1, v_2, \ldots, v_{n-1}, d_1, d_2, \ldots, d_m)$$

Step 2: A checks whether y' and y stored in smart card are equivalent. If they are equal, A can obtain ID and PW. Otherwise, A repeats the steps 1 and 2 until the correct ID and PW pair are obtained.

Case B. The security length of ID is long enough and random with high entropy. In this case, the adversary can not obtain PW using the same attack method presented in case A, since guessing a correct ID is computationally infeasible and the computational complexity is about $2^{-|ID|}$, where $|ID|$ is the length of ID.

Through above discussion, we can see that the conclusion 1 is correct.

Theorem 2. Multi-server authentication schemes using the remote authentication method may resist offline password guessing attack. However, they are vulnerable to the denial-of-service attack caused by wrong password in login phase or password update phase, unless computational cost of detection operation in server or RC is low.

Proof. Because smart card does not contain detection information, password will be detected after sever or RC receives the the login request. Therefore, the adversary cannot achieve password guessing when he obtains the other factors and smart card. It is important to know that this case only happens under the conditions that the adversary is unable to calculate password detection function while he has known the other factors, smart card and messages transmitted through communication channel. Therefore, Multi-server authentication schemes using remote authentication method may resist to the offline password guessing attack.

For the denial-of-service attack, because login request is not detected, server or RC will check the request using wrong password. This operation will increase computation cost of server or RC, especially when ECC point multiplication operation is used. Thus the adversary can achieve the denial-of-service attack. However, if the operation of checking login request in server or RC is very low, such as once hash operation, the computation cost of server or RC is low, so the denial-of-server attack cannot succeed.

Though above discussion, we can see that the conclusion 2 is correct.

We give further comparison results of more authentication schemes in Table 1. From Table 1 and the examples in Sect. 2, we can see that the correctness of conclusion 1 and conclusion 2.

Table 1. Further comparison results of six schemes

Security properties	Relate works					
	Ref. [1]	Ref. [2]	Ref. [3]	Ref. [4]	Ref. [5]	Ref. [6]
Offline password guessing attack	Yes	No	No	Yes	Yes	No
Denial-of-service attack	No	Yes	Yes	No	No	Yes

4   Conclusion

In this paper, we have discussed the limits of resistance to the offline password guessing attack and denial-of-service attack in multi-server authentication schemes. We have analyzed the weakness of He-Wang's scheme to the denial-of-service attack and Odelu et al.'s scheme to the offline password guessing attack. Moreover, we have proved the reason of this case and point out conclusions to solve this problem. At last, the ability of other six recent schemes to against the two attacks has been compared. The comparison has shown that resistant to the two attacks cannot be satisfied at the same time in these schemes. Then we demonstrate that authentication schemes using local authentication method can against the denial-of-service attack, but cannot resistance to offline password guessing attack unless the user identity is long and random enough, authentication schemes using remote authentication method may against the offline password guessing attack, but cannot resistance to denial-of-service attack unless the verification operation of server or RC is too low.Using the conclusion, we can design multi-server authentication scheme not only against offline password guessing attack and denial-of-server attack.

Acknowledgments. This work has been supported by the open research fund of key laboratory of intelligent network information processing, Xihua University (SZJJ2012-032).

References

1. He, D.B., Wang, D.: Robust biometrics-based authentication scheme for multiserver environment. IEEE Syst. J. **9**, 816–823 (2015)
2. Odelu, V., Kumar, A., Goswami, A.: A secure biometrics-based multi-server authentication protocol using smart cards. IEEE Trans. Inf. Forensics Secur. **9**, 1953–1966 (2015)
3. He, D.B., Zeadally, S., Kumar, N., Wu, W.: Efficient and anonymous mobile user authentication protocol using self-certified public key cryptography for multi-server architectures. IEEE Trans. Inf. Forensics Secur. **9**, 2052–2064 (2016)
4. Mishra, D.: Design and analysis of a provably secure multi-server authentication scheme. Wirel. Pers. Commun. **86**, 1095–1119 (2016)
5. Shen, H., Gao, C., He, D.B., Wu, L.B.: New biometrics-based authentication scheme for multi-server environment in critical systems. J. Ambient Intell. Human Comput. **6**, 825–834 (2015)
6. Xiong, L., Wang, K., Shen, J., et al.: An enhanced biometric-based user authentication scheme for multi-server environment in.critical systems. J. Ambient Intell. Human Comput. **7**, 427–443 (2016)

Clustering-Based Self-learning Approach for Security Rules in Industrial Communication Protocol

Ming Wan[✉], Wenli Shang, and Peng Zeng

Key Laboratory of Networked Control System Chinese Academy of Sciences, Shenyang Institute of Automation Chinese Academy of Sciences, Shenyang, China
wanming@sia.cn

Abstract. Modbus/TCP, which is a widely used industrial communication protocol, has serious security flaws because of its openness and simplicity, and developing security mechanisms based on Modbus/TCP is very hot topic. However, it is an onerous task to set rules manually for these security mechanisms. In this paper, we propose a clustering-based self-learning approach for security rules to facilitate the rule setting when carrying out the Modbus/TCP defense. Furthermore, our approach analyzes the address information from Modbus/TCP packets in depth, and automatically learns the address range setting in the white-listing rules by using the K-means algorithm. Our experimental results show that, the proposed approach is very available and effective to generate the white-listing rules for Modbus/TCP.

Keywords: Modbus/TCP · Self-learning · K-means algorithm · White-listing rule

1 Introduction

In recent years, with the development of modern industry, the application area of information communication technology has involved into the traditional industrial control systems. As a consequence, the original closure of industrial control systems has been broken, and the accompanying security issues are increasing exposed [1, 2]. According to the research reports announced by USA ICS-CERT (Industrial Control Systems Cyber Emergency Response Team), the number of security incidents in critical infrastructures has reached 295 in 2015 [3]. Although many kinds of security technologies have been designed in the regular IT system, the application of these technologies has certain limitations in current industrial control system [4].

Therefore, the security states in industrial control systems have attracted extensive attention, and both academia and industry start to research on the vulnerabilities and security mechanisms to protect industrial control systems. In these researches, communication control [5, 6] and intrusion detection [7, 8] are very effective security mechanisms, which have been successfully applied in various critical infrastructures, such as petrochemical plants and water distribution. Furthermore, one of common features is that they may adopt the white-listing rule setting to identify or prevent misbehaviors.

© Springer International Publishing AG 2018
F. Qiao et al. (eds.), *Recent Developments in Mechatronics and Intelligent Robotics*,
Advances in Intelligent Systems and Computing 690, DOI 10.1007/978-3-319-65978-7_40

However, the manual setting of white-listing rules not only is an onerous task, but also requires a lot of professional knowledge or experience.

Modbus/TCP is an open industrial communication protocol, and it has seized a dominant position in various industrial control networks due to its characteristics, including openness, simplicity and ease of development. However, it is also exposing out with security flaws [9, 10] increasingly. In particular, Modbus/TCP is an application layer protocol, which is based on the basic TCP/IP framework [11]. Therefore, it not only inherits the vulnerability of the traditional TCP/IP protocols, but also brings some new security issues, mainly including lacks of authentication, authorization and integrity detection [12]. In order to improve its security, most of the above security mechanisms have provided the defense measures for Modbus/TCP. Moreover, the setting of white-listing rules for Modbus/TCP is one of the most important links.

In order to automatically learn and generate the white-listing rules for Modbus/TCP, this paper proposes a clustering-based self-learning approach for security rules, which facilitates the white-listing rule setting. Furthermore, by analyzing Modbus/TCP packets in depth according to the protocol specification, this approach extracts the key contents and uses the K-means algorithm to learn the address range setting in the white-listing rules. The experimental results show that this approach is available and effective.

2   Modbus/TCP and White-Listing Rule

The basic packet format of Modbus/TCP mainly composed of three parts: MBAP header, function code and data field. Besides, for some frequently-used function codes, the data field also includes address field and value field. In general, when one Modbus master sends an operation request to one Modbus slave, function code is mainly used to distinguish all kinds of operations, and the data field indicates the specific addresses and values to perform one operation.

The common white-listing rule for Modbus/TCP [5] can be shown in Fig. 1, and it can be summarized as two parts: general options and deep options. More exactly, general options refer to three tuples, including source IP address, destination IP address and destination port. Besides, deep options aim to set Modbus application information, primarily including function code and address range. In particular, some more fine-grained white-listing rules may involve the value range setting, and although the proposed approach focus on the self-learning process of address range, it can also be used to the value range setting.

Options:	Source IP address	Destination IP address	Destination port	Function code	Address range
Case:	192.168.1.2	192.168.1.10	502	16	50-100

| | General options | | | Deep options | |

Fig. 1.   Common white-listing rule for Modbus/TCP

3   Clustering-Based Self-learning Approach for Security Rules

When automatically learning the white-listing rules for Modbus/TCP, this approach extracts the key information from captured Modbus/TCP packets, and these information mainly include source IP address, destination IP address, destination port, function code and the following addresses. However, because of the protocol specification, when one Modbus master performs the read or write operation, this operation can work on many addresses at the same time, that is, one Modbus/TCP packet may contain some consecutive addresses. Figure 2 shows an example of many addresses in different packets when the function codes are "01" (Read coils) and "16" (Write multiple registers), respectively. If we learn the white-listing rules for the function code "16", we cannot only generate one rule whose address range is $[R^i_{min}, R^j_{max}]$, but we should generate two rules whose address ranges are $[R^i_{min}, R^i_{max}]$ and $[R^j_{min}, R^j_{max}]$, respectively. Therefore, when generating one white-listing rule, we cannot simply set the address range from the minimum to the maximum under the same function code, because this setting may expand the scope of white-listing rules and leave some unnecessary vulnerability exploited by the attackers. In summary, in order to generate more effective white-listing rules, we use the K-means algorithm [13] to automatically classify the addresses under the same function code, and form different address clusters which represent the responding address ranges in different rules. The clustering-based algorithm can be described below:

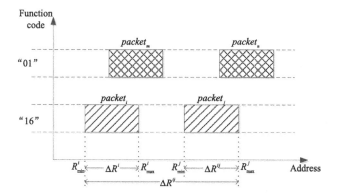

Fig. 2. Example of many addresses under different function codes

Step 1: Estimate the initial number k of white-listing rules under the same function code.

We first rearrange all addresses under the same function code in ascending order, and constitute the address sequence $r_1, r_2, r_3, \cdots, r_n$. And then, we calculate the distances of adjacent addresses, and obtain the distance sequence $d_1, d_2, d_3, \cdots, d_{n-1}$. At last, by calculating the number k of the distances larger than ϕ, we get the initial number of white-listing rules. Here, ϕ is a predefined value according to the actual network traffic. The formula to calculate the distances is the following:

$$d_i = |r_{i+1} - r_i|, \ i \in [1, n-1] \tag{1}$$

Step 2: Select k initial cluster centers according to the above number of white-listing rules.

(1) From the address sequence $r_1, r_2, r_3, \cdots, r_n$, we select the address who has the largest repeat rate as the first cluster center c_1.

(2) Select next cluster center c_j, $j \in [2, k]$. For each address r_i, we calculate the distance $D(r_i)$ between r_i and its nearest cluster center, and obtain the responding selection probability $p_i = D(r_i) \Big/ \sum\limits_{i \in [1,n]} D(r_i)$. Based on the selection probabilities of all addresses, we calculate the cumulative probability P_i of each address. After generating a random value in $(0,1)$, we define r_i as next cluster center if $Rand \in (P_{i-1}, P_i]$. $D(r_i)$ can be calculated as

$$D(r_i) = \left\| r_i - c_j \right\|^2, \ i \in [1, n], j \in [1, k] \tag{2}$$

Here, c_j is the nearest cluster center of r_i.

(3) Repeat the process (2) until all k cluster centers are found.

Step 3: Get the address bound of each rule by using the K-means algorithm.

(1) According to k cluster centers, we calculate the cluster $c^{(i)}$ by

$$c^{(i)} := \arg \min_j \left\| r_i - c_j \right\|^2, \quad i \in [1, n], j \in [1, k] \tag{3}$$

Here, $c^{(i)}$ represents the cluster whose center is nearest to r_i.

(2) For each cluster, we recalculate the corresponding cluster center by

$$c_j := \frac{\sum_{i=1}^{n} 1\{c^{(i)} = j\} r_i}{\sum_{i=1}^{n} 1\{c^{(i)} = j\}}, j \in [1, k] \tag{4}$$

(3) Repeat the above processes until two consecutive cluster centers have very small change. The change can be calculated by

$$E = \sum_{j=1}^{k} (c_{j,b} - c_{j,a})^2, b - a = 1 \tag{5}$$

Here, $c_{j,b}$ represents the current center of the cluster j, and $c_{j,a}$ represents the prior center of the cluster j.

(4) After obtaining all k clusters, the maximum and minimum is the address bound of each rule.

4   Experimental Analysis

In order to evaluate our approach, we build a small control system which is based on Modbus/TCP. As shown in Fig. 3, by sending different Modbus/TCP requests to PLC, the operator workstations can change the states of the valves. In particular, each valve corresponds to one address in PLC, and we simulate 100 valves in this control system. Besides, our approach is applied in one PC, and captures the Modbus/TCP packets from the monitor port of industrial switch. Moreover, we perform 12 experiments, and these experiments can be divided into three groups: the first group (4 experiments) uses contiguous addresses; the second group (4 experiments) uses discrete addresses; the third group (4 experiments) uses the mix of both. By these experiments, we evaluate the availability of the learned white-listing rules.

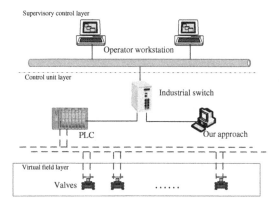

Fig. 3. Simulation control system based on Modbus/TCP

Figure 4 shows different accuracies to generate the white-listing rules in these three groups of experiments. From this figure we can see that, the first group has a very high accuracy, and the second group has the lowest accuracy. In other words, our approach is more specifically suited to learning rules from contiguous addresses. Besides, the average accuracy of these experiments reaches 83.03%, and the consuming time to learn rules from thousands of Modbus/TCP packets is far less than the one consumed by the manual rule setting. In the process of actual application, our approach can generate most of effective and available white-listing rules for Modbus/TCP, and save the working hours of the professionals.

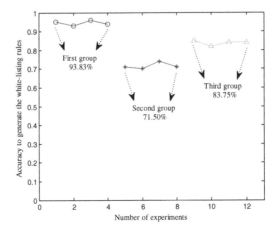

Fig. 4. Different accuracies to generate the white-listing rules under 12 experiments

5   Conclusion

In order to facilitate the Modbus/TCP white-listing rule setting, this paper proposes a clustering-based self-learning approach for security rules. Firstly, we briefly analyze the Modbus/TCP protocol specification, and explain the white-listing rule structure for Modbus/TCP. Secondly, we present the clustering-based self-learning approach in detail. Finally, we evaluate our approach by building a small simulation control system, and the experimental results show that our approach is available and effective to generate the white-listing rules for Modbus/TCP.

Acknowledgment.   This work is supported by the National Natural Science Foundation of China (Grant No. 61501447). The authors are grateful to the anonymous referees for their insightful comments and suggestions.

References

1. Genge, B., Fovino, I.N., Siaterlis, C., Masera, M.: Analyzing cyber-physical attacks on networked industrial control systems. In: Proceedings of 5th IFIP Advances in Information and Communication Technology, NH, USA, pp. 167–183 (2011)
2. Cheminod, M., Durante, L., Valenzano, A.: Review of security issues in industrial networks. IEEE Trans. Ind. Inform. **9**(1), 277–293 (2013)
3. ICS-CERT. ICS-CERT year in review 2015. https://ics-cert.us-cert.gov/Year-Review-2015 (2016)
4. Shao, C., Zhong, L.G.: An information security solution scheme of industrial control system based on trusted computing. Inf. Control **44**(5), 628–633 (2015)
5. Wan, M., Shang, W.L., Kong, L.H., Zeng, P.: Content-based deep communication control for networked control system. Telecommun. Syst. **65**(1), 155–168 (2017)

6. Cheminod, M., Durante, L., Valenzano, A., Zunino, C.: Performance impact of commercial industrial firewalls on networked control systems. In: Proceedings of 2016 IEEE 21st International Conference on Emerging Technologies and Factory Automation, Berlin, Germany, pp. 1–8 (2016)
7. Han, S., Xie, M., Chen, H.H., Ling, Y.: Intrusion detection in Cyber physical systems: techniques and challenges. IEEE Syst. J. 8(4), 1052–1062 (2014)
8. Zhu, B., Sastry, S.: SCADA-specific intrusion detection/prevention systems: a survey and taxonomy. In: Proceedings of the First Workshop on Secure Control Systems (SCS'10), Stockholm, Sweden, pp. 1–16 (2010)
9. Goldenberg, N., Wool, A.: Accurate modeling of Modbus/TCP for intrusion detection in SCADA systems. Int. J. Crit. Infrastruct. Prot. 6(2), 63–75 (2013)
10. Huitsing, P., Chandia, R., Papa, M., Shenoi, S.: Attack taxonomies for the Modbus protocols. Int. J. Crit. Infrastruct. Prot. 1, 37–44 (2008)
11. Modbus-IDA. Modbus messaging on TCP/IP implementation guide v1.0a. http://www.modbus.org/docs/Modbus_Messaging_Implementation_Guide_V1_0a.pdf
12. Wan, M., Shang, W., Zeng, P., Zhao, J.: Modbus/TCP communication control method based on deep function code inspection. Inf. Control 45(2), 248–256 (2016)
13. Singhal, G., Panwar, S., Jain, K., Banga, D.: A comparative study of data clustering algorithm. Int. J. Comput. Appl. 83(15), 41–46 (2013)

Study on Architecture of Wisdom Tourism Cloud Model

Fang Yu and Jiaming Zhong[✉]

Software and Communication Engineering School of Xiangnan University,
Chenzhou, Hunan, China
jmzhongcn@163.com

Abstract. This paper analyzes the connotative features and key technologies of wisdom tourism, illustrates the development status of wisdom tourism, analyzes its demand, and proposes some principles for its construction. Based on this, it designs the general constructing model for wisdom tourism, which can be divided into sensitive layer, IaaS infrastructure layer, PaaS platform supportive layer, SaaS application layer, and service layer. Through information standards and standardized systems, as well as the operation maintenance and safety system, the standard construction and operation maintenance for wisdom tourism can be safeguarded.

Keywords: Wisdom tourism · Information system · Cloud computing · Cloud model

1 Introduction

Tourism industry is the strategic important industry for national economy, while wisdom tourism is the key measure to promote the transforming and upgrading of tourism industry, so tourism industry has become the core of tourism informatization, as well as the popular issue for theoretical study.

Generally speaking, tourism informatization in China has undergone three stages: The first stage is professionalization, where the scenic spot and administration departments established their websites; the second stage is constructing digital tourism and digital scenic spot, where some distributed data integration management functions were achieved. Meanwhile, certain data sharing and service mechanism were established. The third stage is wisdom tourism, and also the intelligent stage. During the process of pushing forward the wisdom tourism, although great achievements have been made, there are also some problems: (1) Development lacks of unified planning. (2) Information lacks of effective sharing. (3) Application lacks of effective integration. (4) Users lack of unified interface.

© Springer International Publishing AG 2018
F. Qiao et al. (eds.), *Recent Developments in Mechatronics and Intelligent Robotics*,
Advances in Intelligent Systems and Computing 690, DOI 10.1007/978-3-319-65978-7_41

2 Basic Concept of Wisdom Tourism

2.1 Connotation of Wisdom Tourism

Based on information communication technology (ICT), wisdom tourism is a systematic and intensive management revolution that can satisfy tourists' individualized demands, and provide quality, as well as highly satisfied service, so as to sharing and effectively utilizing tourism resources and social resources. Viewing from connotation, the essence of wisdom tourism refers to the application of intellectual technology in tourism industry, including ICT. It is a modern project that aims at enhancing tourism service, improving tourism experience, innovating tourism management, and optimizing the utilization of tourism resources, to increase the competitiveness of tourism enterprises, improve the management level of tourism industry, and expand the scale of the industry. Wisdom tourism is a part of wisdom earth and wisdom city [1].

2.2 Features of Wisdom Tourism

Through the information technologies like cloud computing, virtualization, and internet of things (IOT), "wisdom tourism" helps tourism transfer from "traditional tourism" to "digital tourism", and then "wisdom tourism", changing the interactive ways of tourism resources. It integrates the tourism resources and application systems, to enhance the clarity, flexibility, and response speed of the application interaction, thus achieving a tourism model with wisdom service and management. Three core features of wisdom tourism are as follows: (1) Wisdom tourism focuses on providing appropriate individualized services, overall intelligent sensing environment, and comprehensive information service platform to different characters. (2) Wisdom tourism can achieve the interconnection and coordination in tourism service industry, so that the whole tourism, service, and management will be more convenient. (3) Through intelligent sensing environment and comprehensive information service platform, it can offer an interface for mutual exchange and sensing.

3 Demand Analysis for Wisdom Tourism [2, 3]

3.1 Demands from Competent Administrative Department of Tourism

The competent administrative department of tourism hopes to achieve the dynamic management, instant service, and publicity of the industry through information technology. For example, in terms of the industrial management for travel agency, and from the perspective of records management, the tourism bureau needs to understand the tourism groups of travel agencies at real time, such as the number of groups outside, the scheduling of each group, the number of group members, their contact ways, origins, and current position of the group. After mastering this real time information, the tourism bureau can provide favorable data support and help for tourism emergencies, supervision, complaint ruling, general resource scheduling, and urgent information releasing. However, currently, the information of tourism groups from travelling agencies is mostly sent to tourism bureau for records through non-electronic methods, such as fax, and most information is static, so the tourism bureau cannot get real time data. Therefore, when dealing with emergencies or complaints, the tourism bureau is usually passive. Viewing from the management and service of scenic areas, tourism bureau needs to understand the real time information of the scenic areas, such as the calculation of tourists flow during the golden time period, pre warning, and evacuating, as well as the tourist amount and reservation amount of each scenic spot. Nevertheless, the current information system between scenic spot and tourism bureau is relatively independent, so we cannot understand the information timely.

3.2 Demands from Tourism Enterprises

The information demand from tourism enterprises includes publicity and promotion; information exchanges between enterprises at lower reach and upper reach; and internal management of enterprises.

(1) **Publicity and promotion**
   In order to expand the marketing channels, and with the purpose of establishing information communication channels with tourists, publishing and booking tourist product information through the way of E-business can provide tourists with tourism consultation and booking service. Therefore, effective product information publishing channels can be established.

(2) **Information exchange between upper stream and lower stream enterprises**
   Travelling agencies integrate diet, accommodation, outing, travelling, shopping, and entertainment as an integrated product, and finally sell it to tourists. In the whole link, there are much coordination and complicated business links. The information exchange demand between travel agencies and tourists, tourist resource supplies, tour guides, as well as travel agencies is strong. For instance, tourism catering enterprises need to understand the scheduling and number of tourists in time, to arrange the catering and reception of tourism groups, and to reduce the waiting time of tourism groups. However, nowadays, such information is mainly communicated through telephone or fax, so the efficiency is obviously low.

(3) **Information of internal management and operation**
   When making daily operation, such as business management, human resource management, and financial management, enterprises should realize information management. Taking travel agencies as an example, they need to manage aiming at the enrollment of individual tourists or tourist groups, and also send the information of scheduling and group members to tourist guides. During the trip, tourist guides should inquire the information of scheduling and group members in time, while travelling agencies should deal with the complaints from tourists.

3.3 Demands from Tourists

With the skyrocketing development of internet and mobile internet, it is more and more convenient for tourists to obtain information, so the proportion of DIY tour is higher and higher. It has become increasingly popular for tourists to inquire the tourism information or reserve tourism products through internet and mobile apps, so wisdom tourism cloud platform needs to help tourists exchange information during the whole process. The specific introduction is as follows. (1) Information inquiry and reservation; (2) Fast and convenient information acquisition; (3) Tourists' evaluation.

4   Targets and Principles of Wisdom Tourism

4.1   Targets of Wisdom Tourism

The key of wisdom tourism cloud platform is to seek the data sharing and interaction between different systems through technical measures. Moreover, the information exchange between travel agencies can be realized through getting access to wisdom tourism cloud platform. Through this platform, reservation systems can connect with hotel internal management systems and the ticket management systems of scenic spots at real time, to update the information like remaining hotel rooms or tickets. Importantly, tourists can understand the latest information through reserving mobile apps. To share data of different enterprises and systems, the tourism cloud data center needs to solve two kinds of problems, namely the accessibility and comprehensibility. The accessibility can be achieved through constructing shared database, while comprehensibility needs to be achieved through establishing standard systems of data exchange in the system. In this way, the core of constructing tourism cloud data center is to achieve the following four targets: (1) Establish shared database in tourism system; (2) Establish standard system for data exchange; (3) Comprehensive unified management platform; (4) One-stop service demand.

4.2   Principles of Constructing Wisdom Tourism

(1) Principle of unified standards and resource sharing; (2) Principle of openness; (3) Principle of seamless integration; (4) Principle of advancement; (5) Principle of security.

5   General Structuring of Wisdom Tourism Cloud Model

A complete wisdom tourism cloud model should at least include sensitive layer, laaS infrastructure layer, PaaS platform supportive layer, SaaS application layer, and service layer [4–6]. It is shown as Fig. 1.

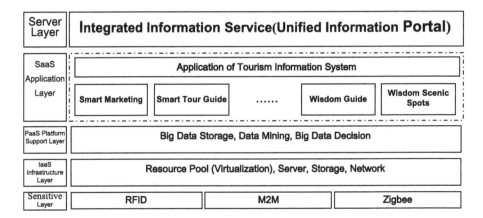

Fig. 1. Architecture of Intelligent Tourism Cloud Model

5.1 Sensitive Layer

Sensitive layer is combined with sensitive objects and sensitive technology. The sensitive objects include: personnel, device, resource; sensitive technology includes: one-card for tourism, real time monitoring, and radio frequency identification. Through applying sensitive technologies, personal life will be connected with wisdom tourism, so that we can clearly understand the dynamics and information of users.

5.2 IaaS Infrastructure Layer

IaaS provides service for hardware infrastructure, which can offer real or virtual computing, storage, and network resources in line with users' demands. During the process of utilizing IaaS layer service, users should provide the configuration information of infrastructure to IaaS service provider, the programming codes for infrastructure, and related users' data. IaaS is the basis of cloud computing. By constructing large scale of data center, IaaS layer can offer a large amount of hardware resources to cloud computing services at higher level.

5.3 PaaS Platform Supportive Layer

As the operating environment of cloud computing application, PaaS can provide deployment and management service to application programs. Through the software tools and development language at PaaS layer, application program developers only need to upload programming codes and data to enjoy the service, but not need to concern about the management of network, storage, and operating system at the bottom level.

5.4 SaaS Application Layer

SaaS is an application program that developed on the basis of cloud computing basic platform. The tourism information problem can be solved through SaaS layer's service, while the application layer mainly includes the individualized applications like wisdom marketing, wisdom tourism guide, wisdom scenic spot, wisdom shopping guide, and wisdom management. Through the application systems with different functions, it can offer individualized applications, which are good for users' work and life.

5.5 Service Layer

Wisdom tourism integrates different application services. When being used by users, it has already been an unified system facing to all users, while different systems are mutually connected and shared. Each user is no longer an "information island". But traditional software services are run independently, so that it has seriously restricted the application and construction of systems, and caused a large number of resource waste. When facing with different users, the application service in wisdom tourism can provide all kinds of intelligent services in accordance with users' demands, so that they can inquire the information such as hotel, scenic spot, business tour, position, transportation, and management through logging in unified portals. Therefore, it has enhanced the management efficiency and service quality.

6   Conclusion

According to the existing problems in the development of the current tourism information, the architecture of wisdom tourism cloud model has been framed through the use of the progressiveness and scalability of cloud computing. The model is divided into Sensitive Layer, IaaS infrastructure layer, PaaS Platform Supportive Layer, SaaS application layer and service layer. The specification of construction and operation maintenance for wisdom tourism can be safeguarded by building information standards and standardized systems, operation maintenance and security systems. It will form a tourism cloud platform and service system to unify management, sales, consumption and so on.

Acknowledgments. Thanks for being supported by the Science-Technology Planning Program of Hunan Province (2014SK3229; 2015SK20083). Natural Sciences Fund Project of Hunan Province (No. 2017JJ2241).

References

1. Zhang, L.Y., Li, N., Liu, M.: Basic concept and theoretical system of wisdom tourism. J. Tour. Trib. **27**(5), 66–73 (2012)
2. Qin, L.J.: Public tourism information service at tourism cloud times. J. Tour.Trib. **27**(2), 7–10 (2012)

3. Sun, C.G., Zhong, K.W.: Landscape pattern changes of coastal wetland in Nansha District of Guangzhou City in recent 20 years. In: Communications in Computer and Information Science. 3rd Annual 2015 International Conference on Geo-Informatics in Resource Management and Sustainable Ecosystem, GRMSE 2015, vol. 569, pp. 408–416 (2015)

4. Zhou, X.B., Ma, H.J., Miao, F.: Study on tourism cloud framework model based on cloud computing. J. Chongqing Normal Univ. (Nat. Sci.) **30**(2), 79–86 (2013)

5. Luo, J.Z., Jin, J.H., Song, A.: etl: cloud computing: system structuring and crucial technology. J. Commun. **32**(7), 3–21 (2011)

6. Wang, P.P.: Study on demands and general framework of wisdom tourism cloud computing service platform. J. Telecommun. Sci. **11**, 61–65 (2014)

Application of Cloud Platform in Nursing Clinical Teaching

Haiyan Deng[1], Zhihui Liu[2], Zhijuan Liu[3(✉)], and Xiaoran Qiu[1]

[1] The School of Nursing, XiangNan University, Chenzhou 423000, Hunan, China
[2] School of Foreign Languages, XiangNan University, Chenzhou 423000, Hunan, China
[3] The School of Public Health, XiangNan University, Chenzhou 423000, Hunan, China
3073143833@qq.com

Abstract. According to the characteristics of *Nursing Clinical Teaching*, the self-directed teaching mode based on the Cloud Platform space is constructed. The experimental shows that this learning method excels over others in terms of improving learning interest, initiative, expression skill, team-work spirit, communication ability, critical thinking ability, ability to solve problems, professional quality and patient satisfaction. At last, this paper comes up with implementation countermeasures for applying self-directed learning to nursing clinical teaching.

Keywords: Cloud Platform Space · Nursing · Clinical teaching · Self-learning

1 Introduction

Nursing clinical teaching is the bond that nursing students from theory to practice, it can improve the operation skills and comprehensive quality of nursing students, shorten the distance between the teaching and clinical, occupies an important position in nursing teaching. But the present actual situation is that pay attention to the knowledge of nursing teaching and despise the cultivation of critical thinking ability, nursing students in clinical, relies too heavily on the doctor's advice, the lack of the independence of the thinking and reasonable doubts, the problem of exposed the weakness of the integrated use of knowledge, the traditional "teacher-centered" and "centered on the books" type of nursing education mode is not conducive to the cultivation of comprehensive ability of nursing students, is not conducive to the development of nursing students [1, 2]. So to explore a new mode of nursing teaching is imperative.

2 Teaching Mode of Nursing Clinical Teaching Based on Cloud Platform

2.1 Creation of Teaching Environments

In the traditional class teaching, teachers and students, between students can conduct face-to-face communication, and teachers can effectively control the teaching process, but also can help students to adjust the learning process [3]. Therefore, making full use

F. Qiao et al. (eds.), *Recent Developments in Mechatronics and Intelligent Robotics*,
Advances in Intelligent Systems and Computing 690, DOI 10.1007/978-3-319-65978-7_42

of the advantages of the traditional classroom teaching and space teaching to create teaching situation.

2.2 Positioning of Teacher-Student Role

Based on the Self-learning of cloud platform space, the role of teachers and students change. Teaching activities, teachers are the developer of teaching resources, teaching activities of the designers, organizers and collaborators, the guidance of student learning, teacher's work focused on how to stimulate students to think and explore [4]. While students from passive knowledge recipients into the master of learning, active exploration, active interaction, collaboration and sharing, to enhance their ability to explore.

2.3 Effective Integration of Teaching and Learning Strategies

Based on the cloud platform space, the Self-learning will integrate the individual teaching strategies, the group teaching strategies and the class teaching strategies effectively. Individualized teaching strategies emphasize students' individual differences, focusing on students' independent inquiry, and the students' self-control, adjustment and evaluation. Group collaborative learning, to motivate students' knowledge desire, and fully form students' learning enthusiasm. Meanwhile, it can cultivate students' interpersonal skills, team spirit, as well as cooperation ability, and help them obtain common goals [5–7].

2.4 Perfect Teaching the Link

Using the cloud platform space rich interactive mode and digital teaching resources, can be very good to improve the class before and after class teaching link, Including preparation before class, internalization in class, and deepening after class.

3 Analysis on the Teaching Effect

This paper secrets 44 people of nursing class 1 of 2013 as the experimental group and 46 people of nursing class 2 as the control group. They are students enrolled based on national college entrance and are randomly divided into different class, so there is no statistical difference between these two group in terms of learning record, sex, age. The experimental group adopts self-directed learning, which encourage students to take advantage of the learning resources around them to carry out self-directed earning based on learning schedule. What's more, each Friday, there will be group discussion with nursing students as the center to focus on relative knowledge and nursing measures of common disease in terms of internal medicine as well as surgery. Furthermore, teachers plays a role in leading the direction of the discussion and creating the atmosphere. After the discussion, the tutor will carry out a brief summary and ask students to reflect on themselves. The control group adopts traditional teaching method. By the way, the tutors, total class hours, weekly hours as well as test contents are same in order to ensure that there is no statistical difference in terms of general information. So, these two groups are comparable.

After the teaching, there will be questionnaires on these two groups, which are collected by secret ballot. 90 questionnaires sent are all effective. The main investigation terms include learning interest, initiative, self-learning ability, expression skill, teamwork spirit, communication ability, critical thinking ability, ability to solve problems, professional quality and patient satisfaction etc., and the result is shown in Table 1.

Table 1. Compared two groups of students learning ability

Items	Interest in learning	Initiative of learning	Ability to Study Independently	Ability of expression	Cooperative spirit of group	Ability to communicate	Ability of critical thinking	Problem solving skills	Occupational qualities	Satisfaction of patients
Experimental group	93.36	90.35	89.71	89.96	90.55	91.78	93.67	96.65	96.33	95.39
Control group	88.92	87.63	87.35	86.22	87.91	88.12	89.11	88.89	91.09	90.81
χ^2 P value	<0.01	<0.01	<0.05	<0.01	<0.01	<0.01	<0.01	<0.01	<0.01	<0.05

Every record of experimental group is obviously higher than the control and the total record is superior.

At the same time, there will be theoretical and operation closed-book exam on experimental group and control group. The written test takes basic theory and application as the testing sites and the topics include subjective and objective ones based on hundred-mark system. All the collected materials will be organized and set into a data base while the statistic software SPSS 19.0 will be adopted to carry out descriptive statistics and t test as well as χ^2 test. The academic record analysis of two methods is shown in Table 2.

Table 2. Compared two groups of students course examination results

Items	Theory test			Operational testing		
	Average score	Excellent rates	Passing rate	Average score	Excellent rates	Passing rate
Experimental group	82.6	39.85	95.12	86.7	41.13	97
Control group	68.3	16.75	92.31	67.5	17.66	93.31
χ^2 P value	<0.01			<0.05		

The score of these two group is in normal distribution and the average score, excellent rate as well as passing rate of the experimental score are higher than the control group.

4   Discussion

The Self-learning based on the network learning space can promote the reform of teaching mode, effectively play the leading role of teachers and students, and has good application value.

4.1 Effects of Cloud Platform Space in Nursing Education

As Cloud Platform Space teaching is featured by intense infectivity and vividness, it can attract students knowledge mind and interest. In this way, it proves solid conditions for the mastering of knowledge and the development of skills.

(1) **Stimulate Interest in Learning, Improve Ability of Cognizance**

Cloud Platform Space can assist teachers presenting case analysis, promoting complicated phenomenon, and organizing different interactive activities to solve puzzles. For example, during the teaching process, pictures and texts could be combined together through applying Cloud Platform Space. Under such circumstance, requirements of politeness, spoken language and behavior, together with operating key points of nurses will be expressed clearly by pictures. Therefore, it tells us that the teaching will be more intuitive and interactive. While appreciating beautiful screens, students would be easier to understand knowledge and accept abstract specification clauses. Not only can it motivate students the interest of learning, but also attracts their curiosity for the exotica of Cloud Platform Space. As a result, it will enhance their learning achievement motivation, increase the ability of cognizance, and contribute to academic improvement.

(2) **Emphasize Key Points, Breakthrough the Difficulty**

The presentation of Cloud Platform Space will let so many abstract and difficult teaching contents lively. Through this platform, complicated matters can be simplified, the time can be shortened, close relationship can be drawn, microscopic things can be magnified, while macroscopic things may be minimized. Such as, in the chapter of "Nursing Process", different graphics and colors can be used to emphasize key points [8]. By doing so, content of courses will be dynamic rather than still, concrete rather than abstract, as well as intuitive but not vague.

(3) **Enhance Interaction between Teaching and Learning**

Cloud Platform Space can assist teachers defining the interaction of online study, and offering more interactive channels of communication along "teaching" and "learning" process. Not being restricted by time, students can gather stored videos of course from teachers or some special content of seminars. Meanwhile, they will give opinions to exchange the knowledge gained ground on the Internet (for example, BBS), and negotiate about study plans in online chat rooms by the means of characteristics, voices and pictures. This technology, under the context of colorful educational surroundings, can provide teachers and students with extensive and in-depth interaction technology platforms by breaking through the limitation of time and space.

(4) **Visualize the Abstract Content**

Cloud Platform Space teaching can present some lively scenes of actual clinical nursing procedures. It also shows interflow communication between nurses and patients and the feeling after making contact with patients to students, and offers a great deal of clinical information. With regard to difficult, uninteresting and nonobjective contents with poor imitation during practical operation, physical pictures and emulational three-dimensional animation demonstration in the teaching videos can be expressed to help students combine theory with practice. As a result, sense experiences will be acquired [9].

(5) **Satisfy the Needs of Personalized Education**
Cloud Platform Space can make sure that students can obtain corresponding research files through the internet while not being influenced by time and space. Students can also learn online and acquire frontier ideas and knowledge from educational field.

4.2 Countermeasures for Applying Self-directed Learning in Nursing Clinical Teaching

4.2.1 Update Ideas

Always keep in mind the concept of people-orientation, comprehensively develop all students. The development of students should not only pay attention to general, comprehensive and proactive prestige but also the difference and continuity. While emphasizing that students should master the basic knowledge and skills, their ability to learn and communicate with other people, their innovative and practical skills should also be paid attention to so as to make them satisfy the requirements of economy and society in the new era. Advocate independent, inquisitive and communicative learning method. The teaching process refers to the interaction communication and mutual development between teachers and students which is able to build a humanistic, harmonious, and equal reaction between them. Teachers are helpers and tutors for students who organize and implement learning activities, always pay attention to the learning process and give guidance as well as feedback. With the continuous assistance of teachers, students can communicative with them based on different approaches, report their results with feedback and raise questions.

4.2.2 Specific Learning Goals

Teachers should help students to make their determination to carry out self-directed learning and come up with overall, hierarchical, dynamic and inspirational goals based on facts so as to improve learning efficiency.

4.2.3 Promote the Deep Integration Between Information Technology and Educational Teaching

Set up efforts on information technology training, improve the information quality of teachers and students. Positively develop, enrich and perfect teaching resources to digitize them and put them on internet to fully apply information technology and internet space into teaching.

4.3 Foster a Good Learning Atmosphere

The teaching environment should be bright, comfortable, simple and liberal while the teaching equipment should be conducive to the communication between teachers and students. What's more, students should be mutually friendly, respectful, supportive, responsible and trust each other so that they can have a positive learning environment with both cooperation and competition.

Acknowledgement. 2013 Hunan Province Colleges and universities teaching reform research project (based on network space "management information system" inquiry teaching research and practice), Twelfth Five-Year Plan topic of Hunan Education and Science (Empirical research on the self-directed learning of nursing clinical teaching XJK012CGD032).

References

1. Li, X., Li, J.: Control and thinking of chinese and foreign nursing teaching. J. Nurs. Sci. **19**(9), 19–21 (2004)
2. Zha, T.: Analysis on the current condition of internal clinical nursing teaching. J. Clin. Exp. Med. **11**(11), 181 (2007)
3. Knowles, M.S.: Self-directed learning: a guide for learners and teachers. The Adult Education Company 9–58 (1975)
4. He, Q., He, J.: Discuss on the concept of self-directed learning. J. Hebei Normal Univ. (Educ. Sci.) **11**(2), 33–36 (2009)
5. Wu, Z., Zhao, L.: On the teaching mode of flipped classroom based on the network learning space. China Educ. Technol. **04**, 121–126 (2014)
6. Zhang, J.: Analysis of the key factors of 'flipped classroom' teaching mode. China Remote Educ. **10**, 59–63 (2013)
7. Zheng, Y., Li, L., Wang, Y.: Construction and application practice of 'Mixed Type – Inquiry'. Mod. Remote Educ. **03**, 50–54 (2010)
8. Liu, Z., Zhong, J.: Application of streaming media technology in nursing teaching. In: 2011 IEEE International Symposium on IT in Medicine and Education (2011)
9. Wang, R., Bai, Q., Hua, Z.: Teaching practice of course based on constructivism. Mod. Educ. Technol. **13**(19), 256–259 (2009)

A Multi-objective Network-Related Genetic Algorithm Based on the Objective Classification Sorting Method

Lianshuan Shi$^{(\boxtimes)}$ and Li Jia

School of Information Technology Engineering,
Tianjin University of Technology and Education, Tianjin, China
`shilianshuan@sina.com`

Abstract. Based on adaptability of the Agent on the environment and the competition and cooperation relations between Agents, and combine with the characteristics of network optimization problem, an improved multi-objective genetic algorithm for the network shortest path optimization problem is given. The chromosome of traditional GA is replaced by agents, to perform the evolutionary operation. By constructing proper encoding and decoding strategy, and use a method based on objective classification sort method to complete the competition behavior of agent. Through the self-learning operations of agents, further improve the performance of genetic algorithm. Finally, through the example test, it shows that the algorithm can obtain better optimal solutions.

Keywords: Genetic algorithm · Multi-objective network optimization · Multi-agent system · Objective value classification sort

1 Introduction

Genetic algorithm is a global optimization algorithm which mimic natural evolution process. Nowadays, and has been widely used in all kind of fields [1]. Genetic algorithm is a very effective algorithm to solve functional optimization problem. Sometimes it maybe does not find the optimal solution, but it can find a better solution [2]. The genetic algorithm tends to be trapped in local optima when the dimension of the function is higher. In recent years, many scholars proposed a genetic algorithm based on the agent to solve the problem of high dimensional function. The genetic algorithm based on the agent is the algorithm which combined the multi-agent technology with the traditional genetic algorithm. It makes each chromosome in the traditional genetic algorithm into the agent which only perceives the surrounding environment and can respond to the change of the surrounding environment [3]. So it has a good performance in solving the function optimization problem with high dimension. In this paper, a shortest path network optimization problem is considered. A multi-objective genetic algorithm based on Agent is given to solve this problem. The two optimization goals, the minimum cost and minimum delay, are considered. Test results shown that the improvement algorithm has good preferment.

© Springer International Publishing AG 2018
F. Qiao et al. (eds.), *Recent Developments in Mechatronics and Intelligent Robotics*,
Advances in Intelligent Systems and Computing 690, DOI 10.1007/978-3-319-65978-7_43

2 Description of the Problem

A network optimization problem also can be described as a multi-objective 0–1 integer programming problem [1, 2, 4]: find an optimum path P_{min}, minimizing the cost C and delay D

$$\min C = \sum_{(i,j) \in E} c_{ij} * x_{ij}$$

$$\min D = \sum_{(i,j) \in E} d_{ij} * x_{ij}, \quad x_{ij} = \begin{cases} 1, (i,j) \in P_{min} \\ 0, (i,j) \notin P_{min} \end{cases}$$

If the node i and the node j is connected, $x_{ij} = 1$, on the contrary, $x_{ij} = 0$. c_{ij} $(c_{ij} > 0)$, d_{ij} $(d_{ij} > 0)$ are the cost and delay corresponding to arc (i, j) are respectively.

3 Algorithm Ideas

In this algorithm, the individual (chromosome) of the traditional GA is replaced by an agent, and the evolutionary operation is used for agents. The agent can exchange information with the environment. At the same time, these agents influence with each other in the evolution process, and can produce cooperation behaviors between the agents, finally form mutual adaptation between agents, as well as the adaptation between agent and environment [6]. In network optimization problem, a chromosome is called an agent, the agent is a feasible path after decoding. The agent contains attributes are: the serial number, the objective 1, ..., the objective m, and the energy value. In the algorithm, the coding scheme is used based on index [7], all path is effective after decoding. The case that $x_{ij} = 0$ is not exist.

4 Algorithm Design

4.1 The Coding Strategy

In this article, a real number encoding strategy is adopted. The coding scheme is based on the index [9]. An agent is generated by generated n random numbers from 1–n, where, n is the number of the network nodes. Although the maximum path chromosome encoding mechanism [1] in genetic algorithm has the completeness and can perform corresponding coding for any path, but it easy produce some illegal individuals by genetic operators, that need to use special operator to adjust. This article uses the encoding scheme based on index, the gene location is used as the index of network nodes. In the process of building path, the genetic value is used decided which node in current candidate nodes is selected as next node of path. By selecting the maximum index of genetic value as a decoding method to generate a valid path. The advantage of this encoding and decoding principle is: (1) any coding string can be decoded to generate a valid path; (2) all the valid path can be achieved by coding; Therefore, compared with other coding way, the feasibility of the proposed coding method is better.

4.2   Objective Value Classification Sort Method

In the improved algorithm, a new method called the objective classification sort (Pareto competition hierarchical) algorithm is used to complete the competition behavior of agent. In the competition, it was assumed that all agents live in the agent grid. When the scope of each agent's perception is 1, there are eight agents in the competition neighborhood of each agent. The purpose of the classification is to form a new different competitive field. The same level agents are at same field, the different level agents have different fields. Through competition, the agent with high energy has higher selection probability in the neighborhood, the agent with lower energy value has lower selection probability. By this intelligent routing strategy, the agents with lower energy are eliminated. For example, a group size = 18, that means the group contains 18 agents. Each Agent has two objectives, cost and delay.

R[18][2] = {{97,88},{86,21},{21,20},{57,80},{43,43},{43,34},{49,81},{96,31}, {49,27},{10,50},{14,38},{29,88},{78,96},{46,77},{59,14},{50,36},{12,87}, {43,52}}.

The main idea of the proposed Pareto competitive classification method is given as follows.

(1) Make an ascending sort according to the value of objective function 1, after sorting, marked the first individual, where denoted as A; also, marked the identical individuals with the first individual; (2) Find irrelevant individuals with the individual A in remaining individuals, these means that find those individuals which objective value 2 is less than objective value 2 of individual A in group except A; If an individual is irrelevant with A, then marked this individual; (3) Continue to compare until the number of remaining individuals equal to zero. The first level non-dominated individuals are obtained. Then loop executes step (1) to step (3), until all individuals in the group have been graded. If there are m objectives, then the rest $m - 2$ objectives also can be used to classified based on the above method.

For the above example with 18 agents, the Pareto competitive grading process is given as follows, where, the number outside the number in the begging. Namely, the agent's serial number (the objective value1, the objective value 2) (Table 1).

Table 1. The agents in group

1 (97,88)	4 (57,80)	7 (49,81)	10 (10,50)	13 (78,96)	16 (50,36)
2 (86,21)	5 (43,43)	8 (96,31)	11 (14,38)	14 (46,77)	17 (12,87)
3 (21,20)	6 (43,34)	9 (49,27)	12 (29,88)	15 (59,14)	18 (43,52)

Firstly, an ascending sort according to the objective 1 (cost value) are given for the group. The sorting result is given in Table 2.

Table 2. The sorting result for group

10 (10,50)	3 (21,20)	6 (43,34)	7 (49,81)	4 (57,80)	2 (86,21)
17 (12,87)	12 (29,88)	18 (43,52)	9 (49,27)	15 (59,14)	8 (96,31)
11 (14,38)	5 (43,43)	14 (46,77)	16 (50,36)	13 (78,96)	1 (97,88)

Mark 10 (10, 50) shown in the Table 3.

Table 3. The irrelative individuals with 10 (10, 50)

10(10,50)	3(21,20)	6(43,34)	~~7(49,81)~~	~~4(57,80)~~	2(86,21)
~~17(12,87)~~	~~12(29,88)~~	~~18(43,52)~~	9(49,27)	15(59,14)	8(96,31)
11(14,38)	5(43,43)	~~14(46,77)~~	16(50,36)	~~13(78,96)~~	~~1(97,88)~~

All irrelative individuals with 10 (10, 50) have been found, which are these individuals in the Table 4 without the strikethrough. Then comparing continue. The first individual 11 (14, 38) is considered, its objective 1 value is 14, not equal to 10, then mark the individual that shown in the Table 4.

According to the above steps to continue, until all the individuals are graded. In the end, all 18 individuals are divided into 7 level. At same level agents meet irrelative, means that the individuals do not dominate each other. The advantage of the proposed Pareto competitive grading sorting method are: (1) when the individuals have the same objective value 1, no matter how about the sort results of individuals base on objective1 in the first time, the final results of the classification are unaffected, which is to ensure the uniqueness of the classification results; (2) no matter the objective value in the sorting process is in the first or second order, the classification results will not be affected, but also to ensure the uniqueness of the classification results; (3) the computational complexity is lower. The algorithm need only compare the individual whether meets irrelevant condition with last individual.

4.3   Selection Behavior

In order to speed up the convergence rate of the population, according to the level of an agent, assign a virtual fitness. The agents in different neighborhood have different virtual fitness values. The roulette wheel selection method is used to select crossover individuals. The first level agents have the maximum virtual fitness. The virtual fitness become smaller and smaller with the level become bigger.

4.4   Crossover Operation

After crossover operation, we hope that the individuals can uniformly distributed in the search space, to improve the quality of the optimization of genetic algorithm. Also, in order to avoid generate infeasible individual, we need consider using the appropriate encoding method and crossover strategy. In the group, it was assumed that an Agent crossover with other agents in the field in the probability Pc. In the process of crossover, we need to avoid generating illegal individuals after crossover operator. Suppose that the agent A crossover with the agent B ∈ neighbors (A). Firstly, the gene sections of parent agent A and B are swapped. In order to guarantee the validity of the offspring agents after crossover, a index mapping relation is established to reconstruct swapping genes sections. It can be seen from the results after crossover operation, the crossover operator can effectively avoid the illegal individuals.

4.5   Hybrid Mutation

Mutation can increase the diversity of population, to obtain the global optimal solution. Two mutation methods are used, one is the inserting method based on the location, another is the exchanging mutation method based on the order. Suppose probability is p_m, then when the Rand $(0, 1) < p_m$, perform mutation operation. If Rand $(0, 1) < 0.01$, then perform the inserting mutation; Otherwise, perform swapping mutation.

4.6   Self-learning Behavior

The purpose of self-learning is to improve agent's energy, to enhance the competitiveness. The self-learning strategy is uses the structure of agent neighborhood to find a better agent in this agent neighborhood. That is a local climbing operation.

5   The Experiment and Analysis

A network optimization problem is given which cost and delay matrixes shown in Tables 4 and 5.

Table 4. The cost matrix   Table 5. The delay matrix

0	11	27	11	0	0	0	0	0	0	0
0	0	18	0	15	21	0	0	0	0	0
0	0	0	10	0	30	24	0	0	0	0
0	0	0	0	0	0	41	0	0	0	0
0	0	0	0	0	0	0	25	0	0	0
0	0	0	0	30	0	0	16	26	36	0
0	0	0	0	0	0	27	0	0	17	0
0	0	0	0	0	0	0	0	28	0	18
0	0	0	0	0	0	0	0	0	39	29
0	0	0	0	0	0	0	0	0	0	40
0	0	0	0	0	0	0	0	0	0	0

0	10	15	3	0	0	0	0	0	0	0
0	0	23	0	4	16	0	0	0	0	0
0	0	0	20	0	6	10	0	0	0	0
0	0	0	0	0	0	19	0	0	0	0
0	0	0	0	0	0	0	20	0	0	0
0	0	0	0	15	0	0	2	30	10	0
0	0	0	0	0	0	20	0	0	19	0
0	0	0	0	0	0	0	0	19	0	5
0	0	0	0	0	0	0	0	0	30	12
0	0	0	0	0	0	0	0	0	0	10
0	0	0	0	0	0	0	0	0	0	0

The index encoding is used to generate Agents. Firstly, n different random numbers from 1 to n are generated, to constitute an Agent. The maximum path decoding method is used to generate a valid path. The algorithm parameters includes: the population size is 300, the number of iterations is 100, the crossover probability and mutation probability are randomly generated in interval [0.7, 0.9] and interval [0.001, 0.05] respectively, and the self-learning probability $p_s = 0.08$. Two optimal paths are obtained as follow: 1-2-6-8-11, the cost and delay [66, 33]; 1-3-6-8-11, cost and delay [91, 28]. When NSGA is used, only above 2 solutions are obtained. The experimental results show the proposed algorithm can obtain optimal solution in solving multi-objective network optimization problems, and it has higher computational efficiency.

6   Conclusion

The improved algorithm is proposed can efficiently solve the multi-objective network optimization problem. The objective classification sort method was adopted to perform the classification sort of agents, so as to determine the agent's energy. The crossover operation based on the index sorting and hybrid mutation algorithm are used. The self-learning strategy is used based on the climbing method, to improve the energy of agents. Finally, a objectives network optimization problem are used to test, the experimental, the results show that the method can obtained optimal solution, and it has higher computational efficiency. In the future, we will do more different kind of examples to test the algorithm.

Acknowledgements. This work was supported by Tianjin Research Program of Application Foundation and Advanced Technology (14JCYBJC15400).

References

1. Ahn, C.W.: A genetic algorithm for shortest path routing problem and the sizing of populations. IEEE Trans. Evol. Comput. **6**(6), 566–579 (2002)
2. Wen, H.Y.: Genetic algorithm-based computation of shortest path in discrete-time networks. J. South China Univ. Technol. **36**(2), 13–16 (2008)
3. Li, Q.: A new adaptive regulation of crossover and mutation probability algorithm. Control Decis. **23**(1), 79–82 (2008)
4. Gen, M., Cheng, R.W., Lin, L.: Network models and optimization, pp. 30–82. Springer, London (2008)
5. Srinva, N., Deb, K.: Multi-objective optimization using non-dominated sorting in GA. Evol. Comput. **1**, 82–87 (1994)
6. Pan, X.: Agent-based multi-objective social evolutionary algorithm. J. Softw. **20**(7), 1703–1713 (2009)
7. Yan, X., et al.: Based the GA network shortest path multi-objective optimization algorithm research. Control Decis. **24**(7), 1104–1109 (2009)
8. Retvari, G., Biro, J.: On shortest path representation. IEEE Trans. Netw. **15**(6), 1293–1306 (2007)
9. Chen, J.: A deque-based transport network shortest path Pallottion optimization algorithm. J. Image Graph **11**(3), 419–424 (2006)
10. Jia, L., Shi, L.: An improved multi-objective genetic algorithm based on agent. In: The 5th International Conference on Intelligent Networks and Intelligent System, pp. 88–91 (2012)

Intelligent Sensor and Actuatorr

Design of Full Temperature Zone Tilt Sensor with the Function of Field Calibration

Na Yu[1(✉)], Zhiyuan Liu[1,2], and Jialong Zhao[1]

[1] China Electronics Technology Group Corporation 49th Research Institute, Harbin, Heilongjiang, China
2624153588@qq.com
[2] Harbin Engineering University, Harbin, Heilongjiang, China

Abstract. The output values of tilt sensor contain the installed initial angle after installed. A method of field calibration is put forward by inputting the corresponding command which makes the sensor output absolute angle or relative angle. The result is clear and visual. In addition, the parameters of semiconductor elements drift along with the variation of temperature which will cause tilt sensor exists measure error and impact accuracy. This paper introduces a temperature compensation method which used in the differential capacitor tile sensor based on acceleration principle. The test results show that this method guarantees the accuracy of tilt sensors up to $0.05\,°C$ in the full temperature zone $(-40\,°C–50\,°C)$. It declines the influence of temperature on tilt sensor effectively and improves its environmental adaptability.

Keywords: Tilt sensor · Field calibration · Temperature compensation

1 Introduction

In the fields of aerospace, military fire control, bridge detection, machine building etc, in order to achieve the goals of condition monitoring and attitude control, requiring the tilt sensor to precisely measure the inclination angle of some plane relative to horizontal plane. The tilt is also needed in the fields of laser instrument horizon, ship sailing posture, tilt monitoring of geological equipment and attitude detection of satellite communication vehicle. However, the sensors are based on absolute zero point when debugging and calibration. The output of tilt sensor is initial installation angle but not zero when installed on the equipment. If the relative angle is need in the field measurement, it need to subtract the initial value from the output value, this brings great inconvenience to measurement. If the sensor has the function of field calibration which make the sensor output the relative angle or absolute angle according to need by inputting corresponding command, it brings great convenience to measurement and the result seems clear and intuitive. Therefore, the function of field calibration is very necessary.

In addition, quite a lot tilt sensor work exposed outdoors for long time, the environmental temperature varies greatly, it will bring the measurement errors because of the thermal output caused by temperature. Especially, the parameters of semiconductor elements drift along with the variation of temperature which will cause tilt sensor exists

© Springer International Publishing AG 2018
F. Qiao et al. (eds.), *Recent Developments in Mechatronics and Intelligent Robotics*,
Advances in Intelligent Systems and Computing 690, DOI 10.1007/978-3-319-65978-7_44

measure error and impact accuracy [1, 2]. At the same time, the temperature also impacts the zero and sensitivity then influences the sensor's static characteristic. So the measures are necessary to decrease the influence on the sensor caused by temperature. We need to do temperature compensation which makes the sensors guarantee accuracy in the full temperature zone and improve the environmental adaptability.

Aiming at the field calibration and temperature drift problem, this paper proposes the design implementation and temperature compensation method, the test result shows the sensor realizes the function and it guarantee accuracy in the full temperature zone. It improves the sensor's practicality and environmental adaptability greatly.

2   Principle and Design

2.1   Measuring Principle

SCA103T series chip of VTI is used in the sensor [3]. The chip is capacitance sensitive chip based on acceleration principle as shown in the diagram. The differential capacitance consists of a pair of fixed plate and a movable plate, the movable plate is connected with the cross beam. When there is a left acceleration signal the Capacitance C_1 increase, C_2 reduction, similarly, when there is a reverse acceleration, the Capacitance C_2 increase, C_1 reduction, two capacitors form differential output then the output signal is converted into a voltage signal through detection circuit and the magnitude of acceleration can be calculated according to this voltage (Fig. 1) [3, 4].

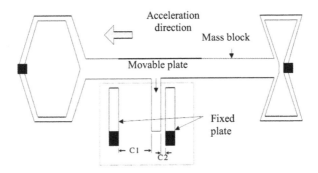

Fig. 1 Principle block diagram of differential capacitance

2.2   Circuit Design

The circuit structure of the sensor is shown in Fig. 2. MSP430 series chip is used which is characterized by low power consumption. The sensor's conditioning circuit consists of power supply, sensitive, A/D, MCU, signal isolation and communication section.

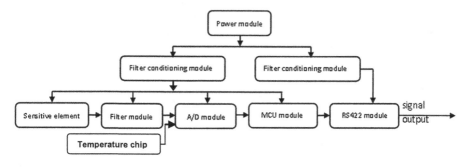

Fig. 2 Block diagram of circuit structure

2.3 Program Design

The program contains AD acquisition, filtering, angle calculation, precision compensation, temperature compensation, etc. They are circulated. The block diagram of the calibration is shown in Fig. 3.

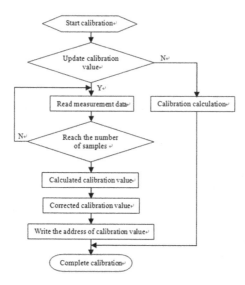

Fig. 3 The program block diagram of the calibration

3   Temperature Compensation Method of Tilt Sensor

Temperature compensation method of tilt sensor is divided into hardware compensation and software compensation. The advantage of hardware compensation is that the hardware parameters are fixed and conducive to mass production but the disadvantage is that the control accuracy is not high, with poor flexibility and complex structure, so it is difficult to achieve the desired results. Software compensation is the error separation and compensation by experiment. With the development of IC technology, online programmable processors can modify parameters and improve accuracy more easily. Therefore, the software method is used for temperature compensation in the tilt sensor.

U is known as the output voltage of the angle sensitive element which exists temperature sensitivity, therefore, measured angle obtained by performing only one dimensional calibration experiment on the angle sensitive element and getting the input and output characteristic curve which will bring large error because the measured A is not unary function of the output value U, if t stands for temperature information, then the angle parameter A can be expressed by U and two-dimensional function of U_t, that is

$$A = f(U, U_t) \tag{1}$$

Similarly, the output voltage U of the angle sensitive element can also be described as the two function of the angle parameter A and the output of the temperature sensitive unit U_t

$$U = g(A, U_t) \tag{2}$$

The A_i is determined by two dimensional coordinates (U_i, U_{it}) on a plane and it can be described by two time surface fitting equation.

$$A = \alpha_0 + \alpha_1 U + \alpha_2 U_t + \alpha_3 U^2 + \alpha_4 U U_t + \alpha_5 U_t^2 + \varepsilon_1 \tag{3}$$

Similarly

$$U = \alpha_0' + \alpha_1' A + \alpha_2' U_t + \alpha_3' U^2 + \alpha_4' A U_t + \alpha_5' U_t^2 + \varepsilon_2 \tag{4}$$

Where, $\alpha_0 \sim \alpha_5$ and $\alpha_0' \sim \alpha_5'$ are constant coefficient, ε_1, ε_2 are high order infinitesimal.

The measured A of the sensitive unit can be calculated when the output value U and U_t are obtained and added them into the formula (3). Therefore, firstly doing the two-dimensional calibration experiment, then determining the constant coefficient $\alpha_0 \sim \alpha_5$ and $\alpha_0' \sim \alpha_5'$ by least square method based on the calibrated input and output.

Determination of n angle mark points in the range of the angle sensitive element and determination of m temperature calibration points in the operating temperature range, thus, the standard value of the calibration value from the angle and the temperature at each standard point is

$$A_i : A_1, A_2, A_3, A_4, \cdots A_n \quad t_j : t_1, t_2, t_3, t_4, \cdots t_m$$

In m different temperature conditions, the input and output characteristic curves of $m$angle sensitive element are obtained by calibration, that is A–U characteristic cluster, at the same time, the n input and output characteristic curves of temperature sensitive units corresponding to different angles are obtained, that is $(t - U_t)$ characteristic cluster.

In order to determine the constant coefficient of the two surface fitting equation represented by the formula (3), usually based on least square method which make the coefficient satisfies the minimum mean square error condition. The calculation method of $\alpha_0{\sim}\alpha_5$ and $\alpha_0' \sim \alpha_5'$ are the same. Take $\alpha_0{\sim}\alpha_5$ for example.

There is a deviation Δ_k between $A\,(U_k, U_{tk})$ and calibration value A_k calculated by the two surface fitting equation, the variance Δ_k^2 is

$$\Delta_k^2 = [A_k - A(U_k, U_{tk})]^2 \qquad k = 1, 2, 3, \cdots, m \times n \tag{5}$$

A total of m × n marked points, the mean square error δ_1 is

$$\delta_1 = \frac{1}{m \times n} \sum_{k=1}^{m\times n} \left[A_k - \left(\alpha_0 + \alpha_1 U_k + \alpha_2 U_{tk} + \alpha_3 U_k^2 + \alpha_4 U_k U_{tk}^2 + \alpha_5 U_{tk}^2\right)\right]^2$$

$$= \delta_1\left(\alpha_0, \alpha_1, \alpha_2, \alpha_3, \alpha_4, \alpha_5\right) \tag{6}$$

Based on the formula above, the mean square error δ_1 is the function of constant coefficient $\alpha_0{\sim}\alpha_5$, in order to meet the minimum mean square error δ_1, according to the limit condition of multivariate function, making the following partial derivative is 0, that is

$$\frac{\partial \delta_1}{\partial \alpha_3} = 0; \quad \frac{\partial \delta_1}{\partial \alpha_4} = 0; \quad \frac{\partial \delta_1}{\partial \alpha_5} = 0, \quad \frac{\partial \delta_1}{\partial \alpha_0} = 0; \quad \frac{\partial \delta_1}{\partial \alpha_1} = 0; \quad \frac{\partial \delta_1}{\partial \alpha_2} = 0;$$

Then the Eq. (7) can be obtained

$$\begin{cases}
\alpha_0 l + \alpha_1 \sum_{k=1}^{l} U_k + \alpha_2 \sum_{k=1}^{l} U_{tk} + \alpha_3 \sum_{k=1}^{l} U_k^2 + \alpha_4 \sum_{k=1}^{l} U_k U_{tk} + \alpha_5 \sum_{k=1}^{l} U_{tk}^2 = \sum_{k=1}^{l} A_k \\
\alpha_0 \sum_{k=1}^{l} U_k + \alpha_1 \sum_{k=1}^{l} U_k^2 + \alpha_2 \sum_{k=1}^{l} U_{tk} U_k + \alpha_3 \sum_{k=1}^{l} U_k^3 + \alpha_4 \sum_{k=1}^{l} U_k^2 U_{tk} + \alpha_5 \sum_{k=1}^{l} U_k U_{tk}^2 = \sum_{k=1}^{l} U_k A_k \\
\alpha_0 \sum_{k=1}^{l} U_{tk} + \alpha_1 \sum_{k=1}^{l} U_k U_{tk} + \alpha_2 \sum_{k=1}^{l} U_{tk}^2 + \alpha_3 \sum_{k=1}^{l} U_k^2 U_{tk} + \alpha_4 \sum_{k=1}^{l} U_k U_{tk}^2 + \alpha_5 \sum_{k=1}^{l} U_{tk}^3 = \sum_{k=1}^{l} U_{tk} A_k \\
\alpha_0 \sum_{k=1}^{l} U_k + \alpha_1 \sum_{k=1}^{l} U_k^3 + \alpha_2 \sum_{k=1}^{l} U_{tk} U_k^2 + \alpha_3 \sum_{k=1}^{l} U_k^4 + \alpha_4 \sum_{k=1}^{l} U_k^3 U_{tk} + \alpha_5 \sum_{k=1}^{l} U_k^2 U_{tk}^2 = \sum_{k=1}^{l} U_k^2 A_k \\
\alpha_0 \sum_{k=1}^{l} U_k U_{tk} + \alpha_1 \sum_{k=1}^{l} U_k^2 U_{tk} + \alpha_2 \sum_{k=1}^{l} U_{tk}^2 U_k + \alpha_3 \sum_{k=1}^{l} U_k^3 U_{tk} + \alpha_4 \sum_{k=1}^{l} U_k^2 U_{tk}^2 + \alpha_5 \sum_{k=1}^{l} U_k U_{tk}^3 = \sum_{k=1}^{l} U_k U_{tk} A_k \\
\alpha_0 \sum_{k=1}^{l} U_{tk}^2 + \alpha_1 \sum_{k=1}^{l} U_k U_{tk}^2 + \alpha_2 \sum_{k=1}^{l} U_{tk}^3 + \alpha_3 \sum_{k=1}^{l} U_k^2 U_{tk}^2 + \alpha_4 \sum_{k=1}^{l} U_k U_{tk}^3 + \alpha_5 \sum_{k=1}^{l} U_{tk}^4 = \sum_{k=1}^{l} U_{tk}^2 A_k
\end{cases} \tag{7}$$

Where $l = m \times n$ stands for the total of marked point. After finishing we can get,

$$\begin{cases}
\alpha_1 l + \alpha_1 E + \alpha_2 F + \alpha_3 G + \alpha_4 H + \alpha_5 I = A \\
\alpha_0 E + \alpha_1 G + \alpha_2 H + \alpha_3 J + \alpha_4 K + \alpha_5 L = B \\
\alpha_0 F + \alpha_1 H + \alpha_2 I + \alpha_3 K + \alpha_4 L + \alpha_5 M = C \\
\alpha_0 G + \alpha_1 J + \alpha_2 K + \alpha_3 N + \alpha_4 O + \alpha_5 P = D \\
\alpha_0 H + \alpha_1 K + \alpha_2 L + \alpha_3 O + \alpha_4 P + \alpha_5 Q = T \\
\alpha_0 I + \alpha_1 L + \alpha_2 M + \alpha_3 P + \alpha_4 Q + \alpha_5 R = S
\end{cases} \tag{8}$$

Where, $E = \sum\limits_{k=1}^{l} U_k$ is the sum of the output values of the angle sensitive element at the

marked point, $F = \sum\limits_{k=1}^{l} U_{tk}$ is the sum of the output values of the temperature sensitive

element at the marked point,

$$G = \sum_{k=1}^{l} U_k^2, \quad H = \sum_{k=1}^{l} U_k U_{tk}, \quad I = \sum_{k=1}^{l} U_{tk}^2, \quad J = \sum_{k=1}^{l} U_k^3, \quad K = \sum_{k=1}^{l} U_k^2 U_{tk},$$

$$L = \sum_{k=1}^{l} U_k U_{tk}^2, \quad A = \sum_{k=1}^{l} U_k, \quad B = \sum_{k=1}^{l} U_k A_k, \quad C = \sum_{k=1}^{l} U_{tk} A_k, \quad M = \sum_{k=1}^{l} U_{tk}^3,$$

$$N = \sum_{k=1}^{l} A_k^4, \quad O = \sum_{k=1}^{l} U_k^3 U_{tk}, \quad P = \sum_{k=1}^{l} U_k^2 U_{tk}^2, \quad Q = \sum_{k=1}^{l} U_k U_{tk}^3, \quad R = \sum_{k=1}^{l} U_{tk}^4,$$

$$S = \sum_{k=1}^{l} U_{tk}^2 A_k, \quad D = \sum_{k=1}^{l} U_k^2 A_k, \quad T = \sum_{k=1}^{l} U_k U_{tk} A_k.$$

According to the standard value of the input marked point A_k and t_k, the output value of two sensitive elements U_k and U_{tk}, we can calculate the value of $A{\sim}D$ and $E{\sim}T$ then solve the matrix Eq. (8). So the constant coefficient $\alpha_0{\sim}\alpha_5$ can be determined. So far, the two surface fitting Eq. (3) is completely determined.

4 Experimental Verification

The operating temperature range of the sensor is $-40\ °C{\sim}50\ °C$, the accuracy of the tilt sensor is measured separately at the lower limit $-40\ °C$, normal temperature $25\ °C$ and the upper limit $50\ °C$, the test results are shown in Table 1.

Table 1 Temperature compensation test results of inclination sensor

Num	Measured angle (°)	Output of sensor (°)		
		$-40\ °C$	$25\ °C$	$50\ °C$
1	−35	−35.02	−35.02	−35.02
2	−25	−25.00	−24.98	−24.98
3	−15	−15.00	−14.99	−14.98
4	−10	−10.02	−10.00	−9.99
5	−5	−4.99	−4.99	−4.99
6	0	0.00	0.00	0.00
7	5	4.99	5.01	5.00
8	10	9.99	9.99	9.99
9	15	14.99	15.01	14.99
10	25	24.98	25.01	24.98
11	35	35.05	35.05	35.04

Verified by experiment, data from Table 1 can be calculated, the accuracy of the sensor are all within $0.05°$ when the temperature at the lower limit $-40\ °C$, normal temperature $25\ °C$ and the upper limit $50\ °C$, that is ensuring the high precision of the sensor in the whole temperature region. The method can effectively reduce the influence of temperature on the sensor accuracy. The environmental adaptability of the sensor is improved.

5   Conclusion

Aiming at the field calibration and temperature drift of tilt sensor, this paper presents a design implementation and temperature compensation method, the experiments show that the sensor can be calibrated by inputting command, the absolute angle or relative angle can be output according to actual needs, the precision can be guaranteed in the full temperature range after temperature compensation and the practicability and environmental adaptability of tilt sensor is improved to meet the needs of the actual environment. The sensor has the advantages of small size, low power consumption and high precision which can satisfy the application in equipment installation, road and bridge construction, automatic level control for robot control, tank and ship gun platform control, aircraft attitude control system. It has broad application prospects [5, 6].

References

1. Wu, Y., Zheng, X., Zeng, Z., et al.: Study on temperature characteristics of tilt sensor. Electron. Meas. Technol. **35**(10), 8–12 (2012)
2. Zhou, Q., Xu, M.: Design of a high resolution tilt sensor with temperature self compensation function. Sens. Microsyst. **31**(1), 107–110 (2012)
3. VTI. The SCA103T differential inclinometer series datasheet. www.vti.fi. Doc.Nr
4. Wang, W., Li, Z., Wang, J., et al.: Inclination sensor based on micro acceleration sensor. Instrum. Tech. Sens. **12**, 12–13 (2010)
5. Lin, D.W.: Design of dual-axis inclinometer based on MEMS accelerometer. In: Proceedings of 3rd International Conference on Measuring Technology and Mechatronics Automation, pp. 959–961 (2011)
6. Yu, X., Wang, H., et al.: Design of high precision angle measuring system based on SOC. Electron. Des. Eng. **18**(12), 34–37 (2010)

Design of Attitude Detection of 2-DOF Ultrasonic Motor Based on Optical Recognition Method

Jian Wang[(✉)], Yang Bai, Zhe Guo, and Jifeng Guo

College of Electrical Engineering, Zhejiang University, Hangzhou, China
usm@zju.edu.cn

Abstract. Attitude detection is a key factor, in applications of multi-degree-of-freedom (M-DOF) ultrasonic motor, such as microrobots, optical instrument and detectors. In order to promote control performance of the two-degree-of-freedom (2-DOF) spherical ultrasonic motor with four stators, a novel non-contact type method of rotor attitude detection is proposed. It's based on optical recognition technology. The construction of 2-DOF spherical ultrasonic motor is introduced, including the optical sensor system. Then mathematical model of the rotor detection is established, to analyze the relationship between relative displacement values and accumulated rotational angles. Finally a solution of detection system is designed. This non-contact method could achieve high accuracy, compactness, and low cost for 2-DOF spherical ultrasonic motor.

Keywords: 2-DOF · Spherical rotor · Ultrasonic motor · Attitude detection · Optical recognition

1 Introduction

Research on attitude detection of multi-degree-of-freedom spherical ultrasonic motor (SUSM) has always been difficult, because the rotor is a sphere. The detection methods for single-degree-of-freedom motor can't be simply copied, it needs to introduce a new mechanism and method.

Currently, attitude detection methods for M-DOF spherical ultrasonic motor could be divided into the contact type and non-contact type. The typical structure of the contact type is slide and bracket system. It consists of two crossing bars and rotary encoders, which was firstly proposed by Toyama, to detect 2-DOF motions [1]. Lee proposed a improvement scheme, which could detect 3-DOF motor [2]. Fu used self-aligning bearing and bar linkage to detect movement of the spherical rotor [3]. The typical structure of the non-contact type is visual recognition system. Lee also utilized a CCD camera to identify the grids, which were sprayed on surface of the rotor [4]. Mashimo presented an attitude sensing system using optical fibers for the SUSM [5].

The contact detection structure has high accuracy, also including increased losses and restricted rotating scope. The non-contact detection mechanism has low loss, simple structure and high recognition. It will gradually become the mainstream of detection.

In this paper, a novel method of attitude detection of 2-DOF spherical ultrasonic motor is put forward. It's utilizing a low power optical mouse sensor, which is based on

F. Qiao et al. (eds.), *Recent Developments in Mechatronics and Intelligent Robotics*,
Advances in Intelligent Systems and Computing 690, DOI 10.1007/978-3-319-65978-7_45

optical recognition. Firstly the structure of 2-DOF motor and optical sensor are introduced. Then mathematical model of the rotor detection is established. Finally a measurement system solution is designed.

2   Motor and Sensor

2.1   2-DOF Spherical Ultrasonic Motor

The construction of the 2-DOF spherical ultrasonic motor mainly consists of a base frame, preload loading mechanism, a spherical rotor and four stators which are distributed around the Z-axis by 90°, as shown in Fig. 1. The axes of four stators are located in the same spherical plane and through the center of sphere.

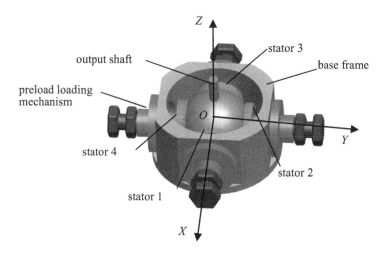

Fig. 1. Sketch of construction of 2-DOF spherical ultrasonic motor

2.2   Sensor Introduction

The Avago chip ADNS-5030 is an optical sensor with low power consumption and small volume. It has the ability to detect high speed motion. It contains an Image Acquisition System (IAS), a internal Digital Signal Processor (DSP), and a four wire serial port. The IAS acquires microscopic surface images via the lens and illumination system. These images are processed by the internal DSP to determine the direction and distance of motion. The internal DSP calculates the Δx and Δy relative displacement values [6].

3   Detection Method and System

3.1   Mathematical Model

This spherical motor could rotate around X-axis and Y-axis. The system detects the attitude of the output shaft of spherical rotor (point O'). According to means of Ref. [7], mathematical model of 2-DOF spherical rotor is built up, as shown in Fig. 2. The ADNS sensor is placed directly below the spherical rotor (point K). In a single sampling period t_s, relative displacement values Δx and Δy can be expressed by detection counts C_x and C_y.

$$\begin{bmatrix} \Delta x \\ \Delta y \end{bmatrix} = \frac{1}{C_p} \begin{bmatrix} C_x \\ C_y \end{bmatrix} \tag{1}$$

Where C_p is the resolution of sensor, 500 and 1000 cpi selectable.

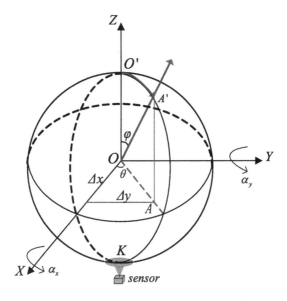

Fig. 2.   Mathematical model of 2-DOF spherical rotor

Assuming the rotor rotates through a trajectory of arc $O'A'$ sensor records straight OA as shown in OXY plane. The relationship between the Azimuth angle θ, zenith angle φ and relative displacement values Δx and Δy can be expressed as

$$\begin{cases} \theta = \arctan \dfrac{\Delta y}{\Delta x} \\ \varphi = \arcsin \dfrac{\sqrt{\Delta x^2 + \Delta y^2}}{r} \end{cases} \tag{2}$$

Where r is the radius of spherical rotor.

Then the angles θ and φ in the spherical coordinate are converted to rotation angles of the X-axis α_x and Y-axis α_y, respectively.

$$\begin{cases} \alpha_x = \varphi \sin \theta \\ \alpha_y = \varphi \cos \theta \end{cases} \tag{3}$$

By substituting Eq. (2) into Eq. (3), the relationship between rotation angles and relative displacement values can be easily obtained.

$$\begin{cases} \alpha_x = \dfrac{\Delta y}{\sqrt{\Delta x^2 + \Delta y^2}} \arcsin \dfrac{\sqrt{\Delta x^2 + \Delta y^2}}{r} \\ \alpha_y = \dfrac{\Delta x}{\sqrt{\Delta x^2 + \Delta y^2}} \arcsin \dfrac{\sqrt{\Delta x^2 + \Delta y^2}}{r} \end{cases} \tag{4}$$

Assumed that the sensor data C_{xi} and $C_{yi}(i = 1, 2,..., n)$ within a certain time $(T = n \times t_s)$, n is samples times. Then the accumulated rotational angels can be expressed as

$$\begin{cases} \alpha_{xn} = \alpha_{x0} + \sum_{i=1}^{n} \dfrac{C_{yi}}{\sqrt{C_{xi}^2 + C_{yi}^2}} \arcsin \dfrac{\sqrt{C_{xi}^2 + C_{yi}^2}}{C_p r} \\ \alpha_{yn} = \alpha_{y0} + \sum_{i=1}^{n} \dfrac{C_{xi}}{\sqrt{C_{xi}^2 + C_{yi}^2}} \arcsin \dfrac{\sqrt{C_{xi}^2 + C_{yi}^2}}{C_p r} \end{cases} \tag{5}$$

Where α_{x0} and α_{y0} describe the starting position of the spherical rotor.

3.2   System Solution

The detection system consists of an optical sensor ANDS-5030, an optical lens, a LED, an external microcontroller and a host PC, as shown in Fig. 3. Lens is disposed between the photosensitive surface of the sensor and a ball rotor. By adjusting the distance of sensors and lens, images could be clearly identified. ANDS-5030 obtains continuous surface images and mathematically determines the direction and scale of action. The micro-controller fetches the Δx and Δy values from the sensor through serial communication. The micro-controller converts the data formats and then sends them to the host PC.

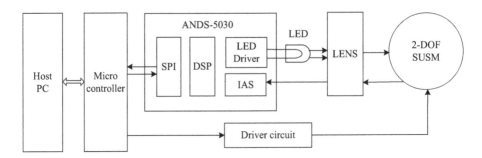

Fig. 3. Schematic diagram of detection system

4 Conclusion

The paper proposes a new method of rotor attitude detection for 2-DOF spherical ultrasonic motor. This is a non-contact type detection system based on optical recognition technology. The sensor system could achieve high accuracy, compactness, and low cost for 2-DOF spherical ultrasonic motor. Further experimental verification will be explained in a following paper.

Acknowledgements. The work was supported by Zhejiang Provincial Natural Science Foundation of China (No. LY15E070002) and the National Natural Science Foundation of China (No. 51475411).

References

1. Toyama, S., Sugitani, S., Zhang, G.Q., et al.: Multi degree of freedom spherical ultrasonic motor. In: Proceedings of the IEEE International Conference on Robotics and Automation, pp. 2935–2940. IEEE, Nagoya (1995)
2. Lee, K.M., Roth, R., Zhou, Z.: Dynamic modeling and control of a ball-joint-like variable-reluctance spherical motor. J. Dyn. Syst. Meas. Control **118**(1), 29–40 (1996)
3. Fu, P., Guo, J.F., Shen, R.J., et al.: Driving circuit and position control system of two degree-of-freedom spherical traveling-wave type ultrasonic motor. Trans. China Electrotech. Soc. **23**(2), 25–30 (2008). (in Chinese)
4. Garner, H., Klement, M., Lee, K.M.: Design and analysis of an absolute non-contact orientation sensor for wrist motion control. In: Proceedings of the 2001 IEEE International Conference on Advanced Intelligent Mechatronics, pp. 69–74. IEEE, Como (2001)
5. Mashino, T., Awaga, K., Toyama, S.: Development of a spherical ultrasonic motor with an attitude sensing system using optical fibers. In: Proceedings of the 2007 IEEE International Conference on Robotics and Automation, pp. 4466–4471. IEEE, Roma (2007)
6. Avago Technologies, ADNS-5030 Datasheet [EB/OL]. http://www.ic37.com/AVAGO_CN/ADNS-5030_datasheet_4869009/. Accessed 14 Apr 2008
7. Lim, C.K., Chen, I.M., Yan, L., et al.: A novel approach for positional sensing of a spherical geometry. Sens. Actuators A Phys. **168**(2), 328–334 (2011)

Target Tracking Algorithm of Wireless Multimedia Sensor Networks

Jing Zhao[1,2(✉)], Zhuohua Liu[1], and Songdong Xue[2]

[1] College of Information Engineering, Guangdong Mechanical and Electrical Polytechnic,
Guangzhou 510515, People's Republic of China
zhaojing_740609@163.com
[2] Division of Industrial and System Engineering, Taiyuan University of Science of Technology,
Taiyuan 030024, People's Republic of China

Abstract. Aiming at the problem of target tracking node control in wireless multimedia sensor networks, a rotating perceptual model is established for sensing nodes, a node control algorithm based on Virtual potential field method is proposed. Through the movement of the target and the virtual force of the Monitoring node, the control monitoring node rotation is dynamically monitored, and the information gain function is established to dynamically select the next monitoring node according to the moving target, so that the target can wake up the node as little as possible to achieve maximum motion trajectory monitoring probability. The simulation results validate the validity of the algorithm.

Keywords: Possible sense area · Target tracking · Wireless multimedia sensor network

1 Introduction

The development of wireless communication, microelectronics and embedded technology has spawned the wireless sensor network technology. Wireless sensor network consists of a large number of low-cost micro sensor nodes [1, 2], complete specific monitoring tasks, wireless sensor networks include the traditional omni-sensing network, multimedia sensor network two categories. The traditional whole-to-sensing network realizes simple environmental data (such as temperature, humidity, intensity, etc. [3]) collecting, transmitting and processing. However, with the complication of monitoring tasks, the more comprehensive and abundant environmental data (such as images, videos, etc.) are needed to accomplish more precise monitoring tasks, thus generating wireless multimedia sensing network [4].

The traditional wireless sensor network tracking algorithms are many, including [5]: aiming at the DWSN node, the target tracking algorithm based on binary detection is proposed by Mechitov et al. [6]. Rabbat and Nowak [7] based on signal strength of distributed positioning and tracking algorithm; Gordon [8] proposes a particle filter algorithm. After the node is randomly deployed, how to dispatch the node to realize the full and efficient coverage of the target motion trajectory is a hot issue that urgently needs to be solved [9]. In the study of WSNs covering problem, the method of virtual

© Springer International Publishing AG 2018
F. Qiao et al. (eds.), *Recent Developments in Mechatronics and Intelligent Robotics*,
Advances in Intelligent Systems and Computing 690, DOI 10.1007/978-3-319-65978-7_46

potential field is often used. The concept of virtual potential field is firstly proposed by Khatib [10] to solve the problem of path planning and obstacle avoidance. The paper [11] applies it to the study of WSNs covering problem, assuming that the node has a virtual field around the field, the exclusion of the other nodes in the entry can be produced, which makes the node spread from the dense area to the sparse area and increases the regional coverage. The force that repelled each other made sensor spread from dense to sparse area. And coverage and scheduling node are very important in WSNs [12, 13].

In this paper, aiming at the application of the target tracking in wireless multimedia sensor network, the method of virtual potential field based on the virtual force of the moving target and the current sensing node is used to control the node rotation to monitor the target, and in the current sensing node of the entire potential sense of the region to establish a warning circle, when the target movement to the alert circle, to produce junction points, according to the target at the junction of the movement direction of the information gain potential area, and the region of all sensor nodes in the information gain calculation, The biggest point of wake-up information gain is the next monitoring node to continue to monitor the target until the target leaves the monitoring area.

2   Algorithm Description

2.1   Idea of Algorithm

When the target randomly enters a specific area, it is perceived by the boundary point, and the virtual gravity of the point of centroid of the target and perceived node can be constructed, and the control node rotates with the motion of the target. Junction points are formed when the target moves to the alert circle of the perceived node. The information gain region is set according to the moving direction of the target at the junction point, and the information gain value is computed for the neighbor node of junction point in the information gain area, the node of the maximal gain is awakened as the next perceptual node, and the current sensing node is converted into hibernation. Duplicate selection perception nodes until the target leaves a specific area.

2.2   Sensing Models

Considering the perceived area of the node and the potential sensing area, the perceptual model is established as follows: using a five-tuple to represent: P, R, ψ, α, v, there P is sensor's location, R is radius of sector' area, ψ is the horizontal angle of the node, it can take ($0°$, $360°$), α represents the visual offset of the video node, i.e. half the angle, v indicates rotational speed, i.e. to monitor the rotational speeds required to move the target.

2.3   Statement

The algorithm includes two parts: node rotation control and perceptual node selection strategy.

2.3.1   Node Rotation Control

When the target randomly enters a specific area, it is perceived by the boundary point, and the virtual gravity of the centroid point of the target and perceived node is established to control the movement of the node with the target.

$$\vec{F}_i = k_1 \times \frac{1}{r_i^2} \times \vec{r_{i0}} \tag{1}$$

where k_1 is density of field; $\vec{r_{i0}}$ is a vector of unit length and describes the direction of force from 'centroid' point of node i to target location; r_i is the distance between the centroid point and the target location of the perceived node i.

When the target is perceived by the node i, the centroid point of the sensing node is rotated by the target's virtual attraction, as shown in Fig. 1, the target at the O point is perceived by the node i, the target is along the dotted MN forward, sensing the direction of the node i as the virtual force of the target is also turned from IO to IP, the final target from the P point leaves the current perception node, the perceived node rotation speed is directly affected by the target motion.

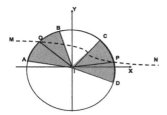 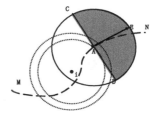

Fig. 1. The perceptual node rotates with the destination

Fig. 2. Information gain area

2.3.2   Sense Node Selection Policy

When the target moves to the edge of the perceived node, the effective selection of the next perceptual node is critical to the monitoring of the target trajectory.

- Determination of junction points
  This article defines the alert circle, and defines the target trajectory and alert circle intersection point, large circle indicates the perception node i the entire potential perception area, small circle for the alert circle, curve MN as the target trajectory, A point is the junction point.
- Determination of information gain region
  Information refers to the length of the target motion trajectory that is not covered by the node; gain means that the target trajectory of the motion is not covered by the

length of the node increases, then the information gain area refers to the adjacent areas of junction points, with the junction point of the target movement direction perpendicular to the diameter of the junction area is divided into two semicircle, then the direction of the target movement of the semicircle points to the information gain area, such as Fig. 2, AE as the target at the junction of the movement direction, CD Vertical AE and divide the adjacent area of junction A, the dark area is the information gain region, which is a forecasting region, and the nodes in the region are most likely to cover the target future trajectory.

2.3.3   Information Gain Computation and the Choice of Perceptual Nodes

In Target Monitoring node control management, we want to use as few nodes as possible to complete the monitoring task, so in the information gain area, we hope that the next perceptual node from the junction a bit more far better, from the line AE closer to the better, so covering the target track to not covered the longest, that is, the node's information gain value is the largest, and there is no covering loopholes, such as Fig. 3, P point in the information gain area of any node, PQ represents P point to line AE distance, AP is P point to the junction of AE points, The area covered by the dotted circle is the range of the predicted point P, and we hope that the larger the AP is better, the smaller the better the PQ, k_2 is controls parameter. The information gain is computed as follows:

$$\inf = k_2 \times \frac{AP}{PQ} \tag{2}$$

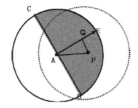

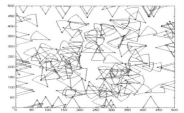

Fig. 3. Predictive sensing node   **Fig. 4.** Initial deployment

3   Simulation

In this paper, the simulation experiment of the Node Control algorithm is carried out, and the experimental results are as follows.

3.1   Simulation Results of Controlling Algorithm

In this paper, the Target Monitoring node control algorithm is simulated, and the sensor is isomorphic to simplify the experiment. Assuming the sensor node's perception radius is 60 m, sensing the sector angle 45°, the specific region is 500 m × 500 m, the number of nodes is 100, and randomly deployed, the target randomly enters the monitoring area, and then moves randomly in the area until it leaves the region, and the algorithm performs the result as Figs. 4 and 5, in which Fig. 4 is the initial deployment of sensor nodes in the monitoring area, the boundary coverage is 59.05%, and Fig. 5 shows that the target is randomly entered the zone from point A, the sensor node wakes randomly moving and leaves from the B point, the small diamond indicates the awakened node, the black dot is the sleeping node, the circular region indicates the potential sensing area of the Wake node, the curve is the target trajectory, the target trajectory coverage is 100%, and the Wake node is minimal. So the target enters the monitoring area randomly moving, finally leaving the area monitored to the probability of 59.05%.

Assuming the perception radius is 60 m, sensing the sector angle 45°, the number of nodes is 50, 100, 150, 200, randomly deployed in the monitoring area, the paper adopts 10 simulation results to average, target trajectory coverage, boundary coverage and target perceived probability as the number of nodes increases with the curves of the change in Fig. 6. It concludes that when the number of nodes is low, the target trajectory coverage is even higher, but due to the lower boundary coverage, so the probability that the target is perceived is lower; as the number of nodes increases, the border coverage increases, the target trajectory increases rapidly, until the node reaches 100, the target trajectory coverage is approximately 1, so that the probability of the target perception is influenced by the boundary probability only.

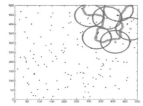

Fig. 5. Sensor scheduling

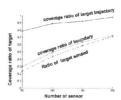

Fig. 6. The effect of node number on target perception probability

4   Conclusions

According to the practical application, the rotational perceptual model is defined and the force model of the target-to-sensing node is established by referring to the method of virtual potential field. The information gain function is defined and the next perceptual node is predicted according to the target leaving the state of the current sensing node and the node information gain, so as to achieve the most probable high target perception probability with as few nodes as possible, and a series of simulation experiments are carried out to validate the validity and correctness of the algorithm. In the future, the target Monitoring node scheduling strategy in 3d state will be researched further.

Acknowledgements. This work is supported by the Shanxi Province Natural Science Foundation under Grant No. 2014011019-2; the school postdoctoral scientific research foundation of China under Grant No. 20142023; The National Science Foundation of China under Grand No. 41272374 and No. 61472269, Shanxi Province Programs for Science and Technology Development under

Grant No. 2015031004, Project from Shanxi Scholarship Council of China under Grant No. 2016-091.

References

1. Akyildiz, I.F., Melodia, T., Chowdhury, K.R.: A survey on wireless multimedia sensor networks. Comput. Netw. **51**(4), 921–960 (2007) ISSN 1389-1286. doi:10.1016/j.comnet. 2006.10
2. Pananjady, A., Bagaria, V.K., Vaze, R.: Optimally approximating the coverage lifetime of wireless sensor networks. IEEE/ACM Trans. Netw. **25**(1), 98–111 (2017)
3. Akyildiz, I., Su, W., Sankara Subramaniam, Y., Cayirci, E.: Wireless sensor networks: a survey. Comput. Netw. **38**(4), 393–422 (2002)
4. Ma, H.D., Tao, D.: Multimedia sensor network and its research progresses. J. Softw. **17**(009), 2013–2028 (2006)
5. Tang, J., Shi, H., Han, Z.: An target tracking algorithm for wireless sensor networks. J. Air Force Eng. Univ. **7**(5), 25–29 (2006)
6. Mechitov, K., Sundresh, S., Kwon, Y., et al.: Agha cooperative tarcking with binary-detection sensor networks. In: Proceedings of the First Internatonal Conference on Embedded Networked Sensor Systems, pp. 332–333 (2003)
7. Rabbat, M.G., Nowak, R.D.: Decentralized source localization and tracking. In: Proceedings of the 2004 IEEE International Conference on Acoustics, and Signal Processing Montreal, Canada, pp. 921–924 (2004)
8. Gordon, N.J., Salmond, D.J., Smih, A.F.M.: Novel approach to nonlinear/non-Gaussian bayesian state stimulation. Radar Signal Process IEE Proc. **140**, 107–113 (1993)
9. Tao, D., Ma, H.D., Liu, L.: Study on path coverage enhancement algorithm for video sensor networks. Acta Electronica Sinica **36**(70), 291–1296 (2008)
10. Khatib, O.: Real-time obstacle avoidance for manipulators and mobile robots. Int. J. Robot. Res. **5**(1), 90 (1986)
11. Zhao, J., Zeng, J.-C.: Sense model and number estimation of wireless multimedia sensor networks. J. Softw. **23**(8), 2104–2114 (2012)
12. Misra, S., Majd, N.E., Huang, H.: Approximation algorithms for constrained relay node placement in energy harvesting wireless sensor networks. IEEE Trans. Comput. **63**(12), 2933–2947 (2014)
13. Djenouri, D., Bagaa, M.: Energy harvesting aware relay node addition for power-efficient coverage in wireless sensor networks. In: IEEE ICC, pp. 86–91 (2015)

Design and Research of Iron Roughneck's Clamping Device

Yongbai Sha[1(✉)], Xiaoying Zhao[2], and Zitao Zhan[1]

[1] School of Mechanical Science and Engineering,
Jilin University, Changchun, China
shayb@jlu.edu.cn
[2] Changchun Vocational Institute of Technology, Changchun, China

Abstract. Analysis and design were carried out to the clamping device of iron roughneck using in the Deep Continental Scientific Drilling Progress. According to the actual working conditions, the static analyses were carried out to the tong dies and the drilling tool. Based on the simulation results and the practical experience, the structure and strength of related components were analyzed, and some suggestions of the tong dies' structure and the selection of drill pipe were given.

Keywords: Iron roughneck · Clamping device · Tong dies · Static analysis

1 Introduction

In order to meet the requirements of deep Continental Scientific Drilling [1], a kind of hydraulic automation drill tools make-up and break-out device - iron roughneck is designed, and the main mechanism is shown in Fig. 1. According to the existing make-up and break-out equipment's questions of injury rod pipe, low automation degree, and high labor intensity of workers, the new device can protect the drilling tool and improve the efficiency in the process of clamping and screwing.

The clamping device is one of the key components of iron roughneck, and clamping effect is the key of drill tools making up and breaking out operation. Under the premise of ensuring the clamping effect, it is necessary to consider the appropriate mechanical structure to reduce the damage to the drilling tool.

In this paper, the iron roughneck clamping device is designed, and the design effect is verified by static simulation. Some suggestions for improvement are made based on the simulation data and the actual situation.

2 Structure and Principle of the Clamping Device

As shown in Fig. 1, clamping and unloading mechanism includes two parts: fixed clamp and screw discharge clamp. Every part includes three clamp monomer which symmetrically arranged, and the two groups of upper and lower clamp together constitute the clamping device, and the upper and lower drilling tools joints are respectively clamped by them to further complete the clamping and screwing action.

© Springer International Publishing AG 2018
F. Qiao et al. (eds.), *Recent Developments in Mechatronics and Intelligent Robotics*,
Advances in Intelligent Systems and Computing 690, DOI 10.1007/978-3-319-65978-7_47

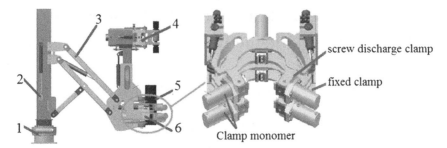

1. slewing mechanism; 2. lifting mechanism; 3. extension mechanism; 4. Screw rotation
mechanism; 5. drilling tool; 6. clamping and unloading mechanism

Fig. 1. Main mechanism of the iron roughneck.

Clamp monomer is a core component of iron roughneck's clamping device, and it can complete clamping, screwing and unloading operation which performance directly determines the iron roughneck's working efficiency. Clamp monomer is mainly composed of clamping cylinder, clamp block, clamp base, and tong dies, as shown in Fig. 2. Each part is connected by a pin shaft and (or) a fastening screw, and this connection mode makes the following installation and maintenance more convenient. The design of clamp base and tong dies is the key of the design of clamp monomer. Figure 3 is the new structure of tong dies and clamp base. Contrast with universal tong dies and clamp base, the new structure has the following advantages:

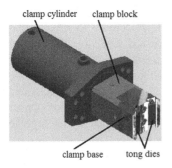

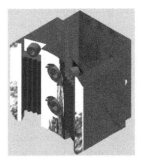

Fig. 2. Clamp monomer structure

Fig. 3. Tong dies and clamp base

1. Clamp base adopts arc transition, and according to the drill diameter, corresponding type clamp base can be installed (in this paper, 127 mm drill pipe is adopted), so that to increase the envelope angle between the drill pipe and the tong dies, and enhance holding effect, and improve the stability.
2. The tong dies are metal matrix composites and triangular groove structure, and the tong dies contact with the drill pipe by multi line, improving the clamping effect,

and the channel can timely discharge the mud on the pipe. This design can effectively alleviate the problem of the drill pipe bite marks caused by the tong dies, and prolong the service life of the drill pipe and the tong dies [2, 3].

3   Simulation Analysis

Combining with the iron roughneck's problem in the drill tools making up and breaking out operation, the tong dies and drill pipe joints are selected to do static analysis, to ensure the components in a predetermined load having reliable strength.

3.1   Mesh Partition

The mesh partition is the key technology to establish the finite element model for static analysis [4]. In order to ensure the accuracy of the finite element model, it is need to divide mesh considering the mesh density and mesh structure combined with the actual complexity, size, stress, and etc.

The oil pipe is mainly composed of drill rod body and the drill pipe joint. Considering the project requirements of the drilling depth (nominal drilling depth is 10000 m), S135 (36CrNiMo4) level which is the highest strength in API series is selected. The drill pipe's diameter is 127 mm, and the wall thickness is 12.7 mm [5]. Compared with the carbon structural steel, the alloy structural steel has the characteristics of high strength, good toughness and wear resistance and good hardenability, so the main material of the tong dies is 20Cr2Ni4.

In the assembly of drill pipe joint and tong dies, in order to ensure the accuracy of static analysis, hexahedral mesh is used. Figures 4 and 5 are the mesh partition graphs of the drill pipe joint and the tong dies, and Fig. 6 is the mesh information.

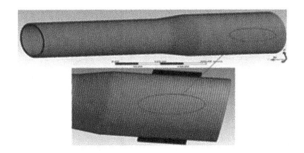   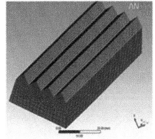

Fig. 4. Mesh of drill pipe   **Fig. 5.** Mesh of tong dies

In order to ensure the quality of the mesh, Element Quality and Skewness are used to characterize the mesh quality in ANSYS Workbench. The element quality is about 0.86, which shows that the element quality is better. The average value of the

Statistics	
Nodes	101280
Elements	79686
Mesh Metric	Element Quality
Min	.744710661004909
Max	.997995027707086
Average	.859923979388132
Standard Deviation	6.03639717153841E-02

Statistics	
Nodes	101280
Elements	79686
Mesh Metric	Skewness
Min	6.93940151697826E-03
Max	.462484067244063
Average	9.25780114534855E-02
Standard Deviation	6.83850058951557E-02

Fig. 6. Mesh information of drill pipe and tong dies

inclination is about 0.46, indicating that the mesh quality is better. The other mesh information of the model is shown in Fig. 6.

3.2 Simulation Analysis

After completing the mesh partition, the constraints are imposed to the above components according to the specific conditions in ANSYS Workbench. Specifically, full freedom constraints are imposed to the upper and lower surfaces of the drill pipe, and normal displacement constraints are imposed to both surfaces of the tong dies; collision constraints with friction are added between the edge of each tong dies and the drill pipe, and the friction coefficient is 0.5. Then, according to the results of the joint simulation, in the assembly of the drill pipe and the tong dies, the resistance torque of 49003215.4 N mm is applied to the drill pipe, along the direction of the drill pipe. The 61.3 MPa pressure is applied to the rear side of the tong dies, and the direction is perpendicular to the side of the drill pipe center.

After completion of the above steps, the simulation and calculation are conducted on the assembly of the drill pipe and the tong dies, and the total deformation and stress nephogram shown in Figs. 7, 8, 9 and 10 are obtained.

For the sake of analysis and comparison, the static analysis results of the key components are shown in Table 1. It can be seen that the maximum stress of the key

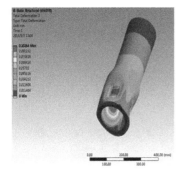

Fig. 7. Total deformation nephogram of drill pipe

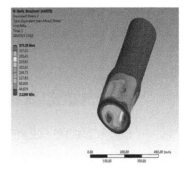

Fig. 8. Stress nephogram of drill pipe

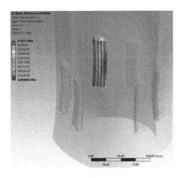

Fig. 9. Total deformation nephogram of tong dies

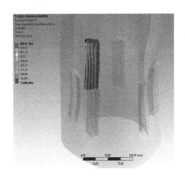

Fig. 10. Stress nephogram of tong dies

components are all in the safe range. The maximum stress of the tong dies appears at the top of the outer edge, and the value is 625 MPa, which is close to the allowable stress of the material; the maximum stress of the drill pipe appears in the inner wall of the lower end, and the value is about 379 MPa. The total deformation of the key components are all about 0.1 mm, that is, deformation is small, and structure is safe.

Table 1. Static analysis results of key components

Result	Maximum strain (mm)	Maximum stress (MPa)	Yield strength (MPa)	Allowable stress (MPa)
Component				
Drill pipe	0.1026	379.23	≥ 931	≥ 621
Tong dies	0.1071	625.07	≥ 1080	≥ 720

Note: Safety factor n = 1.5

4 Improvement Suggestion

According to the requirements of the project, the corresponding super deep well drilling pipe should be developed, which has the advantages of high strength, corrosion resistance, double shoulder design, application of composite materials, internal flush and external upset and so on [6, 7]. Specifically, according to the stress nephogram shown in Fig. 8, it can be considered to increase the external diameter of the drill pipe joint and apply knurling design, so that to increase friction effect and improve the torsional strength of the drill pipe. In addition, from the pipe section stress nephogram shown in Fig. 11, it can be seen that the stress decreases from inside to outside and larger stress appears in the shallow layer of inner wall. According to the rules, the method such as nitriding and high-frequency quenching heat treatment process can be used to strengthen the strength of drill pipe inner wall.

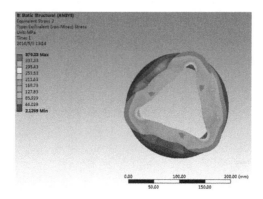

Fig. 11. Stress nephogram of drill pipe section

Figure 10 is the tong dies stress nephogram, and according to the characteristics of the stress concentration in the lateral edge of the external, nitriding, high-frequency quenching heat treatment process can be used to improve its strength; according to the characteristics of the tong dies' triangular groove bearing larger force, without significantly affecting tong dies clamping effect on the drill pipe, it can be considered to polish the serrated edge forming ladder shape type [8], which can reduce the damage to the surface of the drill pipe causing by the tong dies, but also can improve the strength of the tong dies, and reduce the frequency of replacement.

5   Conclusion

(1) An iron roughneck clamping device is designed in this paper, and it can ensure the clamping effect, and reduce the damage to the drill pipe.
(2) Static analysis is carried out on the assembly of the drill pipe joint and the tong dies, and the simulation results verify the performance of the components.
(3) According to the simulation results, combined with the practical situation, some suggestions are put forward to improve the design of the tong dies and the structure of the drill pipe.

Acknowledgment. This paper is funded by the project (SinoProbe-09-05) from the Ministry of Land and Resources and expressing thanks for this.

References

1. Dong, S.W., Li, T.D., Chen, X.H., et al.: Progress of deep exploration in mainland China: a review. Geophy **55**(12), 3884–3901 (2012)
2. Zhu, H., Liu, K., Moli: Tooth profile design of drill pipe slip tooth plate. China Pet. Mach. **36** (11), 68–70 (2008)
3. Wei, L.: Present situation of iron roughneck tong dies. J. Henan Sci. Technol. **11**, 137 (2013)

4. Dai, X., Cui, H., Liu, J., et al.: Hexahedral mesh generation method of swept volume with sub-domain constraints. J. Comput. Aided Des. Comput. Graph. **26**(4), 632–637 (2014)
5. Zhao, H.: Research on a New Type Pneumatic Slip for Oil Drilling, pp. 28–41. Harbin Institute of Technology, Harbin (2006)
6. Li, J., Yu, L., Niu, C., et al.: The production status and development trend of drill pipe. Weld. Pipe **34**(11), 35–38 (2011)
7. Zhao, J., Chen, S., Liu, Y., et al.: Progress on high-performance drill pipe. Oil Field Equip. **40**(5), 96–99 (2011)
8. Fang, Z., Zhu, H., Chen, C.: Discussion of the mechanical model of drilling string setting in the slip. J. Oil Gas Technol. **31**(4), 381–383 (2009)

A Study on Wireless Sensor Network Data Collection Model Based on Real Perception Data Analysis

Yongfa Ling[✉], Xiangxu Xie, Hui Wang, Jie Huang, Chuanghui Li, and Wenya Lai

School of Mechanical and Electronic Engineering, Hezhou University, Hezhou, China
syswangxueleng@163.com

Abstract. In this paper, specific to perception data compression performance and signal recovery quality of a wireless sensor network, a distributed large-scale wireless sensor network data partitioning processing model was presented based on the original perception data analysis. Then, distributed projection data processing implementation was also demonstrated in detail to carry out data recovery performance evaluation and energy dissipation simulation testing. The experimental results indicated that such a model was able to reduce the number of measured values transmitted on the network so as to further bring down costs of network data collection and improve data recovery quality.

Keywords: Data type partitioning · Compressive sensing · Wireless sensor network · Energy dissipation

1 Introduction

A wireless sensor network is an important component of the Internet of Things and has been extensively applied into fields of medical care, environmental monitoring, target tracking and military affairs, etc. It employs large quantities of randomly deployed nodes to cover monitoring area and then form a multi-hop self-organizing network system by means of wireless communication. In this way, automatic monitoring, acquisition and transmission of long hours can be accomplished for environmental data. As an "Intelligent" system that is able to independently complete various monitoring tasks according to environment, the wireless sensor network is an autonomous network system constituted by masses of low-cost and low-power dissipation tiny sensor nodes with sensing, computing and communicating capabilities [1].

Due to limitations of node resources and network resource dissipation imbalance, the sensor network fails to carry out large-scale deployment. Consequently, reduction in the number of data packets transmitted on the network serves as one of the major means of network energy dissipation decrease. Thus, it is clear that network compression technology is an effective method that can be utilized to improve the wireless sensor network size and extend the run time of it. However, the existing data compression techniques basically follow a feature of complex encoding and simple decoding and are inapplicable to wireless sensor network node resources that are under limitations, such as the distributed transform coding and the distributed source coding, etc. [2].

© Springer International Publishing AG 2018
F. Qiao et al. (eds.), *Recent Developments in Mechatronics and Intelligent Robotics*,
Advances in Intelligent Systems and Computing 690, DOI 10.1007/978-3-319-65978-7_48

The compressive sensing theory is a breakthrough of Shannon-Nyquist sampling theorem. As for sparse signals or compressible signals, the former is able to not only bring down sampling frequency greatly, but realize simultaneous execution of sampling and compression. In recent years, the application of compressive sensing in wireless sensor network has become increasingly more extensive. As a result, on one hand, the number of transmitted data packets required by network is effectively reduced provided that there is no need for nodes to execute complex encoding operations, which just conforms to the feature of wireless sensor network node resources under limitations [3].

Currently, research emphases of wireless sensor network based on compressive sensing are concentrated on effective acquisition, fusion and routing of data instead of the improvement of network performance. The corresponding main reasons are as follows. First, transmission of a single measure value requires great costs. Second, the compression performance of real perception data is not ideal. Third, measurement matrix fails to match with real routing preferably [4–6]. Therefore, a study on wireless sensor network data acquisition based on compressive sensing has important practical academic meaning and application value [7–10].

2   Environment Assumption for Wireless Sensor Network

Targeted at the general wireless sensor network structure, the following assumptions are made. In the first place, all sensor nodes are isomorphic and each of them is static. Secondly, a sensor network consisting of n sensor nodes and a single rendezvous point has been deployed within an identical monitoring range. Among them, the latter takes the responsibility to collect data in the whole network and recover sensory data; and, all external manipulations are carried out accompanied with the rendezvous point. Thirdly, passive submission serves as the commit mode of data. Nodes store the data collected by them into their own cache and submit them only when they obtain notifications from such a rendezvous point. Fourthly, type of service is a periodic collection business. To be specific, the nodes collect data regularly from the sensor and the collection of all data completed by the rendezvous point within a single collection cycle is referred to as a round of data collection and the relevant businesses have no particular requirements for transmission delay.

In a wireless sensor network, the number of nodes of the sensor is assumed to be 330. If signals with sparseness of 50 should be collected, 200 measured values are needed at least; then, the number of data packets required transmitting on the network is 200×330. In the case that no data compression is adopted, data packets transmitted in it can also be figured out by $330 \times \log(330)$, where $\log(330)$ is the average hops needed by the transmission of every perception data to the rendezvous point in the case of tree routing. Hence, simply utilizing compressive sensing is unable to reduce communication costs taken by network. In the project, real compressive sensing data should be processed in the first place.

3   Real Perception Data Analysis and Data Type Partitioning

The GreenOrbs large-scale wireless sensor network deployment diagram (330 nodes) is employed to analyze perception data (homogeneous data) collected by the true Green-Orbs system. It can be found that some abnormal data exist in them.

The aim of deploying a wireless sensor network is to monitor the interested physical quantity within a given region and find some abnormalities among them. On this basis, data monitoring from a wireless sensor network is partitioned into data of normal transformation and abnormal data. As for the former, they can be deemed as monitoring data within a certain variation range, while the latter is considered to be perception data going beyond a certain threshold. Therefore, real perception data of this study is divided into parts of normal transformation, abnormal transformation and noise to perform compression and projection processing for diverse data. Hence, the purpose of data compression and recovery performance improvement can be realized further.

4   Idea of Data Type Partitioning Model

How to realize data type partitioning? According to the definition above, perception data with n nodes are n-dimensional signals. The equation $y(t) = [y_1(t), y_2(t), ..., y_n(t)]^T$ refers to the perception data collected by all nodes in network of round t; where, $y_i(t)$ is the perceived value of node i in the round of t. In addition, $y_i(t)$ can be expressed in the following formula.

$$y_i(t) = y_i^*(t) + y_i^e(t)(i = 1, 2 \ldots n) \tag{1}$$

Where, $y_i^*(t)$ is the value with normal variations in $y_i(t)$, containing real monitoring value and sampling noise; and, $y_i^e(t)$ is the value of abnormal variation. In order to further reduce temporal correlation of perception data and improve the compression performance of them, $y_i^*(t)$ can be divided into a predicted value $y_i^p(t)$ and a standardized residual $y_i^p(t)$. Resultantly, the perceived value of the whole network during data collection in the round of T can be denoted as follows.

$$Y(t) = Y^p(t) + Y^q(t) + Y^e(t) \tag{2}$$

Where, $Y^p(t) = [y_1^p(t), y_2^p(t), ..., y_n^p(t)]^T$ is the predicted part of data with normal variations, while $Y^q(t) = [y_1^q(t), y_2^q(t), ..., y_n^q(t)]^T$ is the residue of such data.

With regard to the predicted part of data with normal variation, except data of the first round, the predicted data in other rounds do not require transmitting as long as all nodes posses a forecasting function identical to that of the rendezvous point. Furthermore, the forecasting function can be designed according to data handling capacity of sensor nodes and characteristics of perception data. Thus, rendezvous point of data is substantially reduced. The forecasting function of node i in round t can be written into the equation below.

$$y_i^p(t) = g[y_i(t-1), y_i(t-2), \ldots, y_i(t-T)] \tag{3}$$

Regarding the residue of data with normal variation, 50 4-layer-`haar'wavelet transform coefficients are adopted to recover them, test the peak signal to noise ratio of recovered signals and investigate their compression performances. Through experiments and direct compression with normal variation data, we obtain a significantly preferable precision after data type partitioning. In addition, $y^q(t)$ can be expressed into $Y^q(t) = \Psi X^q(t)$, $\|X^q(t)\|_0 = k^q(t)$, where Ψ refers to sparse orthogonal basis of $n \times n$ and $X^q(t)$ to coefficient variation after conversion.

As for data with abnormal variations, they are seen as sparse signals as only a few of them exist on the network, which is denoted by $Y^e(t) = IY^e(t)$, $\|Y^e(t)\|_0 = k^e(t)$, where I is the unit matrix of $n \times n$.

In this way, after data type partitioning and implementation above, a distributed large-scale perception data partitioning model is preliminarily established based on real perception data analysis.

5   Projection Implementation

The predicted part of normal variation data is only transmitted in the first round and the projection is allowed to be ignored correspondingly.

For the residue of them, they are compressible not sparse data, so that sparse conversion is required here. A sparse random projection method is used and its measurement matrix is defined as follows.

$$\Phi_{ij}^q = \frac{1}{\sqrt{s}} \begin{cases} +1 & withprob. \dfrac{1}{2s} \\ 0 & withprob. 1 - \dfrac{1}{s} \\ -1 & withprob. \dfrac{1}{2s} \end{cases} \tag{4}$$

Where, $i = 1, 2, \ldots, l$ and $j = 1, 2, \ldots, n$. Besides, the sparsity of such a measurement matrix is determined based on parameter s. If $s = n/\log n$, then $l = O(k^q \log n)$ among which k^q is known as the sparsity of residue $Y^q(t)$. During data collection in round t, the measured value ranking i is expressed into $z_i^q(t) = \sum_{j=1}^n \Phi_{ij}^q y_j^q(t)$. The relevant measurement vector specific to the entire residue is,

$$Z^q(t) = \Phi^q Y^q(t) \tag{5}$$

Data of abnormal variation are sparse signals themselves inapplicable for sparse random projection. However, dense random projection can be conducted for them (that is, elements of all measurement matrixes are nonzero). During data collection for residue in round t, the measured value ranking i is denoted to be $z_i^e(t) = \sum_{j=1}^n \Phi_{ij}^e y_j^q(t)$. Then, the compression vector can be written into the following equation overall.

$$Z^e(t) = \Phi^e Y^e(t) \tag{6}$$

Where, Φ^e refers to the matrix of dense random projection. Totally speaking, sparse random projection is used for residue while dense random projection for data with abnormal variation. No matter the measurement matrix or the collection signal is sparse, only a small number of nodes participate into data collection of individual measured values in this process so as to achieve the same data collection effect.

During data recovery, spatial prediction is not allowed for data of the first round which should be collected and processed in a direct manner. Starting from the subsequent rounds, there is no need to transmit the predicted part on the network as long as the sensor nodes are given a prediction mode identical to that of rendezvous point. In other words, the rendezvous point only needs to collect measured values of the residue and the abnormal part. According to the above data recovery process, it can be deferred preliminarily that data processing based on a data type partitioning model is able to enormously bring data communication cost down. In this study, performance assessment is also carried out through simulation testing.

6 Simulation Testing and Conclusions

Based on a data partitioning model, the combination of sparse and dense projections as well as a clustering routing strategy, the experimental tests are conducted from the following two aspects considering that the number of measured values obtained during compressive perception data collection processing not only affects perception data recovery quality, but has an influence on network energy dissipation. On is comparison between perception data recovery qualities in the case of diverse measured values; and, the other is comparison between network energy dissipations under circumstance of diverse routing strategies and data collection processing methods.

While real perception data of GreenOrbs are utilized to carry out such experiments, perception signal recovery is based on an orthogonal basis traceback algorithm. In addition, the peak signal-to-noise ratio serves as an evaluation index of signal recovery. Firstly, normal perception data are tested and compression performances of perception data are partitioned separately. Then, comparative testing is carried out in allusion to data recovery qualities acquired by various perception data partitioning and projection methods. In the end, relevant results are obtained correspondingly.

During network energy dissipation testing, we use OMNeT++ network simulation platform as the energy evaluation platform while bit hop as the energy dissipation model to design three data collection processing strategies respectively, such as the shortest spanning tree random projection, the clustering dense random projection and the clustering sparse random projection, to evaluate network energy dissipation without regard to energies consumed by node calculation.

Ultimately, experimental results indicate that, provided that the number of measured values is identical, the data collection scheme proposed in this paper is able to not only improve data recovery quality, but significantly extend time of network monitoring.

Acknowledgement. This research was supported by the fund project of 1608027; 2017KY0651 (The basic ability promotion project of Young teachers in Guangxi universities)

References

1. Fang, X., Gao, H., Li, J.: A survey of data collection problems in wireless sensor networks. Intell. Comput. Appl **4**(1), 1–6 (2014)
2. Intelligent computer and applications. Wirel. Sens. Netw. **1,** 1–6 (2014)
3. Wu X.: Research on Compressive Sensing Based Data Gathering in Wireless Sensor Networks. University of Science and Technology of China (2013)
4. Yin, H., Liu, Z., Chai, Y., et al.: Survey of Compressed Sensing. Control Decis. **28**(10), 1441–1445 (2013)
5. Yang, H.Y., Huang, L.S., Xu, H.: Distributed compressed sensing in vehicular ad-hoc network. Lasers Eng. **25**, 121–145 (2015)
6. Chen, S., He, Z., Xiong, H., et al.: A reconstruction algorithm of wireless sensor signal based on compressed sensing. Chin. J. Comput. **38**(03), 614–624 (2015)
7. Yang, H., Wang, X.: Data gathering based on regionalization compressed sensing in WSN. Chin. J. Comput. **39**(75), 1–14 (2016)
8. Wang, Y., Wu, X., Li, W., et al.: A reconstruction method based on ALOFGD for compressed sensing in border monitoring WSN system. PLoS ONE **9**(12), 1–17 (2014)
9. Xiang L, Luo J, Vasilakos A. Does compressed sensing improve the throughput of wireless sensor networks. In: 2010 IEEE International Conference on Communications (ICC), pp. 1–6 (2010)
10. Li L., Li J.: Research of compressed sensing theory in WSN data fusion. In: 2011 Fourth International Symposium on Computational Intelligence and Design (ISCID), pp. 125–128 (2011)
11. Huang, C.: Wireless sensor networks data processing summary based on compressive sensing. Sens. Transducers **174**(07), 67–72 (2014)

Analysis and Design of the Warning Process of the Road Slope Monitoring System

Xiao-rong Zhou[✉], Xiang Wang, and Rong-Xin Jin

GuangXi University of Mechanical Engineering, Nanning, China
127580443@qq.com

Abstract. The highway slope disaster for effective monitoring and early warning, to reduce the economic losses caused by highway slope disaster. According to the characteristics of slope monitoring system, the demand of early warning process of highway slope monitoring system is analyzed. To achieve the highway slope monitoring system early warning process analysis and design, so that timely and accurate highway slope disaster early warning disposal. Thus minimizing the prevention of disasters and reducing losses.

Keywords: Road slope · Monitoring system · Warning · Process design

1 Introduction

Guangxi is the southwest of China. The geological conditions are complex, and the terrain is mainly mountainous. Guangxi is one of the seven geological disaster-prone areas in China [1]. Due to the special geological environment and the abundant rain, the highway geological disasters happened frequently. It was seriously affected the road construction and operation safety, and had caused significant loss of life and property.

Because of the special nature of highway geological disasters, it is impossible to carry out engineering treatment for all latent hidden danger points [2]. The traditional technology of disaster prevention and reduction has been difficult to meet the needs of modern development, application of advanced information technology for highway geological disasters, highly effective monitoring and early warning, has become the only way to improve the level of highway geological hazard control [3]. In this paper, the early warning process of slope monitoring system is analyzed and designed in detail. So that the system can be based on the collected monitoring data timely and accurate warning processing, to maximize the prevention of slope disasters.

2 Early Warning System Function Structure Design

The highway slope monitoring system collects, stores, analyzes and processes large slope monitoring data. And then the slope monitoring data into the disaster level and location information. Intuitive presentation to the highway slope user management, disaster inquiry, early warning and tracking monitoring to facilitate emergency response. The road slope monitoring and early warning system covers the slope monitoring

© Springer International Publishing AG 2018
F. Qiao et al. (eds.), *Recent Developments in Mechatronics and Intelligent Robotics*,
Advances in Intelligent Systems and Computing 690, DOI 10.1007/978-3-319-65978-7_49

function module, the query management function module, the slope management function module, the scheduling management function and the system management function. The concrete roadside monitoring and early warning system function diagram is shown in Fig. 1.

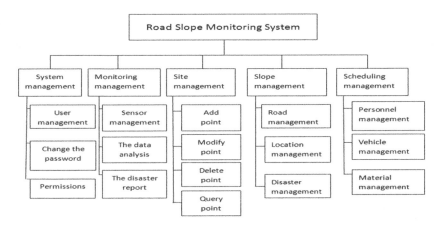

Fig. 1. System function structure diagram

The highway slope monitoring and warning system takes the geological information and climatic information of the highway slope as the input. Based on the analysis and treatment of the slope level of the slope. The system of different levels of early warning processing as output, to achieve highway slope monitoring and early warning processing.

3   Early Warning System Process Design

Highway slope monitoring and warning process: slope sensors to monitor the highway slope, and slope monitoring data transmission to the data acquisition module. The data acquisition module integrates the slope monitoring data. Data acquisition module in the integrated slope data to do data encoding, the encoded data transmission to the GPRS transmission module. GPRS transmission module to upload data to the GPRS receiver module. The GPRS receiving module decodes the encoded data according to the slope information database and sends the monitoring data to the civil analysis module. The civil analysis module analyzes the slope monitoring data, obtains the disaster grade of the monitoring slope, and sends it to the warning server.

Disaster level design for low to high score 4 levels (level 1, level 2, level 3 and level 4), in the road slope monitoring map corresponding to the four disasters were blazing green, blue, orange, red. After landing, users click the corresponding color shining position, the interface can display: sensor number, road section, location, disaster time, disaster level, slope maintenance personnel, maintenance personnel number, contact, notes.

3.1 Level Warning Process

Level 1 (green) is a safe level. The system will continue to monitor automatically. When the level is upgraded, the system will take the corresponding operation.

3.2 Second Warning Process

Level 2 (blue) for the alert level. After the system receives the secondary disaster rating from the civil analysis module, the level color is changed to blue on the map. The system sends a warning message to the slope manager based on the sensor location information. Such as: "Guangxi highway slope monitoring and early warning system to remind: March 15, 2015 05, G3 * 1 National Road K769 + 800-K770 + 000 starting and ending station position H06 sensor shows the disaster rating: Level 2. Please pay attention to the system disaster level changes".

3.3 Level 3 Warning Process

Level 3 (orange) is a dangerous level. After the system receives the Level 3 Disaster Level from the Civil Analysis Module, the color of the displayed level changes to orange on the map. The system sends a warning message to the slope manager based on the sensor location information. Such as: "Guangxi highway slope monitoring and early warning system to remind: March 15, 2015 05, G3 * 1 National Road K769 + 800-K770 + 000 starting and ending station position H06 sensor display Disaster rating: Level 3. Please carry out disaster prevention." Slope managers click on the corresponding warning warning position, pop-up interface display: sensor number, section, location, disaster time, disaster rating, slope maintenance personnel, staff number, contact information. The slope manager sends the sensor number, location, section and grade to the slope engineer to prepare for disaster prevention, as shown in Fig. 2.

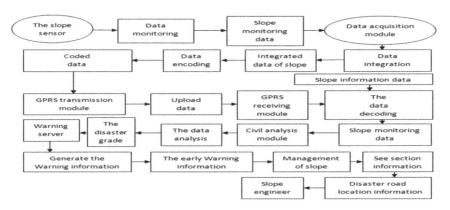

Fig. 2. The third stage early warning flow

4   System Data Process Design

4.1   The Top Data Flow Diagram

The top-level data flow chart of the highway slope monitoring and early warning system shows that the data flow and system outflow of the system. The data flow into the system is the slope sensor, the data outflow is the slope maintenance personnel. The data flow of the top layer of the system is shown in Fig. 3.

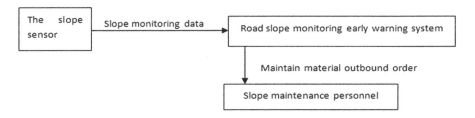

Fig. 3.   The top data flow

4.2   The First Data Flow Diagram

The first layer data flow chart of the highway slope monitoring and early warning system mainly includes four processes: monitoring data uploading, monitoring data analysis, disaster early warning view and maintenance material dispatching. Figure 4 shows the data flow of the highway slope monitoring and early warning system.

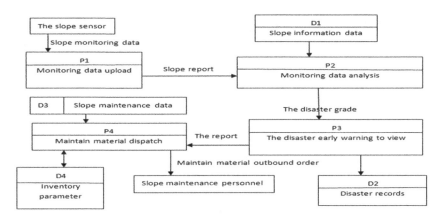

Fig. 4.   The first data flow

5   Conclusion

The functional structure of the early warning system is designed by analyzing the demands of the practical application of the highway slope monitoring system. The early warning and disposal business process and data flow of the highway slope monitoring system are designed. The early warning part of the highway slope monitoring system is enriched and completed. It has practical significance and application value to the overall development of highway slope monitoring system.

Acknowledgements. This work was supported by the Nanning key project of Science and Technology (Granted No. 20143113), the Guangxi key project of Science and Technology (Granted No. 14124004-4-12),the innovation training project of Guangxi college students (Granted No. 201510593165), the Scientific research projects of Guangxi education department (Granted No. 2013YB013).

References

1. Huang, X., Jiang, J., Niu, H.: Guangxi highway geological disaster research. West. Traffic Sci. Technol. **9**, 58–59 (2014)
2. Zeng, Q.: Automatic monitoring system in the application of slope deformation monitoring. J. Technol. Market **21**(6), 179–180 (2014)
3. Luo, J.: The automatic monitoring system based on sensors and the wireless mode in highway slope monitoring the application and development. J. Appl. Technol. **2**, 390–391 (2014)
4. Li, H., Cao, Z., He, S., Zhou, G.: Released meteorological disaster warning system based on beidou satellite design and application. J. Meteorol. Sci. Technol. **5**, 799–802 (2014)
5. Chen, H., Xi, W., Zhou, F., Liu, Y.: Based on the Internet of things technology of landslide monitoring and early warning system design. J. Electron. **5**(2), 279–282 (2014)
6. Zhang, P., Wei, L.: Highway for fog monitoring and early warning system for vehicle design. Mech. Des. Manuf. Eng. **2**, 27–29 (2014)
7. Mandal, D., Chattopadhyay, I.M., Mishra, S: A low cost non-invasive digital signal processor based (TMS320C6713) heart diagnosis system. In: 1st International Conference on Recent Advances in Information Technology I RAIT-2012

Slope Monitoring and Early Warning System Implementation Based on the Internet of Things Technology

Xiaorong Zhou[✉] and Qianghua Shi

School of Mechanical Engineering, Guangxi University, Nanning 530004, China
gxzhouxiaorong@163.com

Abstract. We design a highway slope monitoring and early-warning system suitable for rainfall condition. The system has great significance for evaluation of highway slope disaster grade and the disaster early warning and emergency measures. This paper based on the functional requirements of highway slope monitoring and early warning system, and have a detail designing for the realization of the system.

Keywords: Monitoring and early warning system · The disaster grade · Disaster early warning

1 Introduction

According to the highway slope monitoring and early warning system function requirement analysis, the realization of system function needs the data transmission of network layer and the control and display of application layer. The system that collects field data by sensors will send the data to remote monitoring center by GPRS wireless communication modules. The remote monitoring center receives and processes data as required, and displays the data in a specified data format. Users can inquire various data information conveniently, and realize the monitoring of disaster, early warming and emergency processing of the highway slope. Then it assists with disaster dispatching work [1].

2 Analysis and Design of Data Transmission

According to the characteristics of highway slope sensor (Type JD-05B Tilting Rain Gauge, Type SMS-I-50 Soil Moisture (humidity) Sensor,) and the data transmission requirements, the system uses the GPRS wireless communication modules for data transmission. But it never makes any data confused.

The rendezvous points of wireless sensor network in the system send monitoring data to the remote monitoring center by GPRS. The standard RS232 interface is used to communicate between the two. Then it enters a dormant state. Rendezvous point collects all the data of nodes on time, and sends data in data frame format by GPRS module [2]. The data frame format that is sent by rendezvous point is as follows:

© Springer International Publishing AG 2018
F. Qiao et al. (eds.), *Recent Developments in Mechatronics and Intelligent Robotics*,
Advances in Intelligent Systems and Computing 690, DOI 10.1007/978-3-319-65978-7_50

Time of departure ((date) (month) (year), 12 bytes) + send the IP address (4 bytes)+.

Data acquisition module 1 ID number (4 bytes) + data acquisition module 1 rainfall (3 bytes) +data acquisition module 1 soil moisture content(3 bytes) + data acquisition module 1 slope displacement (3 bytes)+.

Data acquisition module n ID number (4 bytes) + data acquisition module n rainfall (3 bytes) + data acquisition module n soil moisture content(3 bytes) + data acquisition module n slope displacement (3 bytes)+.

3   Requirements of System Design

In allusion to the design of highway slope monitoring and early warning system, performance requirements mainly includes: pertinence, practicability and expansibility [3].

Pertinence. Because the monitor modules of highway slope monitoring and early warning system are the highway side slope, and it can be showed as complexity, uncertainty and confidentiality. Therefore, the designed system should have high pertinence to the object that should be monitored.

Practicability. Highway slope monitoring and early warning system is a system which integrates multi-discipline and a variety of technology. The Designed system should satisfy operation requirements for different users.

Expansibility. Designed operation system can satisfy the requirement of the basic function. But when system used in engineering in the future, there will be shortage in the system. It requests system can have the ability to expand. So we can update and maintain the system in the future.

The function of highway slope monitoring and early warning system is included slope management, fault maintenance, auxiliary tools management and system Settings management [4, 5]. Specific functions are shown in the figure (Fig. 1):

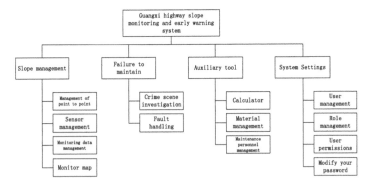

Fig. 1. System function diagram

Slope management modules: this module is the "cornerstone" of highway slope monitoring and early warning system. It needs to grasp the characteristics of highway slope monitoring information, decorate monitoring sensor reasonably, collect the information of monitoring transducer, analysis data and summarize the disaster. Specific functions includes: minority point management, sensor management and so on. It implements the uniform management of highway slope minatory point, minatory slope sensors and minatory data.

Failure maintenance modules: we can get the rainfall, displacement of the slope, soil moisture of the slope by the slope management modules of the system. And then we get the disaster of every monitory point by the formula of slope disaster level of the system.

Auxiliary tool modules: we do the maintenance by this module. We gather statistics about the staffs, goods and materials. Depend on the result of the statistics, we supply goods and materials in time. And we can make sure that there is enough goods and materials if there is a serious disaster. What we should gather statistics are slope disaster road, disaster grade, maintenance personnel, necessary goods etc.

System setup function modules: user management, role management, user permissions and modify your password. The system is a public system. The system users include the average citizen and traffic academy of sciences who manages the slope. We carry out management and set permissions to different users. This can guarantee system works well.

4 System Implementation

4.1 Basic Information Function Realization

The user can enter the main interface of operation procedure by inputting user name and password. You can enter the system if the user name and password is right.

The realization of basic information function of the system includes user management, role management, user permissions and modifying your password. The user management refers to the management of basic information of the users. User permissions is that different types of users have different permissions in system.

The main function of this system is to realize the monitoring and early warning to the highway slope, real-time monitoring of highway slope condition. The first level is a safe level. The second level is warning level. The third level is dangerous level. The fourth level is severity level (Fig. 2).

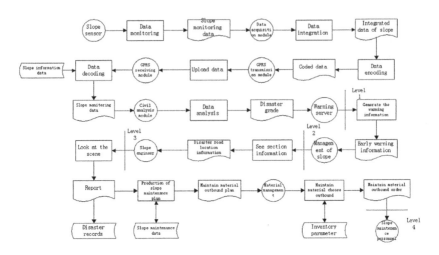

Fig. 2. Slope monitoring early warning process

4.2   System Function Implementation

The Guangxi highway slope monitoring and early warning system mainly includes two aspects: the displacement of highway slope monitoring, rainfall and soil moisture monitoring and maintaining slope disasters. Depend on the actual situation of Guangxi highway, we monitor the monitoring points, connect the monitoring points and management center. We collect the data through the sensor, and send it to the management center. The management center calculates the security level of different monitoring point depend on the slope security level formula, and outputs the data in regulation format (Fig. 3).

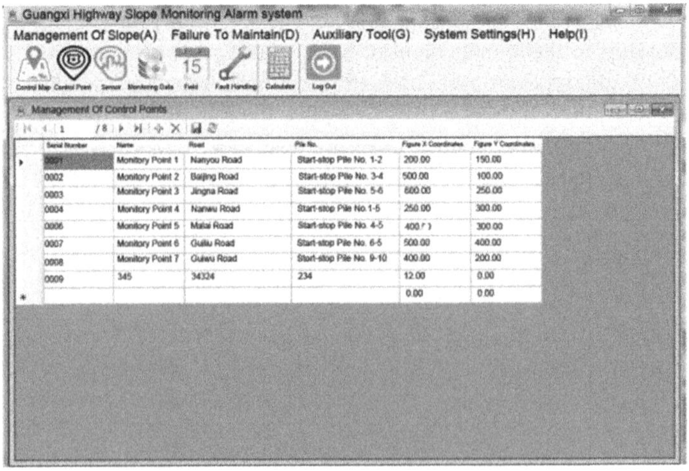

Fig. 3. Management of monitoring point

Another important function of the system is to maintain the slope where have the disaster. If the disaster grade gets to dangerous grade, we take different maintenances according to the disaster grade. We transport goods and material and arrange staffs to maintain the highway slope, and have a feedback of the maintenance information (Fig. 4).

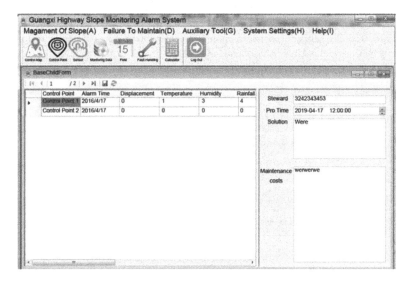

Fig. 4. Slope disaster maintenance and management

The design of remote monitoring and early warning system is to monitor the highway slope and decrease the damage caused by the highway slope disaster. Our system sends the slope displacement, rainfall and soil moisture to management center by SPRS DUT data transmission modules. The management center processes data and presents the intuitive data format (displacement, rainfall, humidity, security level) in the highway slope monitoring and early warning system. This realize the remote control to the slope monitoring [6, 7].

Acknowledgements. This work was supported by the Nanning key project of Science and Technology (Granted No. 20143113), the Guangxi key project of Science and Technology (Granted No. 14124004-4-12), the innovation training project of Guangxi college students (Granted No. 201510593165), the Scientific research projects of Guangxi education department (Granted No. 2013YB013).

References

1. Chubin, L.: Optical fiber heat fire early warning and monitoring system design and application of analysis. J. Guangdong Vocat. Techn. Coll. Hydraul. Electric. Eng. **6**(3), 37–42 (2008)
2. Jifei, H., Jing, J.: The design of the GPRS early warning and monitoring system based on ARM. J. Coal Mine Mach. **32**(5), 218–220 (2011)

3. Li, J.: User-centered monitoring system design and development of optical fiber bragg grating. Xi'an: Northwest University (2008)
4. Xi, F.: Water quality detection system based on ARM and GPRS network design. Wuxi, Jiangnan University (2009)
5. Li, M.: The study of dynamic early warning monitoring system of oilfield well control. Da qing: Northeast Petroleum University (2013)
6. Jiang, S.: Building health monitoring system based on Internet of things technology, the research and design. Xi'an: Chang'an University (2013)
7. Wenzhou, L.: Do network operation monitoring system based on Internet research and implementation. Fudan University, Shanghai (2013)

Numerical Simulation of Underwater Electric Field in Seawater Condition

Zhang Jianchun[1](✉), Wang Xiangjun[1], and Tian Jie[2]

[1] Naval University of Engineering, Wuhan 430033, Hubei, China
hbzjctj@163.com
[2] Wuhan Institute of Technology, Wuhan 430033, Hubei, China

Abstract. In the shallow sea, the source of underwater electric field (UEF) is the corrosion, but in the complex ocean environment, many seawater factors are effect on the UEF, such as seabed conductivity, seawater depth and water conductivity. If there are many damaged parts, the numerical simulation comparability of UEF related to corrosion in these factors is found through multiple horizontal dipoles. The results show that the UEF is easy to be effected by the ocean environment, and it has different effect on the three components of UEF.

Keywords: Shallow sea · Underwater electric field · Corrosion · Multiple horizontal dipoles · Variation trend

1 Introduction

Many researches have shown that the shaft-rate underwater electric field can vary significantly, depending on the environmental conditions during practical measurement, because many factors are combined action, it is hard to get an accurate variation trend when analyzing the vicinity UEF. In the contribution, the variation trend is analyzed depended on some environmental conditions, including the seabed depth, water-conductivity and seabed conductivity that own to environmental conditions, and how they affect the component of shaft-rate electric field. Our conclusion is based on analytical calculations and numerical simulation. The excitation source of corrosion is stemmed from electrochemical reaction, Because of the different extent of corrosion on the hull of vessel, the corrosion current density is also different. In our contribution, the vessel is modeled by multiple horizontal dipoles, which has different potential. By using the mirror image method [1, 2], the expression of electric filed can be derived analytically.

2 Part of Theory Analysis

In this section, the vessel model and computing method will be discussed based on multiple horizontal dipoles.

© Springer International Publishing AG 2018
F. Qiao et al. (eds.), *Recent Developments in Mechatronics and Intelligent Robotics*,
Advances in Intelligent Systems and Computing 690, DOI 10.1007/978-3-319-65978-7_51

2.1 The Vessel Model

Different positions of hull in the sea have different extent of corrosion due to the different painting condition of the proactive hull coating, the pH-value of water and the fouling in the hull, so the polarization curve and the corrosion current density are also different. Multiple horizontal dipoles are closed to model those positions, each dipole is relative and independent, different current density are set to those dipoles combined to the polarization curve [3]. The distance of adjacent dipoles is the same, as shown in Fig. 1(a).

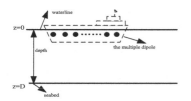   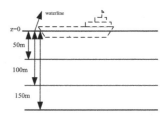

a. The mathematical model of ship   b. The different depth of seabed

Fig. 1. The vessel model

When considering the electric potential at a certain point stemmed from those dipoles, the superposition principle can be considered. In order to achieve it, a theoretical approach called "multipole approximation" can be used, where an arrangement of different current sources can be described by a total electric potential [4].

2.2 The Introduction of Multipole Approximation

The expression as the following:

$$\frac{1}{4\pi\varepsilon} \cdot \sum_{i=1}^{N} \frac{I_i}{|\vec{r}-\vec{r_i}|} = \frac{1}{4\pi\varepsilon} \cdot \sum_{n=0}^{\infty} r^{-(n+1)} \sum_{i=1}^{N} I_i r_i^n F_n(\phi) = \varphi(\vec{r}) \tag{1}$$

The multipole approximation makes it possible to compute the UEF in multiple horizontal dipoles (order n). When there is no one dipole, the total current intensity I_T stemmed from dipole is zero. So the electric potential is the following:

$$\varphi_{n=0}(\vec{r}) = \frac{1}{4\pi r\varepsilon} \sum_{i=1}^{N} I_i = \frac{I_T}{4\pi r\varepsilon} = 0 \tag{2}$$

If there is only one dipole, the dipole will determine the UEF:

$$\varphi_{n=1}(\vec{r}) = \frac{1}{4\pi r^2\varepsilon} \sum_{i=1}^{N} I_i r_i \cdot \vec{e}_r = \frac{\vec{p} \cdot \vec{e}_r}{4\pi r^2\varepsilon} \tag{3}$$

Then the electric field can be got, which is the gradient of electric potential.

When there are multiple horizontal dipoles, the choice of different current intensity I_i is based on polarization diagram of steel in different corrosion condition.

3   Environmental Factors of Influence

In this part the environmental factors of influence to UEF in different values are mainly discussed, if there are any changes, what this change rule should be.

3.1   Seabed Conductivity

It is known that there is a big difference between the rock mass in the seabed and minerals in the continent [5]. Because the seabed soaks in seawater for a long time, material structures have undergone great converts. So it has to be considered a very important factor of influence for UEF signature. The change rule is shown in Fig. 2.

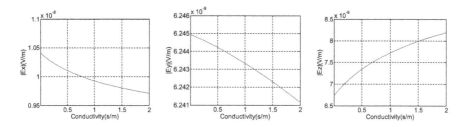

Fig. 2. The influence of seabed conductivity from $0.1\,\mathrm{S}/m$ to $2\,\mathrm{S}/m$

The Fig. 2 is shown that the seabed conductivity has a huge qualitatively and quantitatively impact on E_x- and E_z-component, the smaller the seabed conductivity is, the greater the electric field amplitude of E_x-component is, while the electric field amplitude of E_z-component is opposite. This variation is the same with the corresponding laws presented by Weaver J.T. [6]. For E_y-component, although it also decreases with the increase of the seabed conductivity, it is not obvious. The seabed conductivity has little influence on the E_y-component. The main reason is that the field distribution characteristics of horizontal dipole as shown in Fig. 3.

The Fig. 3 indicates that E_z-component is smaller than other two components. Because in no seabed present the maximum UEF value of horizontal dipole consists E_x- and E_z-component, however, the E_z-component is easy to be effected by seabed conductivity.

3.2   Seawater Depth

As expected, the UEF signature will change in different seawater depth. The seabed conductivity can affect the UEF signature. The distance between vessel and seabed (that

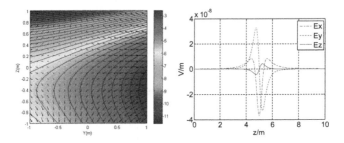

Fig. 3. The distribution of the UEF signature in the vicinity of a low conducting sebed $(\varepsilon_s = 10\%\varepsilon_w)$

is called "seawater depth") will also affect the distribution of UEF signature. So it is another important factor to consider, it is shown in Fig. 1(b).

The UEF signature in different depth is shown in Fig. 4. In this section, we only consider the influence of depth, the transversal distance to ship keeps constant.

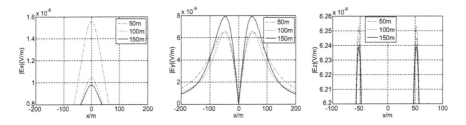

Fig. 4. The amplitude of the UEF signature in different depth

The change in the depth from 50 m to 150 m has a clear impact on the amplitude of the UEF signature in different depth $(\varepsilon_s = 10\%\varepsilon_w)$. The change rule is shown in Fig. 5.

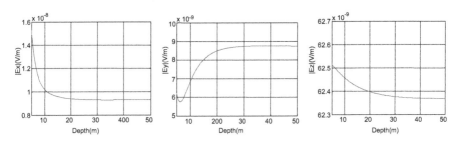

Fig. 5. Influence of shaft-rate underwater electric field with seawater depth

It can be seen from the simulation results that the effect of sea water depth on the three different components of horizontal dipole are different. The field amplitude of E_x- and E_z-component gradually decrease with the increase of seawater depth, while the field amplitude of E_y-component first decreases in a certain depth range and then increases again, which is caused by the characteristic electromagnetic wave in the sea. The depth has great influence on the convert of the E_x- and E_y-component amplitude. Although the E_z-component has decreased, the convert of amplitude range is small, because the seabed conductivity keeps constant, in other word, the depth has a little impact on E_z-component. When the water depth reaches 300 m and above, the three components of the electric field will be stable, which shows that the electric field intensity is limited when the depth tends to infinity. When depth tends to infinity, it has almost impact on the distribution of UEF, so the UEF signature is limited.

3.3 Water Conductivity

As we all know, the water conductivity is a very important factor of influence on UEF signature, it has an empirical formula about water conductivity [7]:

$$\varepsilon_w = \left[A + B\frac{T^{1+k}}{1+T^k}\right]\frac{S^{\lambda-0.8447}}{1+S^h}e^{-\varepsilon S}e^{-\zeta(S-S_0)(T-T_0)} \tag{4}$$

The water conductivity is determined by temperature and saltness. So the water conductivity is different in different sea areas. It is necessary to research the influence of the water conductivity. The UEF signature in different water conductivity is shown in Fig. 6.

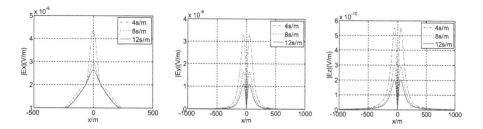

Fig. 6. The UEF signature in different water conductivity

The Fig. 6 displays the water conductivity has a huge effect on the signature peak value, the current intensity of dipoles were set to be independent respectively and water conductivity makes no difference in them, so the change of UEF signature only result from the current intensity in water, that is to say, the change is only caused by the water conductivity. The change rule is shown in Fig. 7.

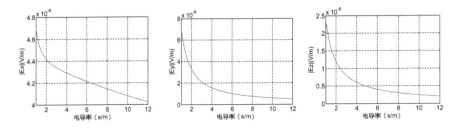

Fig. 7. The variation trend to water conductivity

4   Conclusion

In this paper the variation trend of UEF signature is got in seabed conductivity, seawater depth and water conductivity. The influence of those factors is different, the components are also different in the same factor. The paper demonstrates that the UEF signature is originated not only by the vessel itself but also by the environment factors. Besides, the variation trend of UEF signature in shallow sea is limited.

5   Outlook

In this paper we only research the variation trend of single factor, but in actual measurement the factors are concurrent. So the combined influence in multiple factor should be considered based on the conclusion of the paper in the future.

References

1. Mao, W., Zhou, M., Yu, R.: Electromagnetic fields produced by a moving vertically directed time-harmonic dipoles in the two layer medium. J. Wuhan Univ. Technol. (Transp. Sci. Eng.) **10**, 1081–1085 (2011)
2. Ji, D., Zhu, W., Wang, X.J.: The EMF of a moving time-harmonic VED embedded in the active layer of a n-layer conducting half-space. J. Nav. Univ. Eng. **26**, 71–74 (2014)
3. Hack, H.P.: Atlas of polarization diagrams for naval materials in seawater. Carde-rock Division Naval Surface Warfare Center, Technical report (1995)
4. Schafeu, D., Doose, J., Pichlaier, M.: Comparability of UEP signatures measured under varying, environmental, conditions. In: MARELEC, Conference, Helmut-Schmidt University (2013)
5. Schafeu, D., Remnimgs, A., Erui, D.: Einfluss der orientierung eines dipols auf seine uep signatur (WP#B). Allgemeine und Theoretische Elektrotechnik, Universitat Duisburg-Esseu, Technical report (2011)
6. Weaver, J.T.: The quasi-static field of an electric dipole embedded in a two-layer conducting half-space. Can. J. Phys. **45**, 1981–2002 (1967)
7. Mao, W.: Research on Modeling Method of the Moving Ship SR EM Filed and Its Propagation in Shallow Sea. Naval University of Engineering (2009)

Analysis of Fouling Mechanism of a 10 μm Metal Micro-sieve During Testing with a Medium-Blocking-Type Contamination Tester

Jixia Lu[1(✉)], Fang Zheng[1], Chenyu Fang[2], Ruiyu Guo[1], and Cheng Wang[1]

[1] School of Mechanical, Electronic and Information Engineering,
China University of Mining and Technology, Beijing, China
`lujx1971@126.com`
[2] School of Physics, Huazhong University of Science and Technology, Wuhan, Hubei, China

Abstract. The micro-sieve silting principle is used widely for detecting the solid particle contamination in oil on the spot since it has some distinct merits. DW1 contamination tester is a kind of portable solid particle contamination tester with a metal micro-sieve as its sensor medium. The analysis upon micro-sieve fouling mechanism is useful for establishing the contamination level testing model of the contamination tester and improving the testing accuracy. Based on the fundamental fouling theory of microfiltration, experiments of fouling exponent analysis are carried out at the DW1 contamination tester with a 10 μm-sized metal micro-sieve to study its fouling mechanism, and the fouling mechanisms are also analysed with the blocking mechanism models. The experimental results show that the whole oil contamination testing process at the tester is in the initial blocking moment of the micro-sieve, and the blocking process can be described by one mechanism, which is the standard pore blocking.

Keywords: Solid particulate contamination level · Micro-sieve silting principle · Metal micro-sieve · Fouling mechanism

1 Introduction

Solid particulate contaminants are the most dangerous contaminants in hydraulic systems, which caused 60%–70% of the total contaminant wear failures [1]. In order to promote the reliability and elements' lifespan of hydraulic systems, the concentration of solid contaminants in oil must be measured. DW1 contamination tester is a kind of portable apparatus with a metal micro-sieve as its sensor medium [2, 3], and it tests the levels of solid particulate contaminants in oil according to the micro-sieve silting principle under constant pressure. The operation principle diagram of the DW1 contamination tester is shown in Fig. 1. The measured fluid sample is placed in the sealing chamber. The action pressure can be adjusted by the pressure regulating valve. When the fluid sample flows through the micro-sieve under operational pressure, the contaminants in the fluid block the pores in the sieve, which causes the flow rate decays with time. The rate of decay is related to the degree of contamination. A displacement transducer, which connects with the fluid cylinder, can transfer the information of flow rate into electric

© Springer International Publishing AG 2018
F. Qiao et al. (eds.), *Recent Developments in Mechatronics and Intelligent Robotics*,
Advances in Intelligent Systems and Computing 690, DOI 10.1007/978-3-319-65978-7_52

signal that can be processed by a computer. From the above operational principle of the DW1 contamination tester, it can be concluded that understanding upon the blockage of the metal micro-sieve is the key point for measuring oil contamination levels by this type of tester. Therefore, it is very important to know the filtration process of the medium since the blocking process of a micro-sieve is equal to the process of medium filtration. As for a full filtration process, it can be explained by four kinds of blocking mechanisms, namely, the complete pore blocking, the standard pore blocking, the intermediate pore blocking and the cake filtration. However, different operation conditions result in different fouling mechanisms. So far, many researchers have carried out a lot of work on the fouling mechanisms of the microfiltration process, however they mainly aimed at such fields as medicine, water treatment, chemical engineering and so forth [4–12]. Little has been done to the blocking mechanisms of the metal micro-sieve used in the medium-blocking-type contamination tester to detect the oil contamination level based on the micro-sieve silting principle. Analysis upon the blocking mechanism of the metal micro-sieve is useful for establishing the contamination level testing model of the contamination tester accurately. Therefore, this paper will carried out research work on the analysis of fouling mechanisms of a 10 μm metal micro-sieve that is commonly used in the medium-blocking-type contamination tester.

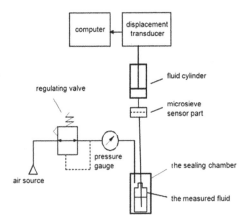

Fig. 1. Operation principle of DW1-type contamination detector

2   Classical Fouling Theory

The metal micro-sieve with pore size of 10 μm is usually used in a DW1 contamination tester, and it intercepts solid particles in oil flow driven by a constant pressure according to the sieving principle like a micro-pore filter membrane, the classical fouling theory of membrane microfiltration is used here to describe the filtration process. The common mathematical form of the fouling theory is [13]:

$$\frac{d^2t}{dV^2} = k\left(\frac{dt}{dV}\right)^n \tag{1}$$

Where t is the filtration time, V is the total filtered volume, k and n are constants related to contamination. The constant n reflects the filtering model [14], with n = 2 for complete blocking, n = 1.5 for standard pore blocking, n = 1 for intermediate pore blocking, and n = 0 for cake filtration. The concrete expressions of blocking mechanisms are shown in Table 1.

Table 1. Concrete expression of blocking mechanisms

Blocking mechanisms	Expression of filtrate flow rate
Complete blocking	$Q = Q_0 - K_b V$
Incomplete blocking	$\dfrac{1}{Q} = \dfrac{1}{Q_0} + K_i t$
Standard blocking	$\sqrt{Q} = \sqrt{Q_0} - (\dfrac{K_s V \sqrt{Q_0}}{2})$
Cake filtration	$\dfrac{1}{Q} = \dfrac{1}{Q_0} + K_c V$

In Table 1, Q, t, and V have the same meaning as the previous section, Q_0 is initial volumetric filtrate flow rate, K_b, K_i, K_s and K_c are constants correspond to different models.

Therefore, the blocking mechanism of the metal micro-sieve during the whole contamination level testing process can be analyzed from two aspects. One is to study the value of the exponent n during the whole testing process, the other is to compare the relationships of the volumetric filtrate flow rate, the total filtered volume and the filtration time with the models in Table 1.

3 Materials and Methods

The micro-sieve sensor part is composed of the micro-pore sieve, the up shell and the down shell. The structure of the micro-sieve sensor part is shown in Fig. 2. The micro-pore sieve made by electro-less plating method is evenly distributed with 10 µm micro pores [15]. ACFTD, a kind of standard test dust, is chosen to represent the actual contaminants in order to implement experiments related to the hydraulic contamination

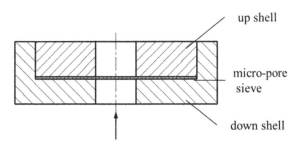

Fig. 2. The structure of the micro-sieve sensor part

control. No. 10 aviation hydraulic oil is selected as the working fluid. Two different concentrations of samples, 4.5 mg/L (NAS10 Level) and 17.85 mg/L (NAS12 Level), were prepared. The experiments were carried out under four kinds of action pressures from 0.3 MPa to 0.6 MPa. The data of NAS 12 are from a 10 μm micro-sieve element with a channel diameter of 2.5 mm and 3000 micro pores in it, and the data of NAS 10 are from a 10 μm micro-sieve element with a channel diameter of 1.2 mm and about 700 micro pores in it.

4   Results and Analysis

4.1   Analysis of Exponent N

By differentiating the logarithm of d^2t/dV^2 with respect to the logarithm of dt/dV, the exponent n can be evaluated analytically through the slope value on the double logarithmic plot [16]:

$$n = \frac{d\left[\log\left(\dfrac{d^2t}{dV^2}\right)\right]}{d\left[\log\left(\dfrac{dt}{dV}\right)\right]} \tag{2}$$

Figures 3 and 4 show the curves of d^2t/dV^2 versus dt/dV on the log-log plot according to data from the oil samples of NAS12 and NAS10 under the action pressures of 0.4 MPa and 0.5 MPa, respectively. From the two figures, all the values of the exponent n drawn from the experiments vary around 1.5, which is correspondent to a standard pore blocking mechanism, that means the main role of solid particles are to block the pores during the testing process, and there are no stacks between them, therefore,t here is no time for forming filter cake, and the detecting process is at the initial fouling moment of the micro-pore filter membrane. The variation in the operation pressure has no effect on the fouling mechanism of 10 μm micro-sieve during the contamination testing process.

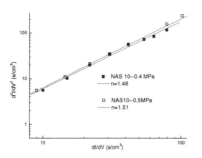
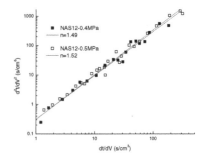

Fig. 3. Fouling exponent of NAS12 sample under different trans-membrane pressures

Fig. 4. Fouling exponent of NAS10 sample under different trans-membrane pressures

4.2   Fouling Mechanism Analysis by Mechanism Models

Figure 5 shows the plots of 1/Q versus t at two kinds of concentrations under four kinds of action pressures. The reciprocal of Q increased rapidly and were nonlinear with time since the flow rate through the blocked micro-sieve declined quickly. The tendency of the curves were the same in despite of different oil concentrations and action pressures. Higher pressure caused faster declining in volumetric flow rate and steeper curve, which means the flow rate decreased to a certain value in a shorter time under a higher action pressure. The curves in Fig. 5 show that the fouling mechanism is not dominated by incomplete blocking since the value of 1/Q versus t are not in linear relationship according to Table 1.

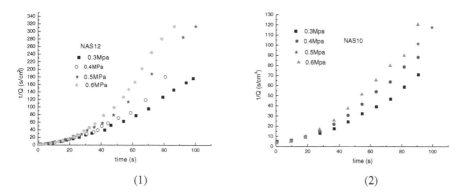

Fig. 5. Curves of 1/Q vs. t under various pressures and different sample concentrations

Figure 6 shows the plots of 1/Q versus V at two kinds of concentrations under four kinds of action pressures. The tendency of the curves were the same in despite of different oil concentrations and action pressures. In the initial period, the total filtered volume through blocked membrane increased quickly, and the flow rate through blocked

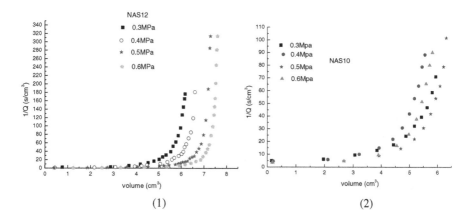

Fig. 6. Curves of 1/Q vs V under various pressures and different sample concentrations

membrane is larger, therefore, the reciprocal of Q increased slowly with volume. With the blockage growing, the increase in volume is much more slowly than the decrease in flow rate, which caused the reciprocal of Q increased greatly with V. Higher action pressure produced larger flow rate and quicker increase in volume, and hence a longer gentle initial period. The behavior is that the curves at lower action pressures came into abruptly increased period earlier than that at higher pressures. Obviously, according to the model in Table 1, the fouling mechanism is also not dominated by cake blocking since the relationship between 1/Q and V are not linear.

Figure 7 shows the plots of Q versus V at two kinds of concentrations under four kinds of action pressures. The volumetric flow rate decreased as the volume growing. In the initial time, the volumetric flow rate declined greatly with the blocking of the membrane pores, but the tendency of the curves became gentle since the declining speed of the flow rate went into slow after a certain time. Higher action pressures corresponded to larger volumetric flow rate. Similarly, according to Table 1, the fouling mechanism is also not dominated by complete blocking since the relationship between Q and V are not linear.

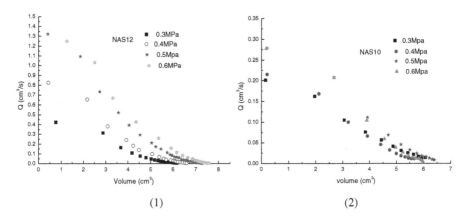

(1)   (2)

Fig. 7. Curves of Q vs. V under various pressures and different sample concentrations

Figure 8 shows the curves of the square root of Q versus V at two kinds of concentrations under four kinds of action pressures. It is clear that the square root of Q is linear with V, which is in good compatible with the standard blocking expression in Table 1. This also indicates that the blocking mechanism of 10 μm metal micro-sieve is standard pore blocking during the whole testing process with DW1 contamination tester.

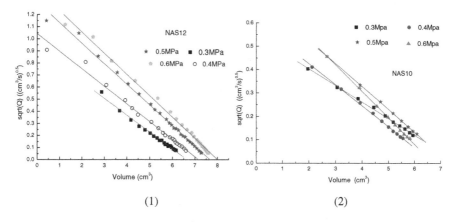

Fig. 8. Curves of sqrt (Q) vs. V under various pressures and different sample concentrations

5   Conclusions

According to the fundamental fouling theory, the fouling exponent of 10 μm micro-sieve are analysed with the data from DW1 contamination testing. Results show that the exponent values vary around 1.5, which means the whole oil contamination testing process with DW1 contamination tester is in the initial pore blocking moment of the micro-sieve, and the blocking process of the metal micro-sieve can be described by the standard pore blocking. The fouling mechanism analysis by blocking mechanism models also show the same conclusion, and the variation in the operation pressures and the channel diameters didn't change the fouling mechanism.

Acknowledgements.   This study was funded by the National Natural Science Foundation of China (Grant No. 51375481).

References

1. Jia, R.: Hydraulic Filtration Technology and Anti-wear Theory. China University of Mining and technology Press, Beijing (1999). (In Chinese)
2. Guo, J.: Study on the testing system of the DW-1 type instrument for measuring fluid contaminant levels. Master's degree dissertation. China University of Mining and Technology, Beijing (2005). (In Chinese)
3. Shi, L.J.: Study on the instrument for monitoring the contaminant level of fluids (DW1). Master's degree dissertation. China University of Mining and Technology, Beijing (2005). (In Chinese)
4. Duclos-Orsello, C., Li, W., Ho, C.-C.: A three mechanism model to describe fouling of microfiltration membranes. J. Membr. Sci. **280**, 856–866 (2006)
5. Hwang, K.-J., Liao, C.-Y., Tung, K.-L.: Analysis of particle fouling during microfiltration by use of blocking models. J. Membr. Sci. **287**, 287–293 (2007)
6. Li, J., Xi, D., Shi, Y.: Resistance distribution and fouling mechanism of dynamic membrane in wastewater treatment. J. Chem. Ind. Eng. (China) **9**, 2309–2315 (2008). (In Chinese)

7. Juang, R.-S., Chen, H.-L., Chen, Y.-S.: Membrane fouling and resistance analysis in dead-end ultrafiltration of Bacillus subtilis fermentation broths. Sep. Purif. Technol. **63**, 518–524 (2008)
8. Li, Z., Youravong, W., Aran, H.: Analysis of fouling mechanism during dead-end microfiltration of tuna spleen extract. J. Anhui Univ. Technol. Sci. **21**, 1–6 (2006)
9. Lee, S.-J., Dilaver, M., Kim, J.-H.: Comparative analysis of fouling characteristics of ceramic and polymeric microfiltration membranes using filtration models. J. Membr. Sci. **463**(3), 97–106 (2013)
10. Ma, J., Xiang, J., Li, J., Li, L.: Fouling machanism of ceramic membranes in pretreating seawater by cross -flow microfiltration. Membr. Sci. Technol. **30**(3), 87–92 (2010). (In Chinese)
11. Iritani, E., Mukai, Y., Tanaka, Y., et al.: Flux decline behavior in dead-end MF of protein solutions. J. Membr. Sci. **103**, 181–191 (1995)
12. Giglian, S., Straeffer, G.: Combined mechanism fouling model and method for optimization of series microfiltration performance. J. Membr. Sci. **417–418**, 144–153 (2012)
13. Hermia, J.: Constant pressure blocking filtration law: application to power law non-Newtonian fluids. Trans. Inst. Chem. Eng. **60**, 183 (1982)
14. Jacob, J., Pradanos, P., Calvo, J.I., et al.: Fouling kinetics and associated dynamics of structural modifications. Colloids Surf. A **138**, 173–183 (1998)
15. Jixia, L., Hui, D., Liang, F., Ruiqing, J.: Method to make metal micro-pore filter membrane through electro-less nickel-plating. J. Mech. Eng. **50**, 171–176 (2014). (In Chinese)
16. Ho, C.C., Zydney, A.L.: A combined pore blockage and cake filtration model for protein fouling during microfiltration. J. Colloid Interface Sci. **232**, 389 (2000)

Three Cameras Car Seat Images Mosaic Based on Improved SIFT Method

Juan Zhu[✉], Xiaoguang Li, and Pingping Xiao

Electronic Information College of Changchun Guanghua University, Changchun, China
zhuj_guanghua@126.com

Abstract. To mosaic the specific application of car seat detection image, improved SIFT method is proposed. Firstly, use the mathematical morphological features of the image to search the similar region between the matching image and the image to be matched. Secondly, calculate the SIFT features in the similar region to match the two images accurately. Last, through the image geometric correction and image fusion to achieve seamless splicing between image sequences.

Keywords: Car seat · Image mosaic · Improved SIFT method · Image fusion

1 Introduction

Automatic monitoring of car back, the main detection back curtain, synchronous rod plastic and backrest spring of three parts, and to detect the back of the whole image, when local detection and global detection are qualified, that car back production of qualified products, or considered unqualified products, need to re processing, placing three the camera for image acquisition of 3 local position in the car behind the backrest, and the global image detection needs formed by 3 independent camera image mosaic, thus directly determines the reliability of the detection of car back.

At present, there are two kinds of algorithms based on gray correlation and feature. The former is unstable to image rotation and scale scaling, and the algorithm based on local invariant feature has a good adaptability to a variety of image transformation and gradually become a research hotspot. However, the biggest drawback of this algorithm is that it is difficult to meet the requirement of real time.

In this paper, an image mosaic algorithm based on improved SIFT feature is proposed, which can realize the seamless splicing of deformation and scale transform images accurately, quickly and stably. This algorithm can reduce the geometric distortion of the image and improve the matching speed of the image.

2 Image Mosaic Framework

The car back images of 3 camera, 3 cameras in different angles, shooting position is different, but the 3 position of the camera is fixed, can roughly determine the actual

© Springer International Publishing AG 2018
F. Qiao et al. (eds.), *Recent Developments in Mechatronics and Intelligent Robotics*,
Advances in Intelligent Systems and Computing 690, DOI 10.1007/978-3-319-65978-7_53

position of the captured image, so in the mosaic of 3 images, mainly according to the following procedure.

The quality of image preprocessing of the image processing to remove noise, remove the uneven illumination interference, followed by a region matching way to find the overlapping area of two images, in the overlap area using the improved SIFT algorithm to extract the feature points to complete image registration. Finally, the image geometric correction and fusion, finally get seamless image mosaic.

Specific diagram is shown in Fig. 1.

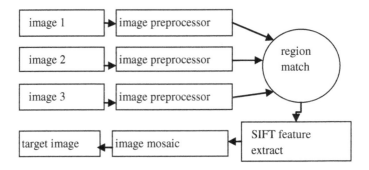

Fig. 1. Image mosaic framework diagram

Three original images are preprocessed respectively. According to the mathematical morphology feature, match the region roughly. Then using SIFT method to get the target image.

3   SIFT Features Extract

The purpose of image matching is to determine the matching image and the corresponding relationship between images matching using a standard measurement, to calculate the transformation parameters of two images, from different perspectives, different sensors with overlapping two or more images into the same coordinate system transform. The SIFT algorithm uses the image scale space theory to extract the feature points, and has good robustness to image rotation, noise and so on.

Generate Scale Space.   In SIFT algorithm, the scale space of a two-dimension image is defined as:

$$L(x, y, \sigma) = G(x, y, \sigma) \times I(x, y) \tag{1}$$

$$G(x, y, \sigma) = \frac{1}{2\pi\sigma^2} e^{-(x^2+y^2)/2\sigma^2} \tag{2}$$

(x, y) is the space coordinate. σ is called scale space coordinate.

To effectively detect the stability features (here the features performance as some key dots). We need Gaussian difference scale space (DoG).

$$D(x, y, \sigma) = (G(x, y, k\sigma) - G(x, y, \sigma)) \times I(x, y) = L(x, y, k\sigma) - L(x, y, \sigma) \tag{3}$$

The construction of scale space could do well with the zooming problem.

Extreme Value Point Detect. Suppose the coordinate of current point is (x,y) in LoG scale space, the corresponding scale space coordinate is σ, compare (x,y) in σ with its every adjacent point: in σ, there are 8 points. They are $(x - 1, y - 1)$, $(x - 1, y)$, $(x - 1, y + 1)$, $(x, y - 1)$, $(x, y + 1)$, $(x + 1, y - 1)$, $(x + 1, y)$, $(x + 1, y + 1)$ respectively. In $\sigma + 1$ and $\sigma - 1$, there are 9 points. They are $(x - 1, y - 1)$, $(x - 1, y)$, $(x - 1, y + 1)$, $(x, y - 1)$, (x, y), $(x, y + 1)$, $(x + 1, y - 1)$, $(x + 1, y)$, $(x + 1, y + 1)$ respectively. When compared with this 26 points, the current point is the maximum value or minimum value, it should be as the extreme value point.

Determine the Precise Extreme Value Points. There are a lot of redundancy and unstable points in the extreme value points. So it is necessary to search the precise extreme value points. First remove the edge point. Introduce Hessian matrix:

$$H = \begin{bmatrix} D_{xx} & D_{xy} \\ D_{xy} & D_{yy} \end{bmatrix} \tag{4}$$

Here

$$
\begin{aligned}
D_{xx} &= L(x + 1, y) - L(x,y) + L(x - 1, y) - L(x,y) \\
D_{xy} &= L(x + 1, y + 1) + L(x - 1, y - 1) - L(x - 1, y + 1) - L(x + 1, y - 1) \\
D_{yy} &= L(x,y + 1) - L(x,y) + L(x,y - 1) - L(x,y)
\end{aligned}
\tag{5}
$$

Define

$$
\begin{cases}
Tr(H) = D_{xx} + D_{yy} \\
Det(H) = D_{xx}D_{yy} - (D_{xy})^2
\end{cases}
\tag{6}
$$

Suppose α is the maximum eigenvalue, β is the minimum eigenvalue. $\alpha = \gamma\beta$. Then

$$\frac{(\alpha + \beta)^2}{\alpha\beta} = \frac{(\gamma + 1)^2}{\gamma} \tag{7}$$

When $\dfrac{Tr(H)^2}{Det(H)} < \dfrac{(\gamma + 1)^2}{\gamma}$, the corresponding extreme value is chosen by our system.

3.1 Distribute the Direction and Generate the Description of Each Feature Points

Distribute the direction for each point using the following formula:

$$\begin{cases} m(x, y) = 2\sqrt{D_x^2 + D_y^2} \\ \theta(x, y) = \text{atan}\dfrac{2D_y}{D_x} \end{cases} \tag{8}$$

Here

$$\begin{cases} D_x = \dfrac{L(x + 1, y) - L(x - 1, y)}{2} \\ D_y = \dfrac{L(x, y + 1) - L(x, y - 1)}{2} \end{cases} \tag{9}$$

$m(x, y)$ is the modulus of the vector, and $\theta(x, y)$ is the angle of the vector.

4 Image Fusion

The relationship between the pixel coordinates of the image acquired by the vehicle seat image sensor in certain motion modes can be determined by the transformation model between the images. The transformation model is:

$$\begin{bmatrix} x' \\ y' \\ 1 \end{bmatrix} = HX = \begin{bmatrix} a_{11} & a_{12} & a_{13} \\ a_{21} & a_{22} & a_{23} \\ a_{31} & a_{32} & a_{33} \end{bmatrix} \cdot \begin{bmatrix} x \\ y \\ 1 \end{bmatrix} \tag{10}$$

According to the matching of the image points, you can find the two images of the transformation matrix H, through the projection transform can achieve the purpose of geometric correction.

5 Experimental Results

This paper mosaics the 3 cameras car seat images. The images have certain rotation angle, a certain zooming. 3 cameras are placed at fixed places. However, due to the impact of the working environment, the 3 original images not always the same places of one seat. So we should do some work to mosaic the three images reliably.

Experimental results show that the mosaic image is stitched. Mosaic effect is as follows:

Here, Figure a are three original images. b are the mathematical morphological rough detect images. The original image is split into four parts. Search the rough match region. Red rectangle shows the match part. c is the SIFT feature. d is the mosaic effect (Fig. 2).

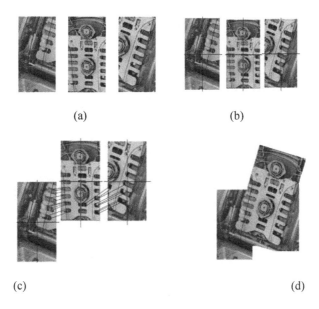

Fig. 2. Three images mosaic effect

When using VC++6.0 to verify the performance of the algorithm, the complexity of the algorithm is as follows: Compared with traditional SIFT method. This method reduce the calculate time. The time cost is shown in Table 1.

Table 1. Time cost of traditional SIFT and this method

Time cost (ms)	Step		
	Preprocessor	Rough detect	SIFT
Traditional SIFT	<1	X	83
This method	<1	<1	20

6   Conclusion

To mosaic the specific application of car seat detection image, improved SIFT method is proposed. Firstly, use the mathematical morphological features of the image to search the similar region between the matching image and the image to be matched. Secondly, calculate the SIFT features in the similar region to match the two images accurately. Last, through the image geometric correction and image fusion to achieve seamless splicing between image sequences. Experimental results show that this method could mosaic 3 cameras car seat images stitched.

Acknowledgement. The work described in this paper was fully supported by Jilin Province Development and Reform Commission (No. 2017C037-2).

References

1. Su, M.S., Wang, L., Cheng, K.Y.: Analysis on multiresolution mosaic images. IEEE Trans. Image Process. **13**(7), 952–959 (2004)
2. Wang, X.H., Huang, W., Ouyang, J.H.: Realtime image registration of the multi-detectors mosaic imaging system. Chin. Opt. **8**(2), 211–218 (2015)
3. Liu, Z.H.W., Liu, D.S.H., Liu, P.: SIFT feature matching algorithm of multi-source remote image. Opt. Precis. Eng. **21**(8), 2146–2153 (2013)
4. Liu, R., Zhang, J.S.H.: Application of the extraction of the image feature points by improved SIFT algorithm. In: IEEE International Conference on Communications, Budapest, Hungary, 9–13 June 2013, pp. 946–949
5. Wang, S., Wang, J.P., Wang, G.T., et al.: Image matching method based on SIFT algorithm. J. Jilin Univ. (Eng. Technol. Ed.) **43**(Suppl), 279–282 (2013)
6. Zhu, J., Wang, S., Meng, F.Y.: SIFT method for paper detection system. In: ICMT 2011, pp. 711–714

Biomechanical Principle of Slip Hip in Shadowboxing

Chun-Min Ma[1(✉)], Jun Guo[1], Yan Li[1], and Kang-Hui Huang[2]

[1] Beijing University of Technology, Beijing, China
machunmin74@126.com
[2] Beijing Sport University, Beijing, China

Abstract. Slip the hip is one of the movement characteristics of Shadowboxing which plays a vital role for the Tai chi enthusiasts relaxing waist and sinking the breath to the dan tian. It can also communicate the up and down parts of body, and support the external force during pushing hands. In this paper, a Taijiquan athletes' attitude of slip hip and cock hip were recorded. By analyze these data using the software AnyBody and Ansys that is famous for human body kinematics analysis and finite element analysis, a conclusion was drawn. The pose slip the hip is twice of pose cock the hip in output force of erector spinae which can equably supports the waist and meets the requirement of Shadowboxing. It also reveals that the pose slip hip is 3 times of the pose cock hip in pressure bearing and force transmission which is important for stimulate the concentrated force of Tai Chi.

Keywords: ShadowBoxing · Biomechanical · Finite element analysis

1 Introduction

Tai Chi person should relax body which need to keep the spine in vertical shape. This kind of pose can make person release force by whole body not part of body. The force like this is called "Integrated force" in traditional way. Many of the old Tai Chi person, after a long practice, whom is over seventy years old can still keep tall and straight bearing [1]. This pose always brings them the feeling of blood smooth and happy spirit. Human being has three physiological bending on the cervical, thoracic and lumbar of spine. A key body position in Tai Chi that can make lumbar spine to straight named slip the hip. It is a basic posture on which Tai Chi person can do shadow boxing and pushing hands better.

In this paper, the human body kinematics analysis software AnyBody and finite element analysis software ANSYS are used to analyze and calculate the kinematics and biomechanics characteristics of the pose slip hip and cock hip.

2 The Kinematics Analysis of the Pose Slip Hip and Cock Hip

Tai Chi is a special exercise. It is very meaningful to analyze the characteristics of the movement and to explore its mechanism [2–4]. It is an effective way to analyze the

© Springer International Publishing AG 2018
F. Qiao et al. (eds.), *Recent Developments in Mechatronics and Intelligent Robotics*,
Advances in Intelligent Systems and Computing 690, DOI 10.1007/978-3-319-65978-7_54

posture of Tai Chi by AnyBody software, especially for the analysis of muscle and ligament [5–7].

2.1 The Difference Between the Pose of Slip Hip and Cock Hip

This topic invite a Taijiquan athletes to do an action named Qi Shi with slip the hip and cock the hip respectively. And recorded his attitude by measuring instrument (Fig. 1).

Fig. 1. Attitude acquisition of slip hip and cock hip

According to the principle of human anatomy [8, 9] and the analysis of the associated human joint angle, found that this gesture can be determined by 5 main angles. The reference system for measuring the obliquity of pelvis is ground. Other angles are the joint angles between the two parts of the human body (Fig. 2, Table 1).

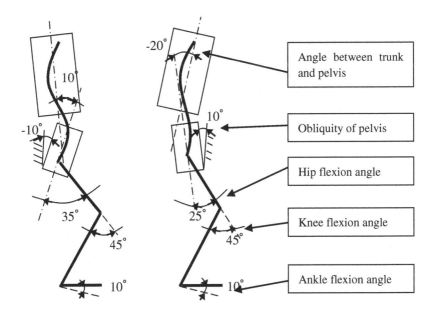

Fig. 2. Pose analysis of slip hip and cock hip

Table 1. Attitude data of slip hip and cock hip

Pose name	Pelvis obliquity	Trunk pelvis angle	Hip flexion angle	Knee flexion angle	Ankle flexion angle
Cock hip	−10°	10°	35°	45°	10°
Slip hip	10°	−20°	25°	45°	10°

Put the angles above into the StandingModel of Anybody [10], the three-dimensional human skeleton shows below (Fig. 3).

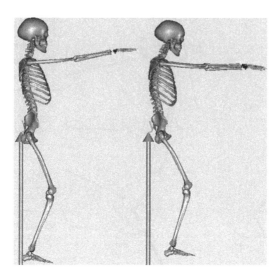

Fig. 3. Skeleton pose of slip hip and cock hip in Anybody

2.2 Muscle Force Research on Slip Hip and Cock Hip

Anybody software has the advantage of calculating human muscle force by reverse kinematics. Which is often used to evaluate the action and analyze the possible damage to human body. Due to erector spinae and vastus lateralis can well reflect the waist and leg force output, this paper focus on the calculation and analysis of these two muscles (Table 2).

Table 2. Muscle force calculate results

Pose name	MusclesSpineLeft.ErectorSpinae.LTptT9S4.Fm	Mus.VastusLateralis.Fm
Cock hip	10.8 N	403 N
Slip hip	25.7 N	419 N

Figure 4 shows that force output of erector spinae under the pose of slip hip is 2.38 times larger than cock hip. Tai Chi person's waist will have a "fill" feeling under the pose of slip hip. This feeling is just the result of the increase of erector spinae output force. During Push Hands the pose of slip hip makes the erector spinae keep larger force which can support the waist around in smooth way. Therefore it plays a very important role in the generation of "Integrated force". Otherwise, it will cause power interruption at the waist which is called "Waist break" in traditional way. In addition, the calculated result shows that output force of vastus lateralis is similar whether it is slip hip or cock hip.

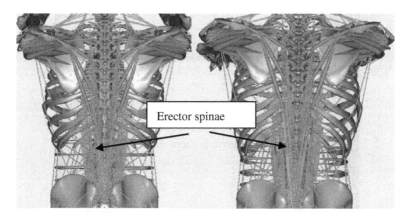

Fig. 4. Difference in erector spinae output force

3 Bone Biomechanics Research on Slip Hip and Cock Hip

For using Ansys software to calculate the spine stress under this two pose, it is needed to establish a mechanical model [11–13]. Many valuable papers can be found [14, 15] on the spine research. But because of such research is aimed at the treatment of certain diseases, the model and loading plan must meet their own needs. This paper focus on the comparative analysis of bone biomechanics between slip hip and cock hip when people practices shadow boxing, therefore a special mechanical model is established.

3.1 Model and Loading Plan

The model is built according to a person whose height is 1.7 m and weight is about 60 kg. Material properties: Young modulus $EX = 12$ (GPa), Poisson ratio $PRXY = 0.3$, Element partition: 10 nodes volume element solid 1,877,348 elements, Loads defined as shown in Fig. 5. Displacement at Sacral = 0, Force at two sides of thoracic vertebra = 300 N (150 N each side).

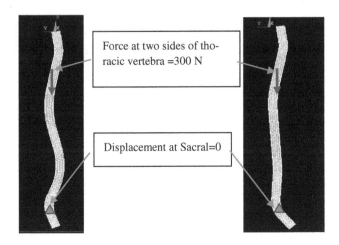

Fig. 5. Load plan of displacement and force

3.2   Calculation Results and Analysis

When Cock Hip (Fig. 6 left), the lumbar stress is 16 MPa, When Slip Hip (Fig. 6 right), lumbar stress is 6 MPa.

(1)   Analysis the different on the carrying capacity. The results above shows that when a person takes the same force under different postures, the stress of the lumbar when slip hip is only about 1/3 of the cock hip posture. Further analysis means that, if the maximum admissible stress is certain for lumbar stability, the carrying capacity under the slip hip posture is about 3 times that of the cock hip posture.

(2)   From the point of view on force block. Because of the assumption that the skeleton is linear elastic, the greater the stress value, the larger the deformation of the bone, the consumption of force is also a linear growth. The results above shows slip hip is 1/3 that of cock hip on the force block. That means slip hip posture can make the force runs through the body better than cock hip during push hands, weather push the opponent or carry the force from opponent. It makes the waist play a pivot role and stimulates the concentrated force of Tai Chi.

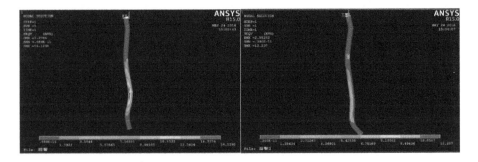

Fig. 6. Stress analysis result on lumbar of slip hip and cock hip

4   Conclusion

In this paper, the joint angle of slip hip and cock hip posture is determined by experiment. Muscle force and bone biomechanics are calculated by Anybody and Ansys software based on the body structure. The results is the pose of slip hip is very important to stimulate the concentrated force of Tai Chi.

References

1. Ma, C.: Wu Style Tai Chi Study Record at South Lake. HuaWen Press, Beijing (2016)
2. Wang, X.: Research progress of physiological and medical effects of Taijiquan. China Sport Sci. Technol. **47**(4), 113–114 (2011)
3. Field, T.: Tai Chi research review. Complem. Ther. Clin. Pract. **17**(3), 141–146 (2011)
4. Wu, D., Liu, H.: Analysis of the biomechanical characteristics of Yang Style Taijiquan main action. J. Beijing Sport Univ. **24**(1), 41–44 (2001)
5. Ji, Z., Li, X., et al.: AnyBody simulation and validation of biomechanical characteristics of lower limb in Tai Chi training. Chin. J. Rehabil. Med. **9**, 799–805 (2014)
6. Liu, S., Yan, S., et al.: Development of human motion modeling method of AnyBodyTM technology based on. Prog. Biomed. Eng. **31**(3), 132–133 (2010)
7. Enguo, C., Yoshio, I., Tao, L.: Estimation of lower limb muscle forces during human sit-to-stand process with a rehabilitation training system. In: Biomedical and Health Informatics, Hong Kong, pp. 1018–1019 (2012)
8. Gao, X.: Human Anatomy. Peking University Medical Press, Beijing (2009)
9. Hao, W.: Biomechanical modeling and computer simulation of human motion. J. Med. Biomech. **26**(2), 97–104 (2011)
10. AnyBody Model System Reference Manual. Aalborg University Biomedical Engineering Research Team, Denmark (2009)
11. Ma, C., Ma, H., Chen, Y.: Preliminary study on the biomechanical principle of the waist and hip of Tai Chi Chuan. Chin. Wushu Res. **3**, 75–79 (2014)
12. He, Y., Zhang, X., Tang, B.: Construction of hip three-dimensional finite element model based on DICOM data. J. Huazhong Univ. Sci. Technol. (Med. Sci. Ed.) **37**(2), 251–254 (2008)
13. Qian, Q., Jia, L., Zhou, W.: Three dimensional finite element analysis of the stress distribution on the sacroiliac joint in different positions. Chin. J. Orthopaed. **20**(3), 173–175 (2003)
14. Qin, J., Yu, W., Pang, X., et al.: A three-dimensional finite element model of the whole lumbar spine and validation of. J. Med. Biomech. **28**(3), 321–325 (2013)
15. Chen, B., Natarajan, N.: Establishment and significance of three-dimensional finite element model of cervical spine. Chin. J. Spine Spinal Cord **120**, 105–107 (2002)

A Real-Time Guiding System Based on ZigBee Network for Highway Traffic Safety in Fog Zone

Hao Li[1], Qiang Wang[1(✉)], and Fang He[2]

[1] School of Mechanical Engineering, University of Jinan, Jinan, Shandong, China
me_wangq@ujn.edu.cn
[2] School of Electrical Engineering, University of Jinan, Jinan, Shandong, China

Abstract. A real-time guiding system for highway traffic safety in fog zone is established in this study. The system integrates heavy fog monitoring system, wireless transformation system based on ZigBee network and warning system. The warning system is composed of sparking lamps and information display system. When receives the signal of heavy fog on highway, the system will send warning messages to drivers in time in case of the occurance of traffic accidents. The information transmission distance between the monitoring system and the display system is about 1 km. To ensure the feasibility of the transmission distance, the research verifies the length of the transmission distance in theory. Besides, amendment formula was established to ensure the correctness of the maximum distance at last.

Keywords: ZigBee network · Guiding system · Fog zone · Transmission distance

1 Introduction

Reminding the drivers driving on highway of real-time information about heavy fog is more and more necessary, since it can rescue drivers' lives in some extent. Traditional methods for reminding drivers to keep speed on highway in fog zone can be divided into following several approaches, such as installing static warning signs, setting tense flanges on the sides of the road, placing sensors for monitoring-distance among vehicles etc. All of the methods can play a role in avoiding the occurance of accidents in some extent, but they can not make the drivers learn the fog information in advance toward their way in time. In recent years, guiding systems for highway traffic safety have taken up and some related studies have been developed. Chen et al. have developed a guiding system based on monitoring and supervisory control, CCTV subsystem, broadcast subsystem etc. [1]. The device sends the warning signals by red and yellow sparking signal lamps. Qian used visibility sensors monitoring the fog weather and sending the messages to CCS [2]. Then, the warning messages will be sent to VMS. According to the event, guidance vehicles will be used to guide the vehicles in heavy fog. Zhou and Sun proposed an early warning discrimination technology and a strategy about load balancing, which decreases the server's pressure when the number of TCP connections is larger [3]. Qian developed an intelligent guiding system of expressway fog zone based on Beidou Satellite-Timing, it can automatically controls the switches and strobes of the

F. Qiao et al. (eds.), *Recent Developments in Mechatronics and Intelligent Robotics*,
Advances in Intelligent Systems and Computing 690, DOI 10.1007/978-3-319-65978-7_55

guidance lamps in bad weather like fog weather [4]. Although these technologies can prevent the inconvenience bring by heavy fog on highway effectively, they are not economical in some extent. Based on the deficiency above, this study created a guiding system that integrates fog monitoring system, alarming system, ZigBee network and display system. With the use of ZigBee modules, the guiding system will be more functional and reliable.

The paper is structured as follows. In Sect. 2, general approaches of the guiding system have been proposed. In Sect. 3, ZigBee network and data acquisition system has been established. In Sect. 3, verification of ZigBee transmission distance has been conducted. The system advantage and function have been discussed in Sect. 4.

2   Guiding System on ZigBee Network

This study established a guiding system mainly composed of fog monitoring system, ZigBee modules and warning system.

Figure 1 shows the intelligent guiding system of fog zone and Fig. 2 demonstrates the basic operational process of the intelligent guiding system. As shown in the two figures, information transmission of the real-time guiding system in the study is based on ZigBee network. The system consists of monitoring system, ZigBee information transformation network and warning system.

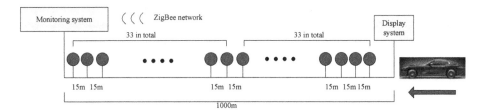

Fig. 1. The intelligent guiding system of fog zone

In the system, fog information signal will be received by fog monitoring system at first. This fog monitoring system is made up of particulate sensors, humidity sensors and data acquisition and processing modules. Firstly, fog information is monitored by fog monitoring system mentioned above. Then, the monitoring results processed by data acquisition and processing modules will be transmitted to display system through ZigBee network in real time. Eventually, the LED display screen will show warning messages to drivers on highway and the warning lamps alongside the highway will be sparking automatically once receives the signal sent by ZigBee network. The color of the warning lamps is orange since it will be conspicuous in fog zone. The distance between the monitoring system and the display system is designed as one kilometers because the distance of one kilometers is enough for drivers passing through the dangerous zone brings by heavy fog. Additionally, the whole system can also be remote controlled by PC through GPRS (General Packet Radio Service) remote transmission.

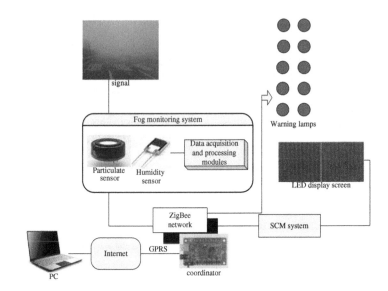

Fig. 2. Operational principle of the guiding system

To make the construction of the whole system clear, ZigBee network, data acquisition system and display system will be specific introduced in Sect. 3.

3   ZigBee Network and Linked System

From Fig. 3, we can clearly see the schematic structure of system networking. The whole system can be divided into three parts, fog monitoring system, ZigBee network and information display system. Schematic structure of the three systems has also been shown in the figure. Basic operational principle of the three subsystems has been introduced in Sect. 2 and concrete connections of hardware will be illustrated in this section.

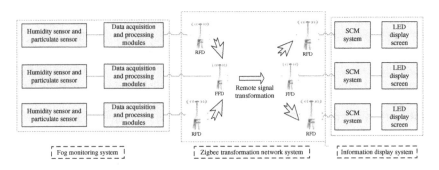

Fig. 3. Schematic structure of system networking

It's easy to see that ZigBee network plays a critical role in the whole system. ZigBee has the advantages like low power consumption, high security, low cost, large capacity of network and low time lag. Low cost can make the system more economical, large capacity of the network can transmit more real-time information and low time lag will accelerate the velocity of messages transmission. In this study, ZigBee network is made up of FFD and RFD. Besides, ZigBee topology choose star topology. The realization of the remote transmission is based on FFD.

The systems link with Zigbee network are data acquisition system and display system. Figure 4 shows data acquisition and processing device connection while Fig. 5 illustrates the information display system of the system.

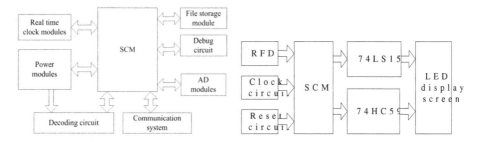

Fig. 4. Data acquisition device connection   **Fig. 5.** Information display system

As shown in Fig. 4, when the system is running, sensors will send acquired data to AD modules to conduct some basic settlement. Then, SCM will settle the data with algorithm. Next, processed data will be deposited into file storage module. Eventually, resolved or acquisition data will be send to FFD through ZigBee network. Besides, the data will also be sent to PC by GPRS to be monitored in real time.

According to the information display system of the system, RFD will send received data to SCM at first. Then, reset circuit provides power signal to power system. When power operates smoothly, the power signal will be revoked. Clock circuit will offer signal to make other units of SCM system operate smoothly. Next, SCM will process the data and send them to decoders. The decoders consists of 74LS154 (line scan driver) and 74HC595 (column scan driver). At last, LED will be successively lightened by the decoders and shows designed Chinese warning messages.

ZigBee networking links the monitoring system and display system like a bridge. To ensure the actual length of the "bridge" can satisfy designed length (1 km), the theoretical verification of ZigBee transmission distance is proposed. Figure 6 shows the classical transmission system and Fig. 7 shows the transmission module of Zigbee based on CC2530 chip of the system.

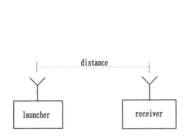

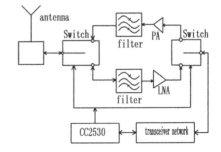

Fig. 6. Classical transmission system

Fig. 7. Transmission module of ZigBee based on CC2530

From Fig. 6, it'll be learned that transmission system is composed of launcher and receiver. In the system, FFD plays the two roles in transmitting information. Besides, CC2530 chip is the core of the transceiver system. Since the limitation of the fog weather, many parameters can't be directly used because the power P_t (power of transceiver), P_{rs} (receiving power on the side of antenna) in the maximum distance formula is referred to transceiver system and antenna connections. The maximum transmission distance formula Eq. (1) is obtained by Kita et al. [5].

$$d_{max} = 10^{\frac{1}{10n}\left[P_t - P_{rs} + A_{ant} + G_t + G_r + P_{L0} + s - 20(k+1)\log(f/5)\right]} \tag{1}$$

n–distance loss correction coefficient
A_{ant}–decay factor
G_t, G_r–the gain of transceiver antenna
P_{L0}–loss under the referenced distance
s–standard division of loss calculation
k–correction factor of frequency
f–frequency of ZigBee networking

Because highway is in outdoor, the value of some parameters proposed by the formula in the context can be confirmed. Concrete value is listed in Table 1.

Table 1. Value of partial parameters

P_{t2530}/dBm	4.5	S_{i4}/dBm	0.2–1.0
P_{rs2530}/dBm	−96	A_{i1}/dBm	6
P_{L0}/dBm	−48.96	n	1.58
A_{ant}/dBm	−3	k	0
S_{i1}/dBm	1–2	s	−3.96
S_{i2}/dBm	0.1–0.5	P_{L0}/dBm	−48.96
S_{i3}/dBm	0.2–2	f/GHZ	2.4

According to ZigBee network in this system, it can obtain the power gain of $PA(A_{o1})$, the power gain of (low noise amplifier) $LNA(A_{i1})$ and the maximum launching power(P_{t2530}). Concrete data of the mentioned parameters can be found out in Table 1.

In the signal transmission process, launching signal will pass through two switches, an power amplifier, a filer and an antenna connection according to Fig. 7. By calculating, value of the power in launching terminal is 16.5 dBm.

Similarly with the transmission process of launching terminal, the signal will pass through an antenna, two switches, a filer and LNA amplifier. the receiving terminal loss and the receiving sensibility of CC2530 can be represented in Eqs. (2) and (3) according to the study Guo et al. have done [6].

$$S_r = S_{i1} + S_{i2} + S_{i3} + S_{i4} - A_{i1} \tag{2}$$

$$P_{rs} = P_{rs2530} + S_r \tag{3}$$

S_{i1}–Transceiver network mismatch loss
S_{i2}–switching loss
S_{i3}–filter loss
S_r–power loss of the receiving side
S_{i4}–Transceiver switching loss
A_{i1}–LNA gain
P_{t2530}–output power of CC2530
P_{rs2530}–receiving power based on CC2530

The gain of Transmitting antenna is supposed as unit gain ($G_t = G_r = 0$).

For the value of parameters into Eq. (1), the maximum distance between ZigBee modules will be evaluated. It's about 9649.50 m. The transmission distance evaluated can naturally satisfy the needed transmission distance (one kilometers).

Considering the special environment in fog weather, correction factors have been added to the Eq. (1). Value of fog humidity $h_{(t)}$ and particulate amount in monitoring zone $p_{(t)}$ are interfered with in this study. Eq. (4) shows the fog influence factor and the Eq. (1) can be transferred into Eq. (5).

$$k_{(t)} = k_1 h_{(t)} + k_2 p_{(t)} \tag{4}$$

k_1–humidity influence factor
k_2–particulate influence factor
$k_{(t)}$–influence factor in fog weather

$$d'_{max} = k_{(t)} d_{max} \tag{5}$$

d'_{max}–amended information transmission distance of ZigBee network in fog zone

$h_{(t)}$ and $p_{(t)}$ will vary with the change of time, so the maximum distance is a variable. Because heavy fog can rarely effect the transmission of ZigBee network, $k_{(t)}$ will be a constant close to one. So the transmission distance of one kilometers can be ensured. For the reason of the limitation of experiment condition, influence factors haven't been estimated so far. Therefore, further study and research should be done in the future.

4 Conclusion

A fog guiding system on highway is established in this study. Different from current approaches, the signal transmission of the system is based on Zigbee network, which makes the system more economical and convenient to launch and receive information. In this case, drivers can receive the warning messages from display screen in real time. Besides, the guiding system can be remote monitored by PC. The research also launches the theoretical verification of information transmission distance based on Zigbee network and raised the amendment formula. Above all, the intelligent guiding system will be helpful for ensuring the traffic safety on highway in fog zone.

References

1. Chen, J.S., Zhou, J.P., Lin, J.B.: Research and application of intelligent electronic guiding system in the fog zone of freeway. Technology Forum **9**, 147–149 (2008). (in Chinese)
2. Qian, J.: The research and exploration of guidance system in the fog area of highway. Technol. Prod. **11**, 109–111 (2004). (in Chinese)
3. Zhou, Y.F., Sun, G.: Research on key technology of expressway fog guidance system. Ind. Control Comput. **26**(12), 66–67 (2013). (in Chinese)
4. Qian, H.F.: Intelligent guiding system of expressway fog zone based on Beidou satellite-timing. Traffic Eng. **22**, 98–101 (2013). (in Chinese)
5. Kita, N., Yamada, W., Sato, A.: Path loss prediction model for the over-rooftop propagation environment of microwave band in suburban areas. Electron. Commun. Jpn. **90**(1), 13–24 (2007)
6. Guo, H.F., Bai, L.N., Guo, Z.H.: Estimation method of the transmission distance for the 2.4 GHz Zigbee application. J. Xidian Univ. **36**(4), 691–695 (2009). (in Chinese)

The Energy Model Does Depend on the Absolute Phase of the Input Images

Li Zhao$^{(\boxtimes)}$

School of Information Engineering, Henan University of Science
and Technology, Luoyang 471023, Henan, People's Republic of China
bcshaust@163.com

Abstract. The energy models for motion and stereo vision is considered as independent of the absolute phase of the input images. However, Zhao and Farell report a simulation result that the energy model may depend on the absolute phase. Here we show that the energy model does depend on the absolute phase of the input images based on a formal theoretical analysis. We further prove that this absolute effect holds no matter quadrature or non-quadrature exists for the receptive fields. Finally we would like to emphasize that, although in the single neuron level phase independence are approximately true for some neurons, in the perception/behavioral level, for complex biological vision systems like human and primates, since there are billions of complex neurons covering the full spectrum of the frequency domain, phase dependence can not be omitted.

Keywords: Binocular complex cell · Energy model · Stereo vision · Motion · Absolute phase

1 Introduction

Stereo and motion vision are crucial functions for human and other primates. These two functions are similar in that both need at least two slightly different images. For stereo vision we need two images (in two eyes) at the same time while for motion we need two images in different times. The neural mechanisms mediates the two phenomena are modeled by variants of energy models [1, 7]. The energy model is built to be independent of the absolute phase of the input stimulus [1] and nobody even doubt about this. (We would like to emphasize that, in Adelson and Bergen [1] paper, there is no indication that the phase independence is only approximately true. On the contrary, as pointed out by Morrone and Burr [5], one may be surprised by the notion that energy model is phase sensitive. This is the start point when we began to do a computational study of stereo vision using energy model in early 2001. However, it [14] was found that the energy model may depend on the absolute phase. This was totally out our expectation, and our result is not generally accepted since the absolute phase effect of the energy model is never been disputed (as pointed out by a reviewer).

Recently Ng, Bharath and Li [6] made us realized that this effect was first pointed out by Morrone and Burr in 1988 [5]. However, their research is so rarely known that we did not know this when began our project. However their result was explanative and

F. Qiao et al. (eds.), *Recent Developments in Mechatronics and Intelligent Robotics*,
Advances in Intelligent Systems and Computing 690, DOI 10.1007/978-3-319-65978-7_56

their result that simple images were independent of the absolute phase is incorrect. As we show later, the phase effect does not depend on whether the images are simple. Here we present a formal theoretical analysis which indicates that the energy model does depend on the absolute phase. The theoretical analysis is without any approximation. Therefore this theoretical result should give us full information of the energy model.

2   An Analytical Proof of the Absolute Phase Effect

Since the energy model for motion can be considered as a special case of the energy model for stereo vision, we present a theoretical analysis of the energy model for stereo vision only. The energy model for stereo vision models a binocular complex cell. This binocular energy model for a binocular complex cell is consistent with most experimental results [2]. In Fig. 1, we can see that the energy model is made of four stereo simple cells. These stereo simple cells make two pairs. The two pairs has a quadrature difference of the phase. Further relations are shown in the figure.

This binocular energy model is nonlinear. To analyze nonlinear relation analytically is generally impossible. This is why the computational methods have been used to investigate nonlinear function. This is also the logic we followed. However, a formal proof was found. This proof corrects previous agreement that model responses are invariant with absolute phase of the input images. Contrary to former theoretical study of binocular energy model [1, 7, 10], our theoretical analysis is without any approximation.

Suppose the two receptive fields of a simple binocular cell in the binocular energy model are $e^{-(ax2+by2)}cos(2\pi f_r x + p_r^1)$ and $e^{-(ax2+by2)}cos(2\pi f_r x + p_r^2)$. a and b are determined by frequency and neurophysiological discovery described in Zhao and Farell [14]. f_r is the optimal spatial frequency of the receptive fields. p_r^1 and p_r^2 are the absolute

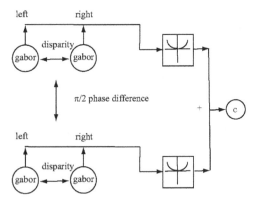

Fig. 1. The energy model of a binocular complex cell. For each eye, there are two pairs (top and bottom, four in total) of binocular simple cells. The two simple cells in each pair correspond to the positive and negative parts of each square function (x^2). The simple cells for negative and positive parts of the square functions inhibit each other. The receptive fields of the two pairs of simple cells differ by 90° phase angle.

phases of receptive fields of the two eyes. Consequently, according to the binocular energy model, the two receptive fields of the other simple binocular cell in the binocular energy model are $e^{-(ax2+by2)}\sin(2\pi f_r x + p_r^1)$ and $e^{-(ax2+by2)}\sin(2\pi f_r x + p_r^2)$.

Moreover suppose the input stimuli to the left and right eyes are $cos(2\pi f_s x + p_s^1)$ and $cos(2\pi f_s x + p_s^2)$ oriented parallel the y axis of the receptive fields. Here f_s is the spatial frequency of the input stimuli. p_s^1 and p_s^2 are the absolute phases of input stimuli to the two eyes.

According to the binocular energy model, the response of the a binocular complex cell is

$$
\begin{aligned}
&\left(\int_{-\infty}^{\infty} \int_{-\infty}^{\infty} e^{-(ax^2+by^2)} \cos(2\pi f_r x + p_r^1) \cos(2\pi f_s x + p_s^1) dx dy \right. \\
&+ \int_{-\infty}^{\infty} \int_{-\infty}^{\infty} e^{-(ax^2+by^2)} \cos(2\pi f_r x + p_r^2) \cos(2\pi f_s x + p_s^2) dx dy)^2 \\
&+ \left(\int_{-\infty}^{\infty} \int_{-\infty}^{\infty} e^{-(ax^2+by^2)} \sin(2\pi f_r x + p_r^1) \cos(2\pi f_s x + p_s^1) dx dy \right. \\
&+ \int_{-\infty}^{\infty} \int_{-\infty}^{\infty} e^{-(ax^2+by^2)} \sin(2\pi f_r x + p_r^2) \cos(2\pi f_s x + p_s^2) dx dy)^2
\end{aligned}
\tag{1}
$$

The variable y can be integrated separately. This yields a constant for the four components. Thus variable y can be omitted from the equation without affecting the conclusion. Therefore the response of the binocular complex cell becomes

$$
\begin{aligned}
&\left(\int_{-\infty}^{\infty} e^{-ax^2} \cos(2\pi f_r x + p_r^1) \cos(2\pi f_s x + p_s^1) dx + \int_{-\infty}^{\infty} e^{-ax^2} \cos(2\pi f_r x + p_r^2) \cos(2\pi f_s x + p_s^2) dx \right)^2 \\
&+ \left(\int_{-\infty}^{\infty} e^{-ax^2} \sin(2\pi f_r x + p_r^1) \cos(2\pi f_s x + p_s^1) dx + \int_{-\infty}^{\infty} e^{-ax^2} \sin(2\pi f_r x + p_r^2) \cos(2\pi f_s x + p_s^2) dx \right)^2
\end{aligned}
\tag{2}
$$

This is a very complicated function. However, it can be significantly simplified by using the fact that the integration of an odd function is 0. Since all the four components of the function are similar, let's first consider the first component.

The first component can be expanded as following

$$
\begin{aligned}
&\int_{-\infty}^{\infty} e^{-ax^2} \cos(2\pi f_r x + p_r^1) \cos(2\pi f_s x + p_s^1) dx \\
&= \int_{-\infty}^{\infty} e^{-ax^2} [\cos(2\pi f_r x) \cos(p_r^1) - \sin(2\pi f_r x) \sin(p_r^1)][\cos(2\pi f_s x) \cos(p_s^1) - \sin(2\pi f_s x) \sin(p_s^1)]
\end{aligned}
\tag{3}
$$

Applying the integration property of the odd function, the above equation can be simplified as

$$
\begin{aligned}
&\int_{-\infty}^{\infty} e^{-ax^2} \cos(2\pi f_r x + p_r^1) \cos(2\pi f_s x + p_s^1) dx \\
&= \int_{-\infty}^{\infty} e^{-ax^2} [\cos(2\pi f_r x) \cos(2\pi f_s x) \cos(p_r^1) \cos(p_s^1) + \sin(2\pi f_r x) \sin(2\pi f_s x) \sin(p_r^1) \sin(p_s^1)]
\end{aligned}
\tag{4}
$$

The other three components can be processed similarly. Therefore, by applying some simple trigonometry equations, the response of the binocular cell is

$$
\begin{aligned}
&(\int_{-\infty}^{\infty} e^{-ax^2} \cos(2\pi f_r x + p_r^1) \cos(2\pi f_s x + p_s^1) dx \\
&+ \int_{-\infty}^{\infty} e^{-ax^2} \cos(2\pi f_r x + p_r^2) \cos(2\pi f_s x + p_s^2) dx)^2 \\
&+ (\int_{-\infty}^{\infty} e^{-ax^2} \sin(2\pi f_r x + p_r^1) \cos(2\pi f_s x + p_s^1) dx \\
&+ \int_{-\infty}^{\infty} e^{-ax^2} \sin(2\pi f_r x + p_r^2) \cos(2\pi f_s x + p_s^2) dx)^2 \\
&= (\int_{-\infty}^{\infty} e^{-ax^2} \cos(2\pi (f_r + f_s)x) dx)^2 \cos^2 \left[\frac{(p_r^1 - p_r^2) + (p_s^1 - p_s^2)}{2} \right] \\
&+ (\int_{-\infty}^{\infty} e^{-ax^2} \cos(2\pi (f_r - f_s)x) dx)^2 \cos^2 \left[\frac{(p_r^1 - p_r^2) - (p_s^1 - p_s^2)}{2} \right] \\
&+ \int_{-\infty}^{\infty} e^{-ax^2} \cos(2\pi (f_r + f_s)x) dx \int_{-\infty}^{\infty} e^{-ax^2} \cos(2\pi (f_r - f_s)x) dx \cos((p_s^1 + p_s^2)[\cos((p_r^1 - p_r^2) \\
&+ \cos((p_s^1 - p_s^2)]
\end{aligned}
$$

$$(5)$$

It is well known that

$$
\int_{-\infty}^{\infty} e^{-\alpha x^2} \cos(\beta x) dx = \sqrt{\frac{\pi}{\alpha}} e^{-\frac{\beta^2}{4\alpha}}
$$

$$(6)$$

Therefore, all integrals in the above equation can be integrated out.

One can see that $(p_r^1 - p_r^2)$ and $(p_s^1 - p_s^2)$ are the relative phases of the receptive fields and input stimuli, thus independent of the absolute phases of the receptive fields and input stimuli. However $(p_s^1 + p_s^2)$ is sensitive to the absolute phase of the input stimuli. Therefore the response of the binocular cell in the energy model is insensitive to the absolute phase of the receptive fields yet sensitive to the absolute phase of the input stimuli.

This result is in sharp contrast with former research. However, our derivation is without any approximation. Therefore our result should override former general agreement regarding binocular energy model. Yet, interesting enough, nature seems build complex cells as phase dependent as the neurophysiological experiments discovered [12].

The case we just analyzed is for the a sinusoid align with the major axis of the binocular complex cell. During the derivation, no approximation is made. Thus the result is a general analytical result which should give every detail of the information of the energy model and be applicable to any admissible situation.

This situation can be further generalized to gabor patches as the input stimuli. The major axis of the gabor patches should parallel to the major axises of receptive fields. Under this situation, it can be proved that Eq. 5 is unaltered except that $cos(p_s^1 + p_s^2)$ is changed to $cos(\text{constant} + p_s^1 + p_s^2)$. Therefore the conclusion still holds.

If the input stimuli is a sinusoid whose direction is not parallel to the major axises of the receptive fields, then the same method used in our derivation can also be applied. Together with the fact that the integrated function can be separated into component x and y functions, the response of the binocular complex cell can be derived analytically. Equation 5 can again be retained, with only difference being the change in term $cos(\text{constant} + p_s^1 + p_s^2)$.

For arbitrary gabor patches as input stimuli, the response of the binocular complex cell modeled by energy model can also be derived analytically with constant difference in some components of the result. We omitted the derivations for these two situations. Interested reader can do the derivation easily.

Theoretically any function can be expanded as a Fourier series, therefore the derivation method applied in this research can also be used to analyze any input stimuli. We expect that the result should also be similar to Eq. 5 with some constant difference in term like $cos(p_s^1 + p_s^2)$. However, such analysis is not the focus of this research and we leave the analysis to future and to interested readers.

The energy model for motion is first introduced by Adelson and Bergen [1]. It is built to be independent of the absolute phase of the input stimuli [1]. Furthermore, the energy model for motion is found to be consistent with the complex cells in cat [3]. Nonetheless the energy model for motion can be considered as a special case of the binocular energy model in which the left and right receptive fields are the same and the inputs to the left and right receptive fields are also the same. Therefore the conclusion for the binocular energy model is also hold for the energy model for motion.

3 Discussion

From Eq. 5, one can see that for both similar frequencies (f_r and f_s) and very different frequencies (f_r and f_s), the response of the binocular complex cell changes when the absolute phase of the input stimuli changes. However, it can be proved that the ratio of change is bigger for very different frequencies (f_r and f_s) than for similar frequencies (f_r and f_s). This indicates that the absolute phase effect is obvious for non-optimal stimulated neurons.

This is also confirmed by a simulation study [14]. In the simulation study, the effect of this term is not obvious and be omitted when the dominant frequency of the input stimuli and the optimal frequency of the neurons' receptive fields of the binocular energy model is similar (e.g. Fig. 2). Only when the dominant frequency of the input stimuli and the optimal frequency of the neurons' receptive fields of the binocular energy model are very different, the effect of the third term is obvious and can not be omitted (e.g. Fig. 3).

One may think that perception will receive little response from neurons with large mismatch input images. However, research in neural decoding [8] found that perception discriminations are not due to the neurons with optimal images. Psychophysical [11, 13, 14] and physiological [9] studies show that perception discriminations are mostly due to the neurons whose optimal input images differ largest with the indeed input images while the neurons with the optimal input images contribute little. Consequently, our

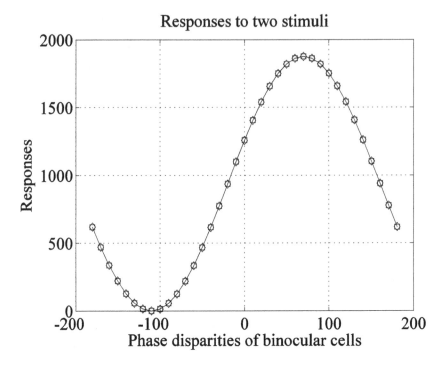

Fig. 2. A group binocular complex cells respond to two vertical 1 c/d sinusoidal gratings with the same disparity but different absolute phases of π/6 (*diamond*) and 0 (*square*). The optimal frequency of the cells is 1.59 c/d.

major contribution of this paper may be important to the perception research and modeling.

One may think that the absolute phase effect may be trivial. However, for envelope decoding mechanism, if the absolute phase of the input images change from 0 (Fig. 4) to π/6 (Fig. 5), the result curves will shift π/6 (Fig. 5). This is a very big change which cannot be ignored.

Furthermore, we would like to point out that the phase dependence cannot be fixed by quadrature. (Some researchers [4] tried incorrectly to fix the phase dependence problem by using quadrature). Even if the quadrature is satisfied, the energy model is still phase dependent. This is actually the main point of the paper by Morrone and Burr [5]. As an example, suppose the two receptive fields of one simple binocular cell in the binocular energy model are $cos\left(2\pi f_r x + p_r^1\right)$ and $cos(2\pi f_r x + p_r^2)$. f_r is the optimal spatial frequency of the receptive fields. p_r^1 and p_r^2 are the absolute phases of receptive fields of the two eyes. Consequently, according to the binocular energy model, the two receptive fields of the other simple binocular cell in the binocular energy model are $sin\left(2\pi f_r x + p_r^1\right)$ and $sin(2\pi f_r x + p_r^2)$. Since the Hilbert transform of $cos(x)$ is $sin(x)$, therefore the two receptive fields are *exactly* quadrature *without approximation*. Moreover suppose the input stimuli to the left and right eyes are $e^{-(ax2 + by2)}cos\left(2\pi f_s x + p_s^1\right)$ and $e^{-(ax2 + by2)}cos\left(2\pi f_s x + p_s^2\right)$

Responses to two stimuli

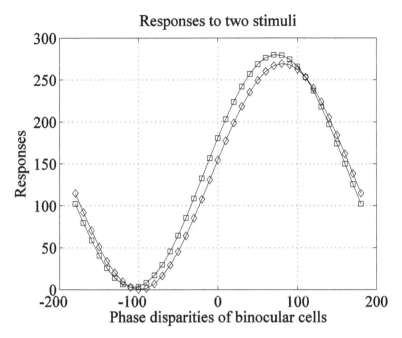

Fig. 3. A group complex cells respond to two vertical 1 c/d sinusoidal gratings with the same disparity but different absolute phases of π/6 (*diamond*) and 0 (*square*). The optimal frequency of the cells is 6.35 c/d.

oriented parallel the y axis of the receptive fields. Here f_s is the spatial frequency of the input stimuli. p_s^1 and p_s^2 are the absolute phases of input stimuli to the two eyes. According to the binocular energy model, the response of the a binocular complex cell is

$$
\begin{aligned}
&(\int_{-\infty}^{\infty}\int_{-\infty}^{\infty} \cos(2\pi f_r x + p_r^1)e^{-(ax^2+by^2)}\cos(2\pi f_s x + p_s^1)dxdy \\
&+ \int_{-\infty}^{\infty}\int_{-\infty}^{\infty} \cos(2\pi f_r x + p_r^2)e^{-(ax^2+by^2)}\cos(2\pi f_s x + p_s^2)dxdy)^2 \\
&+ (\int_{-\infty}^{\infty}\int_{-\infty}^{\infty} \sin(2\pi f_r x + p_r^1)e^{-(ax^2+by^2)}\cos(2\pi f_s x + p_s^1)dxdy \\
&+ \int_{-\infty}^{\infty}\int_{-\infty}^{\infty} \sin(2\pi f_r x + p_r^2)e^{-(ax^2+by^2)}\cos(2\pi f_s x + p_s^2)dxdy)^2
\end{aligned}
\tag{7}
$$

It is easy to prove that the response of the binocular cell is

$$
\begin{aligned}
&(\int_{-\infty}^{\infty} e^{-ax^2}\cos(2\pi(f_r+f_s)x)dx)^2 \cos^2\left[\frac{(p_r^1-p_r^2)+(p_s^1-p_s^2)}{2}\right] \\
&+ (\int_{-\infty}^{\infty} e^{-ax^2}\cos(2\pi(f_r-f_s)x)dx)^2 \cos^2\left[\frac{(p_r^1-p_r^2)-(p_s^1-p_s^2)}{2}\right] \\
&+ \int_{-\infty}^{\infty} e^{-ax^2}\cos(2\pi(f_r+f_s)x)\,dx \int_{-\infty}^{\infty} e^{-ax^2}\cos(2\pi(f_r-f_s)x)dx\cos((p_s^1+p_s^2)[\cos((p_r^1-p_r^2)+\cos((p_s^1-p_s^2))
\end{aligned}
\tag{8}
$$

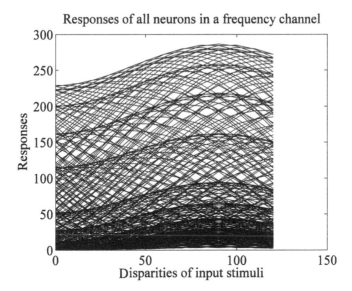

Fig. 4. A group complex cells with optimal frequency 6.35 c/d respond to a 1 c/d input images with absolute disparity 0. Every layer corresponds to different orientations.

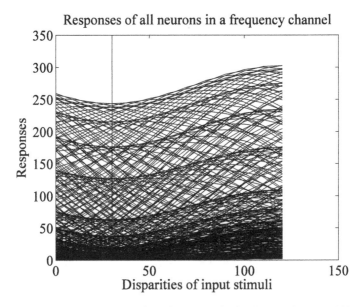

Fig. 5. A group complex cells with optimal frequency 6.35 c/d respond to a 1 c/d input images with absolute disparity $\pi/6$. Every layer corresponds to different orientations.

Since there is a term $(p_s^1 + p_s^2)$ in the result, it is easy to see that the response again depends on the absolute phase of input stimulus *no mater the quadrature relationship holds without any approximation.* Therefore with the paper of Morrone and Burr and result of this paper, phase dependence is clearly the property of energy model no matter quadrature or non-quadrature.

4 Conclusion

The past two decades has witnessed the widely success of the energy model in the visual psychology, systems neuroscience and theoretical neuroscience [1–3, 7]. Nobody even doubts its absolute phase independence property. Indeed, this property is the energy model meant to be and built to be. Nonetheless our result clearly shows that this is not the case.

However, our result does not mean that the former conclusion about absolute phase independence of the energy model is totally wrong in the single neuron level. Instead, our result only shows that the former conclusion about absolute phase independence is approximately valid when the neuron's dominant frequency and the frequency of the input stimuli are similar. The absolute phase dependence only reveals itself where large mismatch between frequencies exists. However, we need emphasize that for a biological system, since there are billions of complex neurons covering the full spectrum of the frequency domain, the phase depence can not be omitted in the perception/behavioral level. As we discussed earlier, the final perception of the biological system depends on the neurons showing phase dependence. The neurons with approximate phase independence contribute little to final perception of the biological system. This is not only consistent with neurophysiological research [12], it also gives us a roughly phase independence model while keeps the subtle phase information which is more important for visual information (review any standard text about Fourier analysis).

Acknowledgement. This work was supported by China National Science Foundation 60475021 and Fund from China Department of Education New Century Excellent Talent NCET-05-0611.

References

1. Adelson, E., Bergen, J.R.: Spatiotemporal energy models for the perception of motion. J. Opt. Soc. Am. A **2**, 284–299 (1985)
2. Cumming, B.G., DeAngelis, G.C.: The physiology of stereopsis. Annu. Rev. Neurosci. **24**, 203–238 (2001)
3. Emerson, R.C., Bergen, J.R., Adelson, E.H.: Directionally selective complex cells and the computation of motion energy in cat visual cortex. Vis. Res. **32**, 203–218 (1992)
4. Lehky, S.R., Sejnowski, T.J., Desimone, R.: Selectivity and sparseness in the response of striate complex cells. Vis. Res. **45**, 57–73 (2005)
5. Morrone, M.C., Burr, D.C.: Feature detection in human vision: a phase-depend energy model. Proc. R. Soc. Lond. B. **235**, 1–17 (1988)
6. Ng, J., Bharath, A.A., Li, Z.: A survey of architecture and function of primary visual cortex. EURASIP J. Adv. Signal Process. **2007**, 221–245 (2007)
7. Ohzawa, I., DeAngelis, G.C., Freeman, R.D.: Stereoscopic depth discrimination in the visual cortex: neurons ideally suited as disparity detectors. Science **249**, 1037–1041 (1990)

8. Pouget, A., Dayan, P., Zemel, R.: Information processing with population codes. Nat. Rev. Neurosci. **1**, 125–132 (2000)

9. Purushothaman, G., Bradley, D.C.: Neural population code for fine perceptual decisions in area MT. Nat. Neurosci. **8**, 99–106 (2005)

10. Qian, N.: Computing stereo disparity and motion with known binocular cell properties. Neural Comput. **6**, 390–404 (1994)

11. Regan, D., Beverley, K.: Postadaption orientation discrimination. J. Opt. Soc. Am. A. **2**, 147–155 (1985)

12. Spitzer, H., Hochstein, S.: Compl-excell receptive field models. Prog. Neurobiol. **31**, 285–309 (1988)

13. Wilson, H.R.: Responses of spatial mechanisms can explain hyperacuity. Vis. Res. **26**, 453–469 (1986)

14. Zhao, L., Farell, B.: The binocular neural mechanism: gnostic and population coding. J. Vis. **2**, 312 (2002)

Assimilation of AMSU-A Microwave Observations Over Land

De Xing[1,2(✉)], Qunbo Huang[1,2,3], Bainian Liu[3], and Weimin Zhang[2]

[1] Institute of Marine Science and Engineering, National University of Defense Technology,
Changsha, China
xd_wony@icloud.com
[2] College of Computer, National University of Defense Technology, Changsha, China
[3] Weather Center of PLA Air Force, Beijing, China

Abstract. Remote sensing satellite microwave observation data has a wide range of coverage and high observation density, which has made the highest proportion of the observed data used in data assimilation. However, due to the uncertainty of the surface emissivity, mostly observations over the sea and observation of high-peaking channels are assimilated, resulting in a large number of land and lower-peaking channel satellite observations are abandoned. Based on the WRF system and the characteristics of AMSU-A microwave observation data, this paper realizes the application of AMSU-A observation data over land in WRFDA assimilation system. The preliminary experiment shows that the addition of AMSU-A observation data over land has a positive effect on the numerical weather forecast.

Keywords: Satellite data assimilation · AMSU-A observations over land · WRF model

1 Introduction

Numerical Weather Prediction (NWP) is numerically solving the basic equations of atmospheric motion based on the known atmospheric initial conditions and boundary conditions, thus predicting the atmospheric state of the future. For numerical weather prediction, the accuracy of the atmospheric initial condition is very important [1].

At present, mostly only the data observed over sea is assimilated because the land surface emissivity cannot be accurately calculated. The premise of assimilating microwave remote sensing data over land is to obtain the more accurate surface emissivity.

The Weather Research and Forecast (WRF) model is a weather prediction system developed by National Center of Atmospheric Research (NCAR), National Centers for Environmental Predictions (NCEP) as well as other research institutes. WRFDA is the data assimilation system of WRF [2].

© Springer International Publishing AG 2018
F. Qiao et al. (eds.), *Recent Developments in Mechatronics and Intelligent Robotics*,
Advances in Intelligent Systems and Computing 690, DOI 10.1007/978-3-319-65978-7_57

The Advanced Microwave Sounding Unit-A (AMSU-A) is a 15-channel cross-track, stepped-line scanning, total power microwave radiometer. The instrument has an instantaneous field-of-view of 3.3° at the half-power points providing a nominal spatial resolution at nadir of 48 km. The antenna provides a cross-track scan, scanning ±48.3° from nadir with a total of 30 Earth fields-of-view per scan line.

For using satellite microwave observation data over land, Karbou et al. retrieved a global monthly averaged land surface emissivity [3]. The global surface emissivity map is called emissivity atlas. Currently, two kinds of emissivity atlases are widely used, namely the TELSEM atlas and the CNRM atlas. The major difference between these two kind of atlases is that they are retrieved from different satellite observations. The TELSEM atlas is calculated using SSM/I observations while the CNRM atlas is calculated using AMSU-A/B observations.

The structure of this paper is listed as follows. Section 1 introduces data assimilation, land surface emissivity and other background information. Section 2 discusses the settings of WRF system and experiment configures. Section 3 gives the results and the evaluations of the experiments. Section 4 gives a brief discussion and conclusion.

2   Method

2.1   Radiative Transfer Model

To directly assimilate microwave observation data, the atmospheric variables must be converted to background radiance by the radiative transfer model [4, 5]. The radiative transfer model used in our experiment is Radiative Transfer for TOVS (RTTOV).

In order to introduce emissivity atlas into RTTOV, data reading interfaces must be set up. Also for matching the position of observation points and the grid points in emissivity atlas, an interpolation method must be implemented. This method can be expressed as follows: for each observation point, set up a 1° multiplies 1° grid centered on it, and then calculated the averaged emissivity in this grid as the emissivity of the observation point. The emissivity atlas being used in experiment is the CNRM atlas.

2.2   Bias Correction

At present, there are mainly two kinds of satellite observation bias correction schemes, namely variational adaptive bias correction scheme and off-line bias correction scheme. The off-line bias correction scheme was first proposed by Eyre [6], which includes bias correction methods depend on scan position and air mass. Variational bias correction scheme introduces the bias correction factor into the objective function and through minimization to get the best estimate [7]. This paper uses variational bias correction method.

2.3 Experiment Settings

In order to show the influence of AMSU-A observations over land to the assimilation result and the prediction result, a set of assimilation and prediction experiments was carried out. The experiments focuses on the 10th typhoon of year 2014, named as 'Matmo', which is a landing typhoon. The experiments perform a cycle prediction, takes July 23nd, 2014, 00:00 (UTC, if no special explanations, the same below) as the start time. The time window is 6 h and the prediction step is 6 h. Typhoon 'Matmo' was formed in west Pacific, and its trace covered west Pacific and Taiwan province, Fujian province, Jiangxi province, Anhui province and Jiangsu province of China, so the assimilation area is confined to 15°S to 45°N, 90°E to 150°E. The number of horizontal grid is 177 × 177, the horizontal distance of the grid is 36 km. The vertical layer number of the pattern is 36. The set of experiments includes 3 different experiments, the detailed descriptions are listed as follows (Table 1):

Table 1. Experiment configure

Experiment number	Experiment name	The data being assimilated
1	CNTL	Conventional observations
2	AMSUA-ATLAS	Conventional observations + AMSUA observations + atlas
3	AMSUA-NOATLAS	Conventional observations + AMSUA observations

3   Result

In order to evaluate the impact of AMSU-A observations over land on assimilation output and prediction accuracy, the result of both assimilation and prediction are listed and discussed.

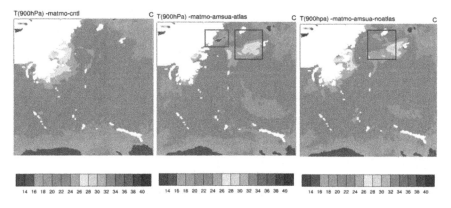

Fig. 1.   Temperature at 900 hPa

3.1 Assimilation Results

We choose the assimilation result of July 24th, 2014, 12:00 to analyze. Figure 1 shows the temperature at 900 hPa, from left to right are of experiment 1, 2 and 3 respectively. From the figures we can see that after introducing AMSUA observations over land, the temperature at 900 hPa has a significant increase (in the blue frame area). Also, using emissivity atlas makes the output temperature more accurate, for example, the temperature of the Bohai sea is significantly colder in experiment 2 than in experiment 3 (in the red frame area).

3.2 Prediction Results

In each experiment, the track of typhoon 'Matmo' is generated. Figure 2 is the track comparing graph, in which there are 4 tracks: track No. 1, real track of typhoon 'matmo'; track No. 2, track generated in exp. 1; track No. 3, track generated in exp. 2; track No. 4, track generated in exp. 3.

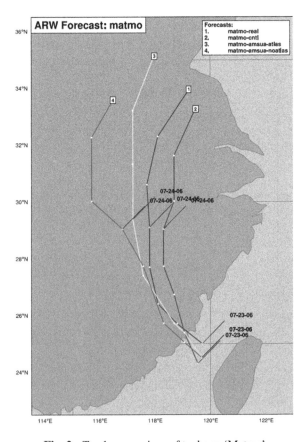

Fig. 2. Track comparison of typhoon 'Matmo'

Figure 3 shows the distances between predicted position of typhoon 'Matmo' in each experiment and the real position of typhoon 'Matmo'. Figure 4 shows the deviation between the real air pressure in the center of typhoon 'Matmo' and that of the experiment predictions.

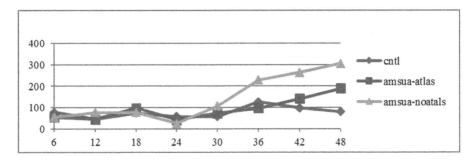

Fig. 3. Forecast position deviation of typhoon 'Matmo'

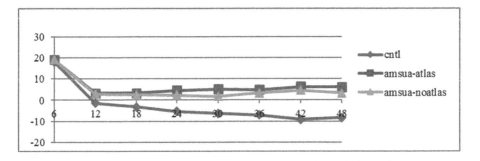

Fig. 4. Forecast air pressure deviation of the center of typhoon 'Matmo'

From Figs. 2, 3, 4 we can see that the introduction of a combination of AMSU-A observations over land and emissivity atlas can significantly improve the accuracy of the typhoon track prediction. It is worth mentioning that the prediction accuracy of using AMSU-A observations without emissivity atlas is not satisfactory. When assimilating AMSU-A observations over land, the introduction of emissivity atlas can increase the accuracy of typhoon 'Matmo' track prediction, but the increase in the accuracy of the air pressure prediction is not very satisfactory.

4   Discussion and Conclusion

Based on WRF system and RTTOV system, this paper focuses on the use of emissivity atlas, implements the assimilation of AMSU-A observations over land in WRFDA system. A set of experiments is carried out and evaluations are made. The paper briefly introduces data assimilation and land surface emissivity in chapter 1 at first, and then comes to the point that when assimilating satellite observations over land, the use of surface emissivity is necessary. To assimilate AMSU-A observations over land using

WRFDA, emissivity atlas must be appropriately used, and the radiative transfer model as well as the WRF system should be correctly set, which is discussed in Sect. 2. Meanwhile, the detailed experiment settings are explained in Sect. 2. Section 3 gives the results and the evaluations of the experiments.

Further work can focus on the following points:

1. Instead of using emissivity atlas, surface emissivity can be calculated dynamically using satellite observations.
2. This paper concerns about the assimilation of the AMSU-A observations over land, other instruments' observations can be discussed later.

Acknowledgements. The work was supported by the Natural Science Foundation of China (Grant No. 41375113).

References

1. Richardson, L.F.: Weather Prediction by Numerical Process. Cambridge University Press, Cambridge (1922)
2. Wang, W., Bruyere, C., Duha, M., et al.: ARW version 3 modeling system User's guide (2016). http://www2.mmm.ucar.edu/wrf/users/docs/user_guide_V3.8/ARWUsersGuideV3.pdf
3. Karbou, F., Prigent, C., Eymard, L., et al.: Microwave land emissivity calculations using AMSU measurements. IEEE Trans. Geosci. Remote Sens. **43**(5), 948–959 (2005)
4. Weng, F., Yan, B., Grody, N.C.: A microwave land emissivity model. J. Geophys. Res. **106**(D17), 20115–20123 (2001)
5. McNally, A.P.: The direct assimilation of cloud affected satellite infrared radiances in the ECMWF 4D-Var. Q. J. Roy. Meteor. Soc. **135**, 1214–1229 (2009)
6. Eyre, J.R.: A bias correction scheme for simulated TOVS brightness temperatures. Tech. Memo. ECMWF **186**, 28 (1992)
7. Dee, D.P.: Bias and data assimilation. Q. J. R. Meteorol. Soc. **131**, 3323–3343 (2006)

Speech Emotion Recognition Based on Deep Learning and Fuzzy Optimization

Jian-cheng Xu and Nan-feng Xiao[✉]

School of Computer Science and Engineering, South China University of Technology,
Guangzhou 510006, China
xiaonf@scut.edu.cn

Abstract. The automatic speech emotion recognition (SER) is a challenging and attractive task. To gain a high accuracy in SER, it is need to extract salient features from the speech. Therefore, in this paper, we first verify whether the deep convolutional neural network (DCNN) can improve the salient of hand-crafted acoustic features, and then based on the conclusion of the first step, considering the fuzziness of emotion, we propose a theoretical approach of fuzzy neural network (FNN) combined with deep learning to deal with the fuzziness of emotion. The first step work is tested on the CASIA Chinese Emotional Corpus, and the second part is introduced in theory and analyzed about its feasibility.

Keywords: Speech emotion recognition · DCNN · Deep learning · FNN

1 Introduction

The most important point in SER is to extract salient feature from speech. In this aspect, many approaches have been proposed, especially in the deep learning field. Zeng et al. [1] proposes to use a deep convolutional neural network (DCNN) to extract features from log-spectrogram and compares them with the hand-crafted acoustic features extracted from the same data set. The result shows that the features extracted by DCNN gain higher accuracy than the hand-crafted acoustic features. There is a consideration that if we try to use the DCNN to extract feature from the hand-crafted acoustic features again, whether the secondly-extracted features can get better performance than that of the features extracted from the log-spectrogram. Considering this problem, we try to test it in the experiments. Besides, It is well known that the emotions contained in the speech are often ambiguous and blended [2]. Because of the fuzziness of emotion, we propose a new fuzzy neural network to deal with it on a theoretical level.

The rest of this paper is organized as follows. Section 2 is an overview of the related work. In Sect. 3, the model of DCNN is described and the experiments are conducted. In Sect. 4, an introduction of new fuzzy neural network is presented on a theoretical level.

© Springer International Publishing AG 2018
F. Qiao et al. (eds.), *Recent Developments in Mechatronics and Intelligent Robotics*,
Advances in Intelligent Systems and Computing 690, DOI 10.1007/978-3-319-65978-7_58

2 Related Work

In SER, to extract the robust and salient features is a challenging problem. In many researches, the speech features can be divided into four categories: acoustic features, linguistic features, context information, and hybrid features [3]. Schuller et al. [4] uses mel-frequency cepstral coefficients (MFCC), perceptual linear prediction (PLP) coefficients etc. as features. As the rapid development of the deep learning in the recent years, SER benefits a lot from it. Mao et al. [3] proposes to learn the affect-salient features for SER using the convolutional neural networks (CNN). Zeng et al. [1] proposes using deep convolution neural network (DCNN) to extract features from log-spectrogram, which results in better performance than that of the hand-crafted acoustic features extracted from the same data set.

When designing a model for speech emotion recognition, the fuzziness in speech should be considered [5]. Wang [6] uses an improved T-S fuzzy neural network for speech recognition which is also a good idea for speech emotion recognition. In [6], the membership function is Gaussian function which is frequently used. However, it is well known that for fuzzy theory, how to determine the membership function is a tough problem, and in most cases, the researcher will determine the membership function according to the previous statistical experience and their subjectivity. To this problem, [7, 8] propose to use a BP neural network to train the membership function of fuzzy theory, and obtain good effect.

3 Model of DCNN and Experiment

3.1 The Structure of DCNN

As for the construction of DCNNs, we can follow [1]. There are two stages of convolution layers, two stages of pooling layers and a fully-connected layer in the DCNN. Figure 1 shows the structure of the DCNN.

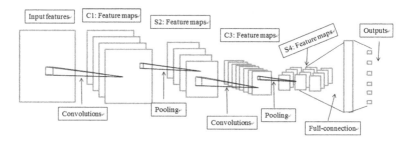

Fig. 1. The structure of DCNN

The input of DCNN: This is a two dimensional convolutional neural network, therefore the input is 2-D matrix.

Convolutional layer: An input is a 2-D map. The training dataset with N inputs make up a 3-D input tensor X, in which the i input map is denoted as x_i. The output of the convolutional layer is also a 3-D tensor with M maps denoted as Y. The j output map is denoted as y_j. The kth convolutional kernel is denoted as W_k, therefore the input maps to the output maps are calculated as follow:

$$y_i = sigmoid(W_k \cdot x_i + b_i) \tag{1}$$

where b_i is the bias.

Pooling layer: After the convolutional layer, the gained feature maps are put into the pooling layer. Here we use mean-pooling for down-sampling:

$$p_i = mean\_pool_{m \times n}(y_i) \tag{2}$$

where $m \times n$ is the size of pooling window, and $mean\_pool(\cdot)$ calculate the average of size $m \times n$ block of the input without overlap. y_i is the output maps of the convolutional layer and is regarded as the input

Fully-connected layer: In this layer, the features gained from the convolutional layer and the pooling layer is transformed into a 1-D feature vector, and this feature vector will be soon put into the classifier to test the emotion recognition accuracy.

3.2 Experiment of DCNN on CASIA

In order to verify whether the secondly-extracted features from the hand-crafted features by the DCNN can get better performance than the features extracted from log-spectrogram or not, we use the CASIA [9] database to test. This database contains audio data from the four professional actors with six kinds of emotions. The labels are respectively: angry, fear, happy, neutral, sad, and surprise. Each actor has 50 utterances of every kind of emotion, and therefore the data set has totally 1200 utterances. The speech data is sampled at 16 kHz.

The experiment training dataset and testing dataset are established in this way. 1200 utterances of six kinds of emotion mean each emotion has 200 utterances. For every emotion, we randomly choose 80% of 200 utterances as the training data while the rest 20% is testing data. Combining all the emotions, the total dataset is completed.

The speech signal is framing with Hamming window of 512 points sliding at 256 points. First, we follow [1] to get the 60×40 PCA-Whitened log-spectrogram, and then extract the MFCC features of 48-dimension including MFCC (13-dimension), delta, delta-delta as well as the mean, median and variance respectively. Delta means the first-order difference of MFCC. Since the input of DCNN is 2-D matrix, we stack 30 frames MFCC features together to form the 48×30 MFCC map as input.

As for the experiment of spectrogram, the DCNN has 2 convolutional layers with respectively 6 and 12 convolutional kernels (size 5×5) and 2 pooling layers with pooling window size 2×2. A softmax classifier is applied to test the performance of the extracted feature. For the MFCC map experiment, the DCNN has 2 convolutional layers with respectively 6 and 12 convolutional kernels of size 7×7 and size 5×5. The number of pooling layer is 2 with window size of 1×1 and 2×2 respectively. A softmax classifier is applied to test the performance of the extracted feature.

The results are showed in Table 1, L_r is the learning rate, S_f is the log-spectrogram and M_f is the MFCC map features.

Table 1. Comparison between log-spectrogram and MFCC using DCNN

L_r	0.1	0.2	0.3	0.4	0.5	0.6	0.7	0.8	0.9	1.0
S_f	0.3388	0.4	0.4187	0.4463	0.4387	0.4563	0.4487	0.4475	0.4472	0.4475
M_f	0.3461	0.3869	0.4048	0.3917	0.3940	0.3917	0.4108	0.4172	0.4024	0.3893

From the result of Table 1, the best accuracy of the log-spectrogram is 0.4563 while the best accuracy of MFCC map features is 0.4172. The results show that the handcrafted feature MFCC after secondly-extracted by DCNN can not get better performance than the PCA-whitened log-spectrogram features extracted by DCNN.

The analysis of the results is showed as follow: the DCNN can gain the useful information from the two dimensional space due to its operating principle. The spectrogram is a two dimensional figure and the data within it is relative to each other. Such relevance can be extracted by DCNN to improve the salient of the feature. As for the MFCC map, it is combined by our subjectivity and thus the relevance between one data point and the other will be quite weak which results in the feature learned by DCNN is not salient enough.

4   A Fuzzy Neural Network Combining with Deep Learning

This section we will introduce a new fuzzy neural network combining with deep learning model. The T-S fuzzy neural network mentioned in [6] is a classical FNN. Since the speech emotion recognition is quite similar to the speech recognition in many aspects and [6] successfully applies its improved FNN model to the speech recognition, we can consider to design a similar approach for SER. In [6], the structure of the improved T-S FNN is showed as Fig. 2. The detail of the improved T-S FNN is showed as follow.

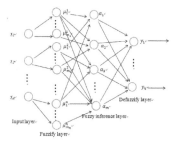

Fig. 2. The structure of FNN

The first layer is input layer, and the neurons of this layer are connected with the input vector $x = (x_1, x_2, \ldots, x_n)$. Each neuron correspond to one component of the input vector.

The second layer is the fuzzify layer. Each neuron represent linguistic variable. It is used to calculated the membership of each input component. Here the membership function is a Gaussian function. The membership is calculated follow this equation:

$$u_{ij} = \exp\left(-\frac{(x_i - c_{ij})^2}{b_{ij}}\right) \tag{3}$$

where, u_{ij} is the membership of each input component ($i = 1,2,\ldots, n; j = 1,2,\ldots,m_i, m_i$ is the number of fuzzy rules) while c_{ij} and b_{ij} is the center and the width of the Gaussian function.

The third layer is the fuzzy inference layer. Each neuron represent one fuzzy rule to deal with the membership from the second layer. The compute method is as follow:

$$a_j = \prod_i^n u_{ij} \text{ or } \quad a_j = \min(u_{ij}) \tag{4}$$

where, $j = 1, 2,\ldots, m; m = m_1 \times m_2 \times \ldots \times m_n$.

The last layer is output layer. It is used to defuzzify the value of the third layer. The calculating equation is as follow:

$$y_i = \sum_{j=1}^{m} w_{ij}a_j, \quad i = 1, 2, \ldots, k \tag{5}$$

The above steps are the process of a classical FNN, and it is trained with the least square method and the back-propagation algorithm together. However, we can find out that the input of the FNN can only accept the 1-dimensional vector. Besides, when the membership function is selected as Gaussian function or other function, although the parameters of the function can be adjusted during the process of training the network, the fixed function expression formula will restrict the performance capability. Considering the above two problems, this paper propose a new fuzzy neural network of which the membership function is the DCNN model but not the conventional function. The new model is denoted as DCNN-FNN. The structure of DCNN-FNN is showed as Fig. 3.

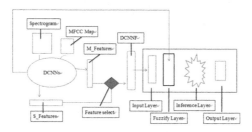

Fig. 3. The structure of DCNN-FNN

The DCNN-FNN uses the DCNN model as the membership function is the main differences compared to the classical FNN. By using the DCNN, this new model is aim to deal with the input sample with two dimensional structure. As the results of the experiments in Sect. 3 and the results from [1], it has been verified that the PCA-whitened log-spectrogram can be extracted salient features by DCNN. Therefore, this new model can be used to recognize the speech emotion using the input of spectro-gram directly. If a classical FNN wants to use the spectrogram for SER, maybe it needs to change the two-dimension matrix into one-dimension vector, during which the relevance of space structure of two-dimension will be destroyed. The feasibility of this model on theoretical level:

First, it has been proved by [10, 11] that when the layers number of the network attains 3 and the activation function of hidden layer is a nonlinear function, such as sigmoid function, the network has the ability to approach to any continuous function with any precision. The DCNN meets all the requirements, and thus it can be regarded as a function without fixed expression formula.

Second, the DCNN can be pre-trained with part of the training set, and during the training stage, it can be adjusted by the back-propagation algorithm. This is the training method of DCNN-FNN model, and it feasible in theory.

By combining the DCNN with the FNN, the new model gain the strong power of extracting feature that the DCNN owns. At the same time, the new model can also deal with the fuzzy component in the features via the fuzzy inference layer.

The DCNN-FNN model is still established on the theoretical level, and the later research will try to achieve it and apply it to the speech emotion recognition.

Acknowledgments. This research is funded by the National Natural Science Foundation (Project No. 61573145), the Public Research and Capacity Building of Guangdong Province (Project No. 2014B010104001) and the Basic and Applied Basic Research of Guangdong Province (Project No. 2015A030308018), the authors are greatly thanks to these grants.

References

1. Zeng, Z., Pantic, M., Roisman, G.I., Huang, T.S.: A survey of affect recognition methods: Audio, visual, and spontaneous expressions. IEEE Trans. Pattern Anal. Mach. Intell. **31**(1), 39–58 (2009)
2. Moataz, E.A., Kamel, M.S., Fakhri, K.: Survey on speech emotion recognition: features, classification schemes, and databases. Pattern Recog. **44**(3), 572–587 (2011)
3. Mao, Q., Dong, M., Huang, Z., Zhan, Y.: Learning salient features for speech emotion recognition using convolutional neural networks. IEEE Trans. Multimed. **16**(8), 2203 (2014)
4. Schuller, B., Steidl, S., Batliner, A., Vinciarelli, A., Scherer, K., Ringeval, F., Chetouani, M., Weninger, F., Eyben, F., Marchi, E., Mortillaro, M., Salamin, H., Polychroniou, A., Valente, F., Kim, S.: The INTERSPEECH 2013 computational paralinguistics challenge: social signals, conflict, emotion, autism. inProc. INTERSPEECH, Lyon, pp. 148–152. ISCA (2013)
5. Emily, M., Narayanan, S.: A framework for automatic human emotion classication using emotion proles. IEEE Trans. Audio Speech Lang. Proc. **19**, 1507–1520 (2011)
6. Wang, P.: Improved T-S fuzzy neural network in application of speech recognition system. Comput. Eng. Appl. **45**(4), 1–10 (2006)
7. Zhang, X.: Using BP neural network to determine the membership function of the input of fuzzy control system. J. Chongqing Univ. **27**(5), 54–56 (2004)

8. Wang, X.: Design and research of fuzzy subset membership function by improved BP neural network. Tech. Autom. Appl. **34**(1) (2015)
9. Institute of Automation Chinese Academy of Sciences, Human Machine Speech Interaction Group. CASIA Chinese Emotional Corpus [DB/OL]. http://more.datatang.com/data/39277 (2012)
10. Hecht-Nielsen, R.: Kolmogorov's Mapping Neural Network Existence Theorem (1987)
11. Kreinovich, V.Y.: Arbitrary nonlinearity is sufficient to represent all functions by neural networks: a theorem. Neural Netw. **4**(3), 381–383 (1991)

Sensitive Information Protection of Power IOT Based on Fuzzy Control Logic

Zheng Wang[1,2(✉)], Yubo Wang[1,2], Yongling Lu[3], Changfu Xu[3],
Zhenjiang Pang[1,2], and Yan Guo[1,2]

[1] State Grid Key Laboratory of Power Industrial Chip Design and Analysis Technology, Beijing
Smart-Chip Microelectronics Technology Co., Ltd., Beijing, 100192, China
wstq5201314@163.com
[2] Beijing Engineering Research Center of High-Reliability IC with Power Industrial Grade,
Beijing Smart-Chip Microelectronics Technology Co., Ltd., Beijing, 100192, China
[3] State Grid Jiangsu Electric Power Company Research Institute, Nanjing, 211103, China

Abstract. With the comprehensive construction of smart grid, Internet of Things technology has been widely applied to the power generation, transmission, scheduling and many other aspects. First, a framework for the information communication system of power Internet of things is proposed Second a method based on fuzzy logic is given to judge the sensitive information. Third, the way to deal with different information is shown according to the sensitivity of the data. The feasibility and correctness of the proposed framework and method are verified by tests.

Keywords: Fuzzy control · IOT · Electric power communication · Sensitive information · Protection strategy

1 Introduction

As an indispensable part of the smart gird information sensing, the Internet of Things has played an important role in power grid construction, safety production management, operation and maintenance, information collection and safety monitoring. Power Internet of Things can improve the perception of each part in smart grid in all aspects including the depth, breadth and densitys [1].

This paper presents a framework for the communication network of power resources, and puts forward a sensitive information protection strategy in this framework. The framework and strategy can improve the security of the data transmission at the network of the Internet of things, so as to protect the sensitive information [2].

2 Information Security Protection Strategy for Network Layer of Power Network

The communication network layer provides a network support platform for the Power IOT, which includes not only routing and switching equipment which supported by network but also security equipment and network infrastructure introduced by security

© Springer International Publishing AG 2018
F. Qiao et al. (eds.), *Recent Developments in Mechatronics and Intelligent Robotics*,
Advances in Intelligent Systems and Computing 690, DOI 10.1007/978-3-319-65978-7_59

protection system construction. The goal of communication network layer security protection is to prevent the business system and the network equipment from malicious attack from the network [3].

2.1 Sensitive Information Classification

The network layer in Power IOT system is composed of three parts: power information communication network, wired private network wireless network, and the Internet.

2.1.1 Non-sensitive Information

Non-sensitive information refers to those who do not cause electricity disaster, and will not cause important information leak, which can be transmitted in the internal and external network.

2.1.2 Semi-sensitive Information

Semi-sensitive information refers to those information who may cause electricity disaster but must be used in the Power IOT system. The transmission of these information to the Internet may cause harm to the power system. So it can only be transmitted in the network but the external network.

2.1.3 Sensitive Information

Sensitive information refers to those who can cause electricity disaster events, can only be viewed by a small number of technicians and executives, in other words, the general staff mustn't access to the information. This information can only be passed in the network of some fixed nodes, can't be transmitted in the external network.

3 Information Sensitivity Division Strategy Based on Fuzzy Control Logic

Fuzzy control is an intelligent control method based on fuzzy set theory, fuzzy linguistic variable and fuzzy logic reasoning. It is an intelligent control method that mimics human fuzzy reasoning and decision making process. Firstly, according to the operator or expert experience to write fuzzy rules, then the real-time signal from the sensor is fuzzed, the fuzzy signal is used as the input of the fuzzy rule to carry out fuzzy reasoning, and finally the output of the reasoning is loaded into the actuator [7].

3.1 Method of Determining Sensitive Information

The method of determining sensitive information is divided into the following steps:

Step 1: Determine the evaluation index u_i ($i = 1,2,3,4,5$), Where u_i corresponds to the five factors in the previous section respectively;

Step 2: Determine the Information Sensitivity Rating Level V = {v1,v2,v3},where v1 denotes non-sensitive information, v2 denotes semi-sensitive information, and v3 denotes sensitive information;

Step 3: Establish the fuzzy relation matrix between evaluation index and comment. Quantify evaluation target from each factor one by one, that is determined from a single factor to evaluate the information on the fuzzy membership subset, and then get the fuzzy relation matrix R:

$$R = \begin{bmatrix} 0.3 & 0.2 & 0.5 \\ 0.5 & 0.1 & 0.4 \\ 0.2 & 0.6 & 0.2 \\ 0.3 & 0.5 & 0.2 \\ 0.6 & 0.1 & 0.3 \end{bmatrix} \tag{1}$$

In the matrix R, the i-th row of the j-th column element rij represents the membership of the evaluation information from the factor ui to the vj-level fuzzy subset. The conditions that must be met in the above matrix are

$$\sum_{j=1}^{3} r_{ij} = 1 \tag{2}$$

Step 4: Determine the weight vector of the evaluation factor A = (a1,a2,a3,a4,a5), which respectively correspond to the five determinants of sensitive information, among them.

The larger the value of ai, the greater the effect of the i-th factor in the determination of sensitive information. In this paper, the semantic judgment of data text has the greatest influence. This group of data is scored by experts for each factor.

Step 5: Synthetic fuzzy comprehensive evaluation result vector.

The fuzzy comprehensive evaluation result vector B could be obtained by using the appropriate operator to synthesize A with R of each evaluated things, which is:

bj represents the degree of membership of things evaluated to the vj-level fuzzy subset as a whole.

Step 6: Anti-fuzzification of fuzzy comprehensive evaluation result vector. Anti-fuzzification generally uses the maximum membership method. The maximum membership method is to take b = max(b1,b2,b3). The value of j corresponds to the level of sensitive information. J = 1 for non-sensitive information, j = 2 for semi-sensitive information, and j = 3 for sensitive information.

3.2 Sensitive Information Access Control

The following access control strategy has been applied to the Power IOT system after using the above method based on fuzzy control theory to determine the information which transmitted at the network layer.

The data judged to be non-sensitive is allowed to be transmitted in the internal and external network of the Power IOT system.

The data judged to be semi-sensitive is not allowed to be transmitted in the internal and external network of the Power IOT system, and only transmitted in the power information communication network and the wired private network/wireless private network.

The data judged to be sensitive is allowed to transmit between fixed secure nodes in the Power IOT system. Data should be intercepted once it intend to transmit to other nodes in the intranet or the Internet.

4   Experimental Results

4.1   Experimental Scenario

Test has been conducted on the Internet of Things system of a provincial grid company. A visit control and sensitive information protection module named IOT-SIRP (internet of things-sensitive information recognition and protection) has been added to the network layer of the Internet of things, the access node including internal employees in the power of the Internet system and users on the Internet.

4.2   Precision and Recall Rate

In order to determine the performance of the IOT-SIRP module designed, the concept of precision and recall has been introduced. The precision and recall rate are the primary indicators for measuring the performance of the IOT-SIRP module. P1 represents the precision rate while P2 represents the recall rate [6].

P1 = the number of sensitive messages that are properly intercepted/the number of sensitive messages for all interception;

P2 = the number of sensitive messages that are properly intercepted/the number of sensitive messages that need to be intercepted;

4.3   Distribution of Test Information

Four sets of data have been tested on a Power IOT system to check the IOT-SIRP module. Each set of data contains 50 information need to intercept, 100 information do not need to intercept.

4.4   Test Results

Following are results after the above four sets of data have been tested in the experimental environment.

According to the above table we can see that the precision rate of IOT_SIRP module is about 95%, and the recall rate is about 93%. It can be seen that the IOT-SIRP module can control the transmission of sensitive information very effectively (Table 1).

Table 1. IOT_SIRP module test results

Test data number	$p_1/\%$	$p_2/\%$
1	0.951	0.942
2	0.943	0.932
3	0.940	0.928
4	0.957	0.947

5   Conclusion

The experimental results show that the access control and sensitive information protection strategy of the network layer of the Power IOT system has better access control performance. It controls sensitive information access comprehensively and accurately which can be very efficient protection of Power IOT system.

Acknowledgements. This work is supported by the Science and Technology Research Project of State Grid Corporation of China (526816160024).

References

1. Wang, Y., Su, B., Zhao, H.: The concept and development trend of electric power network. Telecommun. Sci. **S3**, 9–14 (2010)
2. Shen, X., Cao, M., Xue, W., Li, J., Lu, Y., Zhang, L.: IOT technology of power transmission equipment on-line monitoring research and application of technology. China Southern Power Grid, **10**(1): 32–41 (2016)
3. Tang, L., Wang, H.C., Liu, R.: Design and implementation of information model and communication protocol for power IOT. J. Xi'an Polytech. Univ. **24**, 799–804 (2010)
4. Zou, W., Chen, J., et al.: Risk analysis and security measures of power IOT. Elect. Power Inf. Commun. Technol. **8**, 121–125 (2014)
5. Li, X., Liu, J.: Technology and application of Internet of things for smart grid. Telecommun. Netw. Technol. **8**, 41–45 (2010)

Retrieval of Satellite Microwave Observation over Land Surface by One-Dimensional Variational Data Assimilation

Qunbo Huang[1,2,3(✉)], Bainian Liu[1,2], De Xing[1,2],
Weimin Zhang[1,2], and Weifeng Wang[3]

[1] Academy of Ocean Science and Engineering, National University of Defense Technology,
Changsha, Hunan, China
hqb09@163.com
[2] College of Computer, National University of Defense Technology, Changsha, Hunan, China
[3] Weather Center of PLA Air Force, Beijing, China

Abstract. The data assimilation of satellite microwave observations over the land surface is a worldwide problem. 1D-Var can retrieve the atmospheric parameters by physical constraints, which is a useful technology. In this paper, 1D-Var is used to study the retrieval of atmospheric profile and AMSU-A microwave brightness observations over the land surface. The experimental results show that the retrieved effect of near-surface channel needs to be further improved. However, the retrieved effect of channel 4–15 is very good, which proves that 1D-Var has its advantage in land surface data assimilation.

Keywords: 1D-Var · Land surface · Data assimilation · Retrieval

1 Introduction

The numerical weather forecast (NWP) is a typical initial/boundary values problem. Given the estimation of the current atmospheric state (initial values) and the appropriate surface and boundary conditions, the atmospheric model will be able to simulate/predict the atmosphere in the future. The purpose of data assimilation is to use all existing information to define an atmospheric/ocean state with a maximum possibility [1]. Variational method is an important tool in the field of data assimilation, which based on the basis of the statistical estimation theory in minimizing, solving the cost function to obtain the most accurate analysis of the real state of the atmosphere. Variational data assimilation is often classified as one-dimensional variational data assimilation (1D-Var), three-dimensional variational data assimilation (3D-Var) and four-dimensional data variational assimilation (4D-Var).

1D-Var is often used as a retrieval and pre-processing method due to its algorithmic simplicity and easy to control. Phalippou [2] used 1D-Var method to retrieve the humidity and temperature, surface wind speed, cloud water content and skin temperature. Aonashi [3] described the assimilation of SSM/I by retrieving the TCWV, using the 1D-Var as an intermediary to assimilate new satellite observations, which is in favor

© Springer International Publishing AG 2018
F. Qiao et al. (eds.), *Recent Developments in Mechatronics and Intelligent Robotics*,
Advances in Intelligent Systems and Computing 690, DOI 10.1007/978-3-319-65978-7_60

of eliminating the illusive model precipitation in non-precipitated area. This method has the advantage of better controlling the response of observation operator to the variation of atmospheric state [4]. Huang studied the basic principles of the 1D+4D-Var two-step method and design a one+four-dimensional variational assimilation system process which is used for assimilating the cloud and rain affected Special Sensor Microwave/ Imager (SSM/I) data [5]. Liu did some research about 1D-Var retrieval of 1D+4D-Var system, which builds cloud-effected microwave satellite data retrieval platform by adding super-saturation penalty part in cost function and super-saturation checking in background profiles [6]. Huang et al. analyzed the algorithm which is used to retrieve cloud parameters based on the 1D-Var scheme, this approach suits best for the advanced infrared sounders such as AIRS [7].

This paper presents a method of retrieving satellite brightness temperature, atmospheric temperature and humidity profiles. We focus on the simulation of profile over land surface, investigate the possibility of data assimilation of satellite data on land surface by 1D-Var. Section 2 describes the algorithm process of 1D-Var. Section 3 describes the experiment configure. Section 4 shows and analyzes some experimental results of the brightness temperature, atmospheric retrieval. Finally, conclusions and plans for future work provide in Sect. 5.

2   1D-Var Algorithm

The principle of 1D-Var is similar to other variational data assimilation algorithms (such as 3D/4D-Var), but 1D-Var is generally considered as an inversion problem, and 3D/4D-Var is generally referred to as an analysis process. According to Bayesian theory and assuming that the background error is not dependent to the observation error, and the errors are Gaussian distribution, the solution to one-dimensional variational problem can be expressed as the minimization of the cost function:

$$J(\mathbf{x}) = \frac{1}{2}(\mathbf{x} - \mathbf{x}_b)^T \mathbf{B}^{-1}(\mathbf{x} - \mathbf{x}_b) + \frac{1}{2}\left[\mathbf{y}^0 - H(\mathbf{x})\right]^T \mathbf{R}^{-1}\left[\mathbf{y}^0 - H(\mathbf{x})\right] \tag{1}$$

where $\mathbf{x}_b$ is background profiles, $\mathbf{B}$ and $\mathbf{R}$ are the background error covariance matrix and observation error covariance matrix, $\mathbf{y}$ is satellite observation. $H(\mathbf{x})$ represents the radiance which is simulated by the forward model of RTTOV11, where $\mathbf{x}$ is the input parameters of RTTOV11.

According to the theory of functional analysis, the analytical $\mathbf{x}_a$ value should satisfy the gradient Eq. (2), which is acquired by minimizing the difference between background and observation radiance.$\mathbf{x}$ has the optimal solution when $J(\mathbf{x})$ achieves minimum.

$$\nabla J(\mathbf{x}) \rightarrow 0 \tag{2}$$

A simple linear solution of the 1D-Var is shown in the Eq. (3), where the second term on the right side of the equation exists as a correction term, and it can be seen that the retrieved profile is the sum of the background profile and the correction term. So then the error covariance can be further quantified by 1D-Var retrieval in Eq. (4), which has obviously improved effect on the background field information.

$$\mathbf{x}a = \mathbf{x}b + [\mathbf{HB}]^{T}[\mathbf{HBH}^{T} + \mathbf{R}]^{-1}(\mathbf{y} - \mathbf{H}(\mathbf{x}b)) \tag{3}$$

$$\mathbf{S}_{a} = \mathbf{B} - [\mathbf{HB}]^{T}[\mathbf{HBH}^{T} + \mathbf{R}]^{-1}\mathbf{HB} \tag{4}$$

The whole flow of one-dimensional variational data assimilation algorithm is shown in Fig. 1. It is shown that, before the satellite microwave observations are introduced into the 1D-Var system, the scan bias correction, pre-screening, bias correction and other pre-processing process should be made to revise the observation error, and then the departures between the simulated background field observation and the real observations are obtained. The departures are used for two purposes, one for calculating the analysis field and the other is for the calculation of the adjoint model to obtain the updated model field, and continuing to input to the observation operator until achieves the accuracy condition we pre-set.

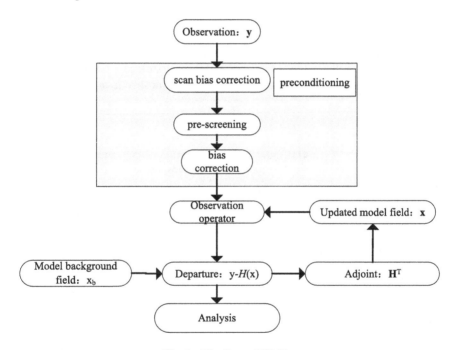

Fig. 1. The flow of 1D-Var

3   Experiment Configure

The data used in this paper are AMSU-A brightness data, the US Center for Environmental Prediction (NCEP) 6-hour forecast data are used as the background grid field. In this paper, we choose the Radiative Transfer Model for TOVS (RTTOV11) as the forward model operator to simulate the brightness temperature. At the same time, US standard atmospheric profile is used to generate true profile with 54 levels. The related data of the variables are shown in Table 1.

Table 1. The variables related to the observation and background field

Variable	Content	Unit
Instrument	AMSU-A	/
First channel	1	/
Last channel	15	/
Date	2014.07.23	/
Surface type	Land	/
Latitude	15.6747S	Degree
Longitude	127.9980E	Degree
Satellite zenith angle	44.650	Degree

The AMSU-A contained 15 channels, and there are 13 temperature detection channels and 2 window/surface channels, Fig. 2 shows the weight functions of AMSU-A channels. Each curve represents the sensitivity of a specific channel to the different levels of atmosphere. It can be clearly seen from the figure that the peak energy contribution layer of the window channels 1–3 and 15 comes from the surface and the peak energy contribution layers of channels 4–14 are the heights of the surface to 60 km, respectively. These curves are calculated by using the US standard atmospheric profile [8].

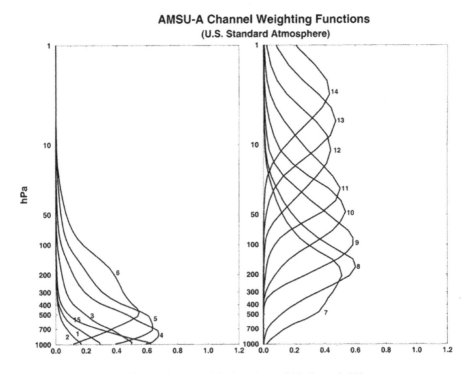

Fig. 2. AMSU-A weight functions of 15 channels [9].

4   Experimental Results

4.1   The Effect of Simulated Observation

We firstly check the effect of forward simulated background brightness temperature and the retrieved brightness temperature by 1D-Var. Figure 3a shows the simulated background brightness temperature (blue line) and 1D-Var retrieved brightness temperature (green line) are very close to the real observation (red line) for the full 15 channels. This suggests that 1D-Var system works well, and the simulation and retrieval accuracy are both high.

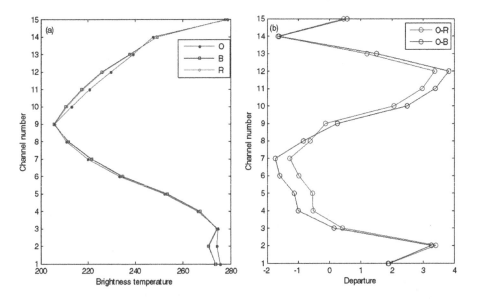

Fig. 3. (a) Comparison among the simulated background brightness temperature (blue line), 1D-Var retrieved brightness temperature (green line) and real observation (red line) for the full 15 channels. (b) Comparison between the background departure (blue line) and retrieval departure (red line). (Color figure online)

Figure 3b gives a comparison of the background departure (observations minus the background brightness, blue line) and retrieval departure (observations minus the 1D-Var retrieved brightness temperature, red line). Retrieval departures are worse than background departures for channels 1–3, mainly because these channels are relatively close to the surface and affected by the surface is relatively large, resulting in background error and observation error is relatively large. While the channels 4–15 retrieved effect is better. As can be seen from Fig. 2, AMSU-A channels 1–5 are more sensitive to the surface, only the channels above channel 5 can be implemented active assimilation in many operational data assimilation centers in the world. This result shows that 1D-Var has the positive potential to assimilate the satellite microwave observations over land surface.

4.2   The Effect of Simulated Atmospheric Profile

Subsequently, we also analyze the background departure (background profile minus true profile) and the retrieval departure (1D-Var retrieved profile minus true profile). Figure 4a and b shows the temperature departure and humidity departure respectively. For the Y axis of the both figures, the larger pressure values are more close to the land surface. In the middle atmosphere (about 700 hPa ~ 300 hPa), the temperature retrieval effect is good because of the influence of land surface is not too large. On the contrary, there is a large difference between the background and the retrieval departure due to the improved surface emissivity leads to a better retrieved atmospheric profile.

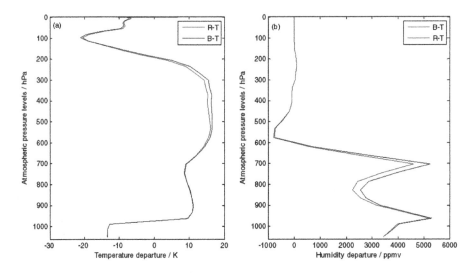

Fig. 4. Comparison the background departure (blue line) and the retrieval departure (red line) of the temperature (a) and humidity (b) profile at 54 atmospheric pressure levels. (Color figure online)

5   Conclusion

Based on the one-dimensional variational data assimilation technology, we try to retrieve the atmospheric profile and microwave brightness temperature over the land surface. It has the advantage of constraining the state variables physically and maintaining physical balance between state variables. The experimental results show that the 1D-Var system works well and the effect of brightness temperature retrieval and atmospheric profile retrieval are ideal. However, because of the difficulty in obtaining accurate surface emissivity, the retrieval quality of brightness temperature of the near surface channels still needs to be further improved. Next, we will conduct global area with various types of land surface to retrieve, and introduce observations of new microwave instruments to test the potential of the 1D-Var system.

Acknowledgements. The work was supported by the Natural Science Foundation of China (Grant No. 41375113).

References

1. Talagrand, O.: Assimilation of observations, an introduction. J. Meteorol. Soc. Jpn. **75**, 191–209 (1997). (Special Issue)
2. Phalippou, L.: Variational retrieval of humidity profile, wind speed and cloud liquid-water path with the SSM/I: potential for numerical weather prediction. Q. J. Roy. Meteorol. Soc. **122**, 327–355 (1996)
3. Aonashi, K., Shibata, A.: The impact of assimilating SSM/I precipitable water and rain flag data on humidity analysis and short-term precipitation forecasts. J. Meteorol. Soc. Jpn. **74**(1), 77–99 (1996)
4. Eyre, J.R., Kelly, G.A., Mcnally, A.P., et al.: Assimilation of TOVS radiance information through one-dimensional variational analysis. Q. J. Roy. Meteorol. Soc. **119**(514), 1427–1463 (1993)
5. Huang, Q.B.: A variational data assimilation technology of the cloud and rain affected satellite microwave data. National University of Defense Technology (2011)
6. Liu, Y.L.: Application research of 1D-Var retrieval technology in cloud- and precipitation-effected satellite microwave observation data assimilation. National University of Defense Technology (2013)
7. Huang, Q.B., Zhang, W.M., Yu, Y.: A cloud parameters retrieval algorithm in the variational assimilation system. In: The 2nd International Conference on Computer Application and System Modeling, pp. 0599–0602 (2012)
8. Ma, G.: A study on assimilation of radiance from FY3's atmospheric vertical sounder system. LanZhou University (2008)
9. Li, J., Wolf, W.W.: Global soundings of the atmosphere from ATOVS measurements: the algorithm and validation. J. Appl. Meteorol. **39**(8), 1248–1268 (2000)

A Discourse Analysis Based Approach to Automatic Information Identification in Chinese Legal Texts

Bo Sun[(✉)]

School of Foreign Languages, Hefei Normal University, Hefei, China
aca_StevenSun@hotmail.com

Abstract. Alternative Dispute Resolution (ADR) is increasingly advocated as an alternative to litigation nowadays. Being one of the possible forms of ADR, eMediation aims to inform the parties of their desired information in legal texts relevant to the current case. A fundamental requirement to achieve this is that information in legal texts should be automatically identified. As a preliminary investigation, this study puts forward a discourse analysis based approach to recognize and present legal information with respect to users' command. First, we make use of the 15 information categories proposed by Discourse Information Theory to describe legal information. Next, through a corpus-based analysis of the hierarchical structure of legal text and the tripartite structure of clause, we formulate a series of processing rules. Finally, an experiment is conducted to examine the efficacy of our approach. Experimental results show that our approach can reach a satisfying accuracy. Moreover, the approach may also provide some insights into the statistics based Natural Language Processing (NLP) techniques.

Keywords: Discourse analysis · eMediation · Processing rules · Support vector machine · Viterbi algorithm

1 Introduction

Litigation has long been notoriously known for its time-consuming and labour-exhausting characteristics. It is reported that ordinary civil cases, on average, can endure 14 months in the U.S., 35 months in Italy, and 30 months in China [1]. Against this backdrop, Alternative Dispute Resolution (ADR), which refers to the proceedings without court hearings, is more and more encouraged.

Among the available schemes for ADR, eMediation may be the most promising one [2]. It aims to help locate former court decisions that are similar to the current dispute. By reviewing similar cases, the parties involved are able to gain an increasing understanding of related legal articles, prospective outcomes and potential risks. In this way, settlement among litigants can be reached more easily and long-running litigation can thus be avoided. However, most of the time, what is needed is not the whole text but merely some of the information contained therein. If an approach can be found to single out the needed information, the efficiency of eMediation can be enhanced.

© Springer International Publishing AG 2018
F. Qiao et al. (eds.), *Recent Developments in Mechatronics and Intelligent Robotics*,
Advances in Intelligent Systems and Computing 690, DOI 10.1007/978-3-319-65978-7_61

To address this issue, knowledge and skills from linguistics can possibly be employed. Linguistic analyses, discourse analyses in particular, have long been stressed in NLP tasks, because they can reveal discourse patterns [3–6].

In this paper, we propose an approach to information identification in legal texts on the basis of Discourse Information Theory [7]. A Discourse Information Theory guided discourse analysis can help yield valuable knowledge of discourse structure and lexical feature of legal texts. With such knowledge, a series of rules can be formulated and the desired information can be captured, so that eMediation can be provided more directly.

The rest of the paper is organized as follows. In Sect. 2, Discourse Information Theory is briefly introduced. In Sect. 3, the proposed method is presented in detail. Legal texts are analyzed in Sect. 4 and processing rules are formulated in Sect. 5. In Sect. 6, the experiment results are given and discussed. Finally, in Sect. 7, the conclusion is drawn.

2 Discourse Information Theory

Discourse Information Theory attaches great importance to the hierarchical structure of discourse and perceives clause as the basic unit, which is called information unit (IU). It holds that the topic of a text can be split up into a series of more detailed IUs at its immediate subordinate stratum, which in turn are developed by their own immediate subordinate IUs. In this recursive manner, the whole text presents a tree structure. IUs can be represented by 15 interrogative expressions: What Attitude (WA), What Basis (WB), What Condition (WC), What Effect (WE), What Fact (WF), What Change (WG), How (HW), What Inference (WI), What Judgement (WJ), When (WN), Who (WO), What Disposal (WP), Where (WR), What Thing (WT), and Why (WY) [7]. An example of information tree can be shown as follows.

Example: 1WB|In accordance with Articles 348 and 357 of the *Criminal Law of People's Republic of China*, the decision is made as follows. 2WJ||The defendant Liu X is guilty of the crime of illegally holding drugs, 3WP|||so he is sentenced to 9 months' imprisonment and 4WP|||fined 3000 Yuan.

There are two sentences in this example. The Arabic numerals represent IUs. "|" means the first level, "||" the second and "|||" the third. The branching nodes (IU1 and IU2) mean that they are developed by their subordinate IU(s) respectively, while the leaf nodes (IU3 and IU4) suggest that discourse development ceases along this branch. IU1 reveals the legal basis of this decision, so it is tagged with WB. IU2, which functions as the legal evaluation of the defendant's behaviour, is attributed to WJ. IU3 and IU4 are the disposals made by the court, so they belong to WP.

3 Proposed Method

The corpus in this research is collected from *China Judgements Online* and *Case Information Disclosure of the People's Procuratorate of the P. R. China*. There are altogether six types of legal texts, i.e. nol pros decisions, civil judgements of second instance, written protests, indictments, reexaminations of criminal petition and criminal judgements of first instance. Through stratified sampling and systematic sampling, 60 texts are included in the training set and 30 texts in the testing set for each type. The training and testing set contain 10962 IUs in all, with an average of about 122 IUs per discourse. The Kappa values of discourse segmentation and information attribution are 0.92 and 0.73 respectively, which shows that the inter-coder reliability is acceptable [8].

In our annotation of the corpus, IU (or clause) is defined as the minimal representative and independent grammatical unit [9]. Therefore, one IU includes at least one predication and expresses at least one proposition. Formally, punctuation marks, such as period, comma, colon and semicolon, provide references for discourse segmentation. Therefore, simple sentence and clauses in complex sentence can serve as IU. As for information attribution, the definition of each information type is referred to [7].

It can be seen that discourse segmentation is the fundamental step in this research. If discourse is wrongly segmented, the following work will turn out to be unreliable. On the other hand, discourse segmentation is not unique to this research. Many discourse parsing tasks such as rhetorical structure tree building also need to tackle this problem [10–12]. The state-of-the-art Chinese discourse segmentation is reported to have achieved an accuracy of 89.2% [11]. In this research, human annotators' segmentation is made use of, in order to eliminate the interference from inaccurate discourse segmentation.

The framework of this identification pipeline is shown as follows (Fig. 1).

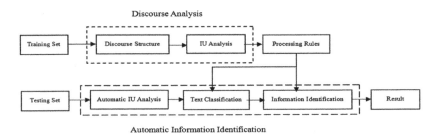

Fig. 1. The framework of automatic information identification

The legal texts in both the training set and the testing set have already been segmented into IUs by annotators. The training set includes annotators' information tagging, while the testing set does not.

The module of discourse analysis is composed of two parts, i.e. analysis of discourse structure and IU. Discourse structure is examined by observing discourse information tree. The first two levels in discourse tree are called upper level, while the other levels are named lower level [7]. Through statistical analysis of the training corpus, information distribution is investigated at both upper and lower levels for each type of legal texts.

The IU analysis is conducted in accordance with Systemic Functional Grammar (SFG), where each clause consists of process, participant and circumstance [13]. Based on the result of discourse structure analysis, the elements in IUs are looked into. From the two steps of discourse analysis, information features can be arrived at, so that processing rules can be formulated.

Regarding testing set, word segmentation and POS tagging will be conducted through NLPIR (http://ictclas.nlpir.org/) for each IU in the testing set. To our best knowledge, no method is reported to identify the three elements in Chinese clauses. In SFG, process is assumed by verb, participant by noun and circumstance by adverbial group and prepositional phrase [13]. Therefore, we treat verb as process, noun as participant and the other words as circumstance. Next, the processing rules are applied to the testing set, going through text classification and information identification. Finally, the identification result is presented and analyzed.

4 Result of Discourse Analysis

4.1 Features of Discourse Structure

To find out the features of discourse structure, we calculate the percentage of overlapping IUs at each level in each type of texts (see Table 1). The overlapping IUs not only belong to the same information category, but also have similar communication function.

Table 1. Overlapping IUs at each level in different types of legal texts

Text class	Upper-level IUs		Lower-level IUs	
	Level 1	Level 2	Level 3	Level 4
1	100%	84.9%	23.37%	7.08%
2	100%	76.92%	18.73%	3.19%
3	100%	84.96%	25.34%	6.45%
4	100%	76%	19.57%	6.61%
5	100%	75.22%	27.91%	4.48%
6	100%	78.18%	27.58%	6.09%

Note: In the "Text Class" column, "1" is nol pros decision; "2" is civil judgements of second instance; "3" is written protest; "4" is indictment; "5" is reexamination of criminal petition, "6" is criminal judgement of first instance.

From Table 1, it is found that, at the first level, all the IUs overlap. With the development of discourse, the overlapping percentage decreases level by level. When it comes to the second level, the overlapping percentage drops a little, ranging from 75.22% to 84.96%. However, at the third level, the overlapping drops sharply, varying from 18.73% to 27.91%. Few IUs overlap at the fourth and deeper levels, with the highest merely 7.08%.

The discourse structure of the same type of legal texts can be characterized as follows.

(1) At the first level, all the IUs are obligatory.
(2) At the second level, most of the IUs are obligatory.

(3) At the third, fourth and deeper levels, few obligatory IUs are discovered. Discourse structure begins to display more variations, as the overlapping IUs decrease sharply.

4.2 Features of Elements within Upper-Level Obligatory IUs

As most of the overlapping IUs are located at the upper level, the upper-level obligatory IUs are probed into. We compare the overlap of process, participant and circumstance elements. The overlapping element means the same word or a synonym (see Table 2).

Table 2. Overlapping elements in upper-level obligatory IUs

Text class	Process	Participant	Circumstance
1	80.44%	58.08%	17.23%
2	73.26%	54.55%	9.01%
3	76.54%	55.37%	16.1%
4	77.81%	63.46%	15.9%
5	80.03%	60.47%	21.51%
6	79.04%	56.76%	17.54%

Note: The numbers in the "Text Class" column mean the same with those in Table 1.

From Table 2, we can find that most of the process elements are shared. Participant elements ranks the second, as more than half of them are observed across texts of the same type. Regarding circumstance elements, very few of them are shared. This phenomenon may be explained by the claim that process is the most central element; participant is close to the centre; circumstance is more peripheral, not directly involved in the process [13].

Accordingly, the feature of elements within obligatory upper-level IUs can be summarized that the probabilities of common process, participant and circumstance rank from the highest to the lowest.

5 Processing Rules

Based on the result of discourse analysis, two groups of information rules can be formulated as follows.

5.1 Text Classification Rules

Legal texts of the same type share a number of obligatory IUs mainly at the upper level. If these IUs can be identified, the text type is naturally determined. To achieve this, elements that constitute the IUs can be relied upon. Therefore, the information rules for text classification are as follows.

(1) Process elements of the obligatory IUs are selected as the primary variable. For example, in "the public prosecution organ charges that", the process "charge" is chosen to represent this IU.

(2) As an alternative, participant elements of the obligatory IUs should be selected on the following occasions.

    (a) Implicit process occurs due to Chinese characteristics. For instance, "the defendant (is) XXX". As Chinese may lack the copular verb be, the participant "defendant" is employed to substitute the implicit process "(is)".

    (b) Process is lexically empty, where process is construed by participant and the relevant process has no substantial meaning. The reason can be attributed to nominalization in legal genre [14]. An example is "file an appeal". What carries the prime semantic meaning is not the process "file" but the participant "appeal".

(3) Circumstance of the obligatory IUs can be discarded owing to its peripheral status.

(4) The linear order of the selected elements in the legal text should be followed. Only in this way can the elements represent the original text to the utmost.

In the above rules, which obligatory IUs are selected is based on the hierarchical structure of discourse; which elements are selected follows the inner structure of IU; the order that the elements should keep is due to the linear structure of discourse.

5.2 Information Identification Rules

The information identification rules fall into two groups, one for upper-level IUs and the other for lower-level IUs.

The rules for upper-level IUs are as follows.

(1) Process and participant elements are combined to serve as the main indicator, because the two make up a proposition. For instance, "this court holds that" can be identified by the presence of the process "hold" and the participant "this court". Simple as it appears, this rule can be expanded into many varieties according to the logical relationship between elements.

    (a) The synonyms of process verbs or participant nouns should be included. For instance, "the public prosecution organ" and "People's Procuratorate" refer to the same party in China. Therefore, both of the two should be taken into account and the relationship between them is "or".

    (b) Where one IU contains more than one common process verb, all these verbs probably need to be considered. This can be exemplified by "the public prosecution organ charges that the defendant has committed the crime of XX", both "charge" and "commit" should be written into the rule. Consequently, their relationship is "and".

(2) Absolute position can be a complementary indicator. This applies only when the target IU lies in a fixed position and lacks explicit process.

After the upper-level IUs have been identified, they can help predict their own lower-level IUs. The rules are listed below.

(1) The obligatory lower-level information categories can be set as the default value. An example can be the criminal fact accused of by the public prosecution organ. The IUs below can be attributed to WF as default.

(2) For those that do not belong to the default category, process and participant can work together as the primary indicator.

(3) Relevant position can serve as an alternative indicator. This rule works when common IU elements cannot be found and absolute position is unavailable.

6 Experiment Results and Analysis

In order to examine the performance of our information rule based approach (IRBA), an SVM based approach (SVMBA) for text classification and a Viterbi based approach (VBA) for information identification are employed for comparison.

6.1 Text Classification

Traditional text classification mechanism consists of preprocessing, feature representation, feature selection and classification [15]. For Chinese text, preprocessing includes word segmentation and stop-word removal. In this research, we make use of NLPIR to segment words and tag their parts of speech. We keep only the nouns and verbs and exclude the other words. We represent these words with Vector Space Model and calculate their *tf-idf* value with the following formula.

$$W(t, \bar{d}) = \frac{tf(t,\bar{d}) \times \log_2(N/n_t+1)}{\sqrt{\sum_{i \in \bar{d}} [tf(t,\bar{d}) \times \log_2(N/n_t+1)]^2}} \tag{1}$$

In order to reduce dimensions, 1000 features with the highest χ^2 values are selected. After that, Directed Acyclic Graph SVMs (DAGSVM) is adopted for this multiclass classification, because SVM is reported to outperform the other widely used classification algorithms [16]. In this study, the machine learning toolkit *scikit-learn* is employed. The performance of the two approaches is shown in Table 3 on the next page.

Table 3. Classification result of IRBA and SVMBA

Text class	IRBA			SVMBA		
	Precision	Recall	F1 measure	Precision	Recall	F1 measure
1	100.00%	96.67%	98.31%	93.75%	100.00%	96.77%
2	100.00%	80.00%	88.89%	100.00%	100.00%	100.00%
3	100.00%	96.67%	98.31%	80.00%	93.33%	86.15%
4	100.00%	90.00%	94.74%	100.00%	100.00%	100.00%
5	100.00%	93.33%	96.55%	100.00%	93.33%	96.55%
6	100.00%	93.33%	96.55%	92.00%	76.67%	83.64%
Macro average	100.00%	91.67%	95.56%	94.29%	93.89%	93.85%

Note: The numbers in the "Text Class" column mean the same with those in Table 1.

With the macro average value of F1 measure, it can be summarized that IRBA achieves better results than SVMBA. In particular, the precision of the former is very stable, which always keeps 100%. In contrast, the two values of the latter are not that stable.

The reason for the different performance may be that IRBA recognizes that discourse, in addition to the surface linear structure, has an inner hierarchical structure. The IUs at the higher levels show more commonalities than those at the lower levels. Therefore, only a few process and participant elements in upper-level IUs can help classify texts. However, SVMBA does not take the hierarchical structure of discourse into account. To a certain extent, this approach may shed some light on text classification techniques in terms of dimension reduction.

6.2 Information Identification

As discourse can be perceived as a linear sequence of IUs, information identification is a de facto decoding problem in Hidden Markov Model. This problem can be solved by Viterbi algorithm, which is formally described as the following expression [17].

$$\lambda = (Q,\ O,\ A,\ B,\ \Pi)$$

In the above expression, λ stands for the Hidden Markov Model, Q for states to be explored, O for observations, A for states transition probability, B for emission probability and Π for initial probability.

In HMM, observations are what is observed from the current sequence. For some problems such as POS tagging, observations are the current word. However, it seems more difficult to choose an observation from IU, as they can be characters, words or even combination of words. In this research, we select the process as the observation, because process is the center of clause [13]. For those IUs containing implicit processes, we set "(be)" as the default. One problem for this treatment is that there are more than 2000 verbs in Chinese, so a process in the training set may not occur in the testing set. The solution can be either smoothing or categorizing the verbs. In this research, we take the latter and adopt the eight categories of process verbs [7].

Table 4. Identification result of IRBA and VBA

Text class	IRBA		VBA	
	Upper-level IUs	Lower-level IUs	Upper-level IUs	Lower-level IUs
1	100%	88.26%	50.31%	62.52%
2	96.78%	86.77%	53.47%	55.26%
3	95.32%	84.45%	41.44%	55.39%
4	91.45%	84.81%	85.03%	82.73%
5	95.57%	83.64%	28.61%	46.21%
6	97.23%	82.39%	47.72%	61.33%
Average	96.06%	85.05%	51.10%	60.57%

Note: The numbers in the "Text Class" column mean the same with those in Table 1.

After that, the five variables of the Viterbi algorithm can be determined. Q denotes the information category, O the category of process verbs, A the probability of information category transition, B the likelihood that a category of process verb occurs in the case of a given information category and Π for initial probability of information category. The performance of the two approaches is listed below (texts wrongly classified are corrected) (Table 4).

It can be seen that the accuracy of IRBA is higher than that of VBA. More specifically, for upper-level IUs, the identification accuracy of IRBA ranges from 91.45% to 100%, while that of VBA can only vary from 28.61% to 85.03%. For lower-level IUs, the identification accuracy of IRBA is lower (from 82.39% to 88.26%), but it is still higher than that of VBA (from 46.21% to 82.73%).

With IRBA, upper-level IUs are more likely to be correctly identified. By contrast, VBA cannot tell exactly IUs at which level can be more correctly identified. For indictments, the identification accuracy of upper-level IUs is higher than that of lower-level ones; while for other text types, the result is in reverse.

Of course, Viterbi algorithm is very successful in NLP tasks. The dissatisfying performance of VBA in this research can be attributed to three reasons. The first is that VBA regards discourse as a linear sequence but ignores its inner hierarchical structure. As first order Markov chain is adopted in this study, the occurring probabilities of two adjacent IUs are calculated. This results in a problem that misidentification can easily occur where a lower-level IU transits to an upper-level one. The second is that the selection of observations is oversimplified. In the identification of IUs, the most ideal observation should be the IU itself, but this treatment is infeasible, as there may be an infinite number of different IUs. On this account, the process verb is selected merely as a makeshift. The third is that we calculate the emission probability not for each process verb but for the eight process categories. In this way, the problem can be avoided that the emission probability of a process verb in testing set is hard to determine if it does not appear in training set. However, this treatment incurs a coarse-grained problem. Different verbs may be ascribed to the same process category, so it may lead to the same emission probability and thus wrong information attribution.

By contrast, IRBA recognizes that discourse is a hierarchical configuration. In identification of information category, it will first explore upper-level IUs. After that, the IUs that lie between two upper-level IUs are automatically regarded as lower-level ones. Therefore, a complete discourse is divided into several parts. If some errors are made in one part, they could have little impact on the others. Besides, to identify the information categories of lower-level IUs, the obligatory information categories are set as the default value. In this manner, most of the lower-level IUs can be correctly identified. Even if some are wrongly attributed, there will not be too many mistakes.

7   Conclusion

In this paper, we aim to develop an information identification approach for Chinese legal texts based on discourse analysis. By employing Discourse Information Theory, we make a series of rules to classify legal texts and identify information categories. The macro average F1 score of IRBA for text classification reaches 95.56%, while its average accuracy for upper and lower level information identification amounts to 96.06% and 85.05%. The experimental results show that our rules are helpful for information identification in legal texts.

Our future work is trying to integrate the rules drawn from discourse analysis into statistics based approaches, so that the information identification approach can be improved and applied to more types of legal texts. What is more, we should also retrieve court decisions by informal case description. All these will lay a good foundation for eMediation in dispute resolution.

Acknowledgments. The author would like to thank the anonymous reviewers for their careful work and valuable suggestions for this paper. This work was supported by Key Project for Social Science Research Project in Universities of Anhui Province (SK2017A0990) and the National Social Science Project of China (Project No.: 13BYY066).

References

1. Yan, J.T.: Choice of alternative dispute resolution – in the perspective of rational economic man. Master thesis in College of Politics and Law, Central China Normal University, May 2012
2. El Jelali, S., Fersini, E., Messina, E.: Legal retrieval as support to eMediation: matching disputant's case and court decisions. Artif. Intell. Law **23**(1), 1–22 (2015)
3. Moens, M.F., Uyttendaele, C., Dumortier, J.: Information extraction from legal texts: the potential of discourse analysis. Int. J. Hum. Comput. Stud. **51**(6), 1155–1171 (1999)
4. Chan, S.W.: Beyond keyword and cue-phrase matching: a sentence-based abstraction technique for information extraction. Decis. Support Syst. **42**(2), 759–777 (2005)
5. Voll, K., Taboada, M.: Not all words are created equal: extracting semantic orientation as a function of adjective relevance. In: Australasian Joint Conference on Artificial Intelligence, pp. 337–346. Springer, Berlin (2007)
6. Taboada, M., Brooke, J., Stede, M.: Genre-based paragraph classification for sentiment analysis. In: Proceedings of the SIGDIAL 2009 Conference: The 10th Annual Meeting of the Special Interest Group on Discourse and Dialogue, pp. 62–70. Association for Computational Linguistics, (2009)
7. Du, J.B.: On Legal Discourse Information. People's Publishing House, Delhi (2014)
8. Altman, D.G.: Practical Statistics for Medical Research. CRC Press, Boca Raton (1990)
9. Xing, F.Y.: A Study on Chinese Grammar. The Commercial Press, China (2016)
10. Feng, V.W., Hirst, G.A.: Linear-time bottom-up discourse parser with constraints and post-editing. In ACL, vol. 1, pp. 511–521, June 2014
11. Li, Y., Feng, H., Feng, W.: Chinese discourse segmentation based on punctuation marks. Int. J. Signal Process. Image Process. Pattern Recognit. **8**(3), 177–186 (2015)
12. Joty, S.R., Carenini, G., Ng, R.T., Mehdad, Y.: Combining intra-and multi-sentential Rhetorical parsing for document-level discourse analysis. In: ACL, vol. 1, pp. 486–496 (2013)
13. Halliday, M.K., Matthiessen, C.: An Introduction to Functional Grammar. Routledge, London (2014)
14. Bhatia, V.: Worlds of Written Discourse: A Genre-Based View. A&C Black, London (2004)

15. Uysal, A.K., Gunal, S.: The impact of preprocessing on text classification. Inf. Process. Manage. **50**(1), 104–112 (2014)
16. Zong, C.Q.: Statistical Natural Language Processing. Tsinghua University Press, Bejing (2013)
17. Feng, Z.W.: A Concise Course of Natural Language Processing. Shanghai Foreign Language Education Press, Shanghai (2012)

Real-Time Serial Communication Data Collection

Jinji Zheng, Wei He[(✉)], Xicai Li, Huafu Li, Lei Xie, and Jiawei Zhang

Yunnan Normal University, Kun Ming, China
he99wei@aliyun.com

Abstract. Serial port communication is a relatively simple form of communication. This thesis introduced how to use Visual Studio 2010 C # programming and finally allows PC to receive information from the serial port. And this can be achieved serial port communication between the computer and the MCU. The serial port data transmission and data reading and writing are carried out through the C # serial port API. The serial port data read is analyzed, and finally the data and analysis result are displayed with visual graph window. This article describes how to read the immunofluorescence data through the C # serial port from MCU and display and draw data curve in real-time. In addition, the design of the reset, scan, save, clear, print and other interfaces, can be perfectly compatible with human-computer interaction.

Keywords: Serial communication · Data acquisition · Visualization

1 Introduction

In order to facilitate the rapid build test platform, to achieve the purpose of flexible operation, you can use the microcontroller to send the sensor information directly through the serial port to the personal computer, and then programmed to processing the receive data on the computer.

In biology, there are some of these substances, which by ultraviolet radiation, absorption of certain wavelengths of incident light. You can send a slightly longer than the incident wavelength of light, when the UV light once the stop, the light also emitted. The emission of light called fluorescence. Due to different molecular structure of the material, can absorb ultraviolet light wavelength and fluorescence emission wavelength is different. The use of this feature can be qualitative analysis of the substance under test, under certain conditions. The higher the concentration of the substance to be tested, the stronger the fluorescence emitted after ultraviolet light irradiation; the lower the concentration. The lower the emitted light, accordingly, the quantitative analysis of the substance can be measured, the technology in medicine the fluorescence immunoassay technology.

At present, there are several types of immunofluorescence detection handheld devices on the market. Most of them are limited to the measurement and display of the machine. Under the pressure of the rapid development of Internet of Things and machine learning, medical instruments can be easily connected with computers, data sharing has become a measure of whether a medical equipment with modern equipment conditions

© Springer International Publishing AG 2018
F. Qiao et al. (eds.), *Recent Developments in Mechatronics and Intelligent Robotics*,
Advances in Intelligent Systems and Computing 690, DOI 10.1007/978-3-319-65978-7_62

of the new logo, has become a medical device development process indispensable to a new function and requirements.

2   Software Functions and Methods of Operation

In recent years, C # programming has helped enterprises solve a lot of industrial problems, while in the field of serial communication has a very good application prospects. After receiving the serial data, but also on the serial port to read the byte string analysis, where the need to use the C # control flow to analyze the contents of the byte string, separated from the bytes they need, then the corresponding variety of sensors Signal value and the actual monitoring of the relationship between the units of the actual industrial information, the software has the following main functions: (1) set the communication port and baud rate. (2) Write the configuration to the default parameters registry. (3) The immune fluorescence detector for gain adjustment. (4) The signal acquisition sensor information to read, sort and curve display. (5) Collect data C value and T value and data point subscript read and display. (6) On the computer to start scanning reset immunofluorescence detector. (7) Control the micro-printer to print the test results. (8) Analysis of the test data to determine the positive and negative results. (9) To read data to copy, save or clear. (10) The equations are mathematically analyzed (Fig. 1).

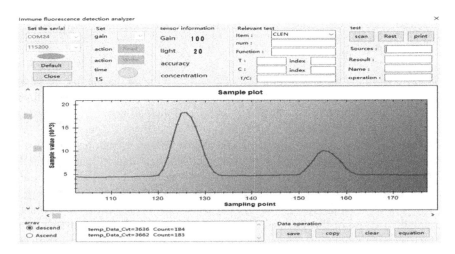

Fig. 1. Software to start scanning interface map

The gain adjustment function is disabled by default. When you need to adjust the gain, you need to input the password in the lower left corner, then open the serial port and select the corresponding gain coefficient. Clicking Read or Write will change the internal gain coefficient or write the gain coefficient into the fluorescence detector.

3 Software Components

3.1 WinForm Development

WinForm is an abbreviation of Windows Form. The default language is C #, developing applications with Visual C # includes building phases, property settings, and code design phases. When you create a new application, you need to define the project's name and physical path location. When you click the OK button, Visual C # will automatically create a new default form FORM1. Based on the C # real-time serial communication data acquisition terminal, the use of WinForm its rich API interface for data reading, curve rendering provides a great convenience, Fig. 2 for the form design interface, the interface can be in the software Each object is set and modified.

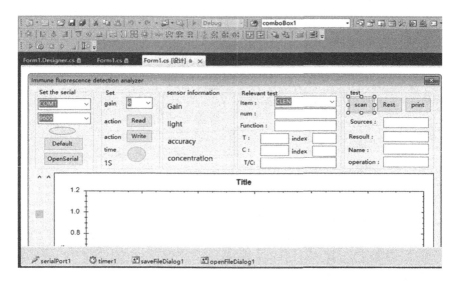

Fig. 2. Schematic diagram of the state transition

3.2 WinForm Drawing Control

The library has a high degree of adaptability, can create almost all the style of the chart, the library is that it provides advanced features of all the default chart, in exchange for the use of simplicity, the library can be drawn according to Curve to select the appropriate range of coordinates also supports switching between different languages, create a chart, you can operate according to the following steps.

(1) Quote GraphPane, this step is equivalent to looking for a professional painter, painter's name is myPane.
(2) Set the title of the chart and the vertical and horizontal coordinates of the logo, set the method to set the chart first, followed by the title set the horizontal and vertical coordinates of the identity, as follows:
    myPane.Title.Text = "My Test Date Graph";

myPane.XAxis.Title.Text = "Date";
myPane.XAxis.Title.Text = "My Y Axis";

(3) Define the middle variables and to generate some random points
    double x, y;
    PointPairList list = new PointPairList();
    for (int i = 0; i < 36; i++)
    {x = (double) new XDate(1995, 5, i+11);
    y = Math.Sin((double) i*Math.PI/15.0);
    list.Add(x, y);}

(4) Use the red curve and the diamond symbol to create a legend named "My Curve".
    CurveItem.myCurve = myPane.AddCurve("My.Curve",list,Color.Red,Symbol-Type.Diamond);
    Set the X axis to the date type.
    myPane.XAxis.Type = AxisType.Date;

(5) ZedGraph automatically refreshes the shape of the curve as the data changes.
    zg1.AxisChange();

After the above six steps can be generated after the following chart, as shown in Fig. 3.

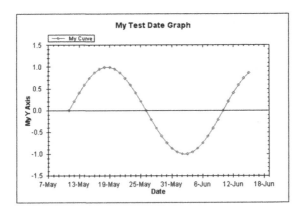

Fig. 3. ZedGraph effect

3.3 Serial Control

Serial communication has several very important parameters, baud rate, data bits, stop bits, parity bit, in the send and receive must be set to the same baud rate and data bits, visual studio 2010 toolbox components The group already has ready-made serial port control, double-click SerialPort, you can call the form code, of course, directly call the form code can achieve the same effect, we can use the introduction of namespace system.io.ports. Figure 4 for the initial control of the serial port control process.

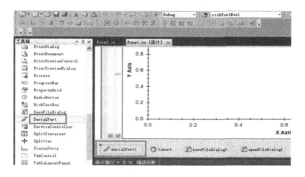

Fig. 4. Serial control call

4 Conclusion

Based on the C # real-time serial communication data acquisition PC control software, combined with Winform powerful serial port control and ZedGraph control, open and close the serial port operation is stable and reliable, with the configuration write configuration table, data read, data write, data saving, data replication, data deletion, sensor status query, gain adjustment curve display and other functions. In addition, the software to follow the standard communication protocol, when the data transmission if the data is wrong, the software can also be adopted by the emergency measures to make the software back to normal, with a certain error correction capability. Through comprehensive debugging and optimization, based on C # real-time serial communication data acquisition PC-side program is stable. In addition, the software has a strong portability according to the need to use the secondary development and upgrading.

Acknowledgment. This project was supported by National Natural Science programs (51267021) and College students' innovative entrepreneurial training programs (2015).

References

1. Qu, X., Wang, B., Fan, M.: Based on the visual C # serial communication program design. Electron. Sci. Technol. **24**(22), 111–112 (2011)
2. Zheng, W., Xiao, B.: Research on new model of serial communication and implementation of C #. Phys. Exp.
3. Zhang, C., Zhao, K.: Development of serial port server system based on STM32. Mod. Electron. Technol. **320**(9), 38–44 (2016)
4. Zhang, J.: C # multi-serial communication strategy research and implementation. J. Liaocheng Univ. **22**(6), 76–78 (2010)

Defect Detection and Recognition Algorithm of Capacitive Touch Panel ITO Circuits

Fan Liao, Yanming Quan[✉], and Changcheng Jiang

South China University of Technology, Guangzhou, Guangdong, China
meymquan@scut.edu.cn

Abstract. As the most widely used human-computer interaction media, capacitive touch panel (CTP) is widely applied to tablets, smartphones and ATM devices, meanwhile, the design of indium tin oxide (ITO) circuits has a decisive impact on the performance of equipment, therefore, the defect detection of ITO circuits is particularly important. With the rapid development of computer technology and digital image processing technology, automatic optical detection technology has been widely used in production and manufacturing, and defect detection algorithm is the core of Automatic Optic Inspection (AOI). Aiming at the pattern of typical CTP ITO circuit, this paper proposed a defect detection and recognition algorithm based on reference comparison and count defect boundary change value. By constructing the corresponding AOI, the paper proved that the algorithm can extract the defects of ITO circuits accurately and identify the extracted defects, which has practical value.

Keywords: ITO · Defect detection · Defect recognition · Algorithm

1 Introduction

With the rapid development of computer technology and digital image processing technology, automatic optical detection technology been widely used in production and manufacturing with its accuracy and high efficiency. ITO is a transparent conductor due to the characteristics of light transmittance and conductivity. Widely used in various types of touch display products [1], ITO circuit pattern has an important impact on the quality of equipment, efficient and accurate automatic detection technology can enhance the quality and production of ITO circuits, and image processing algorithm is the core of automatic detection system [2, 3].

Zongqing Lu proposed a defect detection algorithm for LCD touch screen ITO circuits (shown in Fig. 1(a)) [4]. Firstly, the image to be measured was treated with binary processing, then, the standard template based on the repetition rate of the ITO circuit pattern was reconstructed, and finally the image was compared to the standard image to obtain the defect. However, this algorithm is only applicable to the defect detection of the pattern rule, simple and repetitive small size ITO circuit, and it can only detect defects but cannot identify the defects detected.

© Springer International Publishing AG 2018
F. Qiao et al. (eds.), *Recent Developments in Mechatronics and Intelligent Robotics*,
Advances in Intelligent Systems and Computing 690, DOI 10.1007/978-3-319-65978-7_63

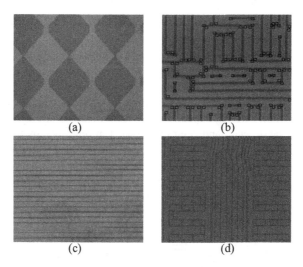

Fig. 1. Several kinds of ITO circuits

Hye Won Kim, Suk i. Yoo proposed a more common defect detection method for PCB and LCD in the non-repetition circuit pattern, namely, feature point matching method [5]. A corner detector was improved so that it can quickly and accurately detect angular points in the image (Fig. 1(b)) and then defect is detected by finding minimum perfect matching of bipartite graph from a complete bipartite graph. Compared with the first algorithm, this algorithm is mainly used in the circuit defect detection of non-repetition pattern which has high accuracy, but if the circuit has large, complex and irregular pattern and more corners, the algorithm has not only low accuracy, but also inefficiency. As with the first algorithm, it only detects defects and does not recognize defects.

Li Changhai proposed a corresponding defect detection and recognition algorithm for the LCD circuits (Fig. 1(c)) [6]. The gray value of the corresponding pixel from the standard image and the image to be measured was subtracted. The residual image is obtained, the defect is searched by contour tracing method and is positioned, and the defects are identified according to the characteristics of the distance variance of the extracted defect and the convex hull of the defect area. Compared with the above two algorithms, this algorithm is suitable for the defect detection of ITO circuits with a linear pattern, which can detect and identify the defects such as missing, scratch, short-circuit, and so on, but it needs a lot of math computation, large amount of data processing and the detection efficiency is low.

The typical capacitive touch panel ITO circuit pattern (Fig. 1(d)) differs greatly from the above three circuit patterns, and its circuits are complex, narrow line spacing, large-sized, more random and no duplication. The first and the second algorithms described in this paper cannot be realized in the current image detection, the third algorithm can only extract some defects and is inefficient. In order to solve this problem, this paper proposed a defect detection and recognition algorithm for the current circuit pattern based on the characteristics of CTP ITO circuit.

2   Detection and Recognition Algorithm of ITO Circuits

The main defects of the ITO circuit pattern in this paper are as follows: short-circuit, protrusion, stain, open circuit, pinhole, notch, among these defects, short-circuit, protrusion and stain can be classified as redundant defects, while open-circuit, pinhole, notch can be classified as missing defects. Common circuit defect detection methods include reference method, non-reference method and mixed method. The pattern of the circuit is complex and large in size, and can only be used in reference method. As the name implies, the reference method is to compare the image to be measured with the standard image by pixel or feature, the advantage of this method is conspicuous which can be widely used, the disadvantage is the high positioning requirement. So before detection, we need to make image registration to obtain high positioning accuracy [7].

The principle of image processing algorithm proposed in this paper is: because the difference of the gray value between the image to be measured and the standard image in the defect area is very large, and the gray value in the no-defect area is similar, the subtraction between the two images only includes defect information. But if there is only one subtraction, only the redundant defects or missing-type defects can be procured, thus, in order to avoid this situation, one more subtraction is needed through exchanging order and a total of two residual images was obtained, and what needs to be stipulated is that the image to be measured minus the standard image obtained is a positive residual image, on the contrary, the result is negative residual image (as shown in Formula 1). The positive residual image contains a missing defect, and the negative residual image contains redundant defects, so that all of the defects can be detected. The specific test steps are as follows:

Firstly, the standard template of ITO circuit is established to obtain positive and negative residual images after the image registration are strictly matched with the standard image and the gray value of the two images is subtracted from the pixel. In order to exclude the disturbance caused by illumination and other factors, the acquisition of standard image and the image to be measured needs to be carried out in the darkroom, using the same illumination system [8].

$$\begin{cases} P_{pos} = P_{ins} - P_{mod} \\ P_{neg} = P_{mod} - P_{ins} \end{cases} \tag{1}$$

where P_{pos} is the positive residual pixel gray value, P_{neg} is the negative residual pixel gray value, P_{ins} is pixel gray value of the image to be measured, P_{mod} is pixel gray value of the standard image, and the residual image of each pixel is treated with binary processing, σ is threshold, which can be configured based on the image gray distribution histogram, the decision rules is shown as the formula (2):

$$\begin{cases} 0 & Pixel \leq \sigma \\ 255 & Pixel > \sigma \end{cases} \tag{2}$$

Then, the residual image is filtered to remove the noise disturbance, also, the defect is searched by using the run-length coding marker in the binary residual image, and the defect is positioned by the center coordinates of the defect.

Finally, transformation numbers fe of defect boundary gray value for each defect are counted to determine the defect, defect boundary schematic diagram is shown in Fig. 2(a), for the acquired defect boundary, the gray value step transformation number is queried clockwise, i.e. fe value. The open circuit is taken as an example to count fe value, as shown in Fig. 2(b):

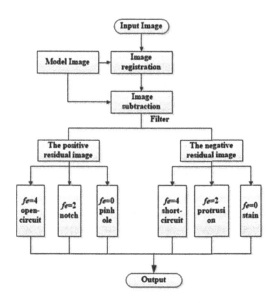

(a)   (b)

Fig. 2. (a) Defect boundary schematic diagram; (b) Open circuit $fe = 4$.

The process of defect detection and identification of ITO circuits as shown in Fig. 3:

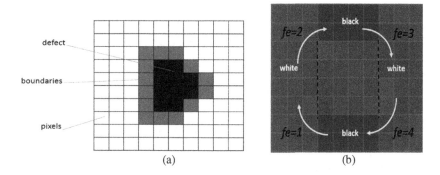

Fig. 3. The flow chart of image processing.

3 Experimental Analysis

For the ITO circuits studied in this paper, the corresponding automatic optical detection system was established, [9] the standard ITO circuit images were produced, the ITO circuit images to be tested were acquired based on the designed AOI, because of the large size of the pattern of ITO circuits, only 420 × 300 pixel images were intercepted for display descriptions to facilitate analysis.

According to the preceding steps, obtaining the image of the positive residual and negative residuals, the filtering denoising and binaryzation processing, and then the defect from the residual image, was located and marked in the image to be measured. For each defect, the fe value is obtained, then the defect is identified according to the fe value and the residual image to which the defect belongs, and the concrete test results are shown in Table 1 (Fig. 4).

Table 1. The results of Defect detection

No	Parameters	
	Location/pixel	Defect recognition
1	(60.4, 329.8)	Notch
2	(71.5, 198.3)	Notch
3	(161.6, 249.2)	Pinhole
4	(182.4, 88.8)	Pinhole
5	(197.5, 159.5)	Open-circuit
6	(210.5, 120.5)	Open-circuit
7	(91.9, 126.4)	Stain
8	(109.6, 308.7)	Stain
9	(164.5, 281.5)	Short-circuit
10	(178.0, 359.8)	Protrusion
11	(248.2, 186.7)	Protrusion
12	(258.5, 305.5)	Short-circuit

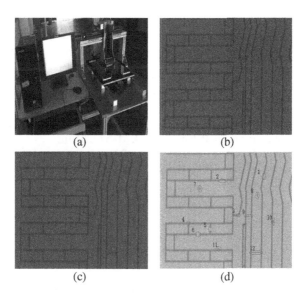

Fig. 4. (a) AOI; (b) The standard ITO circuit; (c) The image to be detected; (d) Detection results.

4 Conclusions

In this paper, an image processing algorithm based on reference comparison and count defect boundary change value is proposed for the detection of the pattern defect of the typical capacitive touch panel. The defect is extracted by pixel by pixel after the image registration with the standard image, and the extracted defect is positioned as well as the defect boundary change value, and then the extracted defects are identified. The method does not need the defect sample, the complex characteristics, or the classifier training to complete the defect classification, because this method is simple and the mathematics is intuitive, therefore it can obtain the higher algorithm efficiency and accuracy rate. Finally, the corresponding AOI system is set up to detect the ITO circuit image, and the experiment proved that the algorithm can extract and identify defects accurately and has practical value.

References

1. Chen, F., Zhao, G., Jiang, S.: High-resolution defect inspection for transparent indium-tin-oxide conductive film. Acta Photonica Sinica **45**(2), 1–6 (2016)
2. Shen, G.S., Yu-Tang, Y.E., Chang-Hai, L.I.: A high resolution defect detection system for ITO lines. Laser Technol. **37**(1), 24–27 (2013)
3. Liu, Y.H., Liu, Y.C., Chen, Y.Z.: High-speed inline defect detection for TFT-LCD array process using a novel support vector data description. Expert Syst. Appl. **38**(5), 6222–6231 (2011)
4. Lu, Z., Peng, Y.: A defect inspection algorithm for LCD touch screen, pp. 1031–1034. IEEE Computer Society (2009)

5. Kim, H.W., Yoo, S.I.: Defect detection using feature point matching for non-repetitive patterned images. Pattern Anal. Appl. **17**(2), 415–429 (2014)
6. Chang-Hai, L.I., Yu-Tang, Y.E., Shen, G.S., et al.: Inspection of circuit defect on LCD panel based on image contour analysis. Laser Technol. **37**(2), 207–210 (2013)
7. Zhang, J., Ye, Y.T., Xie, Y., et al.: Method for fast registration of photoelectric image of appearance detection in printed circuit board. Jiliang Xuebao Acta Metrol. Sinica **36**(3), 238–241 (2015)
8. Lai, W.W., Zeng, X.X., He, J., et al.: Aesthetic defect characterization of a polymeric polarizer via structured light illumination. Polym. Test. **53**, 51–57 (2016)
9. Jiang, C.C., Quan, Y.M., Peng, Y.H.: Design of imaging system for PCTP ITO pattern inspection by using line-scan CCD. Huanan Ligong Daxue Xuebao J. South China Univ. Technol. **42**(11), 19–24 (2014)

Fault Diagnosis of Rolling Bearing Based on Information Fusion

Yanwei Xu[1,2(✉)], Xianfeng Li[2], Tancheng Xie[2], and Yicun Han[2]

[1] School of Mechanical Engineering, Tianjin University, Tianjin 300072, China
xuyanweiluoyang@163.com
[2] School of Mechatronics Engineering, Henan University of Science and Technology, Luoyang 471023, China

Abstract. The rolling bearing is widely used as the mechanical rotating parts, whose state directly determines the performance of the whole machine. Since there exist the problems of low accuracy in traditional single sensor detection and lack of stable diagnostic system, acceleration and acoustic emission can be combined to conduct the study on the fault diagnosis of rolling bearing. The wavelet technology should be used to reduce the noise of the signal obtained by two kind of sensors. Hilbert technique is used to demodulate the signal after noise reduction, and the frequency domain envelope will be got. The eigenvector is obtained by calculating the frequency energy of the fault signal. Then Multi-sensor information fusion system is established by BP neural network in which the suitable samples is selected to train until the error meets the requirement, and the diagnosis of rolling bearing fault is realized.

Keywords: Rolling bearing · Information fusion · Neural network · Fault diagnosis

1 Introduction

Rolling bearing is a kind of mechanical equipment in parts of the world the most widely used, its running state can directly determine the fate of the whole machine [1]; in recent years, the rolling bearing condition monitoring and fault diagnosis has become a hot research at home and abroad. However, the complex working environment of rolling bearings in high speed, high temperature, heavy load and alternating load coupling, its dynamic behavior is influenced by centrifugal force and gyroscopic moment, friction, thermal deformation, external load and other factors, the fault signal is non-stationary and nonlinear characteristics, and the noise is large, the traditional single use sensor fault diagnosis of rolling bearing, weak fault signal is often submerged [2], misjudgment or false. The information fusion technology can integrate the information resources of multi sensors, and combine all kinds of sensors in time and space according to some criteria to achieve the accurate description of the target [3]; The acceleration sensor and the acoustic emission sensor detection technology belongs to the different between the two, with correlation and complementarity, based on this, the two methods are combined by using neural network technology to study the fault diagnosis of rolling bearing.

© Springer International Publishing AG 2018
F. Qiao et al. (eds.), *Recent Developments in Mechatronics and Intelligent Robotics*,
Advances in Intelligent Systems and Computing 690, DOI 10.1007/978-3-319-65978-7_64

2   Information Fusion Fault Diagnosis Method

2.1   Rolling Bearing Fault Vibration Mechanism

When there is a local fault element surface of the rolling bearing surface, fault surface will hit the other element periodically, resulting in uniform force pulse interval, the vibration frequency factors from geometry, bearing speed and fault location; the bearing outer ring fault characteristic frequency is:

$$f_0 = \frac{z}{2}\left(1 - \frac{d}{D}\cos\beta\right)f_s \tag{1}$$

The characteristic frequency of bearing inner ring fault is:

$$f_i = \frac{z}{2}\left(1 + \frac{d}{D}\cos\beta\right)f_s \tag{2}$$

The characteristic frequency of bearing roller fault is:

$$f_f = \frac{D}{2d}\left[1 - \left(\frac{d}{D}\cos\beta\right)^2\right]f_s \tag{3}$$

here: z is rolling body number; β is pressure angle, rad; d is rolling body diameter; D is bearing diameter, m; f_s is rotation frequency of the shaft, Hz.

2.2   Fault Feature Extraction

The principle of rolling bearing fault feature extraction is to find out a kind of characteristic value which can represent different fault types, and the envelope demodulation spectrum of the fault signal of rolling bearing can clearly reflect the distribution of the reactive energy in the frequency domain. Therefore the energy value of the fault characteristic frequency can be taken as the fault characteristic value of the rolling bearing. (1) For noise reduction pretreatment of rolling bearing fault signals: Selecting sym8 wavelet to decompose and reconstruct the signal with 3 layers, and the obtained denoised signal using Hilbert transform, gaining envelope demodulation spectrum of fault signal; (2) Calculating the theoretical characteristic frequency band energy of the demodulated signal: the actual fault characteristic frequency of the bearing and the theoretical fault characteristic frequency error are basically between 2 Hz. Therefore, the theoretical characteristic frequency is chosen as the center, and the interval is $[-2, +2]$ to calculate the energy eigenvalue. If $X(w)$ is the frequency domain signal after fault demodulation, the energy formula of the signal segment is as follows:

$$E = \int_{w_1}^{w_2} |X(w)|^2 dw \tag{4}$$

3 BP Neural Network Information Fusion

In neural networks, the type of feature vector elements need to represent the type of bearing failure. For acceleration signals, respectively the outer ring, inner ring, rolling body of the theoretical characteristics of the frequency of f_0, f_i, f_f as the central frequency, the frequency band of ± 2 Hz, calculate the characteristic energy of different fault types. Setting the characteristic energy is E_1, E_2, E_3, and the characteristic energy of acoustic emission signal is E'_1, E'_2, E'_3, thus the feature vector of the input layer can be combined as follows:

$$\vec{E} = \left[E_1, E_2, E_3, E'_1, E'_2, E'_3, \right] \tag{5}$$

Due to the different units and orders of magnitude, the difference between the acceleration and the acoustic emission sensor is large. In order to reduce the convergence time of the network, it is necessary to normalize the vector [9]. The algorithm is as follows:

$$x_i^0 = 1 - \frac{|x_i - \bar{x}|}{(x_{\max} - x_{\min})} \tag{6}$$

After the above algorithm, all the data in the input vector is between 0 and 1.

The experimental bearing state has four types: normal bearing, outer ring fault, inner ring fault and rolling body fault, so the neural network output layer neurons should be 4, the output vector of the BP neural network is set to the following form:

$$\vec{t} = [\ 1000,\ 0100,\ 0010,\ 0001\ ]^T \tag{7}$$

Among them: 1000 on behalf of the normal bearing, 0100 on behalf of the rolling fault, 0010 on behalf of the inner ring fault, 0001 on behalf of the outer ring fault.

In the previous analysis of the input and output layer, the number of neurons in the input layer has been determined to be 6, and the number of neurons in the output layer is 4. According to the Kolmogorov theorem, the number of neurons in the hidden layer is selected 9, and the neural network of the fault diagnosis of the three layer rolling bearing is established. The structure is 6-9-4. The three layer neural network uses Sigmoid type function as the hidden layer neuron function. The formula is:

$$f(x) = 1 + \frac{1}{1 + e^{\beta x}} \tag{8}$$

Take 0.01 as learning rate, choosing 200 as training step, select 0.001 as learning error, using "train" as training function. Based on the MATLAB neural network toolbox, selecting the appropriate samples to train the network, to achieve the required error range, you can realize the rolling bearing fault diagnosis.

4   Example Analysis

The experiment selects the BVT-5 bearing vibration experiment platform, and the acceleration sensor uses the LC0151T models, matching the Lang LC0201-5 signal conditioning circuit and selecting the Altai PCI8510 acquisition card to collect vibration signals. Acoustic emission system selected model for the PCI-8 acoustic emission acquisition instrument which is the US physical acoustics research and development, and acoustic emission sensor probe selected broadband differential acoustic emission sensor WD. The experimental platform is shown in Fig. 1, and the two sensors start to collect data simultaneously.

Fig. 1. Fault diagnosis of rolling bearing

4.1   Experimental Data

The experiment object selects 40 sets of rolling bearings (including 10 sets of normal bearing and 30 sets of fault bearing), model 6203, which are fault bearing is selected in the production line of quality inspection on the part of the outer ring, inner ring, rolling body unqualified bearing fault. Refer to JB/T 5313-2001 "Rolling bearing vibration (speed) measurement method", the experimental speed selection 1800 r/min, axial loading load of 70 N. According to the theoretical calculation of the frequency characteristics of the rolling bearing, we can get the characteristic frequency of the bearing 6203, as shown in Table 1.

Table 1. The characteristic frequency of rolling bearings 6203

Bearing parts	Outer ring	Inner ring	Rolling body
Characteristic frequency/Hz	92	147	122

4.2   Acceleration Signal Analysis

The envelope is obtained by the Hilbert transform after wavelet denoising. Since the amplitude is higher than 1000 Hz, the low-pass filter (low cutoff frequency is 1000 Hz) is selected to filter the envelope. Figure 2 shows the time domain and the envelope spectra of the Hilbert transform after the noise reduction of the outer ring fault.

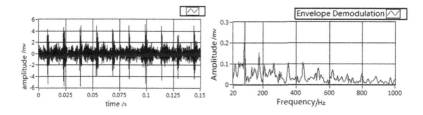

Fig. 2. Time domain chart of outer ring fault and the envelope spectrum

From the envelope spectrum can clearly see the outer ring fault characteristic frequency is 90.8 Hz, closing to the outer ring theory fault characteristic frequency 92 Hz. Figure 3 shows the time domain diagram and its envelope of the wavelet after noise reduction of the bearing inner ring.

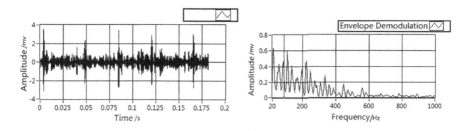

Fig. 3. Time domain chart of inner race fault and the envelope spectrum

The inner characteristic of the inner ring fault Hilbert transform can not see the actual characteristic frequency. This may be due to the fact that the inner ring fault point is far from the measuring point of the sensor and the relative position is not fixed, and the transmission path is more complicated, and the signal has been completely submerged in the background noise and can not pick up, so rolling bearing inner ring with a vibration sensor is relatively difficult to diagnose. Figure 4 is the time domain of bearing rolling body fault wavelet denoising and its envelope figure after Hilbert transform. It can be seen clearly that the characteristic frequency is 122.8 Hz, which is close to the theoretical fault characteristic frequency 122 Hz.

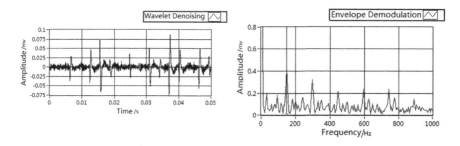

Fig. 4. The time of Inner race fault acoustic emission and the envelope spectrum

4.3 Acoustic Emission Signal Analysis

Acoustic emission is the phenomenon that the elastic energy is released when the material is deformed by external force. Since the spectrum of the acoustic emission signal is wide and the amount of information is larger, it is not susceptible to interference. In this list, the acoustic emission spectrum of the inner ring fault signal, which is difficult to diagnose, is shown in Fig. 4.

It can be seen from the spectrum after demodulation that the characteristic frequency is 147.5 Hz, which is close to the theoretical characteristic frequency of 147. It can be seen that the acoustic emission technology has its unique superiority for recognizing the weak signal of the inner ring. The correct rate of acoustic emission detection is 87.5% by statistics, which indicates that the detection of the inner ring fault of the acoustic emission sensor has obvious advantages over the vibration sensor.

Acknowledgements. This work was financially supported by the National Natural Science Foundation of China (51305127), the youth backbone teachers training program (2016GGJS-057) and scientific research key project fund of the Education Department Henan Province of China (14A460018).

References

1. Tao, M.: Vibration Analysis and Diagnosis of the Gear and Rolling Bearing. Xian: Aviation Power and Thermal Energy Department of Northwestern Polytechnical University, pp. 2–5 (2003)
2. Li, C., Xu, M., Gao, J.: Fault diagnosis of rolling bearing based on independent component analysis. Acad. J. Harbin Inst. Technol. **9**, 1363–1365 (2008)
3. Mei, H.: A Monitoring and Diagnosing System to Vibration of Rolling Bearing, pp. 30–40. China Machine Press, Beijing (1995)

Design and Application of Full-Adder

Jing Zhao[1(✉)], Lan-Qing Wang[2], and Zhi-Jie Shi[2]

[1] School of Physics & Electronic Technology, Liaoning Normal University,
Dalian 116029, China
zjmilk@126.com
[2] School of Mathematics and Computer Science, Shanxi Datong University,
Datong 037009, China

Abstract. In this paper, we focus on the design and application of full-adder. The design is given from the truth table to simplify to logic circuit. The application is given the full-adder implementation of NAND gate, "138" to achieve full-adder, "153" to achieve full-adder, 4-bit parallel adder, full-adder/full-subtractor. Finally design a full-adder application direct on 6 voting system.

Keywords: Full-adder · Truth table · 4-bit parallel adder · 6-position voting system

1 Introduction

Usually the circuit design process is generally:

(1) Through the analysis of the problems to be solved, the truth table is listed. In the analysis we must pay special attention to practical problem how to abstract logic relations between several input variables and several output variables, the existence of constraint relation between the output variable, which lists the truth table or simplified truth Table
(2) By means of the simplification of the graph of Kano or the logical formula, the simplest and the expressions are derived.
(3) According to the actual requirements, the simplest logic expression is drawn (Fig. 1).

© Springer International Publishing AG 2018
F. Qiao et al. (eds.), *Recent Developments in Mechatronics and Intelligent Robotics*,
Advances in Intelligent Systems and Computing 690, DOI 10.1007/978-3-319-65978-7_65

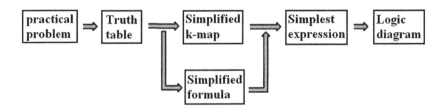

Fig. 1. The circuit design process

2 The Full-Adder Design

A full adder has three inputs: the summand, the addend and carry signals from adjacent low; it has two output: the sum and the adjacent high carry signals.

Therefore, the full-adder is different from the half-adder is that full-adder should consider an adjacent low carry, while half adder does not.

Full adder logical symbol shown in Fig. 2, and its truth table shown in Table 1. C_I is input carry. C_O is output carry. S is sum. A and B are input bits, summand and addend (Fig. 3).

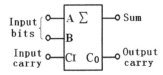

Fig. 2. Logic symbol

Table 1. Truth table of a full-adder

A	B	C_I	C_O	S
0	0	0	0	0
0	0	1	0	1
0	1	0	0	1
0	1	1	1	0
1	0	0	0	1
1	0	1	1	0
1	1	0	1	0
1	1	1	1	1

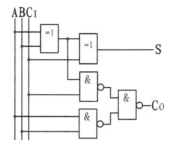

Fig. 3. Full-adder logic circuit

From the truth table, we can write the expression

$$S = \overline{AB}C_I + \overline{A}B\overline{C_I} + A\overline{B}\,\overline{C_I} + ABC_I$$
$$C_O = \overline{A}BC_I + A\overline{B}C_I + AB\overline{C_I} + ABC_I$$

Simplify and transform to:

$$\begin{aligned}
S &= \overline{A}\left(\overline{B}C_I + B\overline{C_I}\right) + A\left(\overline{B}\,\overline{C_I} + BC_I\right) \\
&= \overline{A}(B \oplus C_I) + A\overline{(B \oplus C_I)} \\
&= A \oplus B \oplus C_I
\end{aligned} \tag{1}$$

$$\begin{aligned}
C_O &= \left(\overline{A}B + A\overline{B}\right)C_I + AB \\
&= (A \oplus B)C_I + AB \\
&= \overline{\overline{(A \oplus B)C_I} \cdot \overline{AB}}
\end{aligned} \tag{2}$$

If we are required to achieve full-adder with nine 2- input NOT-AND gate, its circuit is giver in Fig. 4.

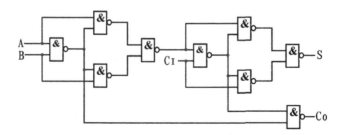

Fig. 4. Full-adder with NAND gate

3 Design a Full-Adder/Full-Subtractor

Now, let's design a full-adder/full-subtractor circuit. It can not only achieve an addition operation, but also achieve a subtraction operation. When the control variable M = 0, the circuit to achieve addition operation; when M = 1, the circuit subtraction operation (Fig. 5).

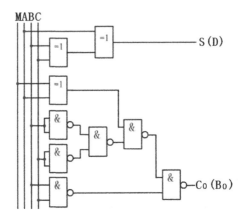

Fig. 5. Full-adder/full-subtractor

Its truth table is given in Table 2.

Table 2. Truth table of a full-adder/full-subtractor

M	A	B	C	S(D)	Co(Bo)
0	0	0	0	0	0
0	0	0	1	1	0
0	0	1	0	1	0
0	0	1	1	0	1
0	1	0	0	1	0
0	1	0	1	0	1
0	1	1	0	0	1
0	1	1	1	1	1
1	0	0	0	0	0
1	0	0	1	1	1
1	0	1	0	1	1
1	0	1	1	0	1
1	1	0	0	1	0
1	1	0	1	0	0
1	1	1	0	0	0
1	1	1	1	1	1

The simplified expression for S(D) and $C_O(B_O)$ is

$$S(D) = A \oplus B \oplus C \quad C_O(B_O) = \overline{\overline{BC} \cdot \overline{\overline{BC}(M \oplus A)}} \tag{3}$$

4 Design a Full-Adder with "138" and "153"

How to realize the function with 138?

First, transformed into a minimum expression. Then, the corresponding output connected with the NAND gate.

$$S = \overline{A}\overline{B}C_I + \overline{A}B\overline{C_I} + A\overline{B}\overline{C_I} + ABC_I = \sum m(1, 2, 4, 7) \tag{5}$$

$$C_O = \overline{A}BC_I + A\overline{B}C_I + AB\overline{C_I} + ABC_I = \sum m(3, 5, 6, 7) \tag{6}$$

The Fig. 6 is given in using "138" and the other logic gates to achieve full-adder.

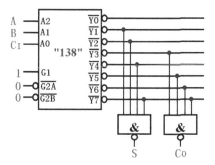

Fig. 6. "138" to achieve full-adder

For the "153" has only 2 address inputs and the full-adder is 3, we can use the Dimension reduction. The truth table is given in Table 3. Figure 7 shows a full-adder with "153" and other logic gates.

Table 3. Truth table of a full-adder with "153"

A	B	C_I	S		Co	
0	0	0	0	C_I	0	0
0	0	1	1		0	
0	1	0	1	$\overline{C_I}$	0	C_I
0	1	1	0		1	
1	0	0	1	$\overline{C_I}$	0	C_I
1	0	1	0		1	
1	1	0	0	C_I	1	1
1	1	1	1		1	

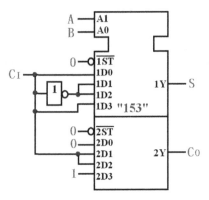

Fig. 7. "153" to achieve full-adder

5   Four-Bit Parallel Adders

A 4-bit parallel adder is composed of four full adders, as shown in Fig. 8. The carry output of the low adder is connected to the carry input of the adjacent high adder. The logic symbol of the 4-bit parallel adder is shown in Fig. 9.

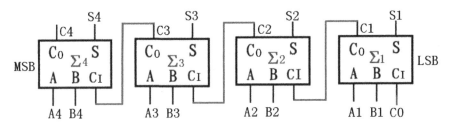

Fig. 8. 4-bit parallel adder

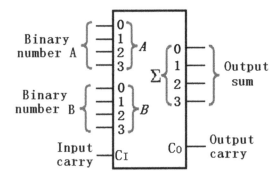

Fig. 9. 4-bit parallel adder logic symbol

6 6-Position Voting System

An example of full adder and parallel adder applications is a basic voting system that can be used to provide both "yes" and "no" votes. The system can be used for a group of people, and can immediately identify opinions (for or against), make decisions, or voting on certain issues or other matters.

The system includes a switch for "yes" or "no" selection at each position in the assembly and a digital display for the number of yes votes and one for the number of no votes. The system is shown in Fig. 10 for a 6-position setup.

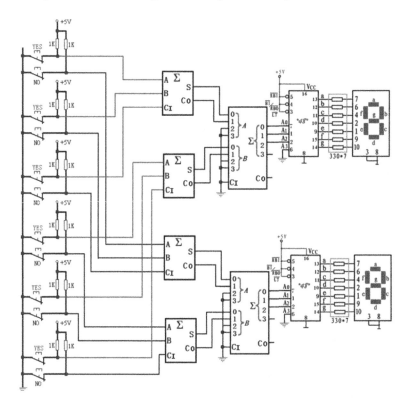

Fig. 10. 6-position voting system

7 Conclusion

In this paper, first, we focus on NAND gate achieving full-adder, and then on "138" and "153" achieving full-adder, 4-bit parallel adder, full-adder/full-subtractor. Finally design a full-adder application direct on 6 voting system.

References

1. Zhao, J.: Design of PLC control system for traffic lights, Mechatronics and Manufacturing Technologies (2016)
2. Zhao, J.: Research on the measurement system of liquid weighing based on PLC. In: International Conference Energy and Mechanical Engineering (2015)
3. Zhao, J.: PLC counter applications technique, Progress in Mechatronics and Information Technology (2013)
4. Floyd, T.L.: Digital Fundamentals (2006)

Safety Device for Stamping Equipment Based on Fuzzy Neural Network Information Combined with Identification Technology

Qingfang Deng[✉]

Department of Mechanical and Energy Engineering, Shaoyang University,
Shaoyang City 422004, China
53730472@qq.com

Abstract. A safety device for stamping equipment was designed to detect the blackness and temperature parameter in the stamping equipment mold area. Based on the fuzzy neural network information identification technology which combined the fuzzy system with neural network, it can be detected whether there is a hand or no hand in the mold area. Feature data were extracted and fused by using Bayes statistics method and the recognition results were output by fuzzy solution. The blackness and temperature detection device of stamping safety equipment was studied to identify the detection signal, providing a theoretical basis for the design of safety device for stamping equipment.

Keywords: Stamping equipment · Safety device · Fuzzy neural network · Identification technology

1 Introduction

Stamping is a kind of technology that uses molds to make the deformation or separation of metal or nonmetal plate on the pressure machine, now widely applied in automobiles, electrical appliances, light industry and other industries [1]. The stamping production in many companies of our country is basically in the state of manual operation, many procedures relying on workers to stretch out into the mold cavity, which can possibly lead to accidents when the mechanical equipment is out of control or operation errors occur. At present, there are many protection devices in the domestic and international, among which the infrared and photoelectric protection device is the most widely used because of its simplicity, practicability and high reliability [2, 3]. Existing infrared, laser, photoelectric safety devices of stamping equipment solve the security problem only to a certain extent and do not have the recognition function of blanks and human hands. Therefore when the blank sender into the screen part blocking the light, the punch will automatically stop, which makes the punch feed discontinuously and causes inconvenience to production, reducing the productivity.

Therefore, in a safe and reliable basis, and under the premise of not affecting the continuous feed, it not only can make the stamping equipment to meet the needs of modern manufacturing enterprises, but also can improve the productivity to a great

© Springer International Publishing AG 2018
F. Qiao et al. (eds.), *Recent Developments in Mechatronics and Intelligent Robotics*,
Advances in Intelligent Systems and Computing 690, DOI 10.1007/978-3-319-65978-7_66

extent to improve the existing safety device, which has intelligent identification function. In the detection and research of stamping safety device, the blackness and temperature are taken as direct detection features according to the theory of computer pattern recognition [4–7]. Based on the fuzzy neural network information combined with recognition technology, the intelligent recognition of man hands and blanks in the production process of stamping equipment is studied to control the interruption and operation of stamping equipment.

2 Basic Principle

Intelligent recognition structure of safety device of stamping equipment based on multipoint infrared thermometry is shown in Fig. 1. Both sides of stamping equipment mold area are arranged with rotatable bracket and slide bar, and 3 infrared probe elements are installed every 4 cm space on each side of the bracket, which are close to the center line to ensure the signal obtain of blank and hand temperature. Infrared temperature measuring element induces surface temperature of object(t) according to the blackness of object itself (ε). The hand temperature is constant, fluctuating in the range of 35–37 °C, whereas the temperature of the punch feeding material (steel, plastic) will change significantly with the environment. Based on the probability statistics, recognition model of blank and man hands can be established by introduction of practical fuzzy neural network algorithm.

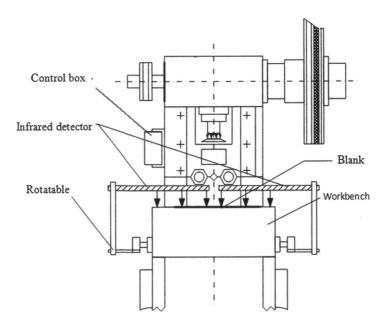

Fig. 1. Intelligent recognition structure of safety device of stamping equipment

In the feed process, the temperature signals in different positions of the stamping equipment mold area are from 6 representative positions on the edge of mold area (covering the hand and blank range), and the 6 point is located in the edge of mold at the same height. For the signals acquired on both sides, using Bayes statistical methods for fusion and extraction of feature signals, the feature signals are identified online based on modern pattern recognition processing method such as fuzzy neural network. The relationship between feature signals and the corresponding mode is analyzed and the corresponding intelligent recognition model is established.

The basic principle of the device is as shown in Fig. 2. The blackness and temperature parameters in the mold area of operating stamping equipment are collected by single chip, and the fuzzy neural network models of hands and blank are stored in the single chip control system in advance. The blackness and temperature parameters are fused by 6 infrared temperature detection devices on the edge of mold area based on Bayes statistics, and they are identified as the input of the fuzzy neural network system. The 3 kinds of identification results, man hand, material, man hand and material are classified as hands and no hand. According to output data of the fuzzy neural network, the single chip controls the clutch of stamping equipment.

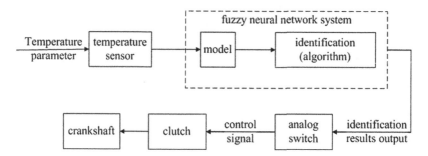

Fig. 2. Schematic diagram of stamping safety device combination of fuzzy system and neural network

Neural network [8, 9] and fuzzy system are important methods of soft computing. Neural network has a network structure of parallel processing and strong self-learning ability [10–12], but the form of knowledge by weight expression is not easy to understand, not making full use of abundant language knowledge of the experts in the field [13–16]. Fuzzy logic can make good use of language knowledge, and the form of knowledge expression is easy to understand [17], but it is weak in self-learning ability and difficult to use numerical information [18, 19]. Therefore, the combination of the fuzzy system and neural network will be more advantageous [20].

Fuzzification, fuzzy inference and fuzzy solution are basic modules of fuzzy system. The multilayer feed forward neural network based on fuzzy inference model is completely equivalent to fuzzy system in the input and output ports, while the internal weights node parameters can be modified by learning. The fuzzy neural network is a multilayer feed forward network which is divided into front, middle and rear layer. The front layer simulates basic fuzzy membership function of premise variables to realize fuzzification. The middle layer constitutes fuzzy logic inference machine. The rear layer realizes the fuzzy solution. The front, middle and rear layer are constituted of monolayer or multilayer node layer, and node number and node weights of each layer can be preset according to the specific modules of fuzzy system, which automatically generates the appropriate shape and fuzzy rules of membership function by certain learning algorithm.

As shown in Fig. 3, when punching operation starts, the test data blackness and temperature are read in from the infrared detector by the A/D converter first and are converted to digital quantity. The saved digital quantity fused and feature extracted by Bayes statistical method is input variable and then fuzzed. The determined input variables are converted to fuzzy sets described by membership degree, which are taken as input layer neurons of neural network. The logic rule layer neurons classifies hands and no hand situation, and then a layer of neurons achieves the fuzzy solution.

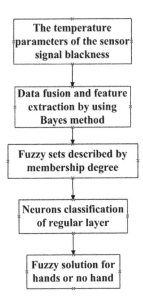

Fig. 3. Recognition block diagram of fuzzy neural network information fusion for stamping safety device

Based on the output of the fuzzy neural network, the hybrid control system in single chip controls timely interruption of stamping equipment when in hand situation, and maintain the normal operation of stamping equipment while it detects no hand and only blank.

3   Conclusion

Safety device for stamping equipment was studied based on fuzzy neural network information fusion identification technology.

(1) It is important for the detection of the blackness and temperature in the feeding area to arrange the number and distribution of the sensors in the mold area in the operation process of stamping equipment reasonably and it has a crucial role in the recognition system.

(2) Based on data fusion and feature extraction by Bayes statistics method, the rule layer neurons classify the fuzzed input variables. Finally the fuzzy solution layer outputs whether there is hand or no hand in stamping area, and the operation system of stamping equipment is controlled by single chip microcomputer system.

Acknowledgements. This project is supported by the Scientific Research Fund of Hunan Provincial Education Department (No. 12C0861).

References

1. Huang, C.: Present situation and development of stamping equipment in China. Machinist (hot working) **4**, 23–25 (2007)
2. Wang, Y., Ji, S., Fei, X.: Research on protection technology of stamping bed based on computer image processing. Mach. Hydraul. **9**, 60–62 (2004)
3. Morales, R., Feliu, V., Sira-Ramírez, H.: Nonlinear control for magnetic levitation systems based on fast online algebraic identification of the input gain. IEEE Trans. Control Syst. Technol. **19**(4), 757–771 (2011)
4. Feng, D., Xie, S.: Fuzzy Intelligent Control, pp. 100–150. Chemistry Industrial Press, Beijing (1998)
5. Culter, C.R., Ramaker, B.L.: Dynamic matrix control: a computer control algorithm. In: Proceedings of Joint Automatic Control Conference, San Francisco (1980)
6. Rouhani, R., Mehra, R.K.: Model algorithmic control: basic theoretical properties. Automatica **18**(4), 401–414 (1982)
7. Chen, C.A., Chiang, H.K., Shen, J.C.: Fuzzy sliding mode control of a magnetic ball suspension system. Int. J. Fuzzy Syst. **11**(2), 97–106 (2009)
8. Hagan, M.T., Demuth, H.B., Beale, M.H.: Neural Network Design. PWS Pub. Co., Boston (1996)
9. Jiaoli, C., Yangshu, Y., Liu, F.: Seventy years beyond neural networks: retrospect and prospect. Chin. J. Comput. **8**, 1698–1716 (2016)
10. Lin, Y.Q., Zhang, T., Zhu, S.H., Yu, K.: Deep coding networks. In: Proceedings of the Neural Information and Processing Systems, Vancouver, Canada, pp. 1405–1413 (2010)
11. Yue, J., Zhao, W.Z., Mao, S.J., Liu, H.: Spectral-Spatial classification of hyperspectral images using deep convolutional neural networks. Remote Sens. Lett. **6**(6), 468–477 (2015)
12. Hui, L., Hong-qi, T., Di-fu, P., Yan-fei, L.: Forecasting models for wind speed using wavelet, wavelet packet, time series and artificial neural networks. Appl. Energy **107**, 191–208 (2013)
13. Li, Y., Fan, R., Jiang, W., et al.: Research on intelligent station area identification method based on BP neural network. Electr. Meas. Instrum. **54**(3), 25–30 (2017)
14. Luo, C., Zhou, L.: Research on neural network model based on improved genetic algorithm. Inf. Mag. **2**, 65–66 (2005)
15. Wang, S.-Q., Chang, J.: Neural network model predictive control for the hydroturbine generator set. In: Proceeding of the Second International Conference on Machine Learning and Cybernetics, Xi'an (2003)

16. Chen, Z., Yuan, Z., Zhang, Y.: Non-liner predictive control based on neural network. Control Eng. **9**(4), 1 (2002)
17. Zhang, N., Yan, P.: Neural Networks and Fuzzy Control. Tsinghua University Press, Beijing (1998)
18. Gu, M.: Research on internal relations among neural network, fuzzy system, support vector machine. Electric and Technology University, Chengdu (2004)
19. Lin, C.J., Lee, C.Y.: Non-linear system control using a recurrent fuzzy neural network based on improved particle swarm optimisation. Int. J. Syst. Sci. **41**(4), 381–395 (2010)
20. Chen, C.H.: Design of TSK-type fuzzy controllers using differential evolution with adaptive mutation strategy for nonlinear system control. Appl. Math. Comput. **219**(15), 8277–8294 (2013)

The Study and Application of High Pressure Water Injection Flow Automatic Control Instrument Based on IOT

Chaodong Tan[1(✉)], Guishan Ren[2], Haoda Wu[1], Guangjun Wu[3], and Xinlun Li[1]

[1] China University of Petroleum (Beijing), Beijing, China
tantcd@126.com
[2] PetroChina Dagang Oilfield Production Technology Research Institute, Tianjin, China
[3] Anhui FirstCon Instrument Co., Ltd., Chizhou, China

Abstract. Aiming at the current situations and the characteristics in water injection wells, the high pressure flow automatic control instrument based on IOT is studied in this paper, which combines the functions of flowmeter, flow regulating valve, intelligent controller and communication technology into one intelligent control device. Through its application in hundreds of water injection wells in Qinghai and Dagang oilfield, it achieves the connections between instruments, the real-time data acquisition and monitor, the intelligent adjustment of water injection, which improves the automatic and intelligent management level for water injection wells.

Keywords: Water injection well · Flow automatic control instrument · IOT · Intelligent adjustment

1 Introduction

For a long time, the water injection is artificially controlled in well site, valve groups and water injection station. Problems existing in conventional water injection system are as follows: big error in water injection rate; poor timeliness; heavy workload of patrolling; difficulties in injected water volume adjustment; no power supply in well site. Therefore, working condition auto-testing system is badly needed to realize fully automatic intelligent adjustment for water injection rate.

The high pressure flow automatic control instrument and its application are studied by Chen Yanhai etc. and it approves that this device can improve the water injection efficiency and adapt to the high pressure environment [1]. The solar steady flow water distribution automatic control technology in water injection well is studied by Wang Yong etc. who put forward the theory of solar energy supply and steady flow valve automatic control [2]. The research and design of new flow control instrument is proposed by Huang Anyi etc. which combined with anti-jamming methods such as isolation and filtering and control technology of PID controller with incomplete derivation [3].

© Springer International Publishing AG 2018
F. Qiao et al. (eds.), *Recent Developments in Mechatronics and Intelligent Robotics*,
Advances in Intelligent Systems and Computing 690, DOI 10.1007/978-3-319-65978-7_67

In recent years, IOT technology is gradually applied to all works of life and plays an important role in accelerating the technology progress in related industries. With the accelerating progress of the intelligent oilfield construction that based on IOT technology, the oilfield has made remarkable achievement in oil, gas and water well intelligent construction [4–6]. The application of real-time acquisition, reliable transmission and intelligent control of oil and water well data laid a relative technology foundation for intelligent construction in the process of oil and gas production.

2   Design of FirstCon High Pressure Flow Automatic Control Instrument

Firstcon Flow Automatic Control Instrument combines the functions of flowmeter, flow regulating valve, intelligent controller and communication technology into one intelligent control device which is shown in Fig. 1. Intelligent controller can compare preset flow value with the flow value detected by flowmeter, when the detected flow value is inconsistent with the preset value, intelligent controller will open flow regulative valve automatically and adjust the flow to the preset value.

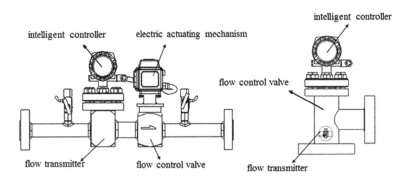

Fig. 1. Structure of Flow Automatic Control Instrument

Intelligent controller can detect upstream pressure and downstream pressure, when the latter is greater than the former, it can automatically close the valve to effectively prevent fluid from flowing backward. Flow Automatic Control Instrument is made up of the flow acquisition and display part, flow control part, pressure acquisition part, data processing part, data transmission part and power supply system. As it shows in Fig. 2.

Fig. 2. Skid-mounted structure of flow automatic control instrument

3 Design of FirstCon Intelligent Water Injection System

FirstCon intelligent water injection system is comprised of sensing layer, control layer, transport layer and application layer as shown in Fig. 3. Users can remotely set daily water-injection rate through the network. Related information can be sent to automatic instrument by mobile web. After comparing current value with preset value, the regulative valve will open automatically to regulate the flow based on PID arithmetic. At the same time the real-time flow data, accumulated flow data and all kinds of pressure data can be sent to RTU. Through public mobile network such as CDMA\GPRS etc., the data can be sent to oilfield intranet. Through analysis and calculation by iWES system in server-side, it can automatically generate related reports and publish in the web by LAN.

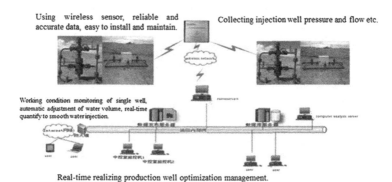

Fig. 3. Frame of FirstCon water injection intelligent system

FirstCon intelligent monitoring system has following characteristics: skid-mounted structure, convenient in installation and replacement; flexible power supply way; standardization of hardware interfaces and communication; reliable equipment performance

4   Analysis of the Application Effect

More than 380 sets of equipment have applied this system in Qinghai oilfield. This system effectively improved the automatic and intelligent management level for the injection well and reduced both the workers' labor intensity and the management cost.

4.1   Real-Time Acquisition of Working Condition

Water injection well: tubing pressure, casing pressure, trunk line pressure, temperature, casing flow, tube flow and electromagnetic valve opening etc.; water allocation station: trunk line pressure, tubing pressure, temperature, flow and electromagnetic valve opening etc. The acquisition and display of production parameters and equipment working parameters can be realized after applying this system.

4.2   Real-Time Analysis and Optimization

The real-time management, real-time analysis, real-time optimization, real-time control and intelligent prediction of production data can be realized.

4.3   Constant Flow can be Realized Under Fluctuant Pressure

System can be run in automatic mode, it can adjust the flow according to preset flow value to realized stable water injection. As it showed in Fig. 4, a pressure fluctuation happened in Yuexin no. 33 water injection well from August 26 to September 2, however, the water-injection curve kept flat, achieving the purpose of constant flow water injection.

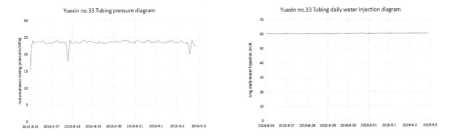

Fig. 4. Graph of constant flow water injection under fluctuant pressure

4.4   Truly Reflect the Condition of Water Injection Well

Daily water-injection curve in Fig. 5 indicated that water injection rate fluctuated sharply on September 20 and October 25 in Yuexin no. 33 well. Through on-site check-up, pump station cut off water supply on September 20 and the well was shut down for testing static pressure on October 25, which means that the water-injection remote control system can truly reflect the condition of the well.

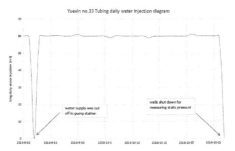

Fig. 5. Graph of daily water-injection rate

4.5 Automatic Adjustment of Valve Opening, Accurately Complete Injection Allocation

In the process of water injection, the current acquired flow rate will be compared with the preset value, if it surpasses the allowable range, the valve opening will be automatically adjusted according to PID arithmetic [7]. The injection rate of Yuexin no. 33 well is 48 square/day, which can be accurately controlled after installing the intelligent regulator control system. The daily water injection is ensured 47.56 square/day with the error rate less than 1% to meet the injection requirement.

5 Conclusion

FirstCon Flow Automatic Control Instrument and Water-Injection Well Intelligent Monitoring System are developed based on IOT technology and combining the reality of water injection wells. This system applies skid-mounted structure in hardware frame, which facilitate the installation. It can support battery power supply as well as photovoltaic power supply to effectively adapt to the current situation of water injection well site without power supply. FirstCon Flow Automatic Control Instrument applies the advanced PID closed-loop control technology, when the real flow is inconsistent with the preset value, the flow regulative valve will automatically start to adjust the real flow, realizing the intelligent allocation, which improved the intelligent level of oilfield water injection and dramatically reduced the workload of daily equipment maintenance.

References

1. Chen, H.: The application of high pressure fiow automatic control instrument in Jilin oil field water injection system. Chem. Enterp. Manage. **26**, 15 (2015)
2. Wang, Y., Shao, D., Lu, H., et al.: Solar steady water injection automatic control technology of water injection well. In: 2013 Digital and Intelligent Gas Fields (International) Proceedings of Conference and Exhibition, Xi'an Petroleum University, Sahnxi Institute of Petroleum, vol. 6 (2013)
3. Huang, A., Li, R.: The research and design of new flow automatic control instrument. Foreign Electron. Meas. Technol. **1**, 57–60 (2017)

4. Tan, C., et al.: Technology research on oil, gas and water well production IOT system iPES. China Pet. Chem. Ind., 64–67 (2012)
5. Wang, Q., et al.: Test application of water injection well remote intelligent control system in Qinghai oilfield. China Pet. Chem. Ind. **5**, 55–57 (2010)
6. Chen, S., et al.: Research and application of real-time analysis and optimization of water injection well. China Pet. Chem. Ind. **9**, 59–61 (2011)
7. Yang, N., Wang, W.: The analysis of new PID function of control algorithm based on LabVIEW. Electron. Meas. Tech. **11**, 74–78 (2015)

The Influence of Driver's Different Degree of Participation on Path Following

Wen-Liang Li, Zhou Wei[✉], Chen Cao, and Xuan Dong

Research Institute of Highway Ministry of Transport,
Xitucheng Road No.8, Haidian District, Beijing 100088, China
c.cao@rioh.cn

Abstract. The vehicle and path following system is established with trucksim. Double lane change and serpentine tests are simulated to analyze the path following effect. The simulation tests include driver controlling the vehicle speed and the steering, driver controlling speed and driver controlling nothing. The speed is from 20 km/h–90 km/h every 10 km/h. The results show that the steering control is the most important factor for path following, and the influence of the path following is different for different condition.

Keywords: ANFIS driver's different degree of participation · Path following · Simulation · Double lane change test · Serpentine test

1 Introduction

In the vehicle performance test, the path following technique can effectively eliminate the influence of driver's variability, can improve the repeatability of the test, which is conducive to the objective evaluation of vehicle performance and the comparative analysis between different vehicles.

The control model of the steering robot is studied, which is used in the vehicle transient response and steady state rotation, but only studies the optimization control technology of the steering, and the research of the path following algorithm is not carried out [1, 2]. A path following algorithm based on two degree of freedom model is proposed, but the two degree of freedom model is not suitable for the condition of large lateral acceleration, which is difficult to meet the requirements of vehicle performance test [3]. Many scholars have studied the application of the vehicle path following technology, which is used to assist or intelligent driving, not to control the vehicle speed [4–10].

In summary, both at home and abroad mainly studied for path following algorithm, and the influence of driver's different degree of participation on path following is not studied, which is important for the construction of key skills of testing ground, is also the main content of this dissertation.

© Springer International Publishing AG 2018
F. Qiao et al. (eds.), *Recent Developments in Mechatronics and Intelligent Robotics*,
Advances in Intelligent Systems and Computing 690, DOI 10.1007/978-3-319-65978-7_68

2 Simulation Test

The vehicle model selected is a two axis bus model (Tour Bus 5.5T/10T) that have been tested by academic paper [11]

Double lane change and serpentine tests are simulated, including driver controlling the vehicle speed and the steering, driver controlling speed and driver controlling nothing. The speed is from 20 km/h–90 km/h every 10 km/h.

(1) Driver controlling vehicle speed and steering. Simulation speed is set according the test speed, the type is 'initial speed, open-loop throttle'.
(2) Driver only controlling vehicle speed. Steering is controlled with driver model, the type is 'steering: Driver path follower, No Offset w/1.5 s. Preview'.
(3) Driver controlling nothing. Simulation speed is set according the test speed, the type is 'constant target speed'. Steering is controlled with driver model, the type is 'steering: Driver path follower, No Offset w/1.5 s. Preview'.

The double lane change and serpentine test path are shown below (Figs. 1 and 2).

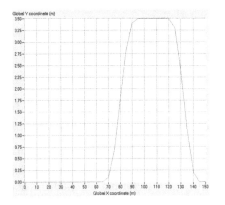

Fig. 1. Double lane change test path

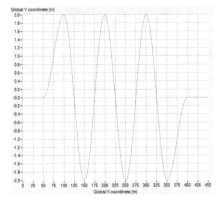

Fig. 2. Serpentine test path

Due to space limitations, the part of the test data is shown as follows (Figs. 3, 4, 5, 6, 7, and 8).

The path following error curves are drew as

As can be seen from Figs. 9 and 10:

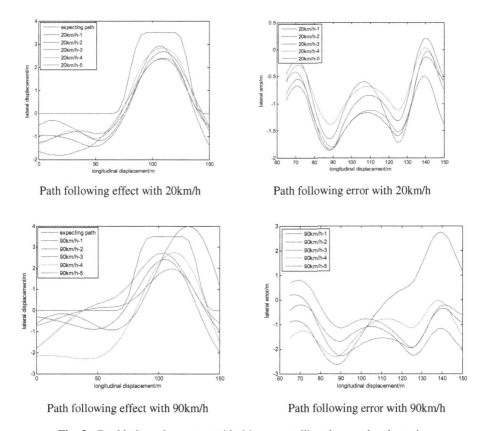

Path following effect with 20km/h   Path following error with 20km/h

Path following effect with 90km/h   Path following error with 90km/h

Fig. 3. Double lane change test with driver controlling the speed and steering

(1) Whether or not the person is involved, the error increases with the increase of vehicle speed, which is consistent with the actual situation;

(2) The control steering is the main source of the error, and the speed of the vehicle has little effect on the error;

(3) The snake following effect is obvious better than the double lane with the path following model. The snake following effect is similar to the double lane with driver control. The error controlled with the driver is 10–50 times of the path following model of snake simulation, and the error controlled with the driver is 1–10 times of the path following model of double lane change simulation.

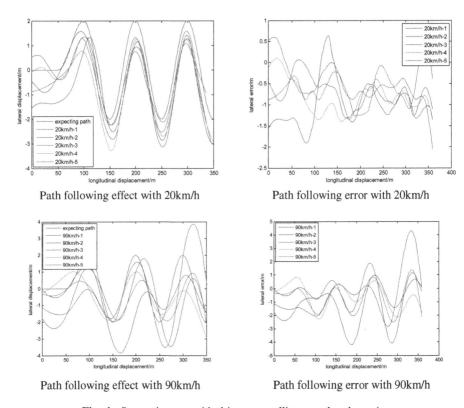

Path following effect with 20km/h   Path following error with 20km/h

Path following effect with 90km/h   Path following error with 90km/h

Fig. 4. Serpentine test with driver controlling speed and steering

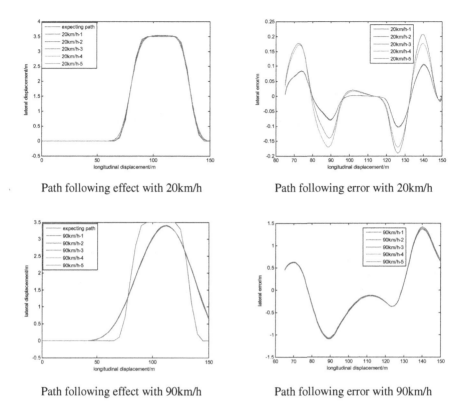

Path following effect with 20km/h   Path following error with 20km/h

Path following effect with 90km/h   Path following error with 90km/h

Fig. 5. Double lane change test with driver controlling the speed

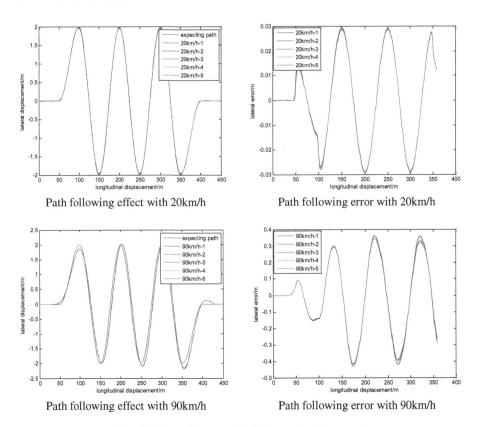

Path following effect with 20km/h   Path following error with 20km/h

Path following effect with 90km/h   Path following error with 90km/h

Fig. 6. Serpentine test with driver controlling speed

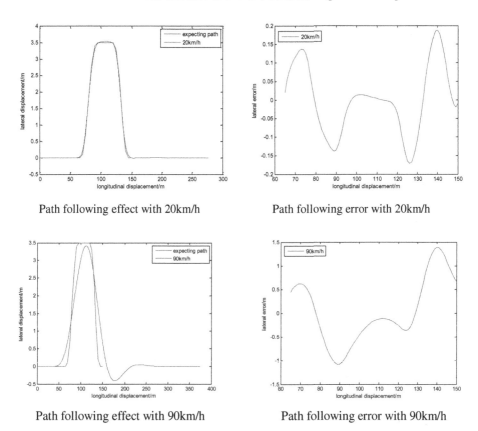

Path following effect with 20km/h   Path following error with 20km/h

Path following effect with 90km/h   Path following error with 90km/h

Fig. 7. Double lane change test without driver

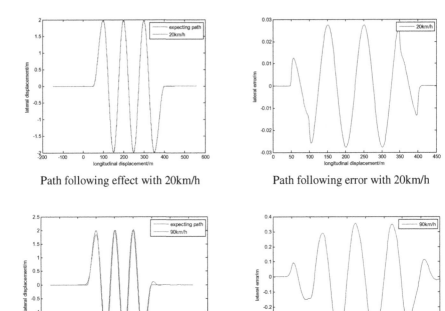

Path following effect with 20km/h   Path following error with 20km/h

Path following effect with 90km/h   Path following error with 90km/h

Fig. 8. Serpentine test without driver

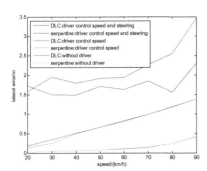

Fig. 9. Path following error with driver participation

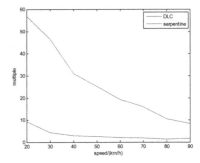

Fig. 10. Contrast path following model with driver

3 Conclusion

Path following model can obviously improve the precision and repeatability of vehicle test. The driver's participation in steering control has the greatest influence on the following effect. For the handling stability test, the driver control speed with steering automatic control mode can basically meet the requirements. Therefore, the test field stability test, only need to buy the steering robot and path following module, no need to buy the throttle robot and brake robot.

Acknowledgements. This work was supported by a Central Public Welfare Scientific Research Institute of basic research funds (2013-9027, 2015-9043).

References

1. Luo, S., Yang, G., Zhu, C., et al.: Research on hybrid control of steering robot in vehicle transient response test. Veh. Power Technol. **128**(4), 12–15 (2012)
2. Luo, S., Zhao, Y., Zhu, C.S., et al.: Control of steering robot in steady static cornering test. Veh. Power Technol. **130**(2), 43–46 (2013)
3. Chen, W.: The Research on Vehicle Path Following Model Based on Driver-In-Loop Simulation Platform. Hunan University (2014)
4. Song, Y., Zhao, P., Tao, X., et al.: UGV robust path following control under double loop structure with μ synthesis. Robot **35**(4), 417–424 (2013)
5. Yan, Z., Chi, D., Zhou, J., et al.: NGPC-based path following control of UUV. J. Huazhong Univ. Sci. Technol. (Nat. Sci. Ed.) **40**(5), 120–124 (2012)
6. Wu, C., Li, L.: A static-feedback control methods for path-tracking of an autonomous vehicle. J. Taiyuan Normal Univ. (Nat. Sci. Ed.) **12**(3), 81–84 (2013)
7. Wang, Q.: Fuzzy preview control method for intelligent vehicle path tracking. Comput. Digit. Eng **287**(9), 1454–1467 (2013)
8. Xiu, C., Chen, H.: Research on path tracking control of autonomous vehicle. Comput. Eng. **38**(10), 128–130 (2012)
9. Zhao, X., Chen, H.: A study on lateral control method for the path tracking of intelligent vehicles. Auto. Eng. **33**(5), 382–387 (2011)
10. Yin, X., Li, L., Jia, X.: Predictive fuzzy control of unmanned vehicle path tracking. J. North Univ. China (Nat. Sci. Ed.) **32**(2), 135–138 (2011)
11. Li, W., Zhou, W., Yu, X., et al.: Methods for On-line monitoring rollover stability of large bus. In: Proceedings of the 3rd International Conference on Machinery, Materials Science and Energy Engineering (ICMMSEE 2015), pp. 248–256, (2015)

Robotics

The Structure and Control Analysis of AMR Automatic Harvesting Robot

Cao Dong, Fuyang Tian$^{(\boxtimes)}$, Xiaoning Dong, Xinqiang Zhao,
and Fade Li

Shandong Provincial Key Laboratory of Horticultural Machineries
and Equipments, College of Mechanical and Electronic Engineering,
Shandong Agricultural University, Taian 271018, China
sdautfy@163.com

Abstract. The rapid development of modern agriculture make the growth of agricultural wheeled robot multibody mechanical system towards Flexible robot. The degree of freedom increases and the corresponding system composition structure becomes more complex. This paper designed and developed the fruit and vegetable automatic harvesting robot AMR1, proposed a generalized dynamic modeling method based on the spatial operator algebra theory. The problem of robot inverse kinematics is solved by ELM neural network. Also, PID-SMC control algorithm of 6 degree of freedom manipulator is designed further to real-time control of the trajectory tracking of robot arm. The simulation results show that the proposed method is practicable and effective.

Keywords: Wheeled robot · Dynamics · Modeling · Simulation

1 Introduction

With the rapid economic development and the continuous advancement of technology, the intelligent agricultural machinery was increasingly sophisticated [1]. The development of precision agricultural intelligent equipment is the mainstream direction of the growth of agricultural machinery. The application and promotion of the agricultural robot is an inevitable choice for modern agriculture towards to precision agriculture [2–4]. The fruit and vegetable automatic harvesting robot is highly intelligent and automated and it is the typical of intelligent agricultural equipment. It includes some advancing front technologies like sensing technology, image recognition technology, automatic navigation technology, communication technology, computer technology, system integration technology and so on. The application of fruit and vegetable automatic harvesting robots will bring another new technology revolution as a new generation of production tool. Based on these points, this paper developed a fruit and vegetable automatic harvesting robot named AMR1, built 3D solid models with *SolidWorks* software and conducted control algorithm simulation in *ADAMS* and *MATLAB* to verify the reasonable modeling of AMR1.

© Springer International Publishing AG 2018
F. Qiao et al. (eds.), *Recent Developments in Mechatronics and Intelligent Robotics*,
Advances in Intelligent Systems and Computing 690, DOI 10.1007/978-3-319-65978-7_69

2   The General Design of AMR1 Vegetable and Fruit Automatic Harvesting Robot

AMR1 vegetable and fruit automatic harvesting robot designed in this paper consists of two parts, intelligent walking part and automatic harvesting part. Intelligent walking part has a four-wheel mobile platform. It drives through the rear two wheels and steers through the nose-wheels. Automatic harvesting part includes a storage box and a 6 DOF vegetable and fruit picking manipulator. Figure 1 shows the structure diagram of AMR1.

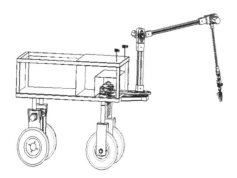
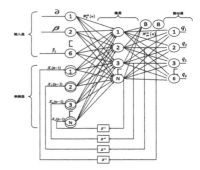

Fig. 1. The structure diagram of AMR1   **Fig. 2.** Inverse kinematics of neural network structure schematic diagram

The distance between the two rear and two front wheel is 500 mm because typically the ridge distance is between 300 mm to 700 mm. AMR1 is designed with independent suspension wheels. The rear two wheels with redundant actuation is driven by servo motors and reducers to achieve a higher position precision. Automatic harvesting manipulator has six degrees, the first three joints and the last three joints are used to identify the position and orientation of manipulator respectively.

3   Inverse Kinematics Algorithm of Agricultural Robot Manipulator Based on ELM Neural Network

The joint variables of Agricultural robot manipulator are denoted $q_i(i = 1, 2, \cdots, q_n)$, these values are getted from output signal of each joint's sensor. By Denavit-Hartenberg method, the kinematics equation is given by the following equation [5]:

$$\begin{bmatrix} {}^n_o R & {}^o P_{no} \\ 0 & 1 \end{bmatrix} = {}^0_n T(q_n) = {}^0_1 T(q_1) {}^1_2 T(q_2) \ldots {}^{n-1}_n T(q_n) \tag{1}$$

where $_n^0R = \begin{bmatrix} n_x & o_x & a_x \\ n_y & o_y & a_y \\ n_z & o_z & a_z \end{bmatrix}$ is the attitude matrix of end-link, $\begin{bmatrix} p_x \\ p_y \\ p_z \end{bmatrix}$ is the position

matrix of end-link, $_i^{i-}T(q_i)$ represents the pose of the link $i - 1$ to link i coordinate frame transformation of the manipulator. Put joint variables $q_i(i = 1, 2, \ldots q_n)$ into the formula (1) for calculating $_n^0T(q_n)$. Since the research manipulator in this paper has six degrees, we define n as 6. Considering $_n^0R$ has only three separate elements, we describe the orientation of the according to the RPY (roll-pitch-yaw) rotations as $R(\partial, \beta, \gamma)$, where ∂, β, γ can be calculated as the following [6]:

$$
\begin{cases}
\partial = A\tan 2(n_y, n_x) \\
\beta = A\tan 2(-n_x, \sqrt{n_x^2 + n_y^2}) \\
\gamma = A\tan 2(o_z, a_z)
\end{cases}
\tag{2}
$$

2000 groups experimental sample data is used to train the network and each group has 6 input vectors and 6 output vectors. Define matrix $P = [\partial, \beta, \gamma, p_x, p_y, p_z]$ and $Y = [q_1, q_2, q_3, q_4, q_5, q_6]$ as the input and output in neural network training. Determination of sample data output vectors are at random in allowable range and grouped in a matrix, input vectors are calculated according to the formula (1). This paper selects ELM neural network for forecasting the angle value of every joint. Figure 2 shows inverse kinematics of ELM neural network structure schematic diagram.

4 The Generalized Dynamics Modelling and Control of AMR1

In order to avoid obstacles, agricultural robot manipulator need a more flexible controlling during the process of picking, so it contains both active and passive hinge. AMR1 also has both active and passive driving wheels in order to select optimal path.

According to literature [7], the dynamics equation of underactuated agricultural robot can be written as:

$$
\begin{pmatrix} M_{aa} & M_{ap} \\ M_{ap}^* & M_{pp} \end{pmatrix} \begin{pmatrix} \ddot{q}_a \\ \ddot{q}_p \end{pmatrix} + \begin{pmatrix} C_a(q_a, \dot{q}_a) \\ C_p(q_p, \dot{q}_p) \end{pmatrix} + \begin{pmatrix} Kq_a \\ Kq_p \end{pmatrix} = \begin{pmatrix} T_a \\ T_p \end{pmatrix}
\tag{3}
$$

where M_{aa} is the mass matrix of active joint system, M_{pp} is the mass matrix of passive joint system, M_{ap} is coupling mass matrix.

Decompose the mass matrix of tree-like mechanical multibody system based on spatial operator algebra theory in formula (3):

$$
\begin{aligned}
M_{aa} &= H_a \phi M \phi^* H_a^* \\
M_{ap} &= H_a \phi M \phi^* H_p^* \\
M_{pp} &= H_p \phi M \phi^* H_p^*
\end{aligned}
\tag{4}
$$

The dynamics differential-algebraic equation of system is calculated as:

$$\begin{pmatrix} T_a(t) \\ \ddot{q}_p(t) \end{pmatrix} = \begin{pmatrix} G_{aa}(t) & G_{ap}(t) \\ -G_{ap}^*(t) & G_{pp}(t) \end{pmatrix} \begin{pmatrix} \ddot{q}_a(t) \\ T_p(t) \end{pmatrix}$$
$$+ \begin{pmatrix} C_a(q_a(t),\dot{q}_a(t),t) + K(t)q_a(t) - G_{ap}(t)\big(C_p(q_p(t),\dot{q}_p(t),t) + K(t)q_p(t)\big) \\ -G_{pp}(t)\big(C_p(q_p(t),\dot{q}_p(t),t) + K(t)q_p(t)\big) \end{pmatrix}$$
$$(5)$$

Where

$$G_{aa}(t) = M_{aa}(t) - M_{ap}(t)M_{pp}^{-1}(t)M_{ap}^*(t) = H_a\left[\left(\Psi - P\Psi^* H_P^* D_p^{-1} H_P \Psi\right)P + P\Psi^*\right]H_a^*$$
$$G_{pp}(t) = M_{pp}^{-1}(t) = [I - H_p \Psi K_p]^* D_p^{-1}[I - H_p \Psi K_p]$$
$$G_{ap}(t) = M_{ap}(t)M_{pp}^{-1}(t) = H_a\left[\left(\Psi - P\Psi^* H_P^* D_p^{-1} H_P \Psi\right)K_P + P\Psi^* H_P^* D_p^{-1}\right]$$
$$(6)$$

Related operators can be found in Ref. [7].

Figure 3 shows the sliding mode control for AMR1 robot manipulator.

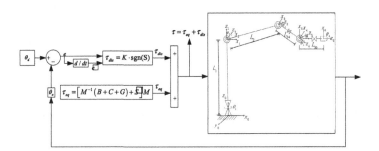

Fig. 3. The sliding mode control for AMR1 robot manipulator diagram

The sliding mode control for AMR1 robot manipulator is designed as:

$$\tau = \tau_{eq} + \tau_{dis} \tag{7}$$

where τ_{eq} is the nominal dynamic system parameter, it can be defined as:

$$\tau_{eq} = \left[M^{-1}(B + C + G) + \dot{S}\right]M \tag{8}$$

τ_{dis} can be defined as:

$$\tau_{dis} = K \cdot \text{sgn}(S) \tag{9}$$

By formula (7–9) the sliding mode control of AMR1 robot manipulator is calculated as

$$\tau = \left[M^{-1}(B + C + G) + \dot{S} \right] M + K \cdot \mathrm{sgn}(S) \tag{10}$$

Furthermore, in order to reduce the chattering phenomenon, this paper designed PID-SMC control algorithm.

where $S = \lambda e + \dot{e} + \left(\frac{\lambda}{2}\right)^2$ in PID-SMC.

5 Simulation Analysis

Use agricultural robot as showed in Fig. 4 to verify the control algorithms in this paper. Import the three-dimensional model into ADAMS, the simplified model of AMR in ADAMS is shown in Fig. 5. Figures 6, 7 and 8 showed joint angular velocity variation, joint angle variation and major joint angular acceleration variation of AMR from Kinematic and Dynamics simulation experiment in ADAMS. From Figs. 6, 7 and 8, angle and angular velocity changed smoothly, angular acceleration fluctuated greatly.

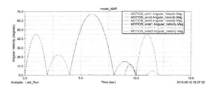

Fig. 6. The joint angular velocity variation of vegetable and fruit automatic harvesting robot AMR

Fig. 4. Practicality picture of agricultural wheeled robot

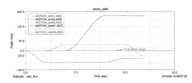

Fig. 7. The joint angle variation of vegetable and fruit automatic harvesting robot AMR

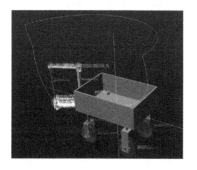

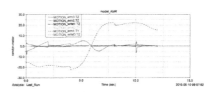

Fig. 8. The major joint angular acceleration variation of vegetable and fruit automatic harvesting robot AMR

Fig. 5. The simplified model of AMR in ADAMS

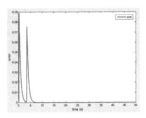

Fig. 9. Tracking control of effector based on sliding mode control

Fig. 10. Error of effector based on sliding mode control

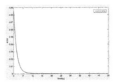

Fig. 11. Tracking control of effector based on PID-SMC

Fig. 12. Error of effector based on SMC-PID

It verified the correctness of the simulation. The methods of SMC, PID-SMC were used respectively to trace the manipulator end-effector trajectory. Results of simulation are shown in Figs. 9, 10, 11 and 12. Simulation experiments are used to verify the validity of the presented algorithm has more efficiency and practicability.

6   Conclusion

In this paper the virtual prototyping model was built by SolidWorks software, the inverse kinematics problem of agricultural robot manipulator is solved by ELM neural network, PID-SMC control algorithm was designed to control the manipulator end-effector. The simulation results show that control scheme has high efficient and the picking efficiency is improved. Experiments show that structure design of the virtual prototype is reasonable. It offers reliable foundation for servo motor selection, reduces product cost and shorten design cycle.

Acknowledgements. This research was financially supported by National Nature Science Foundation under Grant (No. 51205238) and Shandong provincial key research and development plan (No. 2016CYJS05A02).

References

1. Jang-Soo, C., Taw-Won, P., Kim, J.: Dynamic analysis of a flexible multibody system. Int. J. Precis. Eng. Manuf. **6**(4), 21–25 (2005)
2. Zhang, W., Yuan, L.: Design of the control system of fruit and vegetable flexible grasping of agricultural robot based on slip detection international. J. Agric. Mech. Res. **6**, 228–232 (2017)
3. Lei, C., Diyi, C., Xiaoyi, M.: Researching on the fruit and vegetable harvesting robot. J. Agric. Mech. Res. **1**, 224–227 (2011)
4. Wu, X., Guoming, H., Jinxia, L., Bo, X.: Design and experiment of a new type apple picking robot. Sci. Technol. Eng. **2016**(9), 71–79 (2016)
5. Wu, T., Wu, G., Wu, H.: Improvement of kinematics algorithm of 6R industrial robot. J. Mech. Electr. Eng. **7**, 882–887 (2013)
6. Liu, S., Zhu, S., Li, J., Wang, X.: Research on real-time inverse kinematics algorithms for 6R robots. Control Theory Appl. **6**, 1037–1041 (2008)
7. Tian, F., Wu, H., Zhao, Z., Shao, B., Wang, C.: Research on generalized efficient recursive dynamics of flexible macro-micro space robots system. J. Astron. **31**(3), 687–694 (2010)

Design and Implementation of Virtual Instructor Based on NAO Robot

Ni Zeng[✉], Shijue Zheng, Jun Zhou, and Qingyu Cai

School of Computer Science, Central China Normal University, Wuhan, China
2663400144@qq.com

Abstract. With the advancement of science and technology, robot technology has penetrated in all aspects of social production and has rapid development. At the same time, voice technology has become a comprehensive human intelligence in various fields of technology and the mainstream is to realize the human-computer interaction between men while machine natural communication is the mainstream of human-computer. Based on the teaching of campus life and moral education, this paper introduces the speech recognition and synthesis technology and embeds the open artificial intelligence platform of IFLYTEK to NAO robot with voice interchange as the core and designs the program of the voice dialogue database between college instructor and student, which achieves the robot and students intelligent interaction function, promoting the cause of higher education and making further breakthrough in the education industry.

Keywords: NAO robot · Speech recognition and synthesis · Speech database

1 Introduction

With the rapid development of the word "intelligence", especially in the field of voice technology and robotics, people prefer to introduce more in the industrial, commercial, education and other aspects of intelligent development. Robotics as a comprehensive discipline is related to the mechanical manufacturing, automatic control, sensor technology, computer software technology, computer hardware technology and other disciplines.

Based on the integration of campus life and the daily work of instructors, this paper studies the speech recognition and synthesis technology, and embeds the API as the support of the API provided by the IFLYTEK platform using the minimum mean square error estimation method (MMSE) for audio processing, which breaking the traditional mode of human-computer interaction and achieving the NAO robot and students voice dialogue between. It's an increase of human-computer interaction in the fun which greatly reduces heavy work tasks of college instructors engaged in moral education, learning style, day-to-day management, making the work of college counselors and college students more humane.

© Springer International Publishing AG 2018
F. Qiao et al. (eds.), *Recent Developments in Mechatronics and Intelligent Robotics*,
Advances in Intelligent Systems and Computing 690, DOI 10.1007/978-3-319-65978-7_70

2   Related Technology

2.1   Structure and Function of NAO

NAO, manufactured by Aldebaran Robotics, is a biped robot that combines a variety of hardware and software to support 90 min of development using the NAOqi development framework and the DCM (Device Interaction Management System) to control the sensor. NAO can be programmed under Linux, Windows and Max OS X operating systems, supporting C++, Python, C# and net frameworks.

2.2   NAOqiOS Choregraphe Platform

NAOqi OS is the development of the operating system that NAO robot works on. There are a large number of built-in library functions and procedures in the NAOqi OS that can be called and achieved cross-platform under Windows, Linux and MacOS to prepare the program. In the course of the study, we wrote the python program on the Choregraphe platform, connecting the NAO robot, and testing the data changes of the NAO robot through the real-time return of the computer.

Choregraphe's feature is to provide graphical-based programming. Instruction boxes and icons are important elements of Choregraphe, and they all have corresponding charts. Chart-based programming is used when connecting these instruction boxes. Choregraphe contains more than 70 instruction boxes, and users can also according to their own needs to create their own corresponding instruction box.

3   Speech Recognition and Synthesis Technology

The key technology in speech technology is mainly divided into speech recognition technology ASR and speech synthesis technology TTS [5]. The computer can understand the sound of human beings while speech synthesis technology is to make the computer according to the corresponding text. Speech recognition and synthesis technology 'achievement in the computer's intelligent is the information age under the great progress.

4   Proposed Method

4.1   Design of System

NAO robot audio capture system is mainly composed of the head MicroFront, Micro-Rear, MicroLeft, MicroRight four microphones and used to obtain voiceFour microphones can be used to get simple audio, by editing the instruction box, modifying the audio code to get the python code. Due to the limited computational power of the embedded processor on the robot, it is necessary to greatly affect the recognition rate of the speech in view of the fact that there may be background noise and other factors in the real experimental environment. Therefore, we need to transmit the audio to the website Perform step-by-step preprocessing for speech enhancement [4].

Speech enhancement can reduce noise, improve voice clarity and achieve higher speech quality. In this paper, we have adopted Ephraim and Malah proposed minimum mean square error estimation (MMSE) [7, 8] for speech enhancement. Its pure voice spectrum amplitude estimates is

$$\hat{A}_k = \frac{\xi_k}{1+\xi_k} \exp\left(\frac{1}{2} \int_{v_k}^{\infty} \frac{e^{-x}}{x} dx\right) R_k \tag{1}$$

The prior signal to noise ratio is ξ_k and the noisy speech amplitude estimate is R_k. After obtaining the estimated value $\hat{A}_k$, then add the noisy phase and anti-FFT can be enhanced after the sound. Finally the voice can be sent to the IFLYTEK platform for speech recognition, access to more accurate corresponding text.

4.2 Speech Database and System Design

The speech dialogue database is mainly designed for college students in the campus learning and living and other aspects of the school construction. In the paper we established the relevant research activities before the integration of the corresponding data model. The E-R relationship is designed based on the requirements analysis, conceptual structure design, logical structure design, database physical design, database implementation, database operation and maintenance. After sorting out the results of the survey, the problem the high frequency of campus life these questions are written into the database, which will constantly improving the contents. The proposed database is shown with the help of blocks in Fig. 1.

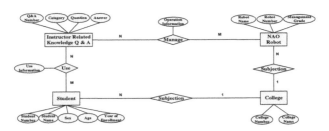

Fig. 1. Speech database E-R diagram

4.3 Voice Call and Implementation

Voice input is mainly used for NAO robot direct dialogue, through the microphone to the audio collection, the original analog signal data into digital signal, into the computer. In order to make the results of the match as accurate as possible, in the voice input stage will be pre-processing. The preprocessing process includes preheat filtering, sampling and quantization, framing, reemphasis, windowing, endpoint detection, etc. [9]. In the computer through the MMSE on the audio again, the pre-processed voice data extraction of the characteristics of the parameters, and then further extract the characteristics of the parameters.

System speech recognition overall framework as Fig. 2.

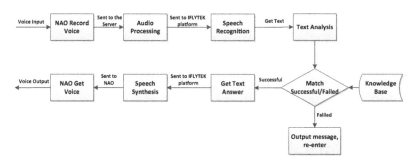

Fig. 2. Speech recognition gram speech recognition system framework

As the NAO robot provides the Audio API for the developers to call, this paper uses the embedded science and technology intelligent voice platform for speech recognition and synthesis. Appid and Key applied to the website in IFLYTEK platform and the information obtained is an important validation condition. The play function in demo.js is passed through this message to obtain a docking with the module, and the audio wav file obtained by the NAO robot is transferred online to KWC and returns the corresponding result.

1. Call the function Recognizer Dialog (Context c, String "appid = " + appid:) to create a speech recognition control. OnResults (ArrayList < RecognizerResult > results, boolean isLast) is used to get the result of the recognition, that is, the speech is converted to text. ArrayList is the result of speech recognition, where isLast is used to determine whether the recognized text is read.
2. The parameters of the information obtained through the network to the voice API analyzer through the microphone input.
3. Return the results into the temporary variables of the results while the results of the corresponding regular expression get the final analysis of the results.
4. When the system API recognizes the corresponding voice, the converted text is passed to the database module to match and the corresponding correct answer is obtained.
5. After matching the answer in the database,call. He function Synthesizer Player. Create SynthesizerPlayer (). In order to output the text in the synthetic database, call the SynthesizerPlayer. Player.playText (String, null, Context) function to synthesize the answer text in the database. Where the play () function is also the main entry. Appid and appkey fill when the parameters can be.

In the Choregraphe platform for dialogue parameter settings, select the language type for the English, as shown in Fig. 3 nine. Through the transmission of parameters, the voice of the corresponding text in the database through the voice output, dang2 set up after the completion of the instruction box connection, the graphical operation, as shown in Figs. 4 and 5 shows

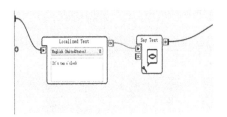

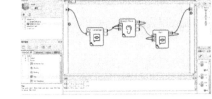

Fig. 3. Select the language type   **Fig. 4.** Choregraphe connection interface

5   Experimental Results

In order to get the accuracy of NAO robot speech recognition, we tested different voice types under certain conditions (bandwidth 10Mbps, input voice length is about 3 s, noise less than 20 dB, standard Mandarin, volume 40 dB or more).There are four types of speech types: words, phrases, short sentences and long sentences. We specify that the length of the Chinese characters is 1–2, which is called the word, the length is 2–5 and consists of at least two words called the phrase, and the sentence is roughly divided into two phrases and long sentences. The test procedure is as follows:

1. Prepare 20 corresponding texts for each voice type;
2. The tester reads the above text in a quieter environment;
3. Record the NAO robot recognition results;
4. Summarize the results into tables;

Table 1 shows the experimental results.

Tables 1. Speech recognition test results table

Number	Voice type	Completely correct	Errors	Unrecognized
1	Words	20	0	0
2	Phrases	19	1	0
3	Shor sentences	19	1	0
4	Long sentences	17	2	1

6   Conclusion

We achieved the system in which the NAO robot and students can have interesting intelligent voice dialogue through the application of speech recognition and synthesis technology and the Choregraphe platform for graphical programming that embedded in the HKD API interface, further promoting the college instructors in college students moral education and physical and mental health to guide the work. In addition, there must be a breakthrough in innovative significance.

Acknowledgements. This work was supported by the key projects China Language committee Research Project NO ZDI135-13; and another is Specific funding for education science research by self-determined research funds of CCNU from the colleges' basic research and operation of MOE NO CCNU16JYKX17.

References

1. Cai, H.: Network Development Based on NAO Service. J. Jinan Univ. (2016)
2. Yin-yu, Yu.: Design and Development Based on Andrews Campus Service Dialogue System, pp. 12–13. Fudan University, Shanghai Shi (2013)
3. Yang, L.: Study on Several Single Channel Speech Enhancement Algorithms. Science and Technology Horizon **26**, 155–222 (2015)
4. Bin, C.A.I., Ying, G.U.O., Hong-wei, L.I., Cheng, G.O.N.G.: AImproved MMSE Speech Enhancement Method. Sig Process **01**, 68–72 (2004)
5. Xiuzhen, Li: Speech recognition algorithm and application technology research, pp. 1–6. Chongqing University, Chongqing (2010)
6. Xiong, F.: Study and Implementation of Embedded Speech Recognition Algorithm, pp. 1–2. Taiyuan University of Technology, Taiyuan (2008)
7. Jiang, J.: Design of Educational Robot Basic Development Platform. J. Suzhou Univ.:1–2
8. Lei, Yu.: Realization of Simple Voice Dialogue System Based on Google Speech-API, p. 1. South China University of Technology, Guangzhou (2012)
9. Jing-ya, Z.: Study on Chinese continuous digital speech recognition based on HMM, pp. 1–3. Suzhou University, Suzhou (2005)

Design and Implementation of an New Embedded Image Tracking System for Micro UAV

Yanfei Liu$^{(\boxtimes)}$, Yanhui He, Qi Tian, and Pengtao Zhao

Xi'an Research Institute of High Technology, Xi'an, China
bbmcu@126.com

Abstract. The current target tracking system of large size, low integration, and high power consumption, mostly use X86 architecture processor as the core. Based on the requirements of target tracking system, this paper designs the circuit of I.MX6Q as the core, integrates power supply, USB interface, expansion interface, video output interface, RTC and so on as the hardware platform, embedded Linux operating system as the software running environment of the target tracking system. In the aspect of hardware, the main devices are selected and designed; In the aspect of software, this paper envisions the development environment which is more humanized, and the embedded Linux system is cut and transplanted. The test results show that the system is not only stable and can track a target in real time, but also meet the requirements of small size and low power consumption.

Keywords: ARM9 · Target tracking · Linux operating system

1 Introduction

Target tracking is one of the most important components of computer vision and is widely used in various areas of science and technology, national defense construction, aerospace, medicine, and national economy [1]. At present, the target tracking system using online mode, captures video images back to the PC through the camera and then complete the tracking target process in PC. The biggest problem of this model is the low integration and high power consumption. Hence, the application of the system has been greatly restricted. The embedded target tracking system based on ARM9 [2] proposed in this paper is a stand-alone system that integrates video capturing and processing into a single system, which is small in size and low in power consumption.

2 System Overall Designs

In this paper, it is proposed that ARM embedded development technology and computer vision technology be combined for the embedded target tracking system design in an attempt for the development a small and real-time embedded target tracking system. Based on the analysis of the functional requirements of the system, the overall

© Springer International Publishing AG 2018
F. Qiao et al. (eds.), *Recent Developments in Mechatronics and Intelligent Robotics*,
Advances in Intelligent Systems and Computing 690, DOI 10.1007/978-3-319-65978-7_71

framework of the system is designed. The entire target tracking system consists of hardware and software components. The hardware section includes circuitry such as processor, memory, peripheral device and interface circuit and power supply. The software part consists of embedded operating system, application program and so on.

3   Hardware Design

The hardware part of the embedded target tracking system mainly includes: processor, SDRAM, FLASH, USB camera, LCD display, power module and interface circuit and RTC circuit. The principle block diagram is shown in Fig. 1.

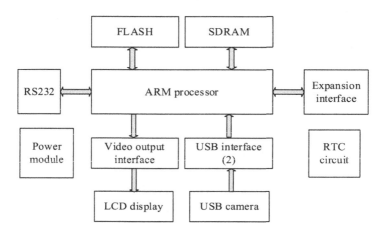

Fig. 1. Block diagram of the hardware platform

3.1   Processor Design

At present, many kinds of embedded processor [3], such as: ARM processor, DSP processor, FPGA processor, the processor type is divided into many types, so choosing a suitable visual image processing and meeting the design requirements of the processor system are very important and complex work. So how can we make the best choice? Under the premise of full analysis of the system requirements, focusing on the following three aspects are needed:

(1)   A comprehensive analysis of the core structure of each candidate processor and the peripheral interface;
(2)   The circulation way of Mastering the understanding of video data in the system;
(3)   Assessing the level of treatment that can be achieved under specified power consumption.

This article chooses I.MX6Q as the core processor based on ARM Cortex-A9 architecture [4], which is a 4-core processor. And the maximum operating frequency up to 1.2 GHZ, the processor internal 64-bit bus structure, can achieve 12000DMIPS

(1.2 billion instructions per second instruction set) high-performance computing power. At the same time, it has 96 k super start ROM, 16 K encryption ROM, up to 128 GPIO interface, 2D/3D graphics and video acceleration processor engine, independent multimedia processor engine IPU and independent video processor unit CPU.

3.2   Storage Devices

The main task is to select the system running memory devices and flash memory devices. The memory device is selected with Samsung produced K4B4G1646D chip, the size of 512 M. Taking into account of the video image acquisition and display process, a large number of data need to be cached, so the use of 4 chips, running a total memory of 2 GB, to meet the requirements of the system. The chip is DDR3 synchronous dynamic random access memory, with higher operating efficiency and lower voltage compared to DDR2, which is currently popular memory products.

Selecting the size of 8 GB eMMC as a flash memory device, is developed by the MMC Association, mainly for mobile phones or tablet PCs and other products, embedded memory standard specifications. An obvious advantage of EMMC is that it integrates a controller in the package, which provides the standard interface and manages flash memory.

3.3   Camera

As the Linux kernel contains UVC (USB video class) module and USB device driver, we can use V4L2 (video for Linux Version2) on the camera to operate, which supports for UVC plug and play USB camera for the system. Here we choose the Logitech USB network camera Pro900, for its maximum support resolution of 720P video capture, and supports auto focus, with good imaging results.

4   Embedded Operating System Transplantation

4.1   Building Development Environment

The first step in the development of embedded systems is to build a development environment.

Installing the windows operating system and VMware virtual machine Linux operating system on a x86 PC at the same time, and then building the cross-compilation environment in the Linux platform. The cross-compilation environment is the most important step in the process of building an embedded Linux system development environment. The block diagram of building development environment is shown in Fig. 2.

Building a cross-compilation environment is the most important step in building an embedded Linux system development environment. So what is cross-compiler? In short, cross-compilation is a platform to compile a program which can run on a different platform with different architectures [5]. For example, in X86 platform, PC can be compiled on the ARM platform to run the program. Now giving a cross-compiler

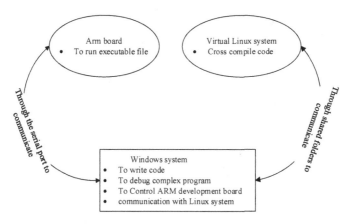

Fig. 2. Block diagram of the development environment

definition: in a computer environment, in order to make the compiler compile another computer platform binary program, the compilation process is called cross-compiled. This compiler that compiles different platform programs is called a cross compiler.

4.2   Configuration and Compile Bootloader and Linux Kernel

Linux operating system transplantation includes three part, Boot Loader, Linux Kernel, and Root File System [6]. Bootloader is a small program that runs after the system starts and runs before the operating system kernel [7]. Bootloader is independent of the operating system that is highly dependent on the hardware, so it must be designed according to the actual situation.

Due to the limited hardware resources of the embedded system [8], we only need to use the Linux Kernel part of the function. So, according to the system's hardware resource and functional requirements of the kernel for cutting and customization. The specific process is as follows:

(1)   To download the Linux Kernel source code.
(2)   To set the cross-compilation environment variable.
        export ARCH = arm
        exportCROSS_COMPILE =/opt/freescale/usr/local/gcc-4.6.2-glibc-2.13-linaro-multilib-2011.12/
        fsl-linaro-toolchain/bin/arm-none-linux-gnueabi-
(3)   To execute *make menuconfig* command into the Kernel configuration interface.
(4)   To compile the Kernel to generate the image file with executing the *make uImage* command.

5 Test and Verification System

In order to reflect more intuitively the system's good real-time and low power performance, the following conducts a comparative test. The test object is the ARM tracking system developed by this paper and a PC. The PC is a configuration i5 series 4-core processor, the maximum operating frequency up to 2.0 GHz, running memory 4 GB, equipped with Windows7 system notebook computer. The setting conditions are as follows: the same target tracking program (via a different compilation environment) is to run on both platforms; the same USB camera is connected; tracking targets and backgrounds are the same; the image resolution is 640 × 480.

Figures 3 and 4 are the 288th and 304th frame of the tracing program running on the development board, respectively. Figures 5 and 6 show the 343th and 520th images of the tracing program on the PC.

Fig. 3. 288th frame   **Fig. 4.** 304th frame   **Fig. 5.** 343th frame   **Fig. 6.** 520th frame

From the results of the above figures and the Table 1, it can be concluded that the average time-consuming to run the target tracking program on the development board is slightly higher than that on the PC, but the power consumption is much lower.

Table 1. Test results

Platform	CPU	Run memory	Operating frequency	Operating system	Power consumption	Average time-consuming
X86 PC	Core i5	4 GB	2.0 GHZ	Win7	15 W	11.28 ms
ARM	I.MX6Q	2 GB	1.2 GHZ	Linux	<3 W	12.54 ms

6 Conclusion

This paper introduces the hardware circuit design and the transplantation of Linux operating system for the target tracking system based on the ARM architecture. The test shows that compared with the X86 architecture target tracking system, under the premise of ensuring the real-time requirement of the target tracking system, the power consumption is greatly reduced, the integration degree is improved, and the application prospect is realized in the field of product realization.

References

1. Smeulders, A.W.M., Chu, D.M., Cucchiara, R., et al.: Visual tracking: an experimental survey. IEEE Trans. Pattern Anal. Mach. Intell. **36**(7), 1442–1468 (2014)
2. Du, C.: ARM architecture and programming. Tsinghua University Press, Beijing (2003)
3. Wang, Z.: Analysis on the characteristics and development of embedded processor. Comput. Knowl. Technol. **22**, 6302–6303 (2009)
4. Shen, J.: ARM processor and embedded system. Microcontrollers Embedded Syst. **11**, 5–7 (2010)
5. Gong, L.: ARM Embedded Linux System Development, p. 2. Tsinghua University Press, Beijing (2014)
6. Fan, Y., Yang, A.: Linux application development technology Xiangjie, p. 2. People Post Press, Beijing (2006)
7. Marshall, W.: Boot with GRUB. Linux J. (2001)
8. Chen, L.: Linux Kernel Code Analysis. People Post Press, Beijing (2000)

Design and Implementation of Platform of Educational Mobile Robots Based on ROS

Xinguo Yu, Lvzhou Gao[✉], Bin He, and Xiaowei Shao

NERCEL, Central China Normal University, Wuhan, China
gaolvzhou@outlook.com

Abstract. This paper develops a ROS-based mobile platform for users to develop their own educational robots. This platform adopts a three-layer architecture and each layer provides one aspect of sharing functions of educational mobile robots. The lowest layer provides the sharing functions of mobility; the middle layer provides the sharing functions of software functions; the upper layer provides the sharing functions of educational clouds. The three-layer architecture can have several good properties: easy of redevelopment, rich of function, and low cost. The experiments show that the platform of educational robots developed in this paper can support the research and development of motion control, vision, and dialogue.

Keywords: ROS · Educational robots · Sensor integration · Software framework

1 Introduction

More and more scholars pay attention to the research of educational robots in recent years [1, 2]. *The NMC Horizon Report: 2016 Higher Education Edition* stated that robot-assisted learning is a typical application in education [3]. Choosing the suitable platform is a top concern when researchers want to develop their own robots. However, there is no widely accepted research platform of educational robots. Hence, researchers end up spending an excessive amount of time with engineering solutions for their particular hardware setup [4]. In recent years, several mobile robotic platforms have emerged such as NAO and Turtlebot. But they are expensive, weak capability of redevelopment, and poor extensibility.

The current situation seriously hinders the application and promotion of educational robot. Hence, this paper designs and implements a mobile robot platform based on ROS with the goal of good versatility, high performance/cost ratio and good extensibility. Then, experiments were carried out on the platform and the performances of experiments show that the platform has the basic functions of video capture, obstacle perception, speech interaction, environment modeling. And this platform supports the application of intelligent education [5].

© Springer International Publishing AG 2018
F. Qiao et al. (eds.), *Recent Developments in Mechatronics and Intelligent Robotics*,
Advances in Intelligent Systems and Computing 690, DOI 10.1007/978-3-319-65978-7_72

2   Introduction to ROS

Robot Operating System (ROS) is a software platform designed with a set of sharing standardized robot functions, which enables every robot designer to use the same platform to develop special robot functions. The existing mainstream operating systems for robots include: Windows, Android, ROS [6], Ubuntu, etc. The differences among them are shown in Table 1.

Table 1. Comparisons among robot operating systems.

Projects	Windows	Android	Ubuntu	ROS
Openness	No open source	Open source	Open source	Open source
Software independence	Poor	Common	Common	High
Difficulty of development	Common	Low	High	Low
Real-time	Fast	Common	Slow	Fast(ROS2.0)
Stability	High	Common	Common	Common
Application areas	Medical field	Commerce	Research, education	Research, education, commerce

As Table 1 shown, compared with Windows, Android and Ubuntu, ROS has the following advantages for educational robots: (a) good scalability. ROS follows the BSD open source license agreement. And all the source code is publicly available. (b) Standardization. MIT Technology Review in 2013 said [7], "ROS has become the de facto standard of robotic software". (c) Loose coupling. Peer to peer communication mechanism is convenient for software transplantation and distributed deployment. (d) Rich functional libraries. Users can get open source libraries through community online. These significant advantages of ROS above can well meet the needs of educational applications [8], which is the main reason for choosing it as the operating system of educational robot.

3   Hardware Framework Design of Educational Robot

To develop cloud service supported educational robots, a mobile robot platform with three-layer architecture was designed and implemented in this paper. Hardware with the drivers in open source is used as much as possible to increase the use range of the system and reduce costs. The hardware framework and platform of educational robots are shown in Figs. 1 and 2. The detailed design of every layer is described as follows:

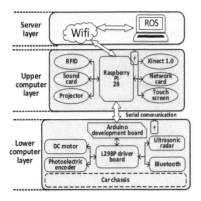
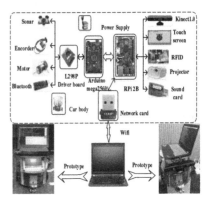

Fig. 1. Three-layer framework of robot   **Fig. 2.** The platform of educational robots

Lower Computer Layer. This layer includes body of vehicle, main control of lower computer, and sensors. The body of vehicle is a stainless frame with rich mounting interfaces for expansion of sensors. As the main controller, Arduino mega2560 [10] integrated sonar, Bluetooth, DC motor, AB phase encoder through L298P two-way motor driver. The platform uses differential drive model and the speed range is 0-0.8 m/s.

Upper Computer Layer. The upper computer layer includes main controller and multimedia equipment. This layer chooses Raspberry Pi 2 Model B as the main controller which is equipped with Broadcom BCM2836 processor. It supports 1 GB LPDDR2 memory and provides USB, GPIO, HDMI and other data interfaces. Raspberry Pi 2 communicates with Arduino through serial port. Educational robot platform integrates Microsoft Kinect 1.0, wireless network card, RFID, projector, sound card, touch screen and other equipment on upper computer to provide support for vision, voice, and touch screen interaction of robot.

Remote Server. The remote server is the cloud brain of robot body, which is connected with robot body through wireless network. With the limited computing resources and storage resources of robot body, the remote server can help robot body to run the complex algorithms on the one hand, can also provide the foundation of the educational cloud-oriented service function of the robot on the other hand. The remote server has no special requirements on the hardware itself, but ROS is to be installed.

4   Design of Robot Software Framework

This paper transplants Ubuntu14.04 kernel into raspberry pi, and then installs ROS on it. In this way the ARM-based ROS development environment is formed [11]. The functional modules of the software framework are deployed into the robot body and the server, respectively. The robot body hosts responsible for such primary function as coordinate transformation, basic motion control and off-line speech processing. However, the server acts as the robot cloud-brain to handle sophisticated algorithms and provide special services of educational cloud [12]. The robot body communicates with the server via Wi-Fi.

The whole software framework of robot can be divided into two parts: ROS-based software framework, bottom software framework. Since both the server and upper computer are in ROS developing environment, they are put in an ROS-based software framework. The lower computer is mainly for the integration of bottom sensors and data collection, and a bottom software framework is designed for the extension of the sensor and stability of data collection.

4.1   Bottom Software Framework

The design of bottom software framework mainly consists of three parts: (1) design of sensor drivers. Bottom computer software encapsulates each sensor driver and provides a unified driver programming interface for adding new sensors. (2) Design of controlling motion error. In order to control the motion of robots accurately, Arduino can obtain error data by encode. Then PID algorithm can be used to control motor speed precisely. (3) Design of basic instruction set. Upper level does not need to care about specific details of lower level, only to send the standard command to lower layer.

4.2   ROS-Based Software Framework

The design of upper framework focuses on providing the interface of programming for users. By setting the ROS_MASTER_URL, a network focusing on ROS Master can be established which connects all processes in ROS. The ROS-based software framework can be shown in Fig. 3.

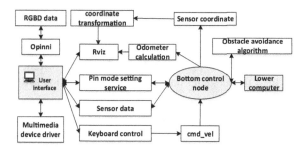

Fig. 3. ROS-based software framework

Users can gain resources from open source community online to build software framework rapidly. This software framework consists of four parts: (1) collection and release of data from the sensor. Bottom control nodes receive the sensor data from lower computer, and release processed data to corresponding topic. Users can obtain data by subscribing the topics which is an important mechanism for the communication between ROS nodes. (2) ROS service interface. Service is another communication mechanism between nodes. Users can write or read information directly by calling the services provided by upper software framework, and can control robot in real time. (3) Coordinate transformation of robots. Robot integrates many sensors. TF library make the setting and converting of coordinate systems much more convenient. This platform uses chassis center of robot as parent frame and releases two child frames that represent coordinate systems of sonar and Kinect respectively.

5   Experimental Results

In order to test the performance of the developed robot platform, a series of robot function development and simulation experiments were carried out. Here are just two examples:

(1) Control of Motion Error. Users can access the corresponding sensor data and odometer data by subscribing to the topics provided by upper software framework. In order to obtain the accurate odometer data which guarantees the accuracy of robot control, the odometer data is calibrated in the environment of wooden floor. We marked a distance of 2 m on the floor, and ran the program to make robots walk 2 m straight, then recorded the actual robot walking distance and error values. In the process, we modified the PID parameters through the command interface provided by the software framework until the distance error value was within 2 cm. After determining the PID parameters, we recorded 1, 3, 5, 7, 9 m of the robot's linear displacements respectively for 20 times and calculated the average velocity of 5 different displacements. Then, the velocity error was obtained by comparing with speed of robot feedback. Besides, we let the robot rotate the specified angle at a fixed angular velocity and recorded the error between actual rotation angle and target rotation angle. Finally, the performance of this platform is evaluated by analyzing the linear velocity and angular velocity error.

Through the comparison and analysis of the experimental data in Table 2, we can conclude that the actual linear velocity of robot is always less than the speed of feedback, but the error is less than 5 mm. Table 3 records the error between actual rotation angle and target rotation angle. The experimental results show that actual rotation angle is slightly larger than the feedback angle, but the angle error is within 1.5°. The experimental data shows that the odometer data can reflect the motion state of this platform and has good stability.

Table 2.   Error statistics of linear velocity of robot 0.3 m/s

Displacement (cm)	Measurement times	Average speed (m/s)	Feedback speed (m/s)	Error (m/s)
100	20	0.301	0.302	0.001
300	20	0.301	0.303	0.002
500	20	0.300	0.303	0.003
700	20	0.300	0.302	0.002
900	20	0.301	0.303	0.002

Table 3. Error statistics of robot 6 rad/s angle

Sequence	Target rotation angle (°)	Actual rotation angle (°)	Feedback angle (°)	Error (°)
1	90	90.8	90.2	0.6
2	180	180.7	180	0.7
3	270	270.9	270.2	0.7
4	360	361.3	360.4	0.9

(2) Navigation. Autonomous navigation of robots based on map is an important solution for robot to move autonomously in teaching scene. The platform machine integrated three commonly used function packages for robotic autonomous navigation: (1) move_base: path planning based on the reference message, and making the mobile robot to reach the designated location. (2) gmapping: map building by using laser data of in-depth data simulation. (3) amcl: positioning according to the existing map. The results of the experiment proved that the platform can satisfy the research of SLAM and navigation algorithm. The effect is shown in Fig. 4.

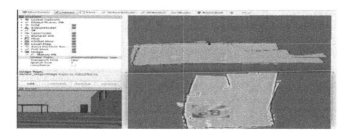

Fig. 4. Robot mapping and navigation

6   Conclusions

This paper has presented a robot platform, which has a three-layer architecture of hardware framework, software framework and cloud support framework. This platform transplanted ROS into a Raspberry Pi, having a good performance and lower price. The experiments of motion control, speech and vision of robot were carried out on the platform. And the experimental results proved that the robot platform can meet the needs of application in education and research of robot, and it has the characteristics of low cost, open source, modularization and multifunction. The platform provides the sharing base for the further study of educational robotics for the cycle of educational robot researchers. We will focus on two aspects in the future: On the one hand, we will optimize the existing educational robot platform. On the other hand, we will focus on the development of the subject-oriented robot solution [13], so that the robot will become a mobile "teacher" assistant.

Acknowledgements. This work is supported by the National Science and Technology Support Program of China (No. 2015BAH33F01).

References

1. Benitti, F.B.V.: Exploring the educational potential of robotics in schools: a systematic review. Comput. Educ. **58**(3), 978–988 (2012)
2. Huang, R., Liu, D.J., Xu, J.J.: The present development status and the trend of educational robot. Mod. Educ. Technol. **1**, 13–20 (2017)
3. Johnson, L., Adams Becker, S., Cummins, M., et al.: NMC Horizon Report: 2016 Higher Education Edition. The New Media Consortium, Austin, TX (2016)
4. Araujo, A., Portugal, D., Couceiro, M.S., Rocha, R.P.: Integrating arduino-based educational mobile robots in ROS. J. Intell. Robot. Syst. **77**(2), 281–298 (2015)
5. Wang, M., Ng, J.W.P.: Intelligent mobile cloud education: smart anytime-anywhere learning for the next generation campus environment. In: 2012 8th International Conference on Intelligent Environments (IE). IEEE (2012)
6. Quigley, M., Conley, K., Gerkey, B.: ROS: an open-source robot operating system. ICRA Workshop Open Source Softw. **3**(3.2), 5 (2009)
7. O'Kane, M.J.: A gentle introduction to ROS, independently published. CreateSpace Independent Publishing Platform (2013)
8. Costa, V., Cunha, T., Oliveira, M., Sobreira, H., Sousa, A.: Robotics: using a competition mindset as a tool for learning ROS. In: Advances in Robotics, pp. 757–766. Springer, New York (2016)
9. Mirats Tur, J.M., Pfeiffer, C.F.: Mobile robot design in education. IEEE Robot. Autom. Mag. **13**(1), 69–75 (2006)
10. Zabala, G., Moran, R., Teragni, M., Blanco, S.: On the design and implementation of a virtual machine for arduino. Robotics in Education, pp. 207–218. Springer, New York (2017)
11. Hodoň, M., Miček, J., Kochláň, M.: Networking extension module for Yrobot—a modular educational robotic platform. Robotics in Education, pp. 159–167. Springer, New York (2017)
12. Casañ, G.A., Cervera, E., Moughlbay, A.A., Alemany, J., Martinet, P.: ROS-based online robot programming for remote education and training. In: 2015 IEEE International Conference on Robotics and Automation (ICRA), pp. 26–30. IEEE (2015)
13. Yu, X., Jian, P., Wang, M., Wu, S.: Extraction of implicit quantity relations for arithmetic word problems in Chinese. In: 2016 International Conference on Educational Innovation through Technology (EITT), pp. 242–245. IEEE (2016)

Practical Secure Certificateless Cryptographic Protocol with Batch Verification for Intelligent Robot Authentication

Qiang Nong[⊠]

School of Computer, Minnan Normal University, Zhangzhou, China
`nong_qiang@163.com`

Abstract. As robots and the mobile internet services have advanced, the security of the robot becomes more and more important for lots of network applications. At present, as an important and serious issue, there are few literatures about the robot authentication techniques. This paper proposes a robot authentication mechanism over a public networks which are vulnerable to several types of attacks. We apply as a new method an authentication protocol based on a certificateless signature scheme which can support batch verification to deal with the specific challenges arising from these applications and to explore possible identification and security mechanisms. The superiority of this method is that few straight-forward cryptology primitives, for instance, message authentication codes and collision-resistance hash functions, are required. It does not rely on secure channels which go with the technological environment of robot security. Detailed analysis and simulation tests show that our scheme is comparatively concise in terms of lower computational overhead and less system cost. It is well suited to applications for practical robot authentication with extremely restricted resources.

Keywords: Robot authentication · Certificateless signature scheme · Hash function · Batch verification · Secure channel

1 Introduction

Robotics have received a great deal of attention from both academia and industry [1, 2]. It can be expected that robotics will improve the efficiency of services and applications aiming at the increase of human-machine interaction and other network environments. Howbeit some critical challenges await to be solved before securable and effective robot applications become available, one of them is practical robot authentication protocol. The robots are connected to local servers via wireless links where information is exposed to various kinds of attacks, for instance, remoted manipulation, replaying or spoofing. Therefore, the security problem is one of the most important challenges to the wide-ranging deployment of robot technology when it is put into practice. Secure information interacting with robots should be well protected, this is mainly because the robots, to deal with an emergency situation, might make critically important judgments based on the communication information. The fast-growing robots relevant

© Springer International Publishing AG 2018
F. Qiao et al. (eds.), *Recent Developments in Mechatronics and Intelligent Robotics*,
Advances in Intelligent Systems and Computing 690, DOI 10.1007/978-3-319-65978-7_73

services requires robots to build reliable and robust security protocols in real application environment. These kinds of problems lead to the research topic of robot secure techniques [3, 4].

Our interest is the cryptographic protocols between a robot and an authentication server. To make use of cryptography in robotics system, the difficulty lies in that the appliance of cryptography demands communication cost and computational overhead that may not be obtainable. We present a secure robot authentication protocol over a public network based on a certificateless signature scheme which can support batch verification. The main purpose of our scheme is to eliminate the inherent problems of certificate management in the public key infrastructure (PKI) which makes the inefficiency of existed schemes. An example application environment for robot authentication is demonstrated in Fig. 1. In this environment, an intelligent robot is connected through a wireless gateway to a local server, which impersonates the role of the authentication server. The intelligent robot has a visible interface which can be conveniently configured by the client. For example, the client can take a phone to sweep the two-dimensional code given by the robot. A robot itself firstly chooses a random value as its secret key. The private key generator (PKG) then extracts a partial private key from the robot's public key which is generated by its secret value. There are several advantages in our scheme. It eases the cost of certifications in PKI and at the same time maintains the original level of security. It provides a solution to the long-standing problem that a secure channel must be used in certificateless-based secure protocol. It also has positive significicance in promoting the research of robot security.

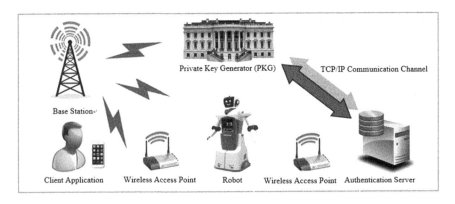

Fig. 1. Application setting for robot authentication

2   Proposed Robot Authentication Protocol

Setup. Given a security parameter k, PKG runs the parameter generator on input k to generate a group G of elliptic curve points with prime order q and determines a generator P in G. Picks a random number $s \in Z_q^*$ and sets $P_{pub} = sP$. Selects five cryptographic hash functions $H_0, H_1, H_2, H_3: \{0, 1\}^* \times G \times G \to Z_q^*$ and $H_4:\{0, 1\}^* \to \{0, 1\}^l$. PKG then publishes the system parameters and keeps the master secret key s secret. The authentication server which acts as a verifier who is designated to perform the batch verification randomly chooses $v \in Z_q^*$, computes and publishes $V = vP$ as his public key.

Robot-Key-Extract. Arobot with identity ID_i selects $x_i \in Z_q^*$, computes $PK_i = x_iP$ and set the secret value/public key as x_i/PK_i. The robot then sends (ID_i, PK_i) to PKG.

Partial-Private-Key-Extract. The robot and PKG perform the following operations.

(1) PKG firstly chooses a random number $t_i \in Z_q^*$, computes $T_i = t_iP$, $U_i = s \cdot PK_i$ and $D_i' = t_i + s \cdot H_0(ID_i, PK_i, T_i) + H_1(ID_i, U_i, T_i)$, outputs (D_i', T_i) as the robot's temporary partial private key. PKG then sends (D_i', T_i) to the robot via a public channel. Note that the value U_i is not required to send to the robot.

(2) The robot then verifies the validity of (D_i', T_i) by checking whether the equation $D_i'P = T_i + H_0(ID_i, PK_i, T_i) \cdot P_{pub} + H_1(ID_i, x_i \cdot P_{pub}, T_i) \cdot P$ holds. It is easy to see that the equation always holds since $x_i \cdot P_{pub} = s \cdot PK_i = U_i$. Finally, the robot computes $D_i = D_i' - H_1(ID_i, x_i \cdot P_{pub}, T_i)$ and sets (D_i, T_i) as its final legitimate partial private key.

Sign. Given message $m_i \in \{0, 1\}^*$, the robot chooses $r_i \in Z_q^*$ and computes $R_i = r_iP$, computes $S_i = D_i + x_i \cdot H_2(ID_i, m_i, PK_i, T_i, R_i) + r_i \cdot H_3(ID_i, m_i, PK_i, T_i, R_i)$ and outputs the signature $\sigma_i = (S_i, T_i, R_i)$. The robots sends σ_i to the authentication server.

Verify. To verify σ_i, the authentication server computes $h_{0i} = H_0(ID_i, PK_i, T_i)$, $h_{2i} = H_2(ID_i, m_i, PK_i, T_i, R_i)$, $h_{3i} = H_3(ID_i, m_i, PK_i, T_i, R_i)$ and checks if the equation $S_iP = T_i + h_{0i} \cdot P_{pub} + h_{2i} \cdot PK_i + h_{3i} \cdot R_i$ holds.

Batch Verify. To verify the validity of n signatures $\sigma_i = (S_i, T_i, R_i)$ for $i \in [1, n]$, the authentication server computes $h_{0i} = H_0(ID_i, PK_i, T_i)$, $h_{3i} = H_3(ID_i, m_i, PK_i, T_i, R_i)$ and $\delta_i = T_i + h_{0i} \cdot P_{pub} + h_{2i} \cdot PK_i + h_{3i} \cdot R_i$ for $i \in [1, n]$, then checks if the equation $H_4(S_1 \cdot V, ..., S_n \cdot V) = H_4(v \cdot \delta_1, ..., v \cdot \delta_i)$ holds.

3   Security Analysis

Firstly, due to the fact that the batch verification is performed by a dedicated authentication server which does not participate in the signing process, our scheme is no longer publicly verifiable. The server verifies the valid of n signatures by adding every individual signature as an input parameter of the hash function H_4. It is quite effective to prevent the coalition attack mentioned above since H_4 is a cryptographic secure collision-resistant hash function. This can guarantee the batch verification that is valid if and only if the validity of every individual signature signed by the designated signer with the robot's public key PK_i. Obviously, the batch verification can only be performed independently by a server who owns a corresponding secret value x_i. Note that the robot's public key PK_i is added as an input of the partial private key generation algorithm. This disallows anyone except the PKG to generate a valid partial private key for every newly generated robot public key. In other words, if there exist two valid signatures of a robot with different public key, the robot can prove the PKG's misbehavior to a third party. The proposed scheme exploits the public key binding technique in [5] and achieve the level-3 security according to Girault's definition [6].

Secondly, our scheme does not require any secure channel used by PKG to deliver the robot' partial private keys. In our improved scheme, PKG is allowed to send the temporary partial private key (D'_i, T_i) to the robot via an open channel. The robot then uses its secret value x_i to compute $D_i = D'_i - H_1(ID_i, x_i \cdot P_{\text{pub}}, T_i)$ and sets (D_i, T_i) as its final legitimate partial private key. Of course, by using the master secret key s, PKG can also compute $D_i = D'_i - H_1(ID_i, s \cdot PK_i, T_i)$ since $s \cdot PK_i = x_i \cdot P_{\text{pub}}$. In other words, the partial private key (D_i, T_i) is co-generated by PKG and the robot. This allows our scheme to restrict the dishonesty of PKG.

4   Performance Evaluation

A comparison of the proposed scheme and the schemes in [7, 8] in terms of the major computational overhead is given below. For ease of analysis, T_p is defined as the execution time of pairing operation, T_{pm} the execution time of pairing-based scale multiplication, T_m the execution time of elliptic curve-based scalar multiplication. The execution times for those cryptology algorithms have been indicated in [9, 10]. In general, the bilinear pairing is defined on the super singular elliptic curve. To evaluate the computational overhead of the proposed schemes, we refer [10] where the cryptology algorithms be simulated by use of MIRACAL, a standard cryptographic library, and their simulation environment is Windows XP OS over an Inter(R) Pentium IV 3.0 GHZ processor with 512-M bytes memory. For the pairing-based scheme, the Tate pairing defined over the super singular elliptic curve $y^2 = x^3 + x$ with embedding degree 2 has been used, where q is a 160-bit prime $q = 2^{159} + 2^{17} + 1$ and p a 512-bit prime satisfying p + 1=12qr. The simulation was run several times, and the results were averaged to compensate for the randomness. We then summarize the result of the performance comparison in Table 1 where n is the total number of individual signature.

Table 1. Performance comparison of [7, 8] and our scheme

Scheme	Signature length	Sign	Time to sign	Verify	Time to verify	Secure channel
[7]	128n	$3nT_{pm}$	19.14n	$3T_p + 4nT_{pm}$	60.12 + 25.52n	Yes
[8]	256n	$4nT_{pm}$	25.52n	$3T_p + 5nT_{pm}$	60.12 + 31.9n	Yes
Ours	60n	nT_m	2.21n	$4nT_m$	8.84n	No

In Fig. 2, we mainly focus on the batch verification time for difference number of individual signature of the proposed schemes. It can be seen from Fig. 2 that the running time of our scheme obviously outperforms schemes [7, 8]. In particular, the secure channel is not required to be used in our scheme. These make our scheme significantly more practical for applications such as robot control systems, where the communication cost and computational overhead that may be constrained.

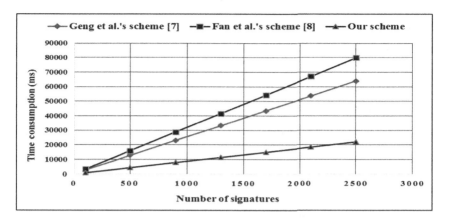

Fig. 2. Comparison of verification time between different schemes

5 Conclusion

Based on the certificateless cryptographical technology, a robot authentication scheme was proposed. Our scheme has some crucial merits. It has neither the key escrow problem nor the huge certificate management overhead. It also achieves better performance compared to related schemes. It has well universal property and it can as well be applied to other similar robot systems with minor modifications. The work is also has positive significance in promoting the research of robot security.

Acknowledgments. This work was supported by the education and research project of the education department of Fujian Province (No. JA15317).

References

1. Shin-ichiro, K., Genci, C.: Human-robot communication for surveillance of elderly people in remote distance. IERI Procedia **10**, 92–97 (2014)
2. Baghernezhad, F., Khorasani, K.: Computationally intelligent strategies for robust fault detection, isolation, and identification of mobile robots. Neurocomputing **171**, 335–346 (2016)
3. Choi, O., Choi, T., Kim, J., Moon, S.: NFC payment authentication protocol for payment agency of service robot. Future Information Technology, Lecture Notes in Electrical Engineering, vol. 309, pp. 65–70. Springer, Berlin (2014)
4. Wael, A.: Mechatronic security and robot authentication. Int. J. Adv. Sci. Technol. **14**, 41–51 (2010)
5. Al-Riyami S., Paterson K.: Certificateless public key cryptography. In: Proceedings of ASIACRYPT 2003, pp. 452–473. Taipei, 30 November–4 December 2003
6. Girault, M.: Self-certified public keys. In: Proceedings of EUROCRYPT 1991, pp. 490–497. Brighton, 8–11 April 1991
7. Geng, M., Zhang, F.: Batch verification for certificateless signature schemes. In: Proceedings of CIS 2009, pp. 288–292. Beijing, 11–14 December 2009

8. Fan, C.I., Ho, P.H., Huang, J.J., Tseng, Y.F.: Secure certificateless signature scheme supporting batch verification. Math. Probl. Eng. (2014). doi:10.1155/2014/854135
9. Cao, X., Kou, W., Du, X.: A pairing-free identity-based authenticated key agreement protocol with minimal message exchanges. Inf. Sci. **180**(15), 2895–2903 (2010)
10. He, D., Chen, J., Hu, J.: An ID-based proxy signature schemes without bilinear pairings. Ann. Telecommun. **66**(11–12), 657–662 (2011)

Research on Hierarchical Intelligent Control-Based Unmanned Vehicle Control System

Cheng-Hui Yang[(✉)]

School of Automation & Electrical Engineering, Lanzhou Jiaotong University, Lanzhou, China
yangchenghui36@163.com

Abstract. Comparing design method of unmanned vehicle, presents a method for designing software architecture based on Intelligent hierarchically control after comparing several traditional design methods. The main task of the underlying control system is to control the actual vehicle speed and direction on the basis of the specific instructions of the planning system, which is required control precision, fast. At the same time, the mathematical model of software architecture is obtained, which provides a reference for coding.

Keywords: Unmanned vehicle · Environmental perception · Planning decision making · Hierarchical intelligent control

1 Introduction

Self-driving Vehicle is a kind of unmanned ground vehicles, which has broad application prospects in intelligent transportation system in the future. Dai Ning' unmanned helicopter control system study based on intelligent stratified hierarchical control [1], the control on the Windows application programming interface (API) and multiple model control, and stratified hierarchical control system structure on the basis of the analysis and research, is an advanced unmanned helicopter flight control system of a kind of intelligent hierarchical control structure, and some of these key technologies are analyzed in this paper. Guo-ming Tang's design of driver-less cars semi physical simulation system [2], the design of the intelligent decision-making system is the most important parts in the unmanned vehicle, whose intelligent decision level directly determine the safety and reliability of the driver-less car. Feng-jiao Li's synthesis of unmanned vehicle obstacle avoidance behavior research and evaluation of unmanned vehicles as a collection of perception, decision-making, control technology in one of the integrated intelligent system, in recent years it has attracted many scholars have devoted to the scientific research work, whether it is theoretical analysis, a virtual experimental study, the actual road test. Intelligent behavior as the unmanned vehicle intelligent level of performance analyzed by people in different ways and gradually formed with the system. With the development of technology, to the intelligent behavior of driver-less vehicles are not limited to roughly judgment, but asked to determine the number of evaluation [3].

© Springer International Publishing AG 2018
F. Qiao et al. (eds.), *Recent Developments in Mechatronics and Intelligent Robotics*,
Advances in Intelligent Systems and Computing 690, DOI 10.1007/978-3-319-65978-7_74

2 Layered Hierarchical Intelligent Control System

2.1 Layered Hierarchical Intelligent Control System

In the '60 s, the development of automatic control theory and technology has gradually matured, and artificial intelligence is still a fledgling emerging technologies. At the Ale University of Canada, Ming Rao professor first put forward the concept of IDIS (Integrated Distributed Intelligent Systems). Technology in many countries, such as the USA and Japan had been a great shock and called the theory of intelligent engineering development milestone. IDIS promote edges cross, emphasizing the new method is on the improvement of the existing methods and complement, the production and management experts experience accumulated in the system software, introducing the process of operation directly, expert system and information management system of quality control system of the integrated [4]. IDIS has succeed in the field of paper making, petroleum, chemical industry and so on in engineering practice, which has achieved obvious economic benefits.

2.2 Stratified Hierarchical Control Substance

Jia-chao Cui's unmanned intelligent vehicle in the obstacle avoidance method in the dynamic environment of [5], environmental perception system is mainly to extract the external environment of the intelligent unmanned vehicles, by car GPS, laser radar and other sensors data fusion to reflect the nature of the smart car current obstacles.

2.3 Distributed Hierarchical Control Structure

As a result of the passengers in and out of the car behavior is uncertain, and the operation of the unmanned vehicle with independent and distributed characteristics. Therefore, adopting distributed hierarchical control structure, putting transportation task decomposition and coordination of the vehicles at different levels to consider. Distributed hierarchical control system uses the level 3 structure, organization, coordination and execution level. In organizational level, the demand of transport to the system is decomposed into associated with each station transport demand; to update the queue waiting for the bus passenger information stations, to seek the recent can transport and generate a vehicle scheduling command. Coordination level according to the state of each vehicle in the system of organization level correction of the vehicle dispatching command, to defuse potential conflict between vehicles. In the executive level, in order to accept the upper control commands and monitor the running status of autonomous unmanned vehicle, the unmanned vehicle behavior based on Petri net model [6]. Based on stratified hierarchical control SAR measurement and control system software architecture design [7], on the basis of comparing the traditional design method, this paper proposes a software architecture based on layered hierarchical control design method.

2.4   Driver-Less Cars of Intelligent Distributed Hierarchical Control Applications

Organization level is stratified hierarchical control structure of the top level, leading thought that the application of artificial intelligence of intelligent decision making, for a given task and command, and provides reasonable and appropriate control mode to control orders for the coordination layer. Coordination level is stratified hierarchical control structure of the middle layer, which is composed of a certain number of the coordinator. The coordinator is derived according to different orders from all sides of a series of reasonable optimization executes instructions, and these instructions are divided into the specific action for executive level operation sequence.

Executive level is at the bottom of stratified hierarchical control structure by classical control theory and modern control theory plays a leading role, which usually consists of multiple controllers, performing a certain action.

3   Realization of the Four Layered Hierarchical Control Structure

3.1   Smart Stratified Hierarchical Control Structure Design

Driver-less cars intelligent stratified hierarchical control structure of the intelligence of driver-less cars, as shown in the stratified hierarchical control structure is divided into coordination layer, control layer, process of three layer structure, coordinating layer can be divided into the upper planning system and the underlying planning system.

3.2   Process Layer Analysis

The underlying control system generally includes three subsystems, and steering control system belongs to the lateral control, throttle and braking control system belongs to the longitudinal control, and within the three subsystems there are respectively independent feedback system. Subsystem receives the instruction, to obtain an expectation, combined with the real information, through a control algorithm, to form a closed loop system, so as to achieve accurate and real-time control of intelligent vehicle.

4   Fuzzy Control Example Analysis

4.1   Membership Degree

Due to the change of E interval [0, 1], if you set the desired E of 0.2, the E change interval for [0.2, 0.8], desirable Ec interval [−1, 1], U interval [1], and then converted into standard interval [−6, 6]. Both input and output quantity is divided into seven membership functions, PL, PM, PS, ZE, NS, NM and NL.

4.2   Determine the Fuzzy Rules

When E is small, control measuring ZE or NS, PS. Fuzzy self-tuning PID parameters is designed to make k_p, k_i, k_d and as the change of the E and Ec and adjust itself, therefore, the relationship between them should be first established. According to practical experience, parameter k_p, k_i, k_d under different E and C of the adjustment should meet the following principles:

(1) The fuzzy rules table of s k_p shown in Table 1.

Table 1. k_p fuzzy rules

ΔK_p		E						
		NB	NM	NS	ZE	PS	PM	PB
Ec	NB	PB	PB	PM	PM	PS	ZE	ZE
	NM	PB	PB	PM	PS	PS	ZE	NS
	NS	PM	PM	PM	PS	ZE	NS	NS
	ZE	PM	PM	PS	ZE	NS	NM	NM
	PS	PS	PS	ZE	NS	NS	NM	NM
	PM	PS	ZE	NS	NM	NM	NM	NB
	PB	ZE	ZE	NM	NM	NM	NB	NB

(2) The fuzzy rules table of k_i, $K_i = K'_i + \{e_i, ec_i\}_i$.
(3) The fuzzy rules table of k_d shown in Table 2.

Table 2. k_d fuzzy rules

ΔK_d		E						
		NB	NM	NS	ZE	PS	PM	NB
Ec	NB	PS	NS	NB	NB	NB	NM	PS
	NM	PS	NS	NB	NM	NM	NS	ZE
	NS	ZE	NS	NM	NM	NS	NS	ZE
	ZE	ZE	NS	NS	NS	NS	NS	ZE
	PS	ZE	ZE	ZE	ZE	ZE	ZE	ZE
	PM	PB	NS	PS	PS	PS	PS	PB
	PB	PB	PM	PM	PM	PS	PS	PB

The fuzzy set theory field of error e and the error rate c is 7, respectively, e = {−6, −4, −2, 0, +2, +4, +6}, c = {−6, −4, −2, 0, +2, +4, +6}; k_p , k_i , k_d Theory of domain is[0,15]. The fuzzy subset is e, Ec, k_p, k_i, k_d = {NB, NM, NS, ZE, PS, PM, PB}. Set E, Ec and k_p, k_i, k_d all obey the normal distribution. Thus getting all the membership degree of fuzzy subsets. According to the membership degree of fuzzy subset assignment and various parameters of the fuzzy control model, fuzzy synthesis reasoning design PID correction parameter fuzzy matrix table found under the fixed parameters into computing [10].

$$K_p = K'_p + \left\{ e_i, ec_i \right\}_p \tag{1}$$

$$K_i = K'_i + \left\{ e_i, ec_i \right\}_i \tag{2}$$

$$K_d = K'_d + \left\{ e_i, ec_i \right\}_d \tag{3}$$

Look-up table, calculating, and completing the PID parameter online setting, parameters are set by fuzzy self-tuning PID controller design. According to the structural characteristics of the design of 9 Fuzzy controller: In the Command prompt, typing the following Command to start the FIS editor: Fuzzy. Named fuzzy inference system (FIS) is fuzzy. In the FIS editor for the system to add two input variables, named E and Ec. Obviously, the system has three output variables, in the editors, which are named k_p, k_i, k_d.

5   Conclusion

In the control of unmanned vehicle model, and intelligent hierarchical structure on the basis of the analysis and research of control system, the paper puts forward a smart stratified hierarchical control structure that is suitable for process control system of unmanned vehicle automated driving, and some of these key technologies are analyzed.

Acknowledgment. This work is supported by 2017 Fundamental Research Funds for the Central Universities of Northwest Minzu University (Grant 31920160003, Grant 31920170079). And the project also supported by Experiment Funds of Northwest University for Nationalities, Northwest University for Nationalities, China. (Grant NO. SYSKF2017035, Grant NO. SYSKF2017036, Grant NO. SYSKF2017037, Grant NO. SYSKF2017043, Grant NO. SYSKF2017044).

References

1. Ning, D.: Hierarchical intelligent control-based unmanned helicopter control system. Aeronaut. Sci. Technol. **2**, 37–40 (2010)
2. Zhong, T.: Electrical inertia simulation system of, Jilin University, a master's degree thesis (2003)
3. Chen, R.Y.: Motion Control System, pp. 33–106. Tsinghua University Press, Beijing (2006)
4. Boshi, C.: Electric Drive Control System, pp. 49–90. Mechanical Industry Press, Beijing (2000)
5. Hu, S.: Automatic Control Theory, 2nd edn., pp. 164–202. Science Press, Beijing (2008)
6. Zhou: Spin "test Tai power inertia brake system control method", Jilin University, a master's degree thesis (2005)
7. Shao, Y., Wang, J.: Intelligent Control, 2nd edn., pp. 25–93. Mechanical Industry Press, Beijing (2009)

8. Tao, Y., Yi-Xin, Y., Ge, L.: New PID Control and Its Application, 1 to 35, pp. 95–128. Mechanical Industry Press, Beijing (1998)
9. Jinkun, L.: Advanced PID Control and MATLAB Simulation, 2nd edn, pp. 94–118. Electronic Industry Press, Beijing (2004)
10. Xiao-Heng, C.: Adaptive fuzzy-tuning PID controller design, Liaoning Technical University, a master's degree thesis (2003)
11. Yang, C., Ren, E., Dang, J.: Analysis research of control method model on automobile brake test rig. Prz. Elektrotech. **88**, 375–378 (2012)

Research on Route Planning Technology of Micro-UAV with SAR

Qiang Liu[(⊠)]

Jiangsu Automation Research Institute, Lianyungang 222061, China
nuaalq@qq.com

Abstract. This paper mainly discuss the route planning of airborne micro SAR data acquisition, and propose a route planning scheme suitable for the coverage of micro-portable SAR unmanned aerial vehicle area. The route design method is described in detail, and the influence of the wind field is introduced. The route planning scheme is modified in accordance with the influence of the wind field, and a simple route planning algorithm based on the change of the wind field is proposed. The design of the route planning algorithm is verified by the design of the route planning software. It is proved that the proposed algorithm can effectively solve the problem of coverage trajectory planning in the polygon region with the presence of wind.

Keywords: Mini SAR · Route planning · Wind field · Micro UAV

1 Introduction

SAR (Synthetic Aperture Radar) is high-resolution imaging radar that can obtain high-resolution radar images which are similar to optical photography, by side-sweep under very low visibility weather conditions. Because of its ability to overcome the restrictions of cloud, fog, night on the implementation of all-day, all-weather, high-resolution reconnaissance detection capabilities, and the ability of anti-interference and reflective optical camouflage, SAR has not only attracted the attention of the military powers, but also been equipped in large and medium-sized unmanned aerial vehicles gradually. In recent years, with the continuous development of its key technologies, SAR equipment continues to be decreased in size and weight, and increased in the imaging resolution and the data transmission rate. Meanwhile, its signal processing capacity continues to be enhanced. SAR can already be installed on the miniature unmanned aerial vehicle, which will greatly enhance the using efficacy of mini-UAV and then all-day, all-weather information access means will be provided for the combat troops of front line. Compared to optical/infrared sensor flight height, however, SAR equipment has a high requirement for communication transmission rate and ground information processing capacity.

For SAR reconnaissance-type unmanned aerial vehicles, the route planning can effectively improve the reconnaissance effectiveness of the aircraft, so as to provide effective remote precision strike. The flight planning of the low-altitude aircraft is essentially to find an optimal or feasible flight route from the starting point to the end point in the planning space and under the given constraints. Because of the difference

© Springer International Publishing AG 2018
F. Qiao et al. (eds.), *Recent Developments in Mechatronics and Intelligent Robotics*,
Advances in Intelligent Systems and Computing 690, DOI 10.1007/978-3-319-65978-7_75

between SAR radar imaging principle and traditional photoelectric load, which is unilateral sweep imaging, its route planning method is essentially different from traditional route planning methods. Based on the full analysis of the working principle of SAR radar, this paper designs a method of UAV route planning based on wind field variation, which effectively solves the problem of UAV with SAR coverage scanning to reconnaissance area.

2 System Components

2.1 Mini SAR System

The mini SAR system consists of SAR loads and ground handling subsystems. The composition of the mini SAR system is shown in Fig. 1.

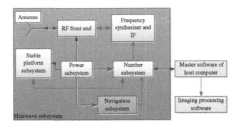

Fig. 1. The composition block diagram of mini SAR system

2.2 UAV System

Mini-UAV consists of the body sub-system, flight controlling sub-system, launching and recovering subsystems, as shown in Fig. 2.

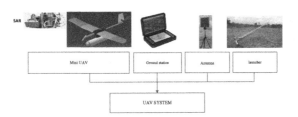

Fig. 2. UAV system components

3 Route Planning Algorithm

SAR radar used in this paper is the side sweeping band imaging one, reconnaissance area is away hundreds of meters from one side of the UAV. Because UAV's flight height, heading changes affect the imaging coverage area, these factors design algorithm should be considered in the route planning algorithm design process. In the route planning process, some navigation parameters are needed (shown in Fig. 3).

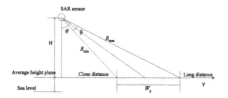

Fig. 3. Airborne SAR data acquisition geometric diagram

1. The High Altitude H: Relative flying height of UAV;
2. Angle of Side View φ: The angle formed in radar side view imaging, the angle is generally between 20 ° $\sim$ 50 °;
3. Directions of Side View: Left or right side view;
4. Mapping Bandwidth Wg: Determined by the beam width and the flight height.

The route is calculated in accordance with aerial parameters. In the aerial photography area, the flight direction of aircraft is from north to south or vice versa, imaged from west to east. First, the height of the reference plane is determined in the route starting position, in order to better cover the area to facilitate the contact between the edges of different areas. Set the starting route to a location that exceeds the aerial range of a survey bandwidth. Then, the next route of the location will be calculated in accordance with the measure of bandwidth and the degree of overlap. So the loop until the entire area is completely covered. The last route is also set to exceed the aerial range of a mapping bandwidth position.

Next, the location of the UAV flight line will be calculated. The plane coordinates (Xp, Yp) of the point P on the route; The location of the UAV on the flight line (Xs, Ys, Zs), according to the geometric relationship of SAR imaging:

$$X_s = X_P - S_{off} = X_p - H * \tan(\varphi) \tag{1}$$

$$Y_s = Y_p \tag{2}$$

The method of plane coordinate re-definition is used to map the space route point to the plane coordinate system in order to facilitate the design of the route. All the waypoint calculation is completed in the plane coordinate system. The planned plane coordinates are converted into geographical coordinates after the route planning is completed.

The plane coordinates are defined by selecting three geographic coordinate points in the reconnaissance area. As shown in Fig. 4, ORG, XX, XY are three definition points. The location data for these three points are entered in the plane coordinate file.

ORG is the origin of the plane coordinate system, and XX is the point on the plane coordinate X axis. XY is the point on the Y-axis side of the plane coordinate, which can be used to determine the direction of the Y-axis and Z-axis.

498   Q. Liu

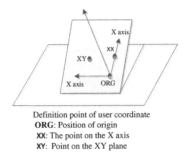

Definition point of user coordinate
ORG: Position of origin
XX: The point on the X axis
XY: Point on the XY plane

Fig. 4. Establishment of plane coordinate system

The implementation algorithm of user coordinate is as follows:

$$L = (XX_x - ORG_x) * (XY_z - ORG_z) - (XY_y - ORG_y) * (XX_z - ORG_z) \quad (3)$$

$$M = (XY_x - ORG_x) * (XY_z - ORG_z) - (XY_x - ORG_x) * (XY_z - ORG_z) \quad (4)$$

$$N = (XY_x - ORG_x) * (XY_y - ORG_y) - (XY_x - ORG_x) * (XX_y - ORG_y) \quad (5)$$

$$L_1 = sqrt(L * L + M * M + N * N) \quad (6)$$

$$d = (XX_x - ORG_x) * (XX_x - ORG_x) - (XY_y - ORG_y)$$
$$* (XX_y - ORG_y) + (XY_z - XX_z) * (XY_z - XX_z) \quad (7)$$

$$L_2 = sqrt(d) \quad (8)$$

The change matrix of coordinate can be expressed as:

$$\begin{array}{lll} a_x = L/L_1 & n_x = (XY_x - XX_x)/L_2 & o_x = a_y * n_z - a_z * n_y \\ a_y = M/L_1 & n_y = (XY_y - XX_y)/L_2 & o_y = a_z * nx - a_x * n_z \\ a_z = N/L_1 & n_z = (XY_z - XX_z)/L_2 & o_z = a_x * n_y - a_y * n_x \end{array} \quad (9)$$

So the transformation matrix of coordinate is as follows:

$$T = \begin{bmatrix} n_x & o_x & a_x & 0 \\ n_y & o_y & a_y & 0 \\ n_z & o_z & a_z & 0 \\ 0 & 0 & 0 & 1 \end{bmatrix} \quad (10)$$

To a certain extent, stable platform reduces the SAR radar's demanding requirements on the stability of aircraft movement and flight attitude. In the specific application, however, under the stable platform's affection to the scope of work in pitching, rolling and azimuth and in the case of a high flight, SAR radar will not work and mission will fail with a heavier wind beyond the working range of the stable platform.

As it is shown in Fig. 6. Aiming at this problem, literature [1] proposed a method of UAV flight planning based on SAR radar imaging. By using this method, it can flexibly cope with the change of wind direction of airborne winds, which is more favorable for SAR radar to complete the task and improve the efficiency of task completion. But however, the algorithm needs to obtain the data of the wind field in real time. The small UAV generally does not configure the sensor to measure the wind field data. Therefore, the method cannot be applied to the mini-UAV with SAR. On the basis of this literature, this paper presents a simple route planning method based on real-time wind direction change.

It can be seen from Fig. 5 that, with the presence of wind, airborne SAR radar imaging fails to complete the mission due to the presence of UAV's drift angle that may make the radar stability platform beyond the scope of work. In order to improve the success rate of airborne SAR radar imaging, this paper presents a simple route planning method of SAR radar area coverage reconnaissance based on the change of wind field. The core idea of this method includes three points:

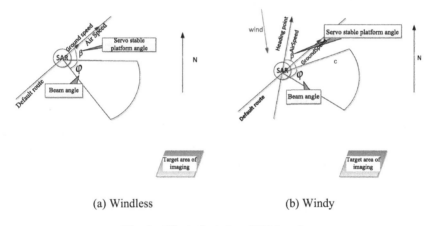

(a) Windless   (b) Windy

Fig. 5. Effect of wind on SAR imaging

1. Selection of Main Scanning Edge: Select the edge with the smallest angle of the wind as the main scanning edge;
2. Considering the transformation of the transformation of coordinate and matrix, as shown in Eq. 11;

$$T = \begin{bmatrix} n_x & o_x & a_x & 0 \\ n_y & o_y & a_y & 0 \\ n_z & o_z & a_z & 0 \\ 0 & 0 & 0 & 1 \end{bmatrix} * \begin{bmatrix} \cos\psi & -\sin\psi & 0 & 0 \\ \sin\psi & \cos\psi & 0 & 0 \\ 0 & 0 & 1 & 0 \\ 0 & 0 & 0 & 1 \end{bmatrix} \tag{11}$$

ψ means the wind direction.

3. Re-planning of Route under The Conditions of Wind: To re-plan the SAR reconnaissance route in the presence of a wind field, by using the route coverage rules
mentioned above, in accordance with the selection of the main scanning edge.

The route planning of SAR radar imaging reconnaissance area coverage under the
conditions of wind is as shown in Fig. 6.

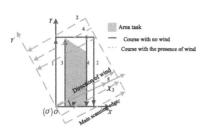

Fig. 6. Re-planning of route under the
conditions of wind

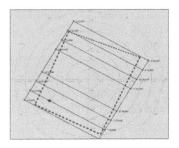

Fig. 7. The results of route planning

4   The Validation of Algorithm

The design of the reconnaissance mission area, the setting of the wind field direction
and the setting of the radar parameters are completed by the navigation parameter
setting page of the route planning software in accordance with the airborne SAR area
coverage reconnaissance route planning algorithm. Based on the route planning
algorithm designed in this paper, the route planning software automatically implements
the coverage route planning of SAR radar reconnaissance area. The results of route
planning are shown in Fig. 7.

It can be seen from Fig. 7 that in order to reduce the effect of unmanned aerial drift
angle on radar imaging in the presence of wind field, the route planning algorithm
designed in this paper will take the initiative to reconstruct the reconnaissance route
according to the direction of wind. It can not only meet the requirements of the
scanning area of the reconnaissance area, but also avoid the radar side view beyond the
work scope of stable platform, thus improving the success rate of radar imaging.

5   Conclusion

The airborne SAR emits pulses by side view. With the flying of aircraft, strip-like
mapping belts are formed. Such imaging method determines the route planning technology that cannot copy the mature optical photogrammetry, but must develop the
planning scheme of route suitable for airborne SAR on the basis of route planning of
the optical photography measurement in accordance with the characteristics of the
airborne SAR itself.

References

1. Guan, H., Li, B., Liu, X.: Route planning of airborne SAR data acquisition in complex area. Sci. Surv. Mapp. **38**, 164–166 (2013)
2. Teng, C., Wan, Y.: An intelligent aerial partitioning algorithm. Bull. Surv. Mapp. **3**, 34–36 (2008)
3. Zhang, C., Wang, H., Xiang, J., Zheng, L.: A Method of UAV Route Planning Based on SAR Radar Imaging. Beijing University of Aeronautics and Astronautics, Beijing
4. Liu, Z., Lv, Y.: Research on imaging algorithm for diving missile-borne SAR based on a curve trajectory. In: International Conference on Computational Problem-Solving (ICCP), Chengdu, pp. 255–259 (2012)
5. Andrew, N.K.: Deployment Algorithms for Mobile Robots Under Dynamic Constrain. University of California, Berkeley (2011)
6. Chen, H., Wang, X., Jiao, Y., Li, Y.: A UAV trajectory planning algorithm for umbrella region. Acta Aeronaut. et Astronaut. Sin. **31**(9), 1802–1807 (2010)
7. Jones, P., Sevanos, G.V., Tang, L.: Multi-unmanned aerial vehicle coverage planner f or area surveillance missions. In: The American Institute of Aeronautics and Astronautics (AIAA), vol. 6453 (2007)
8. Berger, J., Jabeur, K., Boukhtouta, A., et al.: A hybrid genetic algorithm for rescue route planning in uncertain adversarial environment. In: Proceedings of the IEEE Congress on Evolutionary Computation, pp. 1–8 (2010)

Design of Motion Control System for Frog-Inspired Bionic Hopping Robot

Qimin Zhang[1], Jieru Zhao[2], and Shuquan Wang[1(✉)]

[1] Technology and Engineering Center for Space Utilization, Chinese Academy of Sciences, Beijing, China
shuquan.wang@csu.ac.cn
[2] Beihang University, Beijing, China

Abstract. The discrete landing position and the sudden and explosive manner of take off of the hopping movement endow the hopping robot with a strong capability to overcome obstacles and to adapt to various environments. Based on the principle of bionics and imitation of a frog, a new type of bionic hopping robot is presented in this paper. The robot is driven by two elastic telescopic hind legs and adjusted by two electric front legs to realize the dynamic take-off and stable continuous jumping. In this paper, the structure of a hopping robot is discussed, and the control system composed of host computer and embedded main controller is designed and studied experimentally. The jumping experiments verify that the robot has a satisfactory hopping capability by adjusting jumping attitude to meet different distance requirements with a stable landing. This study is valuable for designing the next-stage hopping robot and exploring its application in complex terrain.

Keywords: Bionic · Hopping robot · Control system

1 Introduction

At present, there are two major types of robot motion, wheel-driving and bionic crawling or walking [1]. Wheel-driving mode is applicable in flat terrain, especially in the paved road. Wheel-driving robots are usually difficult to be applied to complex and rugged terrain because of their strictly limited obstacle crossing capability. Crawling robots have redundant degrees of freedom and complex gait algorithm. They can avoid obstacles by changing the planed path. With the increasingly wide range of robot applications, researchers and engineers are developing robots that can adapt to harsh conditions and unstructured environment where human beings cannot penetrate, such as planetary surface exploration, archaeological exploration, military reconnaissance and other fields [2]. The unknown complex environment requires the robot to have a strong capability to adapt to the terrain, efficient movement patterns and autonomous movement capability.

The research of the early hopping robot is based on the application of the extraterrestrial exploration. Due to the strong raggedness of the environment, the wheel-driving robots exploration vehicle is greatly restricted. On the other hand, in many situations

© Springer International Publishing AG 2018
F. Qiao et al. (eds.), *Recent Developments in Mechatronics and Intelligent Robotics*,
Advances in Intelligent Systems and Computing 690, DOI 10.1007/978-3-319-65978-7_76

the gravity coefficient is less than that on the surface of the Earth (such as Mars / earth is 38%, the moon / earth is 17%). Hopping robots suit these situations pretty well because they can jump over obstacles which are several times of their own size. The hopping robot with this capability can easily cross ditch and obstacles, which greatly expanding the scope of activities and improving the ability to adapt to complex terrain.

The hopping robot has been studied extensively in the past three decades. Raibert et al. have significant contributions on studying the motion control principle exampled by their 2D and 3D one-legged hopping robot [3, 4] and even quadruped robot [5]. Zhao et al. proposed a single-motor-actuated miniature steerable jumping robot [6] in 2013, Hamed proposed a 3-D Bipedal Robots [7, 8] in 2014. Reis developed an energy-efficient hopping robot based on free vibration of a curved beam [9] in 2014. Certain animals with outstanding jumping capabilities, such as kangaroos, frogs and rabbits, are good references for designing jumping robot systems.

This paper proposes to design a hopping robot by imitating the movement of a frog. This kind of research is helpful to understand the jumping motion of a frog and reveal its mechanism. Moreover, the extraordinary jumping capability of a frog attracts the researchers' attentions and motivates investigations of mimicking the motion of a frog in designing robots. A frog-inspired bionic hopping robot driven by elastic mechanical legs is investigated in the paper. The motion mechanism and motion control system are studied. The research results will be helpful to explore the practical application of robots in various fields such as extraterrestrial exploration.

2 Mechanical Model of the Hopping Robot

Hopping robots are inspired by animals, frog hopping movement is a good imitation. The frog adjusts its stance by moving the forelegs before taking off, and the energy of the jump mainly comes from the muscular hind legs. According to the characteristics of frog hopping, this paper designs a model to simulate the hopping motion. Schematic diagram of half side of hopping robot is shown in Fig. 1.

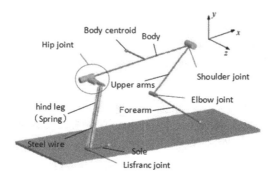

Fig. 1. Schematic diagram of half side of hopping robot

The hip joint, shoulder joint and elbow joint of the frog-inspired hopping robot are simplified as a degree of freedom, and spring is selected as the energy storage element. Two degree of freedom are added in hip joint, rotate around the Z axis and the X axis to adjust the take-off attitude, direction and trajectory. Forelimb function is relatively simple, shoulder joint has a degree of freedom (around the Z axis rotation) which is responsible for balance and adjust the jumping posture. The elbow joint has a passive degree of freedom to support and buffer the robot during the landing phase.

3   Hopping Motion Strategy

There are 4 requirements for the ideal springing mechanism: i. improve the performance of the hopping robot to jump higher and farther; ii. the hopping direction and angle are adjustable; iii. the attitude of hopping mechanism is controllable; iv. smooth landing is controllable. In order to analyze the characteristics of mechanism, we can start with the simplified hopping model. The vertical motion and forward motion are included in the hopping motion, this paper mainly studies the strategy of vertical hopping motion, because the forward motion can be controlled under the condition of adjustable attitude. The coordinates of the robot's hind leg in contact phase is shown in Fig. 2.

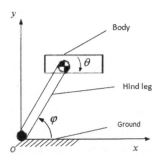

Fig. 2. Coordinate of the robot's hind leg in contact phase

It is necessary to satisfy the constraint condition of the center of mass of hip in coordinate in takeoff phase:

a. h_0 means initial position, h_1 means the position at T_1;
b. v_1 means the velocity at T_1.

The hopping motion can be actualized by control the attitude and velocity in takeoff phase. According to the above constraint conditions, the expression of the vertical direction of the rising center of mass is:

$$y(t) = \lambda_1 t^3 + \lambda_2 t^2 + \lambda_3 t + \lambda_4 \tag{1}$$

The mathematical expression of the above constraint equation:

$$\begin{cases} y(0) = h_0, \dot{y}(0) = 0 \\ y(t) = h_1, \dot{y}(t) = v_1 \end{cases} \tag{2}$$

Figure out λ_i (i = 1,2,3,4) by the constraint conditions, then vertical displacement expression of center of mass of hip in takeoff phase:

$$y(t) = \left[2(h_0 - h_1)/T_1^3 + v_1/T_1^2\right]t^3 + \left[3(h_1 - h_0)/T_1^2 - v_1/T_1\right]t^2 + h_0 \tag{3}$$

The impact force in landing phase has a certain influence on the sensor and mechanical structure, so we need to reduce the impact force by landing motion planning. In the use of "soft landing" ideas, planning of the trajectory rationally to make the mechanical legs be elastic and control the landing impact force in a certain range. It is necessary to satisfy the constraint condition of the center of mass of hip in coordinate in landing phase:

a. h_1 becomes h_0 in the landing phase;
b. $-v_1$ becomes 0 in the landing phase;

Then the constraint equation is:

$$\begin{cases} y(0) = h_1, \dot{y}(0) = -v_1 \\ y(T_2) = h_0, \dot{y}(T_2) = 0 \end{cases} \tag{4}$$

Figure out λ_i (i = 1,2,3,4) by the constraint conditions, then vertical displacement expression of center of mass of hip in landing phase:

$$y(t) = \left[2(h_1 - h_0)/T_2^3 - v_1/T_2^2\right]t^3 + \left[3(h_0 - h_1)/T_2^2 + 2v_1/T_2\right]t^2 - v_1 t + h_1 \tag{5}$$

4   Design of Motion Control System

Hopping robot is able to cross ditches and obstacles, adapt to rugged surface and have a wide range of activities. In order to improve the practicability of the robot, the semi autonomous control mode is adopted, which means mission instructions issued by the operator at the control terminal, the robot actualizes its attitude adjustment and hopping independently and the working state information is fed back to the control system in real time. The communication between host computer and controller adopts Serial Communication Interface (SCI). The motion control system of hopping robot is shown in Fig. 3.

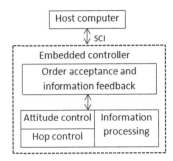

Fig. 3. Structure of motion control system

4.1 Hardware Structure of Motion Control System

The motion control system of hopping robot is composed of host computer PC and embedded main controller. The host computer PC is responsible for human-computer interaction, provide parameters hopping motion and trajectories of joints, issue the motion instructions to embedded main controller through SCI and receive feedback information of hopping motion state. The embedded main controller, DSP-TMS320LF2407, control steering gears of joints and servo motors of hind legs after receiving the task of the upper computer PC and the movement instruction [10]. The embedded main controller accept sensor information in real time.

4.2 Software Structure of Motion Control System

The software structure of motion control system is divided into four layers according to the function level, as shown in Fig. 4. Master plan level is responsible for the initialization of the entire software system and make mission planning according to planning indicators and terrain characters. Attitude plan level is responsible for setting the

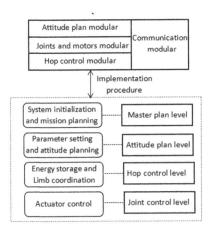

Fig. 4. Software structure of motion control system

parameters of hopping motion according to the mission instruction. Hop control level is responsible for energy storage and limb coordination according to the mission and feedback information from sensors. Joint control level is responsible for actuator control.

The controlled objects in the robot control system is complicated, and the robot software system should have a good expansibility and portability for updating the software and hardware of hopping robot. Therefore, the hopping motion is realized by the combination of modules, which reduce the complexity of the software structure and is propitious to update software system. If more advanced control strategy is applied in the future, modules can be replaced directly, which largely reduces the workload.

4.3 Main Controller and Divers

TMS320LF2407 is selected as embedded main controller and LMD18200 is specially designed for motion control [11]. One TMS controls two LMD, which means 5 steering gears and 2 direct-current motors. Embedded controller is the core of motion control system, which is used to coordinate drives and sensors [12]. Software system is designed in modular, the rotation of the motor, the detection of Hall sensors and the conversion of the A/D are all realized by specific module functions. SCI commands are detected and executed circularly in the main program. In this loop, the state of sensors is queried in real time, and when it changes, SCI instruction is executed. Embedded controller main program flow diagram is shown in Fig. 5.

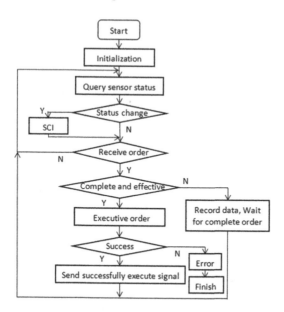

Fig. 5. Embedded controller main program flow diagram

5 Hopping Experiment

According to the mechanical structure and motion control system above, a prototype of bionic hopping robot was designed in this paper. The length is 260 mm, the width is 96 mm, the height is 140 mm and the weight is about 0.7 kg. The horizontal jump distance is about 845 mm, and the maximum vertical jump height is 435 mm in the hopping experiments. The hopping experiments proved that the frog-inspired bionic hopping robot designed in this paper has a satisfactory hopping performance. It can be seen from the vertical displacement curve of the hopping motion that there is a slight vertical fluctuation in the landing phase, which is caused by the landing impact. In view of this situation, a buffer mechanism is added to the forelimb of the robot to have a more stable landing. The curve of mass center position of hopping robot, which is obtained by experiments, is shown in Fig. 6.

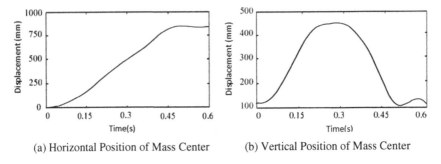

(a) Horizontal Position of Mass Center   (b) Vertical Position of Mass Center

Fig. 6. Displacement of mass center

In the actual hopping motion, two hind limbs of the robot can't be fully synchronized, leads to loss of energy and affects the height and stability of hopping motion. The synchronization is reduced by controlling the lag time of the hind limbs and adjusting the counterweight.

6 Conclusion

Based on the principle of bionics and imitation of a frog, this paper proposes a new type of frog-inspired bionic hopping robot. The mechanism model and control system of hopping robot are designed, and experimental research on hopping motion of robot is carried out. The feasibility of hopping mechanism model and the motion control system is verified by hopping experiments. The frog-inspired bionic hopping robot has a strong capability of obstacle avoidance and high mobility. Meanwhile, the hardware structure and software system presented in this paper are universal, which are easy to transplant to the next-stage hopping robot or other bionic robots.

Acknowledgment. This work was supported by the National Natural Science Foundation 11672294.

References

1. Dragan, A.D., Lee, K.C.T., Srinivasa, S.S.: Legibility and predictability of robot motion. In: ACM/IEEE International Conference on Human-Robot Interaction, pp. 301–308. IEEE (2013)
2. Graichen, K., Hentzelt, S., Hildebrandt, A., et al.: Control design for a bionic kangaroo. Control Eng. Pract. **42**, 106–117 (2015)
3. Raibert, M.H., Tello, E.R.: Legged robots that balance. IEEE Expert. **1**(4), 89–89 (1986)
4. Pratt, G.A.: Legged robots: what's new since Raibert. IEEE Robot. Autom. Mag. **7**(3), 15–19 (2000)
5. Raibert, M., Blankespoor, K., Nelson, G., et al.: BigDog, the rough-terrain quadruped robot. In: World Congress, pp. 10822–10825 (2008)
6. Zhao, J., Xu, J., Gao, B., et al.: MSU jumper: a single-motor-actuated miniature steerable jumping robot. IEEE Trans. Robot. **29**(3), 602–614 (2013)
7. Hamed, K.A., Grizzle, J.W.: Event-based stabilization of periodic orbits for underactuated 3-D bipedal robots with left-right symmetry. IEEE Trans. Robot. **30**(2), 365–381 (2014)
8. Hamed, K.A., Grizzle, J.W.: Robust event-based stabilization of periodic orbits for hybrid systems: application to an underactuated 3D bipedal robot. In: American Control Conference, pp. 6206–6212 (2013)
9. Reis, M., Iida, F.: An energy-efficient hopping robot based on free vibration of a curved beam. IEEE/ASME Trans. Mechatron. **19**(19), 300–311 (2014)
10. Lin, C.K., Liu, T.H., Yu, J.T., et al.: Model-free predictive current control for interior permanent-magnet synchronous motor drives based on current difference detection technique. IEEE Trans. Ind. Electron. **61**(2), 667–681 (2014)
11. Chauhan, S.K., Shah, M.C., Tiwari, R.R., et al.: Analysis, design and digital implementation of a shunt active power filter with different schemes of reference current generation. Iet Power Electron. **7**(3), 627–639 (2014)
12. Aliff, M., Dohta, S., Akagi, T., et al.: Development of a simple-structured pneumatic robot arm and its control using low-cost embedded controller. Proc. Eng. **41**(3), 134–142 (2012)

Automatic Crowd Detection Based on Unmanned Aerial Vehicle Thermal Imagery

Yonghao Xiao[1], Hong Zheng[2(✉)], and Weiyu Yu[3]

[1] Electronic Information School, Foshan University, Foshan, China
[2] Shenzhen Institute of Wuhan University, Shenzhen, China
503948664@qq.com
[3] Electronic and Information Engineering School,
South China University of Technology, Guangzhou, China

Abstract. Automatic crowd detection is one of the most challenging problems in computer vision. Although the thermal image from Unmanned aerial vehicle (UAV) have prominent performance in perspective, but have some disadvantage, such as the platform motion, image instability and so on. In the thesis, the automatic crowd detection system is designed to tackle these disadvantage. The detection system consists of two parts: ROI extraction and SVM classification, circle gradient and geometric filtering are used in ROI extraction, a hybrid feature which combines circle gradient (CG) and histogram of oriented gradient (HOG) is used in SVM classifier. The experimental results prove that the approach performs dense crowd detection from thermal images effectively.

Keywords: Circle gradient · Thermal image · SVM · UAV

1 Introduction

Crowd detection and monitoring play an important role in public safety and so on. Generally, CCTV networks are applied in crowd monitoring. If the crowd in a large square, there need lots of monitors. In addition, it is difficult to learn the overall situation of the crowd rapidly from many surveillance video. Therefore, some research work have been focusing on the detection of dense crowd using surveillance video from UAV. On the UAV surveillance video, the wide range of monitoring can be obtained.

In thermal image, the object pixel's value is got by their temperature or radiated heat, rather than lighting conditions. In fact, crowd usually warmer than other objects in the scene [1, 2]. For the prominent feature, the thermal video is applied to crowd detection in this paper. However, thermal image has its own limitations. First, besides crowd, other no-human objects also produce bright region, which makes it impossible to detect crowd only based on the brightness. Second, thermal image have low resolution and lots of noise due to the limitations in camera technology [3].

Human thermal detection systems usually have two parts structure [4, 5], include learning and detection. For the learning part, the system gathers both negative and positive samples, extracts human's features, then sends the features to a classifier for training the crowd classifier. For the detection part, the regions of ROIs are circled as

© Springer International Publishing AG 2018
F. Qiao et al. (eds.), *Recent Developments in Mechatronics and Intelligent Robotics*,
Advances in Intelligent Systems and Computing 690, DOI 10.1007/978-3-319-65978-7_77

candidates, the features of the ROIs are extracted and transmitted to the crowd classifier, the crowd classifier classifies and locates the Human.

In the thesis, a new approach is proposed to detect crowd thermal imagery of unmanned aerial vehicle. The approach includes two stages, region of ROIs extraction and ROIs classification. The remainder of this thesis is arranged as follows: Sect. 2, some work is reviewed, which related to crowd or pedestrian detection in infrared images. Section 3, the framework of approach is proposed. Section 4, the experimental results are shown. Section 5, this paper is concluded.

2   Literature Survey

The UAV flies higher, the thermal image resolution is lower. Therefore, these crowd objects have insufficient feature. A lot of methods have been proposed for object detection, such as CSM representation [6], background segmentation [7, 8] and intensity contrast [9, 10], these approaches are suited for optical image, but not for thermal image. The blob extraction aims at extraction pedestrian blobs from UAV thermal infrared image [11], it assumes that each pedestrian object presents as small hot blob in thermal image. Blob is extracted with and geometric filtering and the gradient feature of region. The blob extraction approaches can extract single blob of whole pedestrian in sliding window, but if there are multi-blobs in sliding window, the approaches does not work well, especially for dense crowd. In addition, one human maybe present as multi-blobs in infrared image.

3   Proposed Method

The crowd detection method is proposed for aerial vehicle thermal imagery in our research. The method contains two stages, first stage is ROIs extraction, The and extraction approach applies the geometric filter and average circle gradient to get ROIs. The second stage is ROIs classification. In this stage, Histogram of Oriented Gradients (HOG) and circle gradient (CG) feature are combined as a hybrid descriptor and used in SVM classification. The system framework is shown second stage is ROIs classification. The whole system workflow is shown as Fig. 1.

Fig. 1.   Crowd detection system workflow

3.1   ROIs Extraction

The proposed approach is shown as Fig. 2. Thermal image is firstly converted to gray image, then a window with fixed size and step is constructed. Window slides over whole image, at every sliding position, an sub-image is got, CG mean is calculated and compared to threshold T_1, if greater than T_1 then continues the flowchart, the candidate block are segmented. In literature [11], OTSU threshold method is adopt to segment, if the intensity of background is high, the OTSU method is not so well. Here, the region growing method instead of OTSU. The local extremum of sub-image are calculated, and used as seeds. The segmented region are more than one in case of dense crowd. The multi-region centeroid is gotten by all regions centeroid weighted average, the region's weighted coefficient depend on the region's area. Then the distance between centeroid and center of sliding window is computed. When distance less than T_2, major-axis is computed. If the major-axis is less than T_3, the sub-image is ROI. Some key concepts are explained as follows.

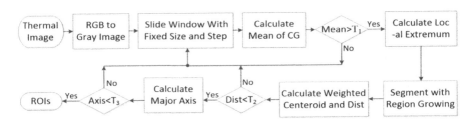

Fig. 2. ROIs extraction flowchart

Circle gradient: In thermal image, the intensity of human body is generally greater (or less) than the background's. Therefore, human body objects could be detected with this type of distinction. Circle gradient is used to detect the distinction. Referring to Fig. 3, here, $I_{i,j}$ is a pixel in sliding window; r_k, given in Eq. 1, is total intensity of all pixels in ring k; circle gradient r_k is defined as Eq. 2; the average of circle gradient c_{mean} is given in Eq. 3.

$$r_k = \sum I_{i,j}, (I_{i,j} \in ring(k)) \tag{1}$$

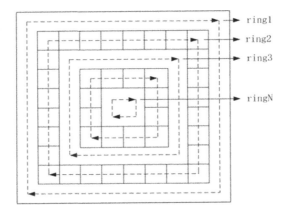

Fig. 3. Circle band in sliding window

$$c_k = |r_{k+1} - r_k| \tag{2}$$

$$c_{mean} = \frac{1}{N-1} \sum_{k=1}^{N-1} c_k \tag{3}$$

Center and centeroid: center is the geometric center of sub-image. Centeroid is center of object region. If there multi-object regions in sub-image, the centeroid is defined for weighting center, presented as Eq. 4. Here, Cd_i is center the i-th region, A_i corresponds to i-th region area, $Cntrd$ is centroid. The distance between center and centroid is Euclidean distance.

$$Cntrd = \sum \left(\frac{A_i}{\sum A_i} Cd_i \right) \tag{4}$$

3.2 ROI Classification

In order to get accurate detection result, a new ROI classifier is proposed here. The classical classifier consists of two parts: one is ROI features extraction; the other is the ROI features matching. In our research, this classical structure is adopted. A feature descriptor combining HOG and CG feature, is first proposed. Then, SVM is applied to distinguish crowd from non-crowd ROI.

Feature descriptor: HOG is a descriptor of feature used for object detection, the features of i-th block are denoted as $F_i = \left[\overrightarrow{f_{i,1}}, \overrightarrow{f_{i,2}}, \cdots, \overrightarrow{f_{i,36}} \right]$. The HOG vector F_{HOG} is obtained by concatenating all blocks histograms, $F_{HOG} = [F_1, F_2, \cdots, F_L]$. CG feature are denoted as $F_{CG} = [c_1, c_2, \cdots, c_N]$, its dimension N depends on the sizes of sliding window. The final combination feature descriptor F_v is constructed with the weighted combination of F_{HOG} and F_{CG}, such as $F_v = [\lambda_1 F_{HOG}, \lambda_2 F_{CG}]$, here, λ_1, λ_2 is weighted factor.

The SVM provides a method to divide objects into two categories in higher dimensional space, it has presented excellent performance on high dimensional space recognition. Here, the SVM is applied to classify crowd and non-crowd objects, Finally, the detected crowd are obtained by constructing the union set of the ROIs which are classified into the crowd category.

4 Experimental Results

In the section, experiments of our approach are conducted. Aerial thermal videos are captured by a thermal infrared camera, which was mounted on a UAVs with flight altitude about 20–70 m. Two measurements are adopted: Precision = TP/(TP + FP), it is the percentage of Correct detection crowd ROIs number over the total detection ROIs; Recall = TP/(TP + FN), it is the percentage of Correct detection crowd ROIs number over the total true ROIs.

There are three video datasets filming in Wuhan, their details are shown in Table 1. They are used to test the proposed approach. In the experiments, some of the frames from the video datasets are chosen as the training set and the others as the test set.

Table 1. Fundamental state of datasets

Datasets	Resolution	Flying altitude	Scenario	Time	Temperature
1	640 × 512	50 m	stadium	2016-12-14	12°C
2	640 × 512	70 m	road	2017-3-18	16°C
3	640 × 512	20 m	road	2016-11-17	18°C

The testing results are shown in Table 2, the proposed detection approach acquired precision above 90% in all datasets, and the recall rate is above 89%. The used parameters are presented in Table 3, here, region_dist is a distance parameter used in region growing method.

Table 2. Detection results

Datasets	Crowd	TP	FP	FN	Precision	Recall
1	1200	1145	48	55	95.98	95.42
2	1600	1474	107	126	93.23	92.13
3	1400	1247	131	153	90.49	89.07

Table 3. Parameters used in the experiments

Datasets	T_1	T_2	T_3	Region_dist	λ_1	λ_2
1	7	3	32	20		
2	6	3	32	20	1	2.5
3	6	4	34	20	1	2.5

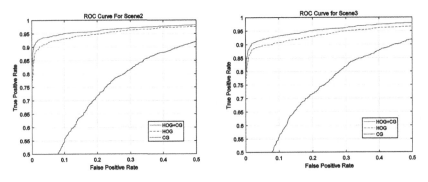

Fig. 4. ROC curves for two scenes

Figure 4 shows the ROC curves for scene 2 and scene 3, the performance of the descriptor (HOG + CG) is better than HOG or CG. Figure 5 shows the crowd detection examples of part images corresponding to three datasets. As a result, the proposed hybrid descriptor achieve a encouraging performance.

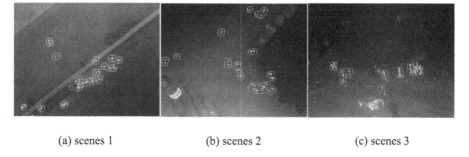

    (a) scenes 1   (b) scenes 2   (c) scenes 3

Fig. 5. Crowd detection sample images from three scenes.

5 Conclusion

In this paper, an new approach to detect crowd from aerial infrared images is proposed. The crowd detection system consists of ROI extraction and SVM classification. The ROI extraction is based on circle gradient and geometric filtering. For complex background, the linear SVM classification is applied to divide the extracted ROI into crowd ROI and non-crowd ROI. A hybrid descriptor which combines HOG and CG feature, is proposed. The experimental results show that the proposed approach performs dense crowd detection from thermal images effectively.

Acknowledgement. This work was supported by the Shenzhen Basic Science & Technology Foundation of China (Grant No. JCYJ20150422150029095) and the Suzhou Industrial Technology Innovation Foundation of China (Grant No. SS201616).

References

1. Xu, F., Liu, X., Fujimura, K.: Pedestrian detection and tracking with night vision. IEEE Trans. Intell. Transp. Syst. **6**(1), 63–71 (2005)
2. Wang, J.T., Chen, D.B., Chen, H.Y., Yang, J.Y.: On pedestrian detection and tracking in infrared videos. Pattern Recognit. Lett. **33**(2), 775–785 (2012)
3. Sun, H., Wang, C., Wang, B., Naser, E.S.: Pyramic binary pattern features for real-time pedestrian detection from infrared videos. Neuro Comput. **74**(11), 797–804 (2011)
4. Ciotec, A.D., Neagoe, V.E., Barar, A.P.: Concurrent self-organizing maps for pedestrian in thermal imagery. Sci. Bull. C **75**(4), 45–56 (2013)
5. Sun, H., Wang, C., Wang, B., EI-Sheimy, N.: Pyramid binary pattern features for real-time pedestrian detection from infrared videos. Neuro Comput. **74**(11), 797–804 (2011)

6. Zhao, X.Y., He, Z.X., Zhang, S.Y., Liang, D.: Robust pedestrian detection in thermal infra-red imagery using a shape distribution histogram feature and modified sparse representation classification. Pattern Recognit. **48**(6), 1947–1960 (2015)

7. Portmann, J., Lynen, S., Chli, M., Siegwart, R.: People detection and tracking from aerial thermal views. In: Proceedings of the IEEE International Conference on Robotics and Automation (ICRA), vol. 93, pp. 1794–1800 (2014)

8. Jeon, E.S., Choi, J.S., Lee, J.H., Shin, K.Y., Kim, Y.G., Le, T.T., Park, K.R.: Human detection based on the generation of a background image by using a far-infrared light camera. Sensors **15**(3), 6763–6788 (2015)

9. Fernandez-Caballero, A., Lopez, M.T., Serrano-Cuerda, J.: Thermal-infrared pedestrian ROI extraction through thermal and motion information fusion. Sensors **14**(4), 6666–6676 (2014)

10. Ge, J., Luo, Y., Tei, G.: Real-time pedestrian detection and tracking at nighttime for driver assistance systems. IEEE Trans. Intel. Transp. Syst. **10**(2), 283–298 (2009)

11. Ma, Y., Wu, X.: Pedestrian detection and tracking from low-resolution unmanned aerial vehicle. Sensors **16**(4), 446–471 (2016)

Research on Networked Collaborative Operations Based on Multi-agent System

Ying Shao[1]([⊠]), Jing Yang[2], and GuangHao Chen[1]

[1] Equipment Academy, Beijing, China
Shao6ll6@163.com
[2] Tawan Street No.77, Shenyang 110035, China

Abstract. The multi-agent simulation method is used to solve the problem of networked collaborative operations. Different from the previous use of communication to achieve collaborative operations, this paper uses shared mental models to achieve network collaborative operations. The influence of shared mental model on cooperative operations is studied by multi-agent model. The results confirm that shared mental model can effectively improve the networked operations effectiveness. The research of this paper provides a new way to realize the networked collaborative operations.

Keywords: Multi-agent system · Shared mental models · Networked operations

1 Introduction

Networked operations command decision-making ability depends on the quality of the information it obtains. The networked operations system has the characteristics of strong communication ability, short combat command chain, fast combat rhythm and high intelligence unit, which make the system has a strong complexity. The battlefield network enhances the information connectivity of the operations system, so that the system unit can join the network equally, every unit has the potential to obtain the battlefield global information. With the support of powerful information capabilities, the power to the edge concept is used in networked operations, makes the unit has a stronger autonomy. The unit in networked operations adopts flexible command and control method, with sharing battlefield information and situational awareness, that can achieve the coordinated operations and enhance the combat effectiveness of the troops. Networked operations require an efficient and robust way to achieve collaborative operations. The current networked collaborative operations mainly through the communication between the units to achieve. The method has following problems: First, communication takes up network resources and affects the efficiency of information transmission; followed by communication occupied the command of the energy and attention, affecting the command effectiveness; third, through the communication between the unit to achieve consistent consumption of a long time, reducing the operations efficiency. We propose a method make the networked operations system units realize the cooperative operations by sharing mental models and form a consistent combat plan. This method provides a new way to further improve the networked operations effectiveness.

© Springer International Publishing AG 2018
F. Qiao et al. (eds.), *Recent Developments in Mechatronics and Intelligent Robotics*,
Advances in Intelligent Systems and Computing 690, DOI 10.1007/978-3-319-65978-7_78

2 Literature Survey

Rouse and Morris define the mental model as a psychological mechanism used to describe system objectives and forms, to interpret system functions, to observe system status, and to predict future state of the system [1]. Johnson-Laird argues that the mental model allows people to speculate and understand current real events and to act accordingly [2]. Endsley argues that the mental model can be used to guide the formation of situational awareness and decision making through the mental model to complete the process of information acquisition to action [3]. The theory of the current mental model includes Simon's proposed Physical Symbol System Hypothesis (PSSH) [4], the Norman model [5], the SOAR model [6] and so on.

In team tasks, members have their own specific mental model, which can be based on the same environment and the facts to make a different conclusion. Cannon-Bowers and Salas extend the mental model from the individual to the team level, which presents the concept of shared mental models, defined as a common knowledge structure by team members, which enables team members to have a team actions and tasks to properly understand and anticipate the ability [7]. The shared mental model among teams has the ability to understand other team roles, plans, information requires, and potential re-planning, as well as predictions and actions for other teams. Shared mental model allows team members to perform tasks with the same reference framework to enhance team communication so that actions by other team members can be predicted. The shared mental model predicts the actions and information requires of other team members when the communication means are limited. Through the sharing of mental models, members can coordinate their own behavior in order to adapt to the team collective combats to meet the needs of other members of the team.

3 Proposed Method

Networked operations have the main characteristics of complex adaptive system; multi-agent modeling and simulation method is to study the effective way of this type of system. In recent years, the method has been widely used in the military field, such as the US Navy Analysis Center developed EINSTein [8], New Zealand Ministry of Defense developed MANA [9], Australian Defense Institute developed cROCADI-LEHl [10] and so on. Compared with the traditional modeling and simulation methods, multi-agent system is more suitable for the simulation of networked operations. Agent has the ability of perception and decision making, which can accurately characterize the networked combat center model. An agent with communication ability can effectively describe the process of networked collaborative operations. Multi-agent system simulation method is used to study the cooperative in networked operations. The effectiveness of networked operations system uses shared mental model method is studied by multi-agent systems. We have established a multi-agent simulation system based on shared mental model network collaborative operations model, which is based on the fact that the existing simulation system only considers the communication between agents and does not consider the shared mental model. Through this simulation model, we can describe how to form a networked collaborative operations process by sharing the mental model.

3.1 Networked Combats Unit Agent Model

The mental model in networked operations shows the role of the commander's cognitive ability in the combat command process. The OODA models proposed by the famous military scientist John Boyd describe the process of military command, which divides the operational command process into four aspects: observation, orient, decide and act. Orient and decision in OODA models occurred in the field of commander cognition, networked operations system unit mental model can describe the two links. According to the physical symbol system hypothesis, the mental model can be described by formalization. In this paper, the networked operations system unit is set as follows, the orient link is described by the situation awareness model, and the decide link is described by the logical decision-making model.

In this paper, Endsley's three-level situation awareness model is used to describe the mental model orient session in networked operations. According to the definition of situation awareness, the orient process is divided into three levels: perception of elements, comprehension of current situation and projection of future status. After the judge, the input battlefield information into a situation awareness product is used in decision-making. In the decide session, the commander refers to the knowledge base and the rule base in the mental model, and establishes the combat plan with the inputted situation awareness product. The networked combat unit agent contains modules such as situation awareness, communication, execution, knowledge base, decision making, and so on. The shared mental model module contains the mental models of other combat units. The shared mental model module contains other combat unit mental models that, when entering a consistent battlefield situation, it can predict the combat plan developed by other units and output them to the decision module. When the decision module has mastered the operational plans of other units, its own operational plans will have a stronger synergy. A networked operations system unit agent with shared mental model module can be show as Fig. 1 below.

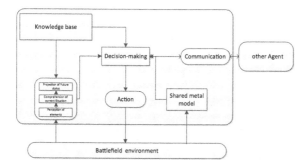

Fig. 1. Networked operations system unit agent with shared mental model function

Networked operations system unit with shared mental model function command process can be showed as Fig. 2:

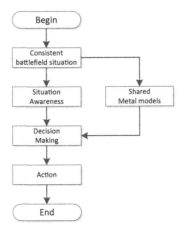

Fig. 2. Shared mental model unit command process

4   Experimental Results

Units A and B involved in networked operations system use the maximum utility function to make decisions. Networked Operations system units have consistent operational situation information. The networked operations system unit with shared mental model can obtain the awareness and combat plan of other units, which reduces the uncertainty of command decision condition. We used the decision table to simulate the process of unit A and B to develop a combat plan, and evaluated the impact of the shared mental model on the combat plan. The command decision conditions show as Tables 1 and 2.

Table 1. Unit A decision table

Income		Units B		
		Action 1	Action 2	Action 3
Unit A	Action 1	7	9	15
	Action 2	2	6	20
	Action 3	3	13	9

The impact of shared mental model on operations effectiveness can be learned by comparing the benefits of a combat plan with a unit that uses a shared mental model or not. When the networked operations system does not use the shared mental model, the

Table 2. Unit B decision table

Income		Units A		
		Action 1	Action 2	Action 3
Unit B	Action 1	7	9	10
	Action 2	12	3	12
	Action 3	6	8	15

unit believes that the possibility of other units performing the combat plan is equal. When the networked operations system uses a shared mental model, the unit will be informed of other units of the combat plan. When acquiring other operations unit plans, the unit will develop a more optimized operational plan, so that the formation of coordination between the units, to obtain higher returns. Shown as Table 3, when the unit can accurately grasp the other unit operations plan, it can develop the maximum benefit of the collaborative operational plan.

Table 3. Operations Unit B decision table

Whether to share the mental model	Choose action		Action		A income	B income	Total income
	A	B	A	B			
No	1	3	1	1	7	6	13
			1	2	9	6	15
			1	3	15	6	21
			2	1	7	8	15
			2	2	9	8	17
			2	3	15	8	23
			3	1	7	15	22
			3	2	9	15	24
			3	3	15	15	30
Yes	1	1	1	1	7	12	19
	1	2	1	2	13	12	25
	1	3	1	3	20	12	32
	2	1	2	1	7	9	16
	2	2	2	2	13	9	22
	2	3	2	3	20	9	29
	3	1	3	1	7	15	22
	3	2	3	2	13	15	25
	3	3	3	3	20	15	35

5 Conclusion

In summary, when the networked operations system uses shared mental model, units can obtain more accurate operations information and formulate more optimal decision. This paper emulates the networked operations unit using the shared mental model to develop the planning process, which proves that the method can effectively improve networked operations effectiveness. The research of this paper opens up a new path for improving networked operations effectiveness.

References

1. Rouse, W.B., Morris, N.M.: On looking into the black box: prospects and limits in the search for mental models. Psychol. Bull. **100**(3), 82 (1985)
2. Johnson-Laird, P.N.: Mental Models. Cambridge University Press, Cambridge (1983)
3. Endsley, M.R.: Toward a theory of situation awareness in dynamic systems. Hum. Fact. **37**(1), 32–64 (1995)
4. Newell, A., Simon, H.A.: Computer science as empirical inquiry: symbols and search. Commun. ACM **19**(3), 113–126 (1976)
5. Norman, D.A.: Some observations on mental models. In: Gentner, D., Stevens, A.L. (eds.) Mental Models, pp. 7–14. Psychology Press, New York (1983)
6. Laird, J.E., Newell, A., Rosenbloom, P.S.: SOAR: an architecture for general intelligence. Artif. Intell. **33**(1), 1–64 (1987)
7. Cannon Bowers, J.A., Salas, E., Converse, S.: Shared mental models in expert team decision making. In: Castellan, N.J. (ed.) Individual and Group Decision Making: Current Issues, pp. 221–246. Lawrence Erlbaum, Hillsdale (1993)
8. Ilachinski, A.: EINSTein: a multiagent-based model of combat. In: Artificial Life Models in Software, pp. 143–185 (2005).
9. Michael, K.L., Stephen, R.T.: Fractals and combat modeling: using mana to explore the role of entropy in complexity science. Fractals **10**(04), 481–489 (2012)
10. Barlow, M., Easton, A.: CROCADILE—an open, extensible agent-based distillation engine. Inform. Secur. **8**, 17–51 (2002)

A New Fast Grouping Decision and Group Consensus of the Multi-UAV Formation

Li Cong[1(✉)], Wang Yong[1], Zhou Huan[1], Wang Xiaofei[1],
and Xuan Yongbo[2]

[1] Air Force Engineering University, Xi'an 710038, China
lealicong@163.com
[2] Beijing Aeronautical Technology Research Center, Beijing 100076, China

Abstract. To solve the problem of formation grouping and group consensus, a Fuzzy C-Means clustering algorithm for grouping decision and a new kind of group consensus are proposed. Firstly, the multi-UAV formation dynamics and grouping process are described by using the graph theory, then, the grouping decision of the multiple unmanned combat aerial vehicles (multi-UAV) formation is made based on the Fuzzy C-Means clustering algorithm considering key elements in the warfare. Meanwhile, a group-consensus with information communication between different sub-formations is put forward. The judgement criterion for the consensus stability is proved by using the Lyapunov theorem. Simulation results demonstrate the superiority and effectiveness of clustering algorithm and group consensus.

Keywords: Multiple unmanned aerial vehicles (multi-UAV) · Grouping decision · Group consensus · Fuzzy C-Means clustering

1 Introduction

The operational requirement of detecting, tracking and attacking cooperatively calls for multiple UCAVs combating in a formation [1]. Compared with one single UCAV, the capability of penetrating, electronic countermeasure, reconnaissance and recognition is highly improved. Now the study of multiple UCAVs formation combat mainly focuses on formation control theory [2], route planning [3] and formation reconfiguration [4], while few published papers discuss the problem of formation split and group consensus of sub-formation. With the development of technology, the scale of formation will continuously extend, thus the cooperative control of formation as a whole is too difficult to achieve. To settle this problem, a large-scale formation can be separated into several sub-formations on the basis of hierarchy theory, after which the cooperative control problem of each sub-formation is investigated respectively. Then the consensus control of a large-scale formation can be achieved.

During the combating process, a large-scale formation should split into several small-scale sub-formations according to operational requirements, after which sub-formations will execute their corresponding task respectively. The above process is called multiple UCAVs formation split. As for the decision making of formation split, the method of community division is widely used in most articles. However, the

© Springer International Publishing AG 2018
F. Qiao et al. (eds.), *Recent Developments in Mechatronics and Intelligent Robotics*,
Advances in Intelligent Systems and Computing 690, DOI 10.1007/978-3-319-65978-7_79

distance between UCAVs is treated as single factor to get results of grouping, which may not match the complicated operational environment. Hence, the key elements concerning formation combat are analyzed to construct UCAV state vector, which serves to the decision making of formation split based on Fuzzy C-Mean clustering algorithm. Multi-agent consensus theory provides a theoretical basis for the study of group-consensus protocol. Many researches have been conducted consensus problems based on the foundation works of Olfati-Saber, Murray, Ren and Moreau. As for the problem of group consensus and group synchronization, [5] discussed the relationship between information transformation and convergence speed of sub-system; [7] proved that the number of sub-system is equal to the number of zero eigenvalue of Laplacian matrix. In the article [8], the necessary and sufficient condition of group consensus is presented based on specific premises. However, the above literatures are all dependent on some conservative assumptions, such as in-degree balance and the sum of adjacent weights from every agent in the same group to all agents in other groups are identical at every time. These special assumptions neglect the interaction between sub-systems, which is not consistent with operational requirements of information sharing among different sub-formations during the process of cooperative combat.

This article is organized as follows. In Sect. 2, we introduce some definitions which will be used for the rest of papers. The decision making of formation split is covered in Sect. 3. A new kind of group-consensus protocol and corresponding criterion are proposed and verified in Sect. 4. Some simulation results are given in Sect. 5. Section 6 concludes the paper.

2   Problem Formulation and Some Definitions

Definition 1. We refer to $F = (G,X)$, $F = (F_1, F_2, \ldots, F_n)$ as a multiple UCAVs formation graph with state x and topology graph G. The sub-formation graph is defined as $F_1 = (G_1, X_1), F_2 = (G_2, X_2), \cdots\cdots, F_n = (G_n, X_n)$.

Definition 2. We say that $G_1 = \{V_1, E_1, A_1\}$ is the sub-formation of $G = \{V, E, A\}$ if and only if (i) $E_1 \in E$ (ii) $V_1 \in V$ (iii) adjacency matrix A_1 inherits A.

3   Decision Making of Formation Split

UCAV state vector which includes key elements concerning combat is defined as

$$X_i = [x_i, y_i, z_i, \alpha_i, v_i] \quad i = 1, 2, \ldots N \tag{1}$$

where (x_i, y_i, z_i), α_i and v_i stands for position, heading angle and velocity of i-th UCAV respectively.

Based on this, the formation center of k-th sub-formation can be defined as

$$C_k = [x_k^0, y_k^0, z_k^0, a_k^0, n_k^0] \; k = 1, 2, \cdots, K \tag{2}$$

In the light of elements of UCAV state vector, the object function of formation split can be represented as (3)–(5).

$$J(X, U, C) = \sum_{i=1}^{N} \sum_{k=1}^{K} \mu_{ik} d_{ik}^2 \tag{3}$$

$$\begin{cases} \mu_{ik} \in [0, 1], & \forall i, k \\ \sum_{k=1}^{K} \mu_{ik} = 1, & \forall i \end{cases} \tag{4}$$

$$d_{ik}^2 = k_d \cdot \sqrt{(x_i - x_k^0)^2 + (y_i - y_k^0)^2 + (z_i - z_k^0)^2} + k_\alpha \cdot \sqrt{(\alpha_i - \alpha_k^0)^2} + k_v \cdot \sqrt{(v_i - v_k^0)^2} \tag{5}$$

where $U = (\mu_{ik})_{n \times K}$ is a membership matrix, and μ_{ik} stands for the degree of membership that i-th UCAV belongs to k-th sub-formation center, the constraint condition of μ_{ik} is expressed as (4). d_{ik}^2 is the square of distance between i-th UCAV state vector and k-th sub-formation center state vector.

Fuzzy C-Mean clustering algorithm is the iteration process of minimizing the object function $J(X, U, C)$ by seeking for suitable membership matrix U and formation center matrix C. The process of seeking for optimal object function is depicted as Fig. 1.

Fig. 1. The flow chart of Fuzzy C-Mean clustering algorithm

4 Group Consensus of Sub-formation

Faced with multiple targets, the entire formation should split into several sub-formations to undertake corresponding combat mission according to task distribution. UCAVs from each sub-formation should reach the same consistent value asymptotically so that they can carry out collaborative operation as a whole. Since the task of each sub-formation is different, there is no need for different sub-formations to reach a same consensus value. Different task distribution among multiple UCAVs results in different consensus value of UCAVs in a formation.

4.1   UCAV Model

In this section, we take a couple-group consensus problem of a formation consisting of $N+M$ UCAVs as the research project. Suppose that sub-formation F_1 consisting N UCAVs and sub-formation F_2 consisting M UCAVs. Consider that each UCAV has the first-order dynamics as follows.

$$\dot{x}(t) = \mu(t), \quad i = 1, 2, \ldots, N+M \tag{6}$$

where $x(t) = \{x_1(t), x_2(t), \ldots, x_{N+M}(t)\}^T$, let $x_i(t) \in R$ be the state of i-th UCAV for $i = 1, 2, \ldots, N+M$ at time t. Besides, $\mu(t) = \{\mu_1(t), \mu_2(t), \ldots, \mu_{N+M}(t)\}^T$ refers to the input of i-th UCAV at time t.

Take first-order model (6) as the research project, we design a new kind of group-consensus protocol to reach two consistent states asymptotically.

Although many articles have studied the consensus problem of multiple UCAVs formation, there are few people taking the group consensus of several sub-formations as the research topic. We refer group consensus as a situation that several UCAVs in a sub-formation reach a consistent value while there is no consensus among different sub-formations.

4.2   Design of Group Consensus Protocol

In this section, we consider a group consensus problem for the first-order dynamics (6) when it is under a certain consensus control protocol. The existing articles are mainly discussing the group-consensus control protocol given by Eq. (9) under constraint condition (7) and (8).

$$\begin{cases} \sum_{j=N+1}^{N+M} a_{ij}=0, \ i \in l_1 \\ \sum_{j=1}^{N} a_{ij}=0, \ i \in l_2 \end{cases} \tag{7}$$

$$\begin{cases} \sum_{j=N+1}^{N+M} a_{ij}=\alpha, \ i \in l_1 \\ \sum_{j=1}^{N} a_{ij}=\beta, \ i \in l_2 \end{cases} \tag{8}$$

$$\mu_i(t) = \begin{cases} \sum_{j\in l_1} a_{ij}(x_j(t) - x_i(t)) + \sum_{j\in l_2} a_{ij}x_j(t), & \forall i \in l_1 \\ \sum_{j\in l_1} a_{ij}x_j(t) + \sum_{j\in l_2} a_{ij}(x_j(t) - x_i(t)), & \forall i \in l_2 \end{cases} \tag{9}$$

where α, β are constant quantities, $l_1 = \{1, 2, \cdots, N\}$ and $l_2 = \{N+1, N+2, \cdots, N+M\}$ are number lists of UCAVs in corresponding sub-formations.

When UCAVs formation combat cooperatively in a dynamic and changeable environment, it is necessary for each sub-formation to be aware of the present situation of other sub-formations so that they can provide each other with cover and support. However, the above two assumptions cannot meet these operational requirements. To solve this problem, in accordance with practical combating problems, a new kind of group-consensus protocol under fixed topological structure is introduced. Group consensus of multiple UCAVs in a sub-formation under general condition can be achieved by the given control protocol Eq. (10).

$$\mu_i(t) = \sum_{j=1;j\neq i}^{N} a_{ij}(x_j(t) - x_i(t)) + \sum_{j=1}^{N} l_{ij}x_{\sigma_j} \tag{10}$$

where a_{ij} is the (i,j) element of adjacency matrix, and $a_{ii}=0$. l_{ij} is the (i,j) element of Laplacian matrix. It is important to note that the roles in a sub-formation can be divided into intra-UCAV and inter-UCAV. The state of intra-UCAV can only be affected by UCAVs from the same sub-formation, while as for inter-UCAV, information change exists not only between UCAVs from the same sub-formation, but also in different sub-formations.

What follows in this article provides the criterion and corresponding proof for the group consensus of multiple UCAVs formation.

Lemma 1. If L has only one simple zero eigenvalue and the rest of eigenvalues of L have positive real parts, then multiple UCAVs formation system can reach group consensus asymptotically.

Proof. (6) can be written as the following linear system in combination with protocol (10)

$$\dot{x}_i(t) = \sum_{j=1;j\neq i}^{N} a_{ij}(x_j(t) - x_i(t)) + \sum_{j=1}^{N} l_{ij}x_{\sigma i} \tag{11}$$

Define the error variables

$$e_i(t) = x_i(t) - x_{\sigma_i}, \quad i = 1, 2, \cdots, N \tag{12}$$

The achievement of group consensus is equal to the establishment of the following conditions

$$\lim_{t\to\infty} |e_i(t)| = 0, \, i = 1, 2, \cdots, N; \, 1 \le \sigma_i \le K \tag{13}$$

which means that UCAVs in each sub-formation can reach a consistent value asymptotically, while there is no consistent value among different sub-formations.

$$\dot{x}_i(t) = \sum_{j=1;j\neq i}^{N} a_{ij}(x_j(t)-x_i(t)) + \sum_{j=1}^{N} l_{ij}x_{\sigma i} = -\sum_{j=1}^{N} l_{ij}(x_j(t)-x_{\sigma i}), \quad i= 1, 2, \cdots, N \tag{14}$$

Let $e(t) = (e_1(t), e_2(t), \ldots, e_N(t))^T$, then (11) can be written as

$$\dot{e}(t) = -Le(t) \tag{15}$$

Define Lyapunov function as

$$V(t) = \frac{1}{2} e^T(t)e(t)a \tag{16}$$

The result of the derivation of (16) is

$$\dot{V}(t) = e^T(t)\dot{e}(t) \tag{17}$$

Plug (17) in the former formula, we can get

$$\dot{V}(t) = e^T(t)[-L]e(t) \tag{18}$$

We say that the error system can achieve stabilization if $V(t) \geq 0$ and $\dot{V}(t) \leq 0$. It is easy to conclude that $V(t) \geq 0$ from (17), the following concentrates on providing the proof of $\dot{V}(t) \leq 0$. When eigenvalues are with non-negative real parts, $L \geq 0$. So that we can get $\dot{V}(t) \leq 0$ from (18), and $\dot{V}(t) = 0$ is achieved if and only if $e(t) = 0$. With the help of Lyapunov global stability theories, we can learn that the error system can achieve stability, $\|e_i(t)\| \to 0$, and it can be said that $x_i(t) - x_{\sigma i} = 0$ is one equilibrium point of the formation, all UCAVs will converge to this point. The above provides proof that formation can achieve group consensus asymptotically by protocol (10).

Compared with results of other papers, Lemma 1 is independent of some special assumptions so that it can serve as a general criterion of group consensus.

5 Simulation Results

In this section, we demonstrate the effectiveness of the group consensus protocol (10) introduced in this article by considering a communication topology graph given in Fig. 2. Consider a formation of five UCAVs, each with dynamic Eq. (6). The UCAVs state vector is shown in Table 1.

5.1 Split of Multiple UCAVs Formation

Assume that the whole formation encounter two unknown areas to be searched, then the whole formation will split into two sub-formations to perform their own reconnaissance task. Combined with values of UCAVs state vector as shown in Table 1, with the help of Fuzzy C-Mean clustering algorithm, the result of formation split is shown as Table 2 and simulation diagram is shown as Fig. 3.

Table 1. Initial values of UCAV state vector

NUMBER STATE	1	2	3	4	5
X(km)	399	409	390	400	410
Y(km)	600	610	580	420	530
Z(km)	3.1	2.9	3.1	3	2.8
a($\circ$)	60	45	30	59	44
V(m/s)	500	400	600	499	601

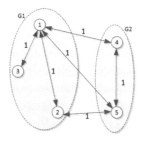

Fig. 2. Communication topology graph

It is easy to learn from Fig. 3 that the formation is divided into two sub-formations from Table 2, we can conclude that 1, 3, 4 UCAV belongs to sub-formation F_1 and 2, 5 UCAV belongs to sub-formation F_2.

Table 2. The result of formation split

Result of formation split	Formation center C(x^0, y^0, z^0, a^0, v^0)	Degree of membership U				
		1	2	3	4	5
Sub-formation F1	(396.3, 533.3, 3.609, 33.5, 425.25)	0.99	0.05	0.97	0.98	0.01
Sub-formation F2	(409.3, 570, 2.853, 52, 449.5)	0.01	0.95	0.03	0.02	0.99

5.2   Group Consensus of UCAV State Vector

The weight of coupling between UCAVs is represented in Fig. 3. Treat the result of formation split that we get in Sect. 4.1 as the research project, and we will verify the effectiveness of protocol (10) based on matrix theory and graph theory.

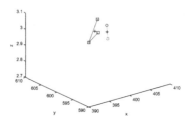

Fig. 3. Multiple UCAVs formation split

From Fig. 3, it is easy to obtain the adjacency matrix, in-degree matrix and corresponding Laplacian matrix as follows.

$$A = \begin{bmatrix} 0 & 1 & 1 & 1 & 1 \\ 1 & 0 & 0 & 0 & 1 \\ 1 & 0 & 0 & 0 & 0 \\ 1 & 0 & 0 & 0 & 1 \\ 1 & 1 & 0 & 1 & 0 \end{bmatrix} \quad D = \begin{bmatrix} 4 & 0 & 0 & 0 & 0 \\ 0 & 2 & 0 & 0 & 0 \\ 0 & 0 & 1 & 0 & 0 \\ 0 & 0 & 0 & 2 & 0 \\ 0 & 0 & 0 & 0 & 3 \end{bmatrix}$$

$$L = D - A = \begin{bmatrix} 4 & -1 & -1 & -1 & -1 \\ -1 & 2 & 0 & 0 & -1 \\ -1 & 0 & 1 & 0 & 0 \\ -1 & 0 & 0 & 2 & -1 \\ -1 & -1 & 0 & -1 & 3 \end{bmatrix}$$

Laplacian matrix has eigenvalues: $\lambda_1 = 0, \lambda_2 = 1, \lambda_3 = 2, \lambda_4 = 4, \lambda_5 = 5$. Because there is only one simple zero eigenvalue and the rest of eigenvalues of L have positive real parts, then we can say that multiple UCAVs formation system can reach group consensus steadily according to Lemma 1. The following simulation will verify the result derived from numerical calculation on the MATLAB platform.

Take the abscissa value x as example. From Table 1, we can learn that initial states of five UCAVs is $x^T(0)= [399, 409, 390, 400, 410]^T$, then the average-consensus value of two sub-formation can be calculated as $x_{\sigma_i} = \begin{cases} 396.3, & i = 1,3,4 \\ 409.3, & i = 2,5 \end{cases}$.

The simulation diagram based on consensus protocol (10) is shown as Fig. 4. From Fig. 4(a) we can learn that the abscissa value x of UCAVs in each sub-formation can reach a same consistent value asymptotically under consensus protocol (10). Figure 4 (b) shows the time evolution of error between the values of x belongs to each UCAV and average-consensus value in corresponding sub-formations. As time goes by, the final error is zero, so that we can say that multiple UCAVs formation achieve average-consensus.

Take $x(t)$ in the protocol (10) stands for y, z, a, v respectively, then we can verify that whether the consensus protocol (10) we introduced in this paper can solve the consensus problem of other elements in the UCAV state vector. Based on the data from Tables 1 and 2, we can get simulation results as shown in Fig. 5(a)–(d).

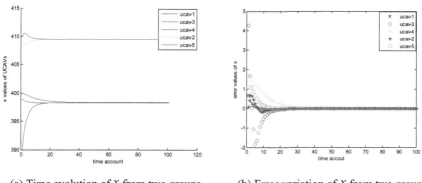

(a) Time evolution of x from two groups   (b) Error variation of x from two groups

Fig. 4. Time evolution of x and error variation of x from two groups

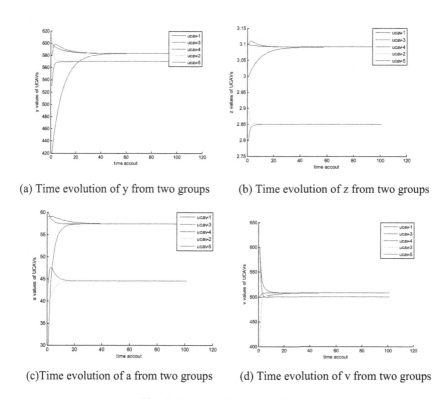

(a) Time evolution of y from two groups   (b) Time evolution of z from two groups

(c)Time evolution of a from two groups   (d) Time evolution of v from two groups

Fig. 5. Time evolution of y, z, a, v

5.3 Verification of Group-Consensus Criterion Through Contrast Experiments

In this section we consider another topology that the weight value between UCAV 1 and UCAV 3 is −1, and then we can obtain the adjacency matrix, in-degree matrix and corresponding Laplacian matrix as follows.

$$A = \begin{bmatrix} 0 & 1 & -1 & 1 & 1 \\ 1 & 0 & 0 & 0 & 1 \\ -1 & 0 & 0 & 0 & 0 \\ 1 & 0 & 0 & 0 & 1 \\ 1 & 1 & 0 & 1 & 0 \end{bmatrix} \quad D = \begin{bmatrix} 4 & 0 & 0 & 0 & 0 \\ 0 & 2 & 0 & 0 & 0 \\ 0 & 0 & 1 & 0 & 0 \\ 0 & 0 & 0 & 2 & 0 \\ 0 & 0 & 0 & 0 & 3 \end{bmatrix}$$

$$L = D - A = \begin{bmatrix} 4 & -1 & 1 & -1 & -1 \\ -1 & 2 & 0 & 0 & -1 \\ 1 & 0 & -1 & 0 & 0 \\ -1 & 0 & 0 & 2 & -1 \\ -1 & -1 & 0 & -1 & 3 \end{bmatrix}$$

The eigenvalues of Laplacian matrix are $\lambda_1 = -1.2548, \lambda_2 = 0.3232, \lambda_3 = 2, \lambda_4 = 4,$ $\lambda_5 = 4.9316$. Because one eigenvalue of L has negative parts, then the formation cannot achieve group consensus according to Lemma 1.

Take the abscissa value x of UCAV state vector as example, the simulation result under protocol (10) is shown in Fig. 6. It is obvious that state trajectory of x is diverging, the value of x cannot achieve group consensus. Compared with the simulation result presented is Fig. 4, the validity of Lemma 1 is further verified.

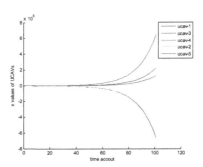

Fig. 6. Time evolution of x from two groups when a13 = a31 = −1

6 Conclusion

The decision making of formation split is investigated by Fuzzy C-Mean clustering algorithm on the basis of UCAV state vector. To deal with the constraint of existing group-consensus protocol presented in other articles, a new kind of group-consensus protocol which concerns the interaction between different sub-formations is introduced and verified. From the simulation result we can learn that the protocol we investigated in this article is effective for sub-formations to achieve group consensus with no need to satisfy existing constraint conditions proposed in existing articles, that is to say, it can be applied in some practical problems. Furthermore, the validity of criterion for group consensus is testified by contrast simulation experiments.

References

1. Masaki, Y., Masahiro, S.: Formation control of SMC with multiple coordinate systems. In: Proceeding of the International Conference on Intelligent Robots and Systems, Sendai, Japan, pp. 1023–1028 (2004)
2. Saboori, K.K.: H∞ consensus achievement of multi-agent systems with directed and switching topology networks. IEEE Trans. Autom. Control **59**(11), 3104–3109 (2014)
3. Schumacher, C.J., Kumar, R.: Adaptive control of UAV in close-coupled formation flight. In: Proceedings of IEEE American Control Conference, Chicago, Illinois, pp. 849–853 (2000)
4. Mellinger, D., Michael, N., Kumar, V.: Trajectory generation and control for precise aggressive maneuvers with quadrotors. Int. J. Robot. Res. **31**(5), 664–674 (2012)
5. Kan, D.A., Shea, J., Dixon, W.E.: Information flow based connectivity maintenance of a multi-agent system during formation control. In: IEEE Conference on Decision and Control, pp. 2375–2380 (2011)
6. Lia, B., Huabga, L.J., Yuana, X.: Analysis of influence of multiple UAVs coordination system. Inf. Comput. Sci. **10**(2), 571–578 (2013)
7. Yu, J., Wang, L.: Group consensus of multi-agent systems with undirected communication graphs. In: Proceedings of the 7th Asian Control Conference, pp. 105–110 (2009)
8. Yi, J., Wang, Y., Xiao, J.: Reaching cluster consensus in multi-agent systems. In: The 2nd International Conference on Intelligent Control and Information Processing, pp. 569–573 (2011)

Research and Implementation of Global Path Planning for Unmanned Surface Vehicle Based on Electronic Chart

Yanlong Wang[1(✉)], Xu Liang[1], Baoan Li[1], and Xuemin Yu[2]

[1] School of Automation Science and Electrical Engineering, Beihang University, Beijing, China
wylloong@163.com
[2] Institute of Information Engineering, Chinese Academy of Sciences, Beijing, China

Abstract. Unmanned Surface Vehicle (USV) is a new type of intelligent surface boat, and global path planning is the key technology of USV research, which can reflect the intelligent level of USV. In order to solve the problem of global path planning of USV, this paper proposes an improved A* algorithm for sailing cost optimization based on electronic chart. This paper uses the S-57 electronic chart to realize the establishment of the octree grid environment model, and proposes an improved A* algorithm based on sailing safety weight, pilot quantity and path curve smoothing to ensure the safety of the route, reduce the planning time, and improve path smoothness. The simulation results show that the environmental model construction method and the improved A* algorithm can generate safe and reasonable global path.

Keywords: USV · Improved A* algorithm · Electronic chart · Global path planning

1 Introduction

Unmanned Surface Vehicle (USV) is a new type of intelligent surface craft that is used in military operations, maritime surveillance cruise, and marine environmental monitoring applications, paper [1] summarizes the history, current situation and development trend of USV. Based on the current situation of domestic and foreign research, combined with "US Department of Defense Unmanned Systems Integrated Roadmap" [2] published by the US military and "The Navy Unmanned Surface Vehicle Master Plan" [3] published by the US Navy, the key technologies involved in the research of USV mainly include automatic route generation and path planning technology, autonomous decision-making and collision avoidance technology, water surface object detection and target automatic identification technology and communication technology.

Automatic route generation and path planning technology mainly studies USV global and local path planning, the method of path planning is mainly grid method, artificial potential field method, genetic algorithm method and ant colony algorithm method. Montes modeled Optimum Track Ship Routing (OTSR) for U.S. Navy using a network graph of the Western Pacific Ocean, a binary heap version of Dijkstra's algorithm determines the optimum route [4]. Zhuang presented a search of shortcut Dijkstra algorithm based on electronic chart, which can generate safety and reasonable routes [5].

© Springer International Publishing AG 2018
F. Qiao et al. (eds.), *Recent Developments in Mechatronics and Intelligent Robotics*,
Advances in Intelligent Systems and Computing 690, DOI 10.1007/978-3-319-65978-7_80

Electronic chart is a digital map of geographical information and maritime information drawn to suit the needs of navigation. The electronic chart in the USV application is in the exploratory stage at present, mainly used in the path planning and collision avoidance.

This paper argues that autonomous global path planning based on the electronic chart, mainly refers to obtaining necessary information such as sea area geographic information and obstacle information from electronic chart documents, rendering the chart information into an environment model that the USV path planning system recognized, and then proposes an algorithm to implement the global path planning of USV automatically.

2 Establishment and Representation of Environmental Model

Radar, camera and other sensor cannot provide the global marine environmental information while USV sails in a wide range of sea, so S-57 electronic chart that can provide detailed and accurate global marine environmental information has become global path planning of USV necessary input, which can guarantee navigation safety.

2.1 Electronic Chart Data Extraction

Electronic chart is mainly composed of marine area elements, which can be expressed in detail as submarine terrain, navigation obstacle, navigation sign, port facility and other elements [6]. The S-57 electronic chart standard package format is ISO /IEC 8211 international standard, so this paper uses ISO8211 open source library to resolve all the electronic chart information according to the package structure of S-57 electronic chart [7]. The S-57 electronic chart package format is shown in Fig. 1 [6].

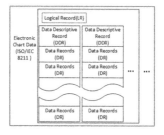

Fig. 1. Package format of S-57 electronic chart

2.2 Establishment of Environmental Model

The establishment of environmental model of USV can be simplified as: USV moves in a limited and arbitrary convex area OS of the sea level, and a limited number of static obstacles $O_i(i = 1, 2, \cdots, n)$ in OS, obstacle shape and distribution is uncertain. The rasterization of the environmental model is done by filling the OS as a rectangle, and treating the padded area as an obstacle area, and the OS is divided into several meshes of equal size by grid method. This paper judges whether there is an obstacle, such as land and islands in the grid according to information extracted in turn, so as to establish navigable areas and non-navigable areas in units of grids.

The original environmental model and the navigable environmental model are shown in Fig. 2 (a) (b), where the shaded portion indicates that it is not navigable.

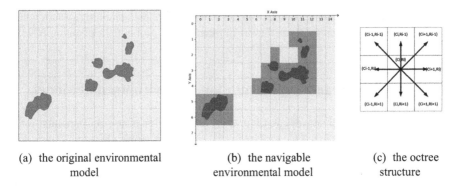

(a) the original environmental
model

(b) the navigable
environmental model

(c) the octree
structure

Fig. 2. Display of the environmental models at different stages

In this paper, the octree structure is used as the extension strategy of route search, so there are eight neighboring nodes around the node i as shown in Fig. 2 (c). USV may deviate from the planned path in the course of avoiding obstacles and temporary mission, and diagonal routes are possible to near non-navigable areas in the octree structure, so this paper considers navigable area near non-navigable area has a potential risk. For the safety of USV, this paper sets sailing safety weight to guide the path planning not to enter potential risk areas. The number of non-navigable grids n in the adjacent eight grids will affect the sailing safety weight of grid (C_i, R_i) as $w(C_i, R_i) = 1 + \frac{1}{2} * 2^{(n-1)}$.

3   The Description and Realization of the Improved A* Algorithm

3.1   Realization of the Improved A* Search Algorithm

In the path planning algorithm, A* algorithm is a classical heuristic search algorithm. The basic idea is to use the preset cost function to calculate the value of each adjacent child node that the current node may reach, and select the minimum cost node to join the search space and expand, and so on until the target point is reached [8].

Under the security constraints, the goal of USV path planning optimization is to minimize sailing cost of the planned route. The sailing cost $D(p_i, p_{(i+1)})$ is the straight-line distance between two points multiplied by sailing safety weight, which is given below.

$$D(p_i, p_{(i+1)}) = \sqrt{(x_{pi} - x_{p(i+1)})^2 + (y_{pi} - y_{p(i+1)})^2} * w(C_{(i+1)}, R_{(i+1)}) \tag{1}$$

There is more than one path of the minimum sailing cost sometimes, and they are all explored although only one is needed, and sometimes the planed route automatically is not the reasonable result because of deviation from the straight-line L_{sg} between the starting and target point. Therefore, this paper introduces a pilot quantity P to guide the planed path close to the straight-line L_{sg}. Set the starting point as (S.x, S.y), the target point as (G.x,G.y), the current node i as (C.x, C.y), we can get angle θi between the starting-target vector and the i-target vector by calculating vector cross product as:

$$\theta i = \arcsin \left(\frac{\sqrt{(C.x - G.x) * (S.y - G.y) - (S.x - G.x) * (C.y - G.y)}}{\sqrt{(C.x - G.x)^2 + (C.y - G.y)^2} * \sqrt{(S.x - G.x)^2 + (S.y - G.y)^2}} \right) \qquad (2)$$

the pilot quantity $p_{(i)}$ of node i can be expressed as

$$p_{(i)} = 3/(4 - \sin(\theta_i)) \qquad (3)$$

In this paper, based on the A* algorithm, the cost function of node i is defined as

$$f_{(i)} = g_{(i)} + h_{(i)} \qquad (4)$$

Where $g_{(i)}$ is the sailing cost function from the starting point to the node i, $h_{(i)}$ is the heuristic function of current node i to the target point [8], this paper takes the distance multiplied by the pilot quantity as heuristic function $h_{(i)}$, which can ensure the route is optimal because $h_{(i)}$ is not greater than the minimum sailing cost from i to target node.

$$h_{(i)} = \sqrt{(x_{pi} - x_{pG})^2 + (y_{pi} - y_{pG})^2} * p(i) \leq \sqrt{(x_{pi} - x_{pG})^2 + (y_{pi} - y_{pG})^2} \leq \min \left(\sum_{j=i}^{G} D_{(j,j+1)} \right) \qquad (5)$$

3.2 Path Curve Smoothing

In the grid environment, if the nodes obtained by the improved A* algorithm are connected in sequence as the planed path of USV, there are sometimes ladder or jagged lines on the route, and it is easy to know that the planned path is not the desired path, so this paper proposes a path curve smoothing method to remove the redundant nodes.

The method of path smoothing is to traverse all the nodes in the planed path, when there is no obstacle on the connection between the node i and $i + 2$, then remove redundant node $i + 1$. Continue above steps until the connection lines between i and j through the obstacles, then take out 3 consecutive nodes as P_{j-1}, P_j and P_{j+1} after node i, continue these steps until you have traversed all the route nodes in the path [9].

4 Simulation Results

In order to illustrate the effectiveness of environment modeling and the improved A* algorithm, the algorithms mentioned above are simulated. Electronic chart analysis base on C++ language, environment modeling and path planning simulates with python.

For a sea area in the South China Sea (regional range of 18.10° N ~ 18.40° N, 109.35° E ~ 109.85° E), given the simulation results of path planning shown in Fig. 3. Figure (a) is the path planning result of the improved A* algorithm. Figure (b) is the path planning result of A* algorithm. Figure (c) is the path planning result of Dijkstra algorithm, where the solid line is the final planned route, the dotted line is the route without path smoothing.

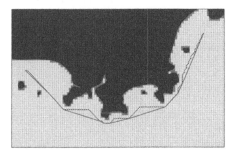

(a) the path planning results of the improved A* algorithm

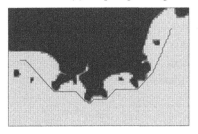   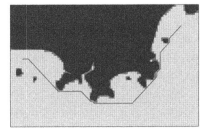

(b) the results of A* algorithm   (c) the results of Dijkstra algorithm

Fig. 3. Simulation results of several path planning algorithms

Table 1 lists the simulation results of path planning algorithm mentioned above, where number of potential hazards is the total number of non-navigable areas is close to the final route within a grid range, and number of turns is the times of turn in the final route.

In the simulation results above, we can get from figure (b) (c) that the number of nodes A* algorithm traversed is less than Dijkstra algorithm because Dijkstra algorithm directly searches the global space without considering the target information [10], so that the route planning efficiency of A* algorithm is much higher than Dijkstra algorithm. It can be seen from figure (a) (b) that the improved A* algorithm can improve the safety of route, reduce redundant nodes, improve path smoothness and shorten sailing distance than A* algorithm without improved, and the navigation node layout is more reasonable.

Table 1. Comparison of simulation results of several path planning algorithms

	Dijkstra algorithm	A* algorithm	Improved A* algorithm
grid accuracy (degrees)	0.005*0.005	0.005*0.005	0.005*0.005
sailing distance (Nautical mile)	35.48	35.52	34.25
number of route nodes	97	97	11
number of nodes traversed	3186	1806	2127
number of potential hazards	32	26	0
route turn times	13	31	6

5   Conclusion

Considering at the characteristics of USV, this paper proposes an improved A* algorithm for global path planning of USV based on S-57 electronic chart, sailing safety weight, pilot quantity function and path curve smoothing method. This algorithm can quickly and accurately generate an optimal collision-free route between any two possible destinations in a given sea area, which can ensure the safety of the planned route.

References

1. Manley, J.E.: Unmanned surface vehicles, 15 years of development. Oceans **2008**, 1–4 (2008)
2. US Department of the Navy. The navy unmanned surface vehicle (USV) master plan (online) http://www.navy.mil/navydata/technology/usvmppr.pdf. Accessed 07 April 2017
3. US Department of Defense.: Unmanned Systems Integrated Roadmap FY2013-2038 (online) http://archive.defense.gov/pubs/DOD-USRM-2013.pdf. Accessed 07 April 2017
4. Montes, A.A.: Network Shortest Path Application for Optimum Track Ship Routing. Master's Thesis Operations Research Department Naval Postgraduate School (2005)
5. Zhuang, J., et al.: Global path planning of unmanned surface vehicle based on electronic chart. Comput. Sci. **38**(9), 211–214 (2011)
6. International Hydrographic Organization, Transfer standard for Digital Hydrogrphic Data (online) http://www.iho.int/iho_pubs/IHO_Download.htm#S-57. Accessed 07 April 2017
7. Bo, H., Feng, L., Yanyong, C.: Research on route planning technology based on electronic charts for long-distance torpedo. Ship Sci. Technol. **37**(7), 116–119 (2015)
8. Duchoň, F., et al.: Path planning with modified a star algorithm for a mobile robot ☆. Proc. Eng. **96**(96), 59–69 (2014)
9. Wang, H., et al.: Mobile robot optimal path planning based on smoothing A~* algorithm. Tongji Daxue Xuebao/Journal of Tongji University **38**(11), 017 (2010)
10. Dijkstra, E.W.: A note on two problems in connexion with graphs. Numer. Math. **1**(1), 269–271 (1959)

Kinect Sensor-Based Motion Control for Humanoid Robot Hands

Bo Hu[(✉)] and Nan-feng Xiao

School of computer science and engineering,
South China university of technology,
University town, Pan yu district, Guangzhou city 510006,
Guangdong province, P. R. China
bohucyjx@foxmail.com

Abstract. This paper presents a Kinect sensor-based control approach for two humanoid robot hands. Firstly, the D-H method is used to establish forward and inverse kinematics models for the humanoid robot hands, establish the 3D models of the humanoid robot hands by 3ds Max, and prove the correctness of the kinematics models by simulations. Secondly, the depth images and the bone joint point information are used to segment the operator's hand gesture from the depth images. Finally, the operator's hand gestures are recognized by DBN, which are converted into a series of instructions to control the humanoid robot hands on time. The experimental results show that recognition accuracy rate is high and can meet the real time requirements.

Keywords: Humanoid robot hands · Hand gesture recognition · Kinematics model · Kinect sensor · Depth image

1 Introduction

The D-H model [1] is a simple method for modeling the robot joints, which can be used to represent any complex robot configuration, commonly used in the humanoid robot kinematics modeling. At present, data glove [2], mouse, keyboard or teaching box are used to control the operation of the humanoid robot hands. In fact, the input control commands by the mouse and the keyboard are unfriendly human-computer interaction, which require the operators to have more special knowledge. Although the data glove can well recognize the operator's hand gestures, the data gloves are very inconvenient and expensive. If the operator's hand gestures can be used to direct control the humanoid robot hands, it will bring much convenience, and it also can significantly reduce the operation difficulty. Therefore, this paper propose a Microsoft Kinect sensor-based approach to control the operations for the humanoid robot hands, no additional equipment is required for the operators, and also does not need to have much professional knowledge to be able to easily control the humanoid robot hands.

The usual methods for the human being's hand gesture segmentation are based on the skin color detection or the Haar features cascade classifier [4]. However, the recognition accuracy is greatly affected by the illumination change or the interference of the objects which are close to the skin color of the hands, especially the human being's hands and faces are often difficult to distinguish, but the hand gesture segmentation

F. Qiao et al. (eds.), *Recent Developments in Mechatronics and Intelligent Robotics*,
Advances in Intelligent Systems and Computing 690, DOI 10.1007/978-3-319-65978-7_81

based on depth images can overcome those problems. The depth image-based hand gesture segmentation usually uses dual threshold or skeletal tracking method, this paper combines the two methods for hand gesture segmentation.

Generally, the gesture recognition methods can be divided into two kinds: static gesture recognition and dynamic gesture recognition, the former mainly studied the hand shapes and postures, the latter mainly studied the hand motion trajectories, and the former usually use RBF, DBN [5], PNN, SVM, HOG features, geometric features, etc., the latter used the methods, for examples, DTW, HMM [6], etc. Zhou Ren [7] propose FEMD (finger-earth mover's distance), it can well distinguish the hand gestures of slight differences, but it only matches the fingers while not the whole hand shape, therefore, it is impossible to judge the direction of the hand. In addition, from the characteristics of the algorithm can be seen, if the camera is not opposite the palm of the hand is likely to lead to the identification of failure or identification errors.

2 Kinematic Model of Humanoid Robot Hand

This paper designed a humanoid robot hand with 15 DOFs (degrees of freedom), a total of 5 fingers, each finger has 4 knuckles and 3 DOFs (each finger's distal joint is coupled with the middle joint motion). In order to facilitate the industrial production and motion control, the 5 fingers have the same structure.

The humanoid robot hand's structure and the coordinate system as shown in Fig. 1, the base joint, the proximal base joint, the middle joint and the distal joint are in the same plane, the base joint is perpendicular to the plane of finger motion (the side swing motion), the rest joints in the finger motion plane (the bending motion). The finger tip in the base joint coordinate system expressed as

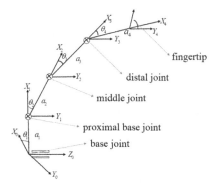

Fig. 1. Finger's structure diagram

$$\begin{bmatrix} x \\ y \\ z \end{bmatrix} = \begin{bmatrix} C_1(a_2C_2 + a_3C_{23} + a_4C_{234}) \\ S_1(a_2C_2 + a_3C_{23} + a_4C_{234}) \\ a_2S_2 + a_3S_{23} + a_4S_{234} \end{bmatrix} \tag{1}$$

where, a_i ($i = 1, 2, 3, 4$) is the length of the i knuckle, θ_i ($i = 1, 2, 3, 4$) is the angle of the i joints, $C_{23} = cos(\theta_2 + \theta_3)$, $S_{23} = sin(\theta_2 + \theta_3)$, $C_{234} = cos(\theta_2 + \theta_3 + \theta_4)$, $S_{234} = sin(\theta_2 + \theta_3 + \theta_4)$.

The finger's kinematic inverse problem is known as the position of the humanoid robot hand in the base coordinate system to solve the various joint variables. The whole solving process is a nonlinear problem, whether or not the solutions depend on whether the hand fingers can reach at the specified position.

According to the physiological structure of human being's hand [2], the distal joint motion of the finger is not independent, when $\theta_4 = 2\theta_3/3$ is more appropriate. θ_i ($i = 1, 2, 4$) is expressed as

$$\begin{bmatrix} \theta_1 \\ \theta_2 \\ \theta_4 \end{bmatrix} = \begin{bmatrix} \tan^{-1}(y/x) \\ \tan^{-1}((-AB \pm \sqrt{A^2 + B^2 - D^2})/(B^2 - D^2)) \\ 2\theta_3/3 \end{bmatrix} \tag{2}$$

where, $A = x/C_1 - a_1$, $B = z$, $D = a_2 + a_3C_3 + a_4C_{34}$. θ_3 is expressed as

$$cos(\theta_3)/a_4 + cos(\theta_3 + \theta_4)/a_3 + cos(\theta_4)/a_2 = \left[(x/C_1 - a_1)^2 + z^2 - a_2^2 - a_3^2 - a_4^2\right]/(2a_2a_3a_4) \tag{3}$$

In Eq. (3), θ_3 can be obtained by the numerical method.

3 Gesture Segmentation and Recognition

3.1 Gesture Segmentation

This paper combines the depth images with the joint point information to segment the gestures, which can effectively remove the background. The median filtering method is used to deal with the noises, and combined with the morphological open and the close operation make the images be more complete and clear.

The watershed algorithm [8] is a morphology segmentation method based on the topological theory, When the rest of the body or the surrounding objects around the robot hand are in the same depth, through the images of the double threshold segmentation, there are not only the gestures and the other parts of the body or the surrounding objects, the watershed algorithm can be used get the gestures. If the calculation results have a number of the segmented regions, the morphology characteristics, the size and the other aspects of the area can be also combined to determine the gesture regions.

Take the left hand control as an example, the system processing steps are as follows: (1) getting the depth images and the skeletal joints information from the Kinect sensor; (2) if the left hand skeletal points are captured, their depth values DL to be gotten. If the left hand points were not caught, but the shoulder joint point be captured, the depth value of the shoulder joint DS is used to minus the threshold threshol1 (according to the actual situation and experience, this experiment takes 390 mm) as the left hand depth DL(DL = DS-threshold1). If the shoulder joints have not been caught, recording Dmin for the minimum depth of the depth images, add Dmin to the depth value threshold

threshold2 (according to the actual situation and experience, this experiment takes 68 mm) as the left hand depth DL (DL = Dmin + threshold2). According to the experiences of taking two threshold D1, D2, depth value in [DL−D1, DL + D2] within the region as a gesture, the depth image is changed into a binary image in this way; (3) the water-shed algorithm and the median filter method are used for step(2) image to segment and processing, finally a 100×100 gesture image is gotten.

3.2 Gesture Recognition

RBM consists of visible layer (input layer) and hidden layer (feature extraction layer). DBN network is composed of a plurality of RBM layers and a translation layer, which is trained from the bottom to the top layer by layer. DBN is trained by the weights between the neurons so that the entire DBN in accordance with the maximum probability generates training datum. In this paper, DBN is composed of 3 RBM layers and 1 softmax layer, and the DBN model structure is shown in Fig. 2. The input vector has 10000 dimension, and K is the number of gestures. The feature vectors obtained from Fig. 2 are small in dimension, it can be quickly identified and processed. The features extracted by the multilayer RBM has better effect than that of the single layer RBM, but not the more layers can get better results, it is found by the experimental comparison that 3 RBM layers can achieve better results, and the training time is not too long. The node numbers in each layer will have a great impact on the final results, therefore it is need to select a reasonable value through the experiments.Through a large number of experimental results, it can be found that when $n_1 = 1000$, $n_2 = 300$, $n_3 = 500$, the best results can be gotten.

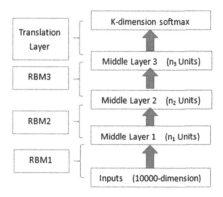

Fig. 2. DBN model structure

Because the different gesture experiments are mutually exclusive, it is suitable to use the softmax regression model [9] for classification. In the softmax regression, the probability that x is classified as a class j is [9]

$$P(y^{(i)} = j | x^{(i)}; \theta) = \exp(\theta_j^T x^{(i)}) / \sum_{l=1}^{k} \exp(\theta_l^T x^{(i)}) \tag{4}$$

where, k is number of gestures, x is the feature vectors extracted by 3 RBM layers. In order to solve the parameter "redundant" set problem in the softmax regression and guarantee a unique solution, the weight decay term is added to modify the cost function, the modified cost function is defined as [9]

$$\mathrm{J}(\theta) = -\frac{1}{m}[\sum_{i=1}^{m}\sum_{j=1}^{k}1(y^{(i)} = j)\log(\exp(\theta_j^{T}{}_{x^{(i)}})/\sum_{l=1}^{k}\exp(\theta_l^{T}{}_{x^{(i)}}))] + \frac{\lambda}{2}\sum_{i=1}^{k}\sum_{j=0}^{n-1}\theta_{ij}^2 \quad (5)$$

where, m represents the training sample number, k is the total category number, n is the dimension of the feature vector x, which corresponds to n_3 in Fig. 2. $\mathrm{J}(\theta)$ is a convex function, the L-BFGS [10] method is used to obtain the optimal solution. In Eq. (5), λ has a significant effect on the results, through many experiments it can be found that when $\lambda = 1.5 \times 10^{-4}$ to achieve a better recognition accuracy rate.

3.3   Gesture Recognition Experiments

This paper uses the data sets collected by our lab., which contains 16 kinds of gestures show as Fig. 3, each gesture 820 images, totally 13120 images (the training set has 9220 images and test set has 3900 imges, size are all 100×100). The depth image resolution is 640×480 by the Microsoft Kinect sensor. The result of DBN, sparse automatic coding machine and CNN are all shown in Table 1. The hardware specifications are dual Intel Core i5-6200U 2.30 GHz CPUs, 8 GB DDR3L 1600 MHz RAM. From the experimental results can be seen that both the accuracy and timeliness can meet the requirements of real-time control.

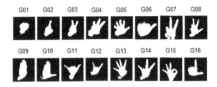

Fig. 3. Gesture image

Table 1. Comparison of the experimental results

Method	Accuracy	Time (ms)
DBN	95.573%	72.169
Sparse auto-encoder	92.318%	81.703
CNN	94.065%	70.521

4   Experiments

In this experiment, the humanoid robot hands (including two arms) have 22 joints, the arms size is designed according to the normal person's arm, 5 fingers are also driven by their respective steering gear, both of the humanoid robot hands have 5 DOFs, throung PC send commands to control the motion and speed of each joint.The operator keeps the distance from Kinect to 2.5 m in the range of 0.8 m,open client and start Kinect, changing gestures, the client will updating color images, depth images, gesture image, gesture recognition results,each joint angle and coordinate in real time, and the robot hand can imitate human hand actions.

For the depth images obtained in accordance with the steps to get the left hand and right hand gestures in Sect. 3.1, then the pre-treatment and identified according to the steps in Sect. 3.2 (the DBN models of the left and the right hands are trained separately), and finally get the left hand gestures and the right hand gestures. Choose 64 kinds of gestures for experiments, and each gesture was performed 4 times, total time consuming 537 s, the accuracy rate is 92.969%. Figure 4 shows the actions of the five gesture combinations, the first column is the RGB images, the second column is the depth images, the third columns and the four columns are the left and right hand gestures respectively. Figure 5 shows the humanoid robot hands gestures.

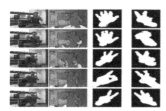

Fig. 4. Motion control

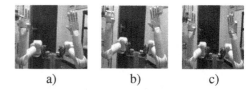

a)   b)   c)

Fig. 5. Humanoid robot hands gestures

5   Conclusions

In this paper, a motion control approach based on the Kinect sensor for the humanoid robot hands is proposed, through the actions of the human being's hand to control the two humanoid robot hand's motion, this kind of human-computer interaction has great development in future.

The main research work and innovation of this paper are as follows: (1)a kinematics model of the humanoid robot hand is established; (2)the control system is also realized, the Kinect sensor is added as an input device, which makes the control of the humanoid robot hands more convenient and simple; (3) the combination of double threshold method and skeleton tracking method for hand gesture extraction; (4)realized the motion control of the two humanoid robot hands.

Although the proposed approach can recognize hand gestures well, there are still also following deficiencies: (1) the gesture recognition type is still limited in the experiments, more types of gestures will be added; (2) the Kinect sensor sometimes can not capture the human being's skeleton points, which will result in a reduction of gesture recognition accuracy, it can be improved by using some other methods, such as combined with the depth images, the skeletal joints information and the RGB image for the positioning human being's hand, so as to obtain more accurate separate the hand gesture; (3) dynamic gesture recognition is not discussed in this paper.

Acknowledgments. This research is funded by the National Natural Science Foundation (Project No. 61573145), the Public Research and Capacity Building of Guangdong Province

(Project No. 2014B010104001) and the Basic and Applied Basic Research of Guang-dong Province (Project No. 2015A030308018), the authors are greatly thanks to these grants.

References

1. Fu, G., Fu, J., Shen, H., Xu, Y., Jin, Y.: Product-of-exponential formulas for precision enhancement of five-axis machine tools via geometric error modeling and compensation. Int. J. Adv. Manuf. Technol. **81**(1-4), 289–305 (2015)
2. Ting, W., Nanfeng, X.: Control system of humanoid robot based on data glove. Computer Engineering and Applications **30**(7), 1707–1711 (2009)
3. Suarez, J., Murphy, R.: Hand gesture recognition with depth images: A review' in 2012 IEEE RO-MAN: The 21st IEEE International Symposium on Robot and Human Interactive Communication, pp. 411–417, Paris, France (2012)
4. Roh, M.C., Lee, S.W.: Human gesture recognition using a simplified dynamic Bayesian network. Multimed. Syst. **21**(6), 557–568 (2015)
5. Santos, D.G., Fernandes, B.J.T., Bezerra, B.L.D.: HAGR-D: a novel approach for gesture recognition with depth maps. Sensors **15**(11), 28646–28664 (2015)
6. Ren, Z., Junsong, Y., Zhengyou, Z.: Robust hand gesture recognition based on finger-earth mover's distance with a commodity depth camera in 19th Inter. Con. on Multimedea 2011, Scottsdale, AZ, USA, pp. 1093–1096, (2011)
7. Smet, P.D., Rui, L.: Implementation and analysis of an optimized rain falling watershed algorithm. Proc. Spie **2**(2), 1116–1117 (2000)
8. Jiang, M., Liang, Y., Feng, X., et al.: Text classification based on deep belief network and softmax regression. Neural Comput. Appl. 1–10 (2016)
9. Liu, D.C., Nocedal, J.: On the limited memory method for large scale optimization. Math. Program. **45**(1), 503–528 (1987)

A New Behavior Recognition Method of Nursing-Care Robots for Elderly People

Xiaojun Zhang, Huanhuan Liu$^{(\boxtimes)}$, and Minglu Zhang

Mechanical Engineering, Hebei University of Technology, Tianjin, China
liuhuanhuan5678@163.com

Abstract. Based on the optical flow technology, a convolutioal neural network (CNN) is proposed for nursing-care robots to perform the behavior recognition task, which considered both static and dynamic information during human motions, thus it is more accurate than the traditional CNN. Firstly, a behavior processing method, combining with Lucas-Kanade optical flow technology, is elaborately designed and tested. In this method, the limitation of static processing method existed in CNN is solved well, then the method is applied to a CNN model for behavior recognition task. Simulation experiment has been carried out, indicating that this method can achieve a higher recognition accuracy and obtain a good recognition effect successfully.

Keywords: Intelligent Nursing-care robot · Target objection · Behavior recognition · Optical flow method · Convolutioal neural network

1 Introduction

With the contradiction among the increase of social pressure, lack of human resources and rising costs of labor is becoming more apparent. It becomes an urgent problem need to be solved that developing nursing-care robots for elderly people. Human behavior recognition technology is the basic technology of nursing-care robots to interact with the elderly people [1]. By testing motion data, the computers capture symbolic motion information and extract the movement characteristics in order to realize the behavior classification process [2]. Generally, human behavior recognition technology has a broad application prospect in the filed of human-computer interaction, machine learning, and intelligent monitoring [3–5]. In addition, it is crucial that this technology also has an important role in improving the work performance, precision and efficiency of nursing-care robots for elderly people.

Convolution neural network (CNN) [6] is one kind of deep model that can achieving a hierarchy of increasingly complicated features and implicit topology structures extracted from training data [7], such as LeNet-5 [8], ImageNet [9]. To effectively comprehend the human behavior information, it is significant to capture discriminative features both spatial dimensions and temporal dimensions in video analysis.

H Mobahi [10] deemed that successive frames contain the same object or objects in video analysis. Karpathy [11] identified mixed-resolution architectures that include a low-resolution context and a high resolution fovea stream, as a way of speeding up

F. Qiao et al. (eds.), *Recent Developments in Mechatronics and Intelligent Robotics*,
Advances in Intelligent Systems and Computing 690, DOI 10.1007/978-3-319-65978-7_82

CNN without sacrificing accuracy. However, current methods ignore the input data optimization problem. How to use the plenty of sequential information from motion images to analyze human behavior pattern, which makes nursing-care robots can judge the intent and purpose of the elderly in real-time, is very worthy of study.

In this work, from the viewpoint of capturing static information and dynamic information during human motion, we established a convolutional neural network model based on optical flow technology. These characteristics are used for training the neural network until achieving the goal of human behavior recognition. The experimental result shows that our planning method can implement human target detection, and simultaneously ameliorate the limitation of traditional identification methods.

This paper is organized as follows. Section 2 illustrates the method of behavioral treatment based on the Lucas-Kanade optical flow technology [12]. The convolutional neural network model is established in Sect. 3, including model scheme designing and model training. In Sect. 4, we show the experiment result and concluding remarks are drawn in Sect. 5.

2 Behavior Processing Method

Optical flow is explained as the instantaneous pixel velocity of moving object on the observation surface, playing an important role in target detection and location. In this paper, we use Lucas-Kanade optical flow technology to deal with a series of motion postures generated in human motion. In the sequence of human motion images, the optical flow, which to calculate the latter frame's target position, is extracted by calculating the time-varying displacement of pixels between consecutive images.

When dealing with the sequential images $I(u, v, n)$ of motion, we assume that each two consecutive images take place in a very short time and have extremely small change in gray variance, pose, and surrounding clutter. The assumption can be established as

$$I(u, v, n) = I(u + \xi, v + \eta, n + 1) \tag{1}$$

$$d = (\xi, \eta)^T \tag{2}$$

where, (u, v) represents the position values of different pixels in the images, generated by human body continuous motions. n is the number of sequential images, d represents the instantaneous velocity of pixel gray values among human continuous motions. The above formula expresses that the image $n + 1$ can be switched from image n by applying a moving vector d, as we called the flow of light, to image n.

Let $J(x)$ and $I(x + d)$ be the description of motion target in two consecutive frame images $I(u, v, n)$ and $I(u + \xi, v + \eta, n + 1)$ respectively. Based on basic theory of error minimum method, a version of error energy function is defined to estimate the optical flow, shown as follows

$$E = \sum [I(x+d) - J(x)]^2 \qquad (3)$$

According to the principle of Taylor's formula $I(x+d) = I(x) + I(x)'d$, d is obtained by minimizing E. After calculating the optical flow field, then the feature points' coordinates corresponding to congregate points can be obtained in latter frame. With plenty of reiterative iterations, human body track can be implemented gradually. The iterative formula is defined as follows [13]

$$d_0 = 0, \quad d_{k+1} = d_k + \frac{\sum I'(x+d_k)^T [J(x) - I(x+d_k)]}{\sum I'(x+d_k)^T I'(x+d_k)} \qquad (4)$$

3 The Convolutional Neural Network Model

3.1 Overall Model Design Scheme

The convolutional neural network consists of one preparation layer, three convolutional layers, two subsampling layers and one output layer. The preparation layer is used to encode prior knowledge and usually leads to better performance, compared to the random initialization. The behavior processing method we proposed takes place in preparation layer. Subsequently, convolution is performed at the convolutional layers to obtain features from the local neighborhood on feature maps in previous layers. Formally, the value of the j th feature map in the lth convolutional layer, denoted as x_j^l, is given by [14]

$$x_j^l = f\left(\sum_{i \in M_j} x_i^{l-1} * k_{ij}^l + b_j^l \right) \qquad (5)$$

where, f(.) is the activation function, M_j is a collection of input feature maps in the current layer, "$*$" is the convolution operation, k_{ij}^l is the kernel of ith feature map in M_j connected to the jth feature map in current layer, b_j^l is the bias of feature map. In subsampling layers, the resolution of feature maps is reduced by pooling over local neighborhood on the feature maps in previous layer, thereby reducing the computational complexity and enhancing robustness of network. The feature map in subsampling layer is expressed by

$$x_j^l = f(\beta_j^l down(x_j^{l-1}) + b_j^l) \qquad (6)$$

where $down(\cdot)$ represents the down-sampling function β_j^l and b_j^l are the multiplicative bias and additive bias of the jth feature map in the lth layer respectively.

In this architecture, shown in Fig. 1, we first apply the behavior processing method to generate multiple feature maps from the input frames in preparation layer, including the gray pixel feature maps, gradient feature maps along vertical and horizontal in spatial dimension, and optical flow feature maps along the horizontal and vertical in temporal dimension. This results in feature maps of which size is 28×28 in preparation layer.

Then, we apply a convolution with one kernel size of 5×5 on each feature maps separately, resulting in 6 feature maps of which size is 24×24 in C_2 layer. In the subsampling layer S_3, we use 2×2 subsampling unit to each feature maps in the previous layer, leading to 24 feature maps of which size is 12×12. The next convolution layer C_4 is obtained by applying a convolution kernel with kernel size of 5×5, resulting in 12 feature maps of which size is 8×8. The S_5 is obtained through applying 2×2 subsampling to each feature map in C_4 layer, leading to the same number of feature maps of which size is 4×4. In the C_6 layer, we use a convolution with kernel size of 4×4, so that C_6 layer contains 192 feature maps of which size is 1×1, connecting to the feature maps in S_5 layer.

After the convolution layers and subsampling layers, the input video has been changed into one 192 dimensional feature vector that captures the critical motion information in input data. The units of output layer contains is same with the actions, and each unit is fully connected to the 192 units in C_6 layer. Finally, we apply the softmax classifier on output layer for action classification.

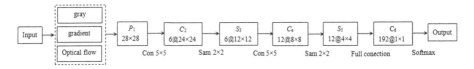

Fig. 1. The CNN model for action recognition. The model contains one preparation layer, three convolution layers, two subsampling layers, and one output layer.

3.2 Model Training

In the training stage, we set label for each video, and take the videos which have same label as a training database of one kind of behavior category. According to the back propagation algorithm, we use the error between output vector and corresponding label to update calculation parameters until meeting the requirement. The each unit of output layer is fully connected to the C_6 layer, so the error is defined as the difference between output vector and label vector. The error value in output layer l, denoted as δ^l, is expressed as

$$\delta^l = f'(w^l x^{l-1} + b^l) \circ (y^n - t^n) \tag{7}$$

where, w^l is the weight of output layer l, y^n and t^n are the ideal value and actual value of output corresponding to the nth input, "$\circ$" is dot product.

When calculating the error value of convolution layers, in order to make the sampling dimension equal to convolution layer, we convert the feature maps in the following subsampling layer with up-sampling operation. Thus the error value in the lth convolution layer, denoted as δ^l, is given as

$$\delta_j^l = \beta_j^{l+1}(f'(\sum_{i \in M_j} x_i^{l-1} * k_{ij}^l + b_j^l) \circ up(\delta_j^{l+1})) \tag{8}$$

where, $up(\cdot)$ is the up-sampling function. So we get

$$\frac{\partial E}{\partial k_{ij}^l} = \sum_{u,v}(\delta_j^l)_{uv}(p_i^{l-1})_{uv}, \quad \frac{\partial E}{\partial b_j^l} = \sum_{u,v}(\delta_j^l)_{uv}, \quad \frac{\partial E}{\partial \beta_j^l} = \sum_{u,v}(\delta_j^l \circ down(x_j^{l-1}))_{uv} \tag{9}$$

where, $(p_i^{l-1})_{uv}$ is the patch convolved with k_{ij}^l in x_i^{l-1}. At each iterative training procedure, the values of convolution kernel, multiplicative bias and additive bias are update in accordance with the following formula

$$k_l = k_l - \alpha\frac{\partial E}{\partial k_l}, \quad \beta_l = \beta_l - \frac{\partial E}{\partial \beta_l}, \quad b_l = b_l - \frac{\partial E}{\partial b_l} \tag{10}$$

where a is the learning rate. At test time, we manage a video with the same method in training stage. When the output vector is obtained, we match it with the label vector and determine the category of this video according to the closet matching label information.

4 Experimental Results and Consideration

Four types of daily behaviors for elderly people are considered as follows: walking, getting up, rolling over and sitting down. The videos of each category are recorded by four persons imitating the elderly motions. In the course of experiment, we take three persons as the training database, and the last one as the test database.

Results of behavior processing method in preparation stage are shown in Figs. 2, 3, 4 and 5 respectively. Figure 2 shows the feature maps of walking category, from left to right is gray feature maps, gradient feature maps and optical flow feature maps successively. Similarly, Figs. 3, 4 and 5 represent feature maps of the rest categories: getting up, sitting down and rolling over. In general from the result, we conclude that the behavior processing method we proposed can track human body effectively and obtain the main motion characteristics accurately.

Fig. 2. Feature maps of walking category   **Fig. 3.** Feature maps of getting up category

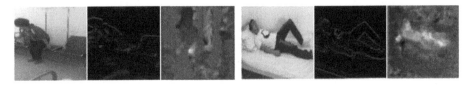

Fig. 4. Feature maps of sitting down category   **Fig. 5.** Feature maps of rolling over category

After training the model, recognition accuracy has been calculated, as shown in Table 1. We can see that the CNN model has a better recognition accuracy in "getting up" and "rolling over", which have achieved 90%. Relatively lower, the recognition rate of "walking" is at 81%. The recognition rate of "sitting down" achieves at 85%. And the average recognition accuracy of this model is 86.5%, which reflects the model has achieved a good result. Figure 6 is the mean square error curve of the CNN model on the database. The curve shows that the identification error is stably tend to 0, and the recognition performance of the method is proved effectively.

Table. 1. Behavior recognition accuracy

Behavior categories	Walking	Getting up	Sitting down	Rolling over	Average
Recognition accuracy	81%	90%	85%	90%	86.5%

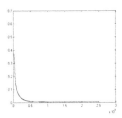

Fig. 6. The mean square error of curve

5   Conclusion

In this paper, we have proposed a convolutional neural network model based on Lucas-Kanade optical flow technology to perform human behavior recognition tasks. Compared to the processing method of ordinary CNNS, we pay attention to the sequential postures that human behavior produced. By using the temporal and spatial characteristics between consecutive motion postures, our model has already effectively realized human body detection. Based on the presented simulation results, the model performs well.

In a future work, we will spread this method to more general recognition tasks and combine this proposed method with other methods. Also, we will research other effective recognition methods which result in less computation costs and a higher accuracy.

References

1. Cheng, L.Y., Sun, Q., Su, H.: Design and implementation of human-robot interactive demonstration system based on kinect. In: Proceedings of the 24th Chinese Control and Decision Conference, pp. 971–975. IEEE (2012)
2. Gu, J.X., Ding, X.Q., Wang, S.-J.: A survey of activity analysis algorithms. J. Image Graph. **14**(3), 377–387 (2009)
3. Aggarwal, J.K., Ryoo, M.S.: Human activity analysis: a review. ACM Comput. Surv. **43**(3), 16 (2011)
4. Du, Y.T., Chen, F., Chen, W.L.: A survey on the version-based human motion recognition. ACTA Electron. Sin. **35**(1), 84–90 (2007)
5. Weinland, D., Ronfard, R., Boyer, E.: A survey of vision-based methods for action representation, segmentation and recognition. Comput. Vis. Image Underst. **115**(2), 224–241 (2011)
6. Lecun, Y., Bottou, L., Bengio, Y.: Gradient-based learning applied to document recognition. Proc. IEEE **86**(11), 2278–2324 (1998)
7. Lecun, Y., Bengio, Y., Hinton, G.: Deep learning. Nature **521**, 436–444 (2015)
8. Lecun, Y., Boser, B., Denker, J.S.: Back propagation applied to handwritten zip code recognition. Neural Comput. **1**, 541–551 (1989)
9. Krizhevsky, A., Sutskever, I., Hinton, G.E.: Imagenet classification with deep convolutional neural networks. In: Advances in Neural Information Processing Systems, pp. 1097–1105 (2012)
10. Mobahi, H., Collobert, R., Weston, J.: Deep learning from temporal coherence in videos. In: Proceedings of the 26th International Conference on Machine Learning (2009)
11. Karpathy, A., Toderici, G., Shetty, S.: Large-scale video classification with convolution neural network. In: 2014 IEEE Conference on Computer Vision and Pattern Recognition (CVPR) (2014)
12. Lucas, B., Kanade, T.: An iterative image registration technique with an application to stereo vision. IEEE Trans. Pattern Anal. Mach. Intell. **8**, 565–593 (1986)
13. Zhang, Y.L., Lu, H.Z., Gao, J.: Improvement of Lucas-Kanade method for optical flow estimation. J. Nav. Aeronaut Astronaut. Univ. **24**(4), 443–446 (2009)
14. Xu, K.: Study of Convolutional Neural Network Applied on Image Recognition. Zhejiang University, Zhejiang, China (2012)

Dynamic Force Modeling for Robot-Assisted Percutaneous Operation Using Intraoperative Data

Feiyan Li[1], Yonghang Tai[1,2(✉)], Junsheng Shi[1], Lei Wei[2],
Xiaoqiao Huang[1], Qiong Li[1], Minghui Xiao[3], and Min Zou[3]

[1] Institute of Color and Image Vision,
Yunnan Normal University, Kunming, China
taiyonghang@126.com
[2] Institute for Intelligent Systems Research and Innovation,
Deakin University, Geelong, Australia
[3] Yunnan First People's Hospital, Kunming, China

Abstract. Percutaneous therapy is an essential approach in minimally invasive surgery, especially of the percutaneous access built procedure which without represent neither visual nor tactile feedbacks through the actual operation. In this paper, we constructed a dynamic percutaneous biomechanics experiment architecture, as well as a corresponding validation framework in surgery room with clinical trials designed to facilitate the accurate modeling of the puncture force. It is the first time to propose an intraoperative data based dynamic force modeling and introduce the idea of continuations modeling of percutaneous force. The result demonstrates that the force modeling of dynamic puncture we proposed based on our experimental architecture obtained is not only has a higher fitting degree with the biological tissue data than previous algorithms, but also yields a high coincidence with the intraoperative clinic data. Further proves that dynamic puncture modeling algorithm has a higher similarity with the medical percutaneous surgery, which will provide more precise and reliable applications in the robot-assisted surgery.

Keywords: Dynamic force · Percutaneous · Intraoperative data · Robot-assisted surgery

1 Introduction

Minimally invasive surgery (MIS) gradually becomes the primary choice for surgeries and patients' treatment with the characteristics of minimal damages, less suffering for patients, short postoperative recovery time as well as lower incidence rate. Especially in the cancer diagnosis, generally, the first procedure needs to extract the lesion of tumor samples for cytopathological examination, which normally achieved by the surgical incision or the percutaneous biopsy [1]. As the traditional surgery treatment is harmful to patients with excessive cost, thus allowing most of the patients choose puncture biopsy. With the navigation of intraoperative images, residents usually rely on their clinical experience, to operate puncture instrument on the internal body target to extract

F. Qiao et al. (eds.), *Recent Developments in Mechatronics and Intelligent Robotics*,
Advances in Intelligent Systems and Computing 690, DOI 10.1007/978-3-319-65978-7_83

the lesion sample in the percutaneous biopsy. In this paper, we constructed a dynamic percutaneous biomechanics experiment architecture, as well as a corresponding validation framework in surgery room with clinical trials designed to facilitate the accurate modeling of the puncture force. It is the first time to propose an intraoperative data based dynamic force modeling and introduce the idea of continuations modeling of percutaneous force.

2 Literature Survey

Miller et al. utilized the super elastic constitutive equation to model the deformation behavior of the liver and kidney tissue [2]. Frick's experiment of sheepskin and sheep tendon are performed in the same manner, the force data showed that the sheepskin needs more force than the sheep tendon when the needle penetration [3]. Simone et al. classified the forces into three components: capsule stiffness, friction and cutting [4]. O'Leary, et al. conducted needle penetrates bovine livers under the CT imaging, described the effects of friction force and needle deformation during the robotic needle puncture [5]. Maurin et al. conducted a series of experiments based on different organs and different condition and discussion [6]. Barbé et al. proposed a method to online estimate the forces in vivo experiment with a low velocity of the needle tip [7]. Carra studied the needle insertion through several layers and finally reach to the liver, focused on finding an optimal model to describe the force profile [8]. Asadian proposed a nonlinear dynamics method based on the modified LuGre model to analyze the relationship between the force and the depth [9]. Ng et al. in order to find an appropriate experimental object to replacement of the animals, multilayer gelatin was designed to simulate the needle insertion into soft tissue [10]. Most of the above studies have been performed on the static puncture of soft tissue in vitro, and it is difficult to implement in vivo experiment. As mentioned in the previous section, considering the body's breathing, the actual needle puncture surgery is a complex dynamic process. To obtain the experimental data in vitro is similar to in vivo, we present a novel method to perform the test.

3 Materials and Methods

3.1 Model of Insertion Force

According to Fung [11], the mechanical properties of most soft tissues are characterized by inhomogeneity, nonlinearity, anisotropy and viscoelasticity. According to the mechanical properties of the soft tissue and the insertion stage, the insertion forces is consisted of three components: hardness, friction and cutting force. The total force equation as follow:

$$f(x) = f_{hardness}(x) + f_{frition}(x) + f_{cutting}(x) \tag{1}$$

Where $f(x)$ is the insertion force, x means the tip's position.

As far as we know, the insertion procedure is a complex process with continuity, uncertainty. Here, we take a puncture process as the research object, according to the idea of periodic extension, transform the non-periodic function into periodic function. Therefore, the insertion force can be considered as a periodic function. According to the principle of Fourier transform, any periodic function can be formed by superposition of trigonometric functions, which is the Fourier series. The equation as follow:

$$f(t) = a_n \cos(k\omega t) + b_n \sin(k\omega t) \tag{2}$$

According to Euler formula $e^{it} = \cos(t) + i \sin(t)$ the trigonometric function can be transformed into an exponential form. Which is:

$$f(t) = \sum_{k=-\infty}^{\infty} c_k e^{ik\omega t} = a_0 + + \sum_{k=1}^{\infty} a_k \cos(k\omega t) + \sum_{k=1}^{\infty} b_k \sin(k\omega t) \tag{3}$$

In the case the mechanics model is unknown, the approximation method is used to fit the experimental data. It has been proved mathematically, the first N term of the Fourier series and the original function is the best approximation under a given energy.

$$\lim_{N \to \infty} \int_0^T \left| f(t) - \left[a_0 + \sum_{k=1}^{N} (a_k \cos(k\omega t) + b_k \sin(k\omega t)) \right] \right|^2 dx = 0 \tag{4}$$

Based on this, we employed a finite number of triangular periodic signals to approximate the periodic function, a mechanical model is obtained:

$$f(t) = a_0 + + \sum_{k=1}^{N} a_k \cos(k\omega t) + \sum_{k=1}^{N} b_k \sin(k\omega t) \qquad 1 \le N \le 8 \tag{5}$$

Where, a_0 is the DC component, a_k, b_k are named Fourier coefficient, N is the number of terms of the Fourier series. As most of the energy concentrated in the low-frequency, we set N from 1 to 8.

3.2 Experimental Setup

In this paper, two experimental systems were designed to evaluate abovementioned algorithm, one is a biomechanics experiment architecture established in laboratory with needle insertion, another one is built in the operating room for data verification. The lab system mainly consists of a surgical instrument (trocar needle), force sensor acquisition system (ATI Nano-17), stepper motor drive system, container for soft tissue and computer. The entire experimental devices are shown in Fig. 1. Clinical trials setup is conducted in the urology department of Yunnan First People's Hospital in Kunming, all the procedures related to the patients and operation are designed and constructed by the professional surgeons according to the safety and ethical rules. The clinical system in the operating room is utilized to clinical data collection, after that, interoperated data are compared with the experimental data to verify whether the algorithm we proposed for the dynamic puncture has a high fitting degree with the modeling value. The detailed system implementation in the operation room is demonstrated in Fig. 2. During the actual percutaneous therapy, a professor surgeon hold the percutaneous instruments to finish the percutaneous surgery and the data is recorded and saved through our system.

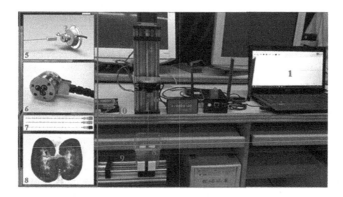

Fig. 1. The entire simulation system of biomechanics experiment: 1-Acquisition software, 2-Multichannel data conversion card, 3-Servo Motor, 4-Linear Guideway (vertical), 5-Connector, 6-Force sensor (ATI Nano-17), 7-Surgical trocar needle, 8-Biopsy sample, 9-Linear Guideway (horizontal), 10-Power Supply.

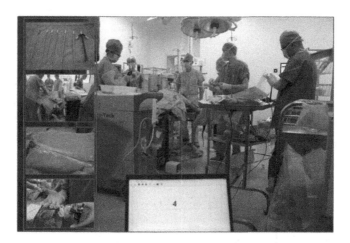

Fig. 2. Clinical trials setup: 1-Surgical instruments, 2-Force sensor and Connector, 3-Data conversion line, 4-Acquisition software, 5-Patient, 6-Puncture devices integration

4 Experimental Results

In this paper, all the experimental data are analyzed by MATLAB2016. The nonlinear least squares are used to fit the whole insertion force by using the Fourier model proposed in Sect. 3 to determine the mechanical model of the complete insertion force. The force data of pre-puncture is analyzed, as shown in Fig. 3, the Fourier model is compared with the fitting function proposed by the predecessors. Table 1 shows the accuracy of each model to fit the insertion. It can be seen from the table that the Fourier model has a higher fitting degree with the biological tissue data than previous algorithms (0.995).

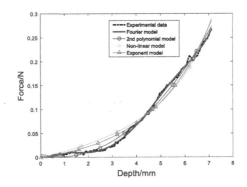

Fig. 3. Comparison of four fitting models with experimental data: Black-Experimental data, Red-Fourier model, Blue-2nd Polynomial model, Cyan-Non-linear model, Magenta-Exponent model.

Table. 1. The fitting degree cooperation of each mechanical model

	Fourier	2nd polynomial	Non-linear	Exponent
Fitting degree	0.995	0.982	0.953	0.965

Based on the Fourier model, our algorithm also yields a high coincidence with the intraoperative clinic data, and the whole expression of the insertion force model can be obtained:

$$
f(x) = \begin{cases} a_0 + \sum_{k=1}^{4} a_k \cos(k\omega x) + b_k \ \sin(k\omega x) & N = 4 \\[2mm] a_0 + \sum_{k=1}^{8} a_k \cos(k\omega x) + b_k \ \sin(k\omega x) & N = 8 \\[2mm] a_0 + \sum_{k=1}^{3} a_k \cos(k\omega x) + b_k \ \sin(k\omega x) & N = 3 \end{cases} \tag{6}
$$

Where, N = 4 is the pre-puncture stage, N = 8 is the post-puncture, N = 3 is withdrawal. Figure 4 shows a comparison of the experimental data of the dynamic with the mechanical model.

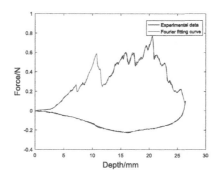

Fig. 4. The complete model of dynamic force

5 Conclusions

In this paper, we proposed a novel force modeling methodology for the percutaneous therapy considering with the human respiration. The dynamic percutaneous biomechanics experiment architecture, as well as a corresponding validation framework in surgery room with clinical trials are designed to facilitate the accurate modeling of the puncture force. It is the first time to propose an intraoperative data based dynamic force modeling and introduced the idea of continuations modeling of percutaneous force. The principal contribution of this paper is to implement and verify the modeming method, with obtaining the clinical data of needle puncture, compare with the experimental data of dynamic, which illustrated that the dynamic insertion can effectively simulate the real operation. The future work will include: the experiment equipment needs to optimize, such as the simulation of human respiratory needs more accurate. More experiments need to be carried out, such as the velocity of needle, the size of needle, the type of needle as well as the features of various soft tissue.

References

1. LiYing, G.: Study on the typical biological soft tissue cutting for needle biopsy. Shangdong University (2012)
2. Miller, K.: Constitutive modelling of abdominal organs. J. Biomech. **33**(3), 367–373 (2000)
3. Frick, T.B., Marucci, D.D., Cartmill, J.A., et al.: Resistance forces acting on suture needles. J. Biomech. **34**(10), 1335–1340 (2001)
4. Simone, C., Okamura, A.M.: Modeling of needle insertion forces for robot-assisted percutaneous therapy. In: IEEE International Conference on Robotics and Automation, vol. 2, pp. 2085–2091 (2000)

5. O'Leary, M.D., et al.: Robotic needle insertion: effects of friction and needle geometry. Proc. IEEE Int. Conf. Robot. Autom. **2**, 1774–1780 (2003)

6. Maurin, B., et al. In vivo study of forces during needle insertions. In: Proceedings of the Medical Robotics Navigation and Visualization Scientific Workshop, pp 415–422 (2004)

7. Barbé, L., et al.: Needle insertions modeling: identifiability and limitations. Biomed. Signal Process. Control **2**(3), 129–134 (2006)

8. Carra, A., Avila-Vilchis, J.C.: Needle insertion modeling through several tissue layers. In: 2010 2nd International Asia Conference on Informatics in Control, Automation and Robotics (CAR), vol. 1, pp. 237–240 (2010)

9. Asadian, A., Kermani, M.R., Patel, R.V.: A compact dynamic force model for needle-tissue interaction. In: 2010 Annual International Conference of the IEEE Engineering in Medicine and Biology Society (EMBC), pp. 2292–2295 (2010)

10. Ng, K.W., et al.: Needle insertion forces studies for optimal surgical modeling. Int. J. Biosci. Biochem. Bioinform. **3**(3), 187 (2013)

11. Fung, Y.C.: Biomechanics: mechanical properties of living tissues. J. Appl. Mech. **61**, 1007 (1994)

Trajectory Planning of Robot Based on Quantum Genetic Algorithm

Guo Qingda, Quan Yanming$^{(\boxtimes)}$, Liu Peijie, and Chen Jianwu

South China University of Technology, Guangzhou, Guangdong, China
meymquan@scut.edu.cn

Abstract. There are many possible trajectories between the given points in Euclidean geometry space for industrial robots. Under constraints of robotic kinematics or dynamics, the optimization of robotic running time is required. In this paper, a trajectory planning method of robot based on quantum genetic algorithm is proposed. Firstly, cubic B-spline function and constraint condition of robotic kinematics were introduced. Then, the chromosome and evolutionary update strategy in quantum genetic algorithm were described, and the variable fitness function of trajectory planning was also defined. Finally, the methods and procedures were introduced in detail, and the first three joint trajectories and the quantum genetic algorithm were programmed on MATLAB platform. The simulation experiments showed that trajectories of joint displacement, velocity, acceleration and jerk could be obtained effectively. The results verified the reliability and practicability of the method.

Keywords: Robot · Trajectory planning · Quantum genetic algorithm · B-spline

1 Introduction

Industrial robot is important factory automation equipment in modern manufacturing industry, which involves many subjects, such as machinery, electronics, controlling, computer, sensor and artificial intelligence. Under the given control algorithm, the industrial robot can perform the specified actions repeatedly. The industrial robot is capable of performing conventional PTP (point-to-point) motion and simple CP (continuous path) movement (such as linear and circular movement). However, there are many possible trajectories between the two given points in the Euclidean geometry space. In the process of robot motion, it is necessary to meet constraint conditions of robotic kinematics and dynamics.

The trajectory planning of industrial robot means that the running time is the shortest and the working efficiency is improved under the constraint conditions of kinematics or dynamics. Lin et al. [1] proposed the cubic spline function to plan the trajectory of the robot and took velocity, acceleration and jerk as constraints. Valero et al. [2] proposed a trajectory planner to obtain efficient trajectories from the dynamic point of view in workspaces with obstacles and discussed the two cases results based on robot Puma 560. Chettibi et al. [3] solved the problem of minimum cost trajectory planning consisting of linking two points in the operational space; the generic optimal

© Springer International Publishing AG 2018
F. Qiao et al. (eds.), *Recent Developments in Mechatronics and Intelligent Robotics*,
Advances in Intelligent Systems and Computing 690, DOI 10.1007/978-3-319-65978-7_84

control problem is transformed into a non-linear constrained problem which is treated then by the Sequential Quadratic Programming (SQP) method.

Trajectory planning algorithm includes genetic algorithm, artificial neural network and chaos optimization algorithm, etc. [4, 5]. Narayanan et al. [6] proposed the quantum genetic algorithm by combining the genetic algorithm and quantum theory. Han et al. [7] used qubits and quantum gates to update chromosome encoding. Compared with the traditional genetic algorithm, the quantum genetic algorithm is very different in terms of expression and evolution mechanism. It has many advantages such as good population diversity, fast convergence speed and strong global search ability, etc.

This paper presents a method for trajectory planning of industrial robots based on quantum genetic algorithm. Through establishing value points of cubic B-spline, the method used quantum genetic algorithm to plan the time optimal trajectory under the constraint of dynamic velocity, acceleration and jerk of the robot. In order to adapt to trajectory planning of the robot, variable fitness function was used in quantum genetic algorithm. In the simulation experiment, PUMA560 is used as the test object, the effective trajectory planning results are obtained after iteration convergence and show the effectiveness of the proposed method.

2   Methodology

2.1   Model of Cubic B-spline

Cubic B-spline curve has C^2 order continuous, Matrix form of cubic B-spline is written as

$$\theta_i(u) = \frac{1}{6} \begin{bmatrix} 1 & u & u^2 & u^3 \end{bmatrix} \begin{bmatrix} 1 & 4 & 1 & 0 \\ -3 & 0 & 3 & 0 \\ 3 & -6 & 3 & 0 \\ -1 & 3 & -3 & 1 \end{bmatrix} \begin{bmatrix} V_{i-1} \\ V_i \\ V_{i+1} \\ V_{i+2} \end{bmatrix} \tag{1}$$

Where $i = 1, 2, \ldots, m - 1$; m is knot number of joint space; V are control points.

2.2   Object Function and Constraint Conditions

For object function of B-spline trajectory optimization, we choose the minimum working time of robots as the optimization function, which is written as

$$T = \min \sum_{i=1}^{m-1} h_i \tag{2}$$

Where T is total time of the robot to run through the entire B-spline curve; $h_1, h_2, h_3, \ldots, h_{m-1}$ are independent time of each B-spline curve; m is the total number of internal knots.

Set joint velocity $\dot{\theta}_i(u)$ of the B-spline curve numbered i of the robot as

$$\dot{\theta}_i(u) = \frac{q'}{t'} \tag{3}$$

Velocity constraint of the robot manipulation is described as

$$\dot{\theta} \leq \dot{\Theta} \tag{4}$$

Where $\dot{\Theta}$ is the maximum velocity value of selected joint.

Constraint of acceleration and jerk are obtained by using the same method. Set joint acceleration $\ddot{\theta}_i(u)$ and jerk $\dddot{\theta}_i(u)$ of the B-spline curve numbered i of the robot as

$$\ddot{\theta}_i(u) = \frac{d\dot{\theta}_i(u)}{dt} = \frac{d\dot{\theta}_i(u)}{du}\frac{du}{dt} \tag{5}$$

$$\dddot{\theta}_i(u) = \frac{d\ddot{\theta}_i(u)}{dt} = \frac{d\ddot{\theta}_i(u)}{du}\frac{du}{dt} \tag{6}$$

Acceleration and jerk constraints of the robot manipulation are described as

$$\ddot{\theta} \leq \ddot{\Theta};\ \dddot{\theta} \leq \dddot{\Theta} \tag{7}$$

Where $\ddot{\Theta}$ is the maximum acceleration value of the selected joint; $\dddot{\Theta}$ is the maximum jerk value of the selected joint.

3 Quantum Genetic Algorithm

3.1 Quantum Bit Coding

Quantum bits differ from classical bits in that they can be superimposed on two quantum states at the same time, such as:

$$|\varphi\rangle = \alpha|0\rangle\beta|1\rangle \tag{8}$$

Where (α, β) are the two amplitude constants, and they also satisfy the following constraint $|\alpha|^2 + |\beta|^2 = 1$; $|0\rangle$ and $|1\rangle$ are respectively spinning-down state and spinning-up state. The genes that encode the number m parameters using multiple qubits are as follows:

$$q_j^t = \begin{pmatrix} \alpha_{11}^t & \alpha_{11}^t & \cdots & \alpha_{mk}^t \\ \beta_{11}^t & \beta_{11}^t & \cdots & \beta_{mk}^t \end{pmatrix} \tag{9}$$

Where q_j^t represents chromosome of the number t generation and the number j individual; k is quantum bits of each encoding gene; m is the number of chromosomes.

3.2 Quantum Revolving Door

Quantum gate is used as the executing mechanism of evolutionary operation. According to the characteristics of quantum genetic algorithm, we choose quantum revolving door. The update process is as follows:

$$\begin{bmatrix} \alpha_i' \\ \beta_i' \end{bmatrix} = U(\theta_i) \begin{bmatrix} \alpha_i \\ \beta_i \end{bmatrix} = \begin{bmatrix} \cos(\theta_i) & -\sin(\theta_i) \\ \sin(\theta_i) & \cos(\theta_i) \end{bmatrix} \begin{bmatrix} \alpha_i \\ \beta_i \end{bmatrix} \tag{10}$$

Where represent probability amplitude before and after the update of the number i qubit revolving door; θ_i is revolving door and its size and symbol are determined by the previously designed adjustment strategy.

The adjustment strategy of quantum revolving door is shown in Table 1. Where x_i is current chromosome numbered i; $best_i$ is current best chromosome numbered i; $f(x)$ is fitness function; $s(\alpha_i, \beta_i)$ is direction of revolving angle; $\Delta\theta_i$ is revolving angle. The adjustment strategy is to compare the fitness $f(x)$ of current individual q_j^t with the fitness $f(best_i)$ of current best individual. If $f(x) > f(best_i)$, then adjust the corresponding bit qubit of q_j^t, so that the probability amplitude (α_i, β_i) evolves along the most favorable emergence direction of x_i; Otherwise, If $f(x) < f(best_i)$, then adjust the corresponding bit qubit of q_j^t, so that the probability amplitude (α_i, β_i) evolves along the most favorable emergence direction of $best_i$. In order to meet dynamic constraints of the robot manipulation, a variable fitness function is proposed to trajectory planning. A variable fitness function is defined as

$$f = \begin{cases} 1/\sum_{i=1}^{m-1} h_i, & \dot\theta \leq \dot\Theta \,\&\, \ddot\theta \leq \ddot\Theta \,\&\, \dddot\theta \leq \dddot\Theta \\ -1/\sum_{i=1}^{m-1} h_i, & others \end{cases} \tag{11}$$

Table 1. The adjustment strategy of quantum revolving door

x_i	$best_i$	$f(x) > f(best)$	$\Delta\theta_i$	$s(\alpha_i, \beta_i)$			
				$\alpha_i\beta_i > 0$	$\alpha_i\beta_i < 0$	$\alpha_i = 0$	$\beta_i = 0$
0	0	FALSE	0	0	0	0	0
0	0	TRUE	0	0	0	0	0
0	1	FALSE	0.01π	+1	−1	0	±1
0	1	TRUE	0.01π	−1	+1	±1	0
1	0	FALSE	0.01π	−1	+1	±1	0
1	0	TRUE	0.01π	+1	−1	0	±1
1	1	FALSE	0	0	0	0	0
1	1	TRUE	0	0	0	0	0

4   Simulation Experiment

In parameter settings of quantum genetic algorithm, the number of iterations is 300, the size of the population is about 40, the length of each chromosome is 20, each time interval of B-spline is [0.1 10].

4.1   Simulation Object

PUMA560 is a six degree-of-freedom robot and all joints rotatable. We need to create and link robot PUMA560 object using the link and robot functions in Robotics toolbox. In the simulation process, the joint space kinematic constraints of the first three joints are shown in Table 2, and internal points of the first three joints are shown in Table 3 (note: the boundary condition is V0 = V1, V8 = V9.).

Table 2. Kinematic constraints

	$\dot{\theta}(\circ/s)$	$\ddot{\theta}(\circ/s^2)$	$\dddot{\theta}(\circ/s^3)$
1	100	45	60
2	95	40	60
3	100	75	55

Table 3. Given internal points of the first three joints

Number (°)	1	2	3	4	5	6	7	8
joint1	10	60	75	130	110	100	−10	−50
joint2	15	25	30	−45	−55	−70	−10	10
joint3	45	180	200	120	15	−10	100	50

4.2   Simulation Results

Kinematics of the first three joints and quantum genetic algorithm were programmed on MATLAB platform, the optimal time of each joint were solved, as shown in Table 4. Convergence trajectories of the first three joints fitness were given and the corresponding iterative optimal trajectories convergence in the 200 step, as shown in Fig. 2; the optimal trajectory planning of each joint was shown in Fig. 3(a); trajectory results of velocity, acceleration and jerk were shown in Fig. 3(b) (c) (d) (Fig. 1).

Table 4. Optimization results

	h_1	h_2	h_3	h_4	h_5	h_6	h_7	T
1	2.5767	2.015	1.9639	1.9606	1.9628	2.5845	2.2919	15.3554
2	1.3596	1.8103	2.5771	1.9568	2.5776	2.6023	2.0375	14.9212
3	2.5766	1.9566	2.578	2.5765	2.3466	2.5775	2.7565	17.3683

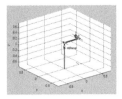

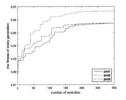

Fig. 1. PUMA560 robot model

Fig. 2. Convergence trajectory of the best fitness of the first three joints

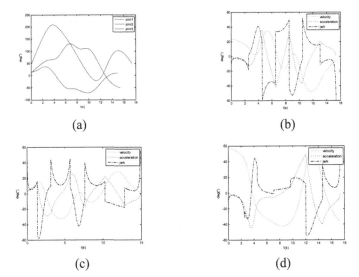

(a)

(b)

(c)

(d)

Fig. 3. (a) The optimal trajectory planning of the first three joint; Trajectory results of the first three joint's velocity, acceleration and jerk ((b) joint1; (c) joint2; (d) joint3).

5 Conclusions

In this paper, quantum genetic algorithm is used as an optimization tool for trajectory planning of industrial robots. Because of the fast convergence speed and strong global search ability of the quantum genetic algorithm, we could obtain the optimal results. In this method, the optimal time of trajectory of the robot was taken as an object function, and the variable fitness function was defined under kinematics constraints of the robot to adapt to the quantum genetic algorithm. The simulation results showed that PUMA560 robot was used as an example to simulate three trajectories in MATLAB platform, and the reliability and practicability of the method were verified.

References

1. Lin, C.S., Chang, P.R., Luh, J.Y.S.: Formulation and optimization of cubic polynomial joint trajectories for industrial robots. IEEE Trans. Autom. Control **28**(12), 1066–1074 (1983)
2. Valero, F., Mata, V., Besa, A.: Trajectory planning in workspaces with obstacles taking into account the dynamic robot behavior. Mech. Mach. Theory **41**(5), 525–536 (2006)
3. Chettibi, T., Lehtihet, H.E., Haddad, M., et al.: Minimum cost trajectory planning for industrial robots. Eur. J. Mech. A. Solids **23**(23), 703–715 (2004)
4. Gasparetto, A., Zanotto, V.: A new method for smooth trajectory planning of robot manipulators. Mech. Mach. Theory **42**(4), 455–471 (2007)
5. Qingda, G.U.O., Chuanheng, W.A.N., Buhai, S.: Planning and simulation of industrial robot time optimal trajectory based on genetic algorithm. Compu. Measur. Control **22**(4), 1240–1242 (2014)
6. Narayanan, A., Moore, M.: Quantum-inspired genetic algorithms. In: IEEE International Conference on Evolutionary Computation, pp. 61–66. IEEE Xplore (1996)
7. Han, K.H., Kim, J.H.: Quantum-inspired evolutionary algorithm for a class of combinatorial optimization. IEEE Trans. Evol. Comput. **6**(6), 580–593 (2002)

Experimental Analysis on Motion Energy Consumption for the Caudal Fin of Robotic Fish

Hao Si, Yan Shen[✉], Qixin Xu, and Jie Zhang

School of Control Engineering, Chengdu University of Information Technology,
ChengDu 610225, China
sheny@cuit.edu.cn

Abstract. As a typical mobile platform, the energy consumption of motion of the caudal fin of robotic fish is more than those of communication, sensing and processing since the robotic fish have limited battery. In order to implement the goal of saving energy and prolong the working time of the robotic fish, the current measurement equipment is installed to obtain the current of the servo when the robotic fish swims straight and turns in the robotic fish. According to the experimental data, a mathematical model of the motion energy consumption is established with respect to the swing parameters of the caudal fin. The experimental results show that the model is effective.

Keywords: Robotic fish · Caudal fin · Motion energy consumption

1 Introduction

Unlike terrestrial robots, robotic fish work in the aquatic environment, and its working time is limited due to the battery capacity.Therefore, robotic fish can't be easily charged.If we want the robotic fish to complete the task more efficiently, the researchers must fully consider its energy consumption.However,most of the researches on the energy consumption are focused on terrestrial robots. For example, Amol Deshmukh et al. proposed the idea of energy management system for the robots [1]. Pratap Tokekar et al. studied energy-optimal velocity profiles for car-like robots [2]. Sun Z et al. studied on finding energy-minimizing paths on terrains [3]. In the aforementioned studies,there is relatively little attention to the energy consumption of biomimetic robotic fish. Shen Yan et al. proposed an energy-saving task assignment for robotic fish sensor network [4], and energy-efficient cluster-head selection with fuzzy logic for robotic fish swarm [5]. Max C-K Wang et al. designed a real-time energy monitoring and management system for robotic fish [6]. Paul Phamduy et al. studied the autonomous underwater charging system for robotic fish [7]. However, the studies above did not establish the motion energy consumption model.

The energy consumption of the robotic fish can be divided into device energy consumption and motion energy consumption. The device includes the controller, sensor, wireless module. Motion energy consumption mainly comes from the motor

© Springer International Publishing AG 2018
F. Qiao et al. (eds.), *Recent Developments in Mechatronics and Intelligent Robotics*,
Advances in Intelligent Systems and Computing 690, DOI 10.1007/978-3-319-65978-7_85

since the caudal fin is driven by the motor [8]. The motion energy consumption is an important part of the total energy consumption of robotic fish. Therefore, this paper focuses on the motion energy consumption of robotic fish actuated by an oscillating caudal fin and models the motion energy consumption. Meanwhile, its model is validated.

2   Modeling of Motion Energy Consumption

2.1   The Beat Rule of the Caudal Fin

A prototype of robotic fish is shown in Fig. 1. The caudal fin is a single joint structure with crescent shape. The motion of the robotic fish can be realized through the servo motor. The servo motor is controlled by formula (1) as follows:

$$a(t) = a_0 + a_A sin(wt) \tag{1}$$

where a_A, w and a_0 denotes the amplitude, frequency and bias of the caudal fin. The oscillating of the caudal fin is related to its three swing parameters. The motion energy consumption of the caudal fin is directly related to the three parameters above.

Fig. 1. The prototype of robotic fish   **Fig. 2.** Current measurement equipment

2.2   Motion Energy Consumption Model of Caudal Fin

In order to get the motion energy consumption model, current measurement equipment is designed to install in the robotic fish, as shown in Fig. 2. The current and voltage passing through the caudal fin may be obtained by making use of this equipment under different swing parameters. The swing of the caudal fin is periodic, so that its motion energy consumption can be obtained by calculating the motion energy consumption of the caudal fin in the unit time and its swing time. The total motion energy consumption of the caudal fin is the sum of the motion energy consumption of the robotic fish under the conditions of swimming straight and turns. The total motion energy consumption of the caudal fin is expressed as

$$E_{SS} = E_{SSL} + E_{SSC} \tag{2}$$

where $E_{SSC} = t_C \cdot E_{unitT-C}$, $E_{SSL} = t_L \cdot E_{unitT-L}$, t_C and t_L are the swing time when the fish turns and swims straight, $E_{unitT-C}$ and $E_{unitT-L}$ are the motion energy consumption of the caudal fin in the unit time when the fish turns and swims straight, E_{SS} is the total motion energy consumption for the caudal fin of robotic fish.

The experimental motion energy consumption data of the caudal fin in unit time are obtained by changing the swing parameters when robotic fish swims straight and turns. Through multi-variable polynomial fitting method, when robotic fish swims straight, straight motion energy consumption model in the unit time is written as

$$
\begin{aligned}
E_{unitT-L} = {} & P_{00} + P_{10}f + P_{01}a_A + P_{20}f^2 + P_{11}a_Af + P_{02}a_A^2 + P_{30}f^3 + P_{21}a_Af^2 \\
& + P_{12}a_A^2f + P_{03}a_A^3 + P_{40}f^4 + P_{31}a_Af^3 + P_{22}a_A^2f^2 + P_{13}a_A^3f
\end{aligned} \tag{3}
$$

where f is the frequency of the caudal fin, P_{00}, P_{10}, P_{01}, P_{20}, P_{11}, P_{02}, P_{30}, P_{21}, P_{12}, P_{03}, P_{40}, P_{31}, P_{22}, P_{13} are the coefficient of the equation.

By making use of the same fitting method, when robotic fish turns, turning motion energy consumption model in the unit time is expressed as

$$E_{unitT-C} = Q_{00} + Q_{10}f + Q_{01}a_A + Q_{20}f^2 + Q_{11}a_Af + Q_{02}a_0 \tag{4}$$

where Q_{00}, Q_{10}, Q_{01}, Q_{20}, Q_{11}, Q_{02} are the coefficient of the equation.

According to the two models above, the motion energy consumption of the caudal fin in the unit time increases with the frequency, bias and amplitude of the caudal fin.

2.3 Calculation of the Swing Time

Assume that the distance which the robotic fish cruise is known, the cruising time t_L is written as

$$t_L = S/V_L \tag{5}$$

where S and V_L are the cruising distance and the cruising velocity, respectively.

The experimental velocity data of the robotic fish are obtained by changing the swing parameters under the conditions of swimming straight and turns. Through multi-variable polynomial fitting method, the cruising velocity model is expressed as

$$
\begin{aligned}
V_L = {} & R_{00} + R_{10}f + R_{01}a_A + R_{20}f^2 + R_{11}a_Af + R_{02}a_A^2 + R_{30}f^3 + R_{21}a_Af^2 \\
& + R_{12}a_Af^2 + R_{03}a_A^3 + R_{40}f^4 + R_{31}a_Af^3 + R_{22}a_A^2f^2 + R_{13}a_A^3f
\end{aligned} \tag{6}
$$

where f and V_L are the frequency of the caudal fin and straight speed, respectively, and R_{00}, R_{10}, R_{01}, R_{20}, R_{11}, R_{02}, R_{30}, R_{21}, R_{12}, R_{03}, R_{40}, R_{31}, R_{22}, R_{13} are the coefficient of the equation.

When the robotic fish turns, the turning time t_C is gotten as

$$t_C = L_{arc}/V_C \tag{7}$$

where L_{arc} and V_C are the arc length and the linear velocity.

By making use of the same fitting method, the linear velocity model is written as

$$V_C = M_{00} + M_{10}a_0 + M_{01}a_A + M_{20}a_0^2 + M_{11}a_0a_A + M_{02}a_A^2 + M_{30}f + P_{22}f^2 + M_{03}a_Af \tag{8}$$

where f is the frequency of the caudal fin, M_{00}, M_{10}, M_{01}, M_{20}, M_{11}, M_{02}, M_{30}, M_{22}, M_{03} the coefficient of the equation.

The turning radius of the robotic fish is mainly affected by the bias of the caudal fin. The experimental turning radius data of the robotic fish are obtained by changing the bias of the caudal fin when robotic fish turns. Through multi-variable polynomial fitting method, the turning radius model is expressed as

$$R = P_1 \cdot a_0^2 + P_2 \cdot a_0 + P_3 \tag{9}$$

where R and P_1, P_2, P_3 are the turning radius of the robotic fish and the coefficient of the equation, respectively.

Given the starting point $A(x_0, y_0)$ and the end point $B(x_1, y_1)$, the distance between two points is L_{AB}. The rotational angle is $\Omega = 2arcsin(L_{AB}/2R)$. The rotational radian γ in the time t_C is $\gamma = \Omega \cdot \pi/180$. The arc length L_{arc} is written as

$$L_{arc} = (\pi/90)(P_1a_0^2 + P_2a_0 + P_3)arcsin(L_{AB}/2(P_1a_0^2 + P_2a_0 + P_3)) \tag{10}$$

3 Experimental Model Validation

The parameters of robotic fish are listed as follows: $S = 0.06$ m^2, $m_b = 2.5$ kg, $L = 0.09$ m and $d = 0.025$ m, where S and m_b are wetted surface area and body mass respectively, L and d are the length and width of caudal fin respectively. The servo motor (HS-5646WP from Hitec) is used to drive the caudal fin of robotic fish.

In order to verify the motion energy consumption model for the caudal fin of the robotic fish, the new motion energy consumption data of the caudal fin are measured when robotic fish swims straight for cruising time $t_L = 20$ s and turns for turning time $t_C = 20$ s under the same experimental conditions, respectively. And the motion energy consumption data of the model are obtained. Finally, the relative error is obtained.

When robotic fish swims straight for cruising time $t_c = 20$ s, the results of motion energy consumption model validation are shown in Fig. 3.

When robotic fish turns for turning time $t_C = 20$ s, the results of motion energy consumption model validation are shown in Table 1.

By comparing the experimental motion energy consumption data with the model motion energy consumption data, it can be seen that, the relative errors of the two models are within reasonable range. Based on a large number of experiments, the two

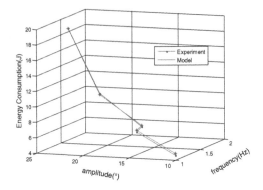

Fig. 3. Comparison of experimental and model motion energy consumption when robotic fish swims straight

Table 1. Motion energy consumption model validation data when robotic fish turns

Bias (°)	Frequency (Hz)	Amplitude (°)	Experimental. energy consumption (J)	Model energy consumption (J)	Relative error
20	1.5	10	5.983	5.746	4.13%
20	1.75	10	6.803	7.086	3.99%
20	1	10	4.112	4.086	0.64%
25	2	10	8.589	8.988	4.44%
30	1.5	15	8.805	9.246	4.77%

models can well describe the change of motion energy consumption of caudal fin, when the robotic fish turns and swims straight respectively. Therefore, the motion energy consumption model for the caudal fin of the robotic fish is effective.

4 Conclusion

The motion energy consumption is an important part of the total energy consumption of robotic fish, and the amplitude, frequency and bias of the caudal fin beat directly affect the motion energy consumption. In this paper, according to experimental data, the motion energy consumption model for the caudal fin of robotic fish actuated by an oscillating caudal fin is established, and the experimental results show that the motion energy consumption model is effective.

Acknowledgements. The author wishes to thank National Natural Science Foundation of China, Applied Basic Research Programs of Sichuan Province, the Research Programs of Sichuan Department of Education and Chengdu university of information technology for proving support under Grant. No. 61472050, 2014JY0257, 17ZA0057 and No. J201406.

References

1. Deshmukh, A., Vargas, .P.A., Aylett, R., et al.: Towards socially constrained power management for long-term operation of mobile robots. The 11th conference towards autonomous robotic systems, UK, Plymouth (2010)
2. Tokekar, P., Karnad, N., Isler, V.: Energy-optimal velocity profiles for car-like robots. IEEE International Conference on Robotics and Automation. IEEE, pp. 1457–1462 (2011)
3. Sun, Z., Reif, J.H.: On finding energy-minimizing paths on terrains. IEEE Trans. Rob. **21**(1), 102–114 (2005)
4. Yan, S., Jie, Z., Bing, G., Hao, S.: Energy-saving task assignment for robotic fish sensor network based on artificial fish swarm algorithm. In: Proceedings of IEEE 12th International Conference on Electronic Measurement & Instruments, July 16–18, Qingdao, China, pp. 536–540. (2015) ISBN:978-1-4799-7618-8
5. Shen, Y., Guo, B.: Energy-efficient cluster-head selection with fuzzy logic for robotic fish swarm. International Conference on Fuzzy Systems and Knowledge Discovery, IEEE, pp. 513–518 (2015)
6. Wang, M.C., Wang, W., Xie, G., et al.: Real-time energy monitoring and management system and its application on bionic robot fish. IEEE (2014)
7. Phamduy, P., Cheong, J., Porfiri, M.: An autonomous charging system for a robot fish. IEEE/ASME Trans. Mechatron. **21**, 2953–2963 (2016)
8. Mei, Y., Lu, Y. H., Hu, Y.C., et al.: A case study of mobile robot's energy consumption and conservation techniques. International Conference on Advanced Robots, 2005. Icar '05. Proceedings. IEEE 492–497 (2005)

Motion Energy Modeling for Tail-Actuated Robotic Fish

Qixin Xu, Yan Shen[(✉)], Hao Si, and Jie Zhang

School of Control Engineering, Chengdu University of Information Technology,
Chengdu 610225, China
sheny@cuit.edu.cn

Abstract. The energy consumption generated by the motion of tail-actuated robotic fish has become one of principal factors for restricting the cruising capability of the robotic fish. Under the robotic fish keeping uniform velocity, the relationship between the frequency and amplitude of the caudal fin beat and the energy consumption of motion is analyzed based on the dynamic model of the tail-actuated robotic fish. And then the energy consumption of motion is modeled with the frequency and amplitude of the caudal fin beat. Finally, the compared tests between the simulation results estimated by this motion model and those of experiment are executed. The results show that the error between the calculation by using the model and experimental results is less than 5%, and the proposed model is effective.

Keywords: Robotic fish · Motion energy · Modeling

1 Introduction

Tail-actuated robotic fish have various applications, such as military operations, ocean sampling, and aquaculture inspection, due to its speed, efficiency, and agility [1]. However, since the robotic fish are powered by the batteries, the cruising ability of the robotic fish is limited. Therefore, it is an important significance to reduce the energy consumption and prolong the life of the robotic fish.

The existing studies on the energy consumption of the robotic fish are mainly focused on the communication energy consumption [2], the sensing energy consumption [3] and the processing energy consumption [4]. However, compared with the sensing and processing energy consumption, the motion of the robotic fish consumes more energy. There is little research on the energy consumption of the robotic fish's motion. For example, Shen [5] proposed energy-saving task assignment for robotic fish which focused on task assignment instead of motion energy model. Hu [6] built a motion energy model base on IPMC propelled robotic fish rather than tail-actuated robotic fish. A precise and reliable motion energy model which could help the robotic fish selecting the low energy consumption motion mode and extend its working life is vital [7].

Through the analysis on the dynamic model, we reduce the influence of oscillation and assume that the robotic fish keeps constant velocity. Therefore, the model of the motion energy consumption for the tail-actuated robotic fish is constructed with the motion control parameters. The inputs of the model are the motion control parameters

© Springer International Publishing AG 2018
F. Qiao et al. (eds.), *Recent Developments in Mechatronics and Intelligent Robotics*,
Advances in Intelligent Systems and Computing 690, DOI 10.1007/978-3-319-65978-7_86

and the running time, the output is the energy consumption value. Compared experimental data with the simulation results, the model can predict the energy consumption under different amplitude and frequency of the caudal fin beat effectively.

The paper is structured as follows. In Sect. 2, the dynamics model is briefly described. Section 3 analyzes the relationship about velocity, propulsion and motion control parameters, and set up the motion energy consumption model. The results are discussed in Sect. 4. The conclusions and future works are carried out in Sect. 5.

2 Dynamic Model

The motion state of the tail-actuated robotic fish in the 2D plane is shown in Fig. 1. The propulsion by the tail swinging drives the robotic fish, the velocity and propulsion of the robotic fish is determined by the state of the caudal fin.

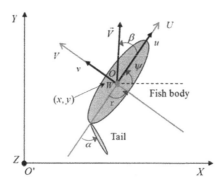

Fig. 1. Fish motion in plane

Therefore, the motion control parameters α_0, α_A, ω_A of the caudal fin swing are described as the input of the model. It is written as:

$$\alpha(t) = \alpha_0 + \alpha_A \sin(\omega_A t) \tag{1}$$

In the hydrostatic environment, the classical caudal fin dynamic model is obtained by using the Lighthill slender body theory without considering the force at the end of the caudal fin [8]. It described by:

$$
\begin{cases}
\dot{x} = \cos\psi \cdot u - \sin\psi \cdot v \\
\dot{y} = \sin\psi \cdot u + \cos\psi \cdot v \\
\dot{\psi} = r \\
\dot{u} = vr + f_u(u,v) - k_1 \ddot{\alpha}\sin\alpha \\
\dot{v} = -ur + f_u(u,v) + k_1 \ddot{\alpha}\cos\alpha \\
\dot{r} = -k_3 r^2 \mathrm{sgn}(r) - k_2 \ddot{\alpha}\cos\alpha - k_4 \ddot{\alpha}
\end{cases}
\tag{2}
$$

where $k_1 = mL^2/(2m_b)$, $k_2 = L^2mc/(2\,J)$, $k_3 = K_D/J$, $k_4 = Lm^3/(3\,J)$,

$$f_u(u, v) = -\frac{1}{2m_b}\rho SC_D u\sqrt{u^2 + v^2} + \frac{1}{2m_b}\rho SC_L v\sqrt{u^2 + v^2}\arctan\left(\frac{v}{u}\right),$$

$$f_v(u, v) = -\frac{1}{2m_b}\rho SC_D v\sqrt{u^2 + v^2} - \frac{1}{2m_b}\rho SC_L u\sqrt{u^2 + v^2}\arctan\left(\frac{v}{u}\right).$$

Determining the motion control parameters of caudal fin, the instantaneous velocity, propulsion and position of robotic fish can be obtained by the dynamic model. And some notations in the model are described in Table 1.

Table 1. Notations used throughout the paper

Notation	Description
(x, y)	Robotic fish coordinate
ψ	Angle between the head direction and the X-axis
t	Time of caudal fin swing
$\alpha_0, \alpha_A, \omega_A$	Bias, amplitude and frequency of the caudal fin beat
u, v, r	Speed of vertical, horizontal, flip
ρ	Fluid density
S	Wetted surface area
C_d	Drag coefficient
C_l	Lift coefficient
m_b	Body mass of robotic fish
m	Virtual mass density of caudal fin
d	The width of caudal fin
L	The length of caudal fin
c	The distance between the caudal fin and the center of the fish
J	Moment of inertia
K_D	Resistance moment coefficient
V, F, E	Velocity, force, energy consumption of robotic fish
β	The angle between the velocity and the U-axis

3   Motion Energy Model

Based on the dynamic model above, the relationship between the frequency and amplitude of the caudal fin beat and the energy consumption of motion is analyzed. By reducing the influence of oscillation, we build the model among velocity, propulsion and the motion control parameters. Finally, motion energy model is established.

3.1 Velocity Model

It is necessary to build a functional relationship between the motion parameters of the caudal fin and the velocity of the robotic fish from dynamic model. During the cruising process, fish move from stationary state to the accelerated motion state. Following this process, it will maintain the uniform motion cruising. As shown in Figs. 2 and 3, during the accelerated process, instantaneous velocity varies with time. But in the uniform motion process, velocity does not change with time. It is noticeable that velocity will oscillate slightly which caused by the periodic changes of α.

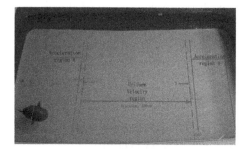

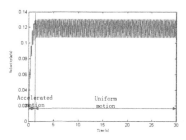

Fig. 2. Fish motion process   **Fig. 3.** velocity of fish

According to formula (2), combined fish coordinate system with the inertial coordinate system, we obtain $v = V \sin \beta$, $u = V \cos \beta$, and rewrite formula as

$$
\begin{cases}
\dfrac{du}{dt} = V \sin \beta \cdot r - \dfrac{1}{2m_b}\rho SC_D V^2 \cos \beta + \dfrac{1}{2m_b}\rho SC_L V^2 \sin \beta \cdot \beta - k_1 \sin \alpha \cdot \left(-\omega_A^2 \alpha_A \sin \omega_A t\right) \\
\dfrac{dv}{dt} = -V \cos \beta \cdot r - \dfrac{1}{2m_b}\rho SC_D V^2 \sin \beta - \dfrac{1}{2m_b}\rho SC_L V^2 \cos \beta \cdot \beta + k_1 \sin \alpha \cdot \left(-\omega_A^2 \alpha_A \sin \omega_A t\right)
\end{cases}
\tag{3}
$$

By using total differential equation, the accelerated equation of the robotic fish can be obtained as

$$
\frac{dV}{dt} = -\frac{1}{2m_b}\rho SC_D V^2 + k_1 \omega_A^2 \alpha_A \sin \omega_A t \sin (\alpha - \beta)
\tag{4}
$$

Given the virtual mass density as $m = (\pi \rho d^2)/4$. When $dV/dt = 0$ and $\alpha_0 = 0$, fish move in the uniform motion process. Therefore, velocity is written as

$$
V = \sqrt{\frac{\pi d^2 L^2}{4SC_D}\omega_A^2 \alpha_A \sin \omega_A t \sin \left(\alpha_A \sin \omega_A t\right)}
\tag{5}
$$

To simplify the model, we use the velocity value in $t = \pi/(2\omega_A)$ to replace the velocity in whole time line, and then the influence of oscillation has been remove. Here c_1 is the effective coefficient of velocity and velocity can be obtained by:

$$V(\omega_A, \alpha_A) = c_1 \sqrt{\frac{\pi d^2 L^2}{4SC_D} \omega_A^2 \alpha_A \sin \alpha_A} \tag{6}$$

3.2 Propulsion Model

Acceleration equals zero when the robotic fish maintain uniform motion. And on the basis of Newton's Second laws of motion, the resultant force of propulsion and resistance is zero value. The resistance of fish, according to Newton's Third laws of motion, which compares with the propulsion from caudal fins swing equal in magnitude and opposite in direction, can be divided into the resistance of fish body and the resistance of caudal fin under fish coordinate system. It is written as

$$
\begin{aligned}
F_u &= F_{hu} + F_{ku} = -\frac{1}{2}m\ddot{a}L^2 \sin\alpha - \frac{1}{2}\rho SC_D u\sqrt{u^2+v^2} + \frac{1}{2}\rho SC_L v\sqrt{u^2+v^2}\arctan\left(\frac{v}{u}\right)^2 \\
F_v &= F_{hv} + F_{kv} = \frac{1}{2}m\ddot{a}L^2 \cos\alpha - \frac{1}{2}\rho SC_D v\sqrt{u^2+v^2} + \frac{1}{2}\rho SC_L u\sqrt{u^2+v^2}\arctan\left(\frac{v}{u}\right)^2
\end{aligned}
\tag{7}
$$

In the uniform motion process, the propulsion makes fish body have slightly periodic oscillation. Compared velocity value in V-axis and U-axis, it can be given $v = 0$, $u = V$. Here propulsion is written as:

$$F = \sqrt{F_u^2 + F_v^2} = \sqrt{\frac{1}{4}m^2\ddot{a}^2 L^4 + \frac{1}{2}m\ddot{a}L^2 \sin\alpha\rho SC_D V^2 + \frac{1}{4}\rho^2 S^2 C_D^2 V^4} \tag{8}$$

By making use of the same method as it in velocity processing, the oscillatory effect of propulsion model is removed. The propulsion model is expressed as formula (9), here c_2 is effective coefficient of propulsion.

$$F(\omega_A, \alpha_A) = c_2 \left(\frac{1}{8}\rho\pi d^2 L^2 \omega_A^2 \alpha_A (1 + \sin\alpha_A)\right) \tag{9}$$

4   Motion Energy Model

In the uniform motion process, motion energy consumption is converted to the displacement of fish. After modeling the velocity and propulsion, it is easy to construct the motion energy model as

$$E(\omega_A, \alpha_A, t) = \int_{t_0}^{t_1} F(\omega_A, \alpha_A)V(\omega_A, \alpha_A)dt \tag{10}$$

5   Experimental Results

The parameters of tail-actuated robotic fish in experiment as following, $\rho = 1000$ kg/m^3, $S = 0.06$ m^2, $L = 0.09$ m, $c = 0.07$ m, $d = 0.025$ m, $m_b = 2.5$ kg. Due to the performance of the servo motor which is connected with the tail and the limited of fish body coun-terweight, it is necessary to limit the motion control parameters within a reasonable range to ensure fish cruise stably.

In the same experimental conditions, the velocities of the robotic fish are measured under different motion control parameters, and then compare with the corresponding velocity model. As shown in Fig. 4, good coincidence is achieved between the model results and the experimental measurements, where the solid line represents the results of the experiment, and the dashed line represents the results of the velocity model.

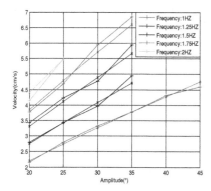
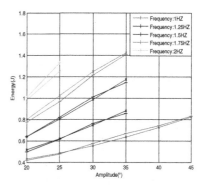

Fig. 4. Velocity value   **Fig. 5.** Energy value per unit time

The effectiveness of motion energy model is verified in the same way, as shown in Fig. 5. The errors between the calculation by using the model and experimental results are within ±5%, and the proposed model is effective.

6 Conclusion

The main work of this paper is to model the motion energy consumption of the tail-actuated robotic fish under uniform velocity cruising, which builds a relationship between the frequency and amplitude of the caudal fin beat and the energy consumption of motion. Good coincidence between the model results and the experimental measure-ments has proved model effective. Model provided a theoretical basis for energy saving motion.

Acknowledgments. The author wishes to thank National Natural Science Foundation of China, Applied Basic Research Programs of Sichuan Province, the Research Programs of Sichuan Department of Education and Chengdu university of information technology for proving support under Grant No. 61472050, 2014JY0257, 17ZA0057 and No. J201406.

References

1. Ryuh, Y.S., Yang, G.H., Liu, J.: A school of robotic fish for mariculture monitoring in the sea coast. J. Bionic Eng. **12**(1), 37–46 (2015)
2. Shen, Y., Guo, B.: Energy-efficient cluster-head selection with fuzzy logic for robotic fish swarm. In: International Conference on Fuzzy Systems, pp. 513–518. IEEE (2015)
3. Mei, Y., Hu, Y.C.: Energy-efficient motion planning for mobile robots. In: 2004 Proceedings of the IEEE International Conference on Robotics and Automation, Vol. 5. IEEE (2004)
4. Wang, M.C.C., Wang, W., Xie, G., Shi, H.: Real-time energy monitoring and management system and its application on bionic robotic fish. In: Proceeding of the, World Congress on Intelligent Control and Automation, pp. 3256–3261 (2014)
5. Shen, Y., Jie, Z.: Energy-saving task assignment for robotic fish sensor network based on artificial fish swarm algorithm. In: International Conference on Electronic Measurement & Instruments (ICEMI), Vol. 1. IEEE (2015)
6. Hu, Q., Zhou, H.: IPMC propelled biomimetics robotic fish energy consumption model construction and its application to energy-saving control. In: IEEE International Conference on Robotics and Biomimetics, pp. 2151–2156. IEEE Xplore (2010)
7. Dabirmoghaddam, A., Ghaderi, M., Williamson,C.: Energy-efficient clustering in wireless sensor networks with spatially correlated data. In: INFOCOM IEEE Conference on Computer Communications Workshops, pp. 1–2. IEEE Xplore (2010)
8. Wang, J., Tan, X.: Averaging tail-actuated robotic fish dynamics through force and moment scaling. IEEE Trans. Rob. **31**(4), 906–917 (2015)

Detection and Maintenance Technology for Oil and Gas Pipeline Based on Robot

Zhengyu Liu[1], Xulong Yuan[2], Maosheng Fu[1], and Shuhao Yu[1(✉)]

[1] School of Electronic and Information Engineering, West Anhui University,
Lu'an 237012, China
yush@wxc.edu.cn
[2] School of Marine Science and Technology, Northwestern Polytechnical University,
Xi'an 710072, China

Abstract. China has huge amounts of oil and gas transportation pipelines, high costs of routine testing and maintenance becomes one of the most important factors in pushing up oil and gas production costs. With the development and application of robot technology, it is expected to substantially reduce the cost of testing and maintenance of oil and gas pipelines. This paper reviewed the key technologies and development of robotic of oil and gas pipelines in recent years, and provide a reference for the development and application of relevant units.

Keywords: Oil and gas pipeline · Robot · Test system · Detection · Maintenance

1 Introduction

As a reliable, environmentally friendly and an efficient carrier, pipe undertook the task of most of the world's oil and gas transportation. At present, the total length of existing oil and gas pipelines more than 120,000-kilometer [1]. Due to developing status of natural gas of green energy, increasing demand for crude oil and refined oil, as a cost-efficient mode of transport, oil and gas pipelines will play a very important role in the future. During the long-term use, corrosion, blockage, leakage, cracks and other phenomenon may cause damage to pipes. Pipeline maintenance artificial excavation detection methods are often used in the past, which means a great spend of human resources and material resources, and high maintenance costs [2].

Pipeline robot is king of robot which able to carry a variety of sensors, crawling along the inner walls of pipelines for pipeline inspection, maintenance and cleaning. As an important part of the robot industry, this could be a great opportunity and has a strategic meaning for China's high-tech industry to enhancing national competitiveness. The core carrier of robot is movement device which has different driving modes, it could carry CCD cameras, ultrasonic distance sensors, Eddy-current sensor, gesture sensors, in order to monitor the corrosion on the inside wall of the pipes; It also carry clean blades to dredge and clean the pipe wall; and carry manipulator, [3], sealing of cracks or repairs on the pipeline.

© Springer International Publishing AG 2018
F. Qiao et al. (eds.), *Recent Developments in Mechatronics and Intelligent Robotics*,
Advances in Intelligent Systems and Computing 690, DOI 10.1007/978-3-319-65978-7_87

2 Related Researches

United States and the United Kingdom, and France, and Japan and other developed countries started the research of pipeline robot in 1950 of 20th century; the major product is a fluid-driven non-powered pipe cleaning equipment [4] there was a rapid development after 80s of the 20th century. Domestic robot research and development started later than them, in the 1980 of the 20th century, Deng zongquan who from Harbin Institute of Technology start the research of mobile mechanism of pipeline robot at first [5], after that the National Defense Science and Technology University, Harbin Institute of technology, Shanghai Jiao Tong University, China University of petroleum, Guangdong University of technology, Zhejiang University and the Dalian University of technology and other units have carried out research work in this area. In this paper, in accordance with the non-active power or not, the pipeline robots are divided to the pipeline robot for autonomous driving pipeline robot and fluid pipeline robot driven categories, detailed information about the current stage of development are also included.

2.1 Autonomous Driving Pipeline Robot

Wheeled, tracked, retractable inchworm, crawling and other types of pipeline robot driven by independent power, they have a great autonomy, easy to control and carrying several sensors to detect or repair equipment. But most of such robots have complex machinery, and limited scope of activities because of the battery limits. In recent years, the representatives of the research are:

(1) Wheeled pipeline robot. Currently, most of the autonomous wheeled pipeline robots are characterized by using of multiple sets of supporting wheels, so that support oil and gas pipe wall. Segon Roh, has been achieved by Sungkyunkwan University [6] development of the fifth generation of differential in-pipe robot MRINSPECT V, made up with the fuselage, drive modules and CCD components. It could adapt diameter φ 85–109 mm variable diameter pipes, self-0.7 kg, 9.8 N, towing, and its travel speed up to 0.15 m/s. Such as Dewei Tang [7, 8] developed the three-shaft differential pipeline robot. Based on the common wheeled mobile robot. It solved the problem of interference during robot turns. Drive in the middle section based on the work unit and control unit connected by Hooke, towing than 240 N, walking speed of it up to 0–10 m/min, and it can adapt to the diameter φ 310 mm pipe.

(2) The Crawler Pipeline Robot. Crawler pipeline robot driven system is made up of positive wheels, derived wheels, tracks, it has great traction, but that means increasing fuselage volume, and reduction of its flexibility. Korea's Yonsei University Jungwang Park [9] PAROYS-II crawler pipeline robot developed by the normal force control and attitude control, pipe diameter, tilting, bending, and other states, it has a strong climbing ability. The weight of the robot is 7.8 kg, maximum drive speed is: 2.5 M/min, available in different diameter φ 400–700 mm movement.

(3) The Retractable Creeping Pipeline Robot. Retractable creeping pipeline robot is mainly suitable for small diameter, but larger bending radius. Its movement mechanism derives for imitation earthworms, inchworm and other lower organisms. It

crawls at low speed, has properties of flexibility and low energy consumption. Jinwai Qiao [10] who invented self-locking pipeline robot prototype called RSIRobot and PSIRobot, supported by two groups composed of institutions. High pressure produced by the pump gas into the telescoping mechanism at both ends of the cylinder. Through control solenoid open circuit, achieving cylinder telescopic movement. Its driving force up to 15.2 N of maximum velocity 13.72 mm/s, diameter ф 16 mm above sports.

(4) Crawling Pipeline Robot. Crawling pipeline robot imitates insects such as the rules spiders' crawling, DC servo-driven joints, as well as connecting foot, various types of maneuvers. Germany Munich Technology University Zagler [11] eight-legged MORITZ developed robot can climb along with different angles of the pipes, the robot is 0.75 m, self-weight 21 kg, four-legged, and middle joints, motor-driven, it can adapt ф 600–700 mm the diameter of the pipes.

2.2 Fluid Pipeline Robot

Fluid pipeline robot driven itself without power mainly depends on both ends of the pipeline fluid pressure to provide thrust, and self-propelled. Infiltration at run time the body in a fluid medium, in the presence of fluid, overcome and pipe wall friction and move forward. Carrying steel brushes, pipe cleaning functions can be achieved. It can be to walk long distance operations, features such as low-energy supply can be adapted to the different bore sizes of pipe, a walking distance to the center of 300 km, and no need to carry batteries or external cable [12], but its movement is less controllable, over speed will cause damage to the pipeline.

Typical of foreign engineering applications are: Brazil Sao Paulo University Okamoto [13] developed for detecting corrosion of pipeline robot. Robots rely on fluids to provide auxiliary power, 16 sets of ultrasonic transducer by mechanically connected with the equipment cabin, between them and the wall by the fluid acts as a lubricant, measurement of ultrasonic transducer by a set of reflective wave transmission time and concrete corrosion of the pipe can be detected. When the robot stops moving, the meter turn no signal output, ultrasonic transducer to stop working. The movement speed of the device up to 0.5 M/s; corrosion detection resolution is 0.04 mm.

In recent years, including Harbin, Shanghai University, Guangzhou University of technology, Dalian University of technology and other units involved in the project of development of a prototype and laboratory tests. such as Zhao Hui [14], was developed using a medium flow-driven, adjustable speed, and online storage of pipeline robot prototype. Prototype driven mainly by online access to devices, and the speed control device, connected device, adjustable support system and the control system components, gross weight of about 2 kg. Φ 80 mm Plexiglass tube inner diameter, maximum design speed 10 m/min. Wenfei Wang studied the fluid-driven in-pipe robot driven characteristics were. Its drive unit from vibration reduction system, the throttle control system and auxiliary support system. Fluid-driven pipeline robot can be applied to ф 400 mm diameter pipelines, theoretical speed for 3 M/s [15].

2.3 Performance Comparison

In general, positive and passive driving robot has its own characteristics. Because of the costs of their manufacturing are different, the salient points of the performance itself, can be applied to different areas. Different driving performance contrast as shown in Table 1.

Table 1. Comparison of different driving performance

Drive mode	Velocity	Volume	Controllability	Crossing ability	Control Distance	Complexity	Working Diameter
Wheel	Low; stable	General	Easy	Strong	General	General	Relatively big
Crawler	Low; stable	Big	General	Strong	Short	Complex	Big
Retractable creeping	Low; stable	Small	Easy	Strong		General	Small
Crawling	Low; stable	Big	General	Strong		Complex	General
Fluid-driven	Unstable	Small	Relatively hard	General	Long	Simple	Variety

3 Pipeline Corrosion Inspection Technologies

Pipeline on-line corrosion inspection mechanism can be divided into magnetic flux leakage testing, ultrasonic testing, Eddy current testing, mechanical testing; different detection techniques can be applied to different types of corrosion tests. Magnetic flux leakage testing applied due to internal or external corrosion of metal loss inspection, ultrasonic testing applicable to manufacturing, fatigue and stress corrosion cracking or weld defects detection [16], eddy current testing applies to pipe inner surface crack detection, mechanical testing applies to local concave deformation detection of pipeline. The principle of magnetic flux leakage testing is: pipeline robots arranged in a permanent magnet, and import the magnetic circuit piping to form a closed loop. If there are internal and external pipeline corrosion, its thickness changes lead to changes in magnetic flux, the wall around the size of Hall element signal can be measured. The schematic is shown in Fig. 1.

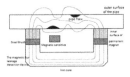

Fig. 1. The principle of magnetic flux leakage testing

Ultrasonic testing principle is: the lateral arrangement of robot ultrasonic sensor, a pulse signal is fired into the wall, lateral reflection of the signal reach of Guan Binei. By measuring the time of ultrasonic reflected twice, combined with the probes coordinates and location information; get the specific location of the corrosion and serious situation. The schematic is shown in Fig. 2.

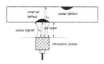

Fig. 2. Ultrasonic testing principle

Eddy current test principle is: the lateral arrangement of pulsed Eddy current sensor of robot, sensors inside a pipe, by the principle of electromagnetic induction, the sensor coil generates an alternating magnetic field, pipe swirl on the surface. If the pipe has cracked, Eddy current distribution changes cause the magnetic field, the magnetic induction coil impedance, voltage and have changed. Measure the actual output voltage distribution curve that crack distribution. The schematic is shown in Fig. 3.

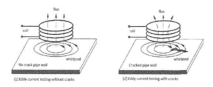

Fig. 3. Eddy current test principle

Mechanical testing principle is: the lateral arrangement of displacement sensor of robot, by way of inertial motion, along the pipeline runs for a week. In vertical direction, through the length of the sensor probe changes, combined with the meter wheel record location information, get the location of the pipeline within local concave and defects. The schematic is shown in Fig. 4.

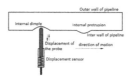

Fig. 4. Mechanical testing principle

Due to the inner wall of oil pipeline wax [17] it is difficult to thoroughly clean, detection means single, low precision, anti-jamming ability, so the current pipeline corrosion inspection errors. Future tests will move towards a more accurate, faster and efficiency of multi-sensor combination direction. Tube imaging [17] is the future direction of corrosion inspection, at present has been successfully developed to carry TV monitor system of pipeline robot. Therefore, future technicians will use pipe cleaning, corrosion detection, integration in one dimensional imaging of pipeline robot in pipe inspection work carried out.

4   Pipeline Cleaning Technology

The pipes which after long-term use, some impurities of oil deposited on the inner wall, they can affect the efficiency of oil transportation. Therefore, after a while, it is necessary to clean the pipes. Conventional pipeline cleaning chemical cleaning and physical cleaning method is generally used two ways [18].

In this case, the method is using the robot for medium mainly for physical cleaning. The main principle is by the mechanical action of the robot, split pipe inner surface impurities or attachment, and with the water discharge pipe, achieve the purpose of cleaning. Routine cleaning methods are retrofitting various types of in-pipe robot head drills, knife or a wire brush [18]. Similar working model of shield machine, robot in pipeline when cleaning device play a role under the physical effects such as extrusion, tearing, grinding, cutting the tube impurities, eventually followed the cleaning jets with discharge pipe.

The method of cavitating water jet cleaning, which use the principle of cavitation erosion for surface scaling. Work mode: under the effect of in-pipe robot driven by fluid flow, in the pipe run. Gap between robot and inner walls of pipelines, jet may break [19], the local low-voltage, formation of vacuoles. Generation, growth and collapse of cavitation and cavitation of the formation of phenomena, can strip the pipe inner wall of impurities. The Jet flow scouring impurity discharge pipe.

5   Conclusions

The Robot technology, related to mechanical engineering, automatic control, computer, sensors, artificial intelligence and electronic technologies, and other subjects. High integration, technology covers a wide range, irreplaceable and unique characteristics, is an important part of robots in the future; countries from all over the world will make more effort in this area. At present, the domestic robot is still in the initial stages of development, market demand and insufficient production capacity, and large dependence on imported products.

With the development of the oil industry and the special robot technology advances, pipes in the future robots will exceed the following difficulties: (1) The ability of turning and adaption to variable diameter: for large angle bends in the pipeline, most of pipeline robots performed not so well, often stuck in the tube which limited the length, the size of pipeline robot, and has greater constraints on its development. Under the trend of technology development in the future, not only need the universal joints connecting solve the length issue, but also more to design all kinds of high accuracy deformation, improve its passing ability. (2) Flexibility and reliability: a except for fluid-driven and other types of robot mechanical structure are more complex, especially the crawler and crawling, in order to improve sports performance, they give up the flexibility and reliability. With the development of artificial intelligence technology in the future, there will be more improvements in this area. (3) Operation distance and energy supply: most active pipeline robots' battery carrying space is limited, partly by towing, energy supply cannot be guaranteed. Dalian University of technology to develop long-range, fluid storage robot solve the energy problem, but it still needs break through to make it more stable and reliable. (4) Sensors: most of domestic pipeline robot research on the stage of lab prototype, during the thinking of movement and performance in the pipes, sensor interface are not enough, how to combine the detection, cleaning, three dimensions imaging, without losing its properties, that is the reason and development needed to be solved in the future.

Acknowledgements. The authors gratefully acknowledge the support of the Natural Science Foundation of the Higher Education Institutions of Anhui Province (No. KJ2015ZD44) and the Universities Excellent Young Talents Foundation of Anhui Province (No. gxyqZD2016249).

References

1. Gao, P., Wang, P., Luo, T., Wang, H.: China's oil and gas pipeline construction in 2014. Int. Petrol. Econ. **23**(03), 68–74 (2015)
2. Wang, H.: The examination method and safety evaluate of oil and gas transmission pipeline. Yanshan University (2014)
3. Zhang, S., Mei, X., Wang, G., et al.: Oil/gas pipeline maintenance and emergency repair method and its progress. Gas Storage Transp. **2014**(11), 1180–1186 (2014)
4. Zhang, T.: The research on the robot of measuring the internal diameter of the pipeline online. Changchun University of Science and Technology (2004)
5. Deng, Z., Sun, X., Liu, C.: A research of robot mechanism moving inside a pipe. Robot **11**(6), 45–48 (1989)
6. Roh, S.G., Kim, D.W., Lee, J.S., et al.: In-pipe robot based on selective drive mechanism. Int. J. Control Autom. Syst. **7**(7), 105–112 (2009)
7. Tang. D., Li, Q., Liang, T., et al.: Mechanical self-adaptive drive technology of triaxial differential pipe-robot. J. Mech. Eng. 44(9):128–133 (2008)
8. Li, Q.: Research on the tri-axial differential pipeline robot and robot property. Harbin Institute of Technology (2011)
9. Park, J., Hyun, D., Cho, W.H., et al.: Normal-force control for an in-pipe robot according to the inclination of pipelines. IEEE Trans. Ind. Electron. **58**(12), 5304–5310 (2012)
10. Qiao, J.: Mechanism design and property study of in-pipe robot based on self-locking. National University of Defense Technology (2012)
11. Zagler, A., Pfeiffer, F.: "MORITZ" a pipe crawler for tube junctions. In: 2003 Proceedings of the IEEE International Conference on Robotics and Automation, ICRA 2003, vol. 3, pp. 2954–2959. IEEE (2003)
12. Cui, G., Jiang, S.: Development status of fluid driven pipeline robot. Metall. Ind. Autom. **S1**, 661–663 (2012)
13. Okamoto Jr., J., Adamowski, J.C., Tsuzuki, M.S.G., et al.: Autonomous system for oil pipelines inspection. Mechatronics **9**(7), 731–743 (1999)
14. Zhao, H.: Pipelines robot of fluid drive and self-obtain energy. Dalian University of Technology (2009)
15. Wang, W.: Research on driving characteristics of a fluid-driven pipeline robot. Harbin Institute of Technology (2011)
16. Li, T., Pu, L.-Z., Su, Y.-J., et al.: Analysis of foreign pipeline magnetic flux leakage testing techniques. Pipeline Tech. Equip. **4**, 17–18 (2014)

17. Liu, H.F., et al.: Current state and development trend of internal corrosion detection technology for oil and gas pipeline. Pipeline Tech. Equip. (2008)
18. Wen, W., Yin, X.: A summary of oil & gas pipeline cleaning technology at home and abroad. Pipeline Tech. Equip. (1), 34–37 (2000)
19. Jiang, Y.: Numerical simulation and experimental study on cleaning oil pipes utilizing cavitating water jets. Harbin University of Science and Technology (2012)

Non-rigid Image Feature Matching
for Unmanned Aerial Vehicle
in Precision Agriculture

Zhenghong Yu[1(✉)] and Huabing Zhou[2]

[1] Guangdong Polytechnic of Science and Technology, Zhuhai, China
honger1983@gmail.com
[2] Wuhan Institute of Technology, Wuhan, China

Abstract. In this paper, we propose a novel feature matching method for agricultural UAV based on probabilistic inference when they undergo non-rigid transformations. Firstly, a set of putative correspondences was generated based on the similarity of features. Then, we focus on eliminating outliers from that set meanwhile estimating the transformation. This procedure is formulated upon a maximum likelihood Bayesian model. We impose three effective regularization techniques on the correspondence, which helps to find an optimal solution. The problem is finally addressed rely on the EM algorithm. Extensive experiments on real farm images shows accurate results of our method, which is superior to the current state-of-the-art methods.

Keywords: Unmanned aerial vehicle · Precision agriculture · Image feature matching

1 Introduction

Real-time, accurate crop growth information extraction is of great significance for precision agriculture [1]. Although high-resolution satellite imagery has been using to reach this goal for a long time, the availability, timing, and interpretation of its data limit its applications. Recently, unmanned aerial vehicle (UAV) are proved to be a potential alternative, thanks to its low-cost, high spatial and temporal resolution, and high flexibility in image acquisition programming [2]. From automated planting to real-time monitoring, UAV has been playing more and more important role in the world of agriculture.

Image feature matching, whose objective is to find a reliable correspondence between two sets of features, is one of the most important fundamental technologies embedded in UAV. It is indispensable for many tasks, such as image stitching [3], image-based navigation [4], specific object localization [5], etc.

A commonly used strategy for feature matching problem is the two-stage way. Firstly, a set of putative correspondences is generated by using a similarity constraint. This stage only requires that the points should match to the points which has the similar local descriptors (such as SIFT) [6]). The second stage mainly focus on eliminating the outliers by using a geometrical constraint. This stage asks the matches must meet the

© Springer International Publishing AG 2018
F. Qiao et al. (eds.), *Recent Developments in Mechatronics and Intelligent Robotics*,
Advances in Intelligent Systems and Computing 690, DOI 10.1007/978-3-319-65978-7_88

underlying geometrical requirement. One typical example of the above strategy is the RANSAC algorithm [7]. It attempts to remove the outliers while do global transformation estimation. Although this method achieves success in many cases, it is prone to fail when the constraints are nonparametric, like non-rigid case. It is also easy to degrade dramatically when the proportion of outlier within the hypothetical set comes high. To solve these problems, several new nonparametric methods such as GS [8] and ICF [9] have been proposed. Unfortunately, the matching accuracy of these methods is still unpromising in the case of large outlier percentage, especially for matching agricultural images. Because the UAV images usually contain local distortions caused by imaging viewpoint changes, and image blurring because of camera motion. That means these images cannot be precisely matched by simple parametric model, like rigid or affine transformation used in most existing methods. In addition, the complexity of UAV images including noise, repeated structures and occlusions often leads to a large number of negative matches, which has great influence on the determination of the transformational model. This paper presents a probabilistic method to solve these problems. The method can deal with the non-rigid deformation in the UAV image pair, and is robust to a large number of outliers.

2 Our Proposed Algorithm

Given two feature point sets $\bar{A} = \{\mathbf{x}_i\}_{i=1}^{N}$ and $\bar{B} = \{\mathbf{y}_i\}_{i=1}^{N'}$ of two images, we can establish a set of putative correspondences $S = \{(\mathbf{x}_i, \mathbf{y}_i)\}_{i=1}^{M}$, where $\mathbf{x}_i$ and $\mathbf{y}_i$ are the two dimensional coordinates, $M \leq \min \{N, N'\}$ is the number of correspondence. Our task is to find out the underlying inliers that fitting a transformation f robustly from the outliers and has the relationship that for any inlier correspondence $(\mathbf{x}_i, \mathbf{y}_i)$, $\mathbf{y}_i = \mathbf{x}_i + f(\mathbf{x}_i)$.

To have a robust estimation, we build a model both taking the inliers and outliers into account. Assume the noise subject to Gaussian distribution with zero mean and uniform standard deviation σ and the outlier is supposed to be uniform distribution $1/b$. b is the area of the image. We then associate the i-th correspondence with a latent variable $z_i \in \{0, 1\}$. $z_i = 1$ means the correspondence $(\mathbf{x}_i, \mathbf{y}_i)$ is an inlier and $z_i = 0$ points to an outlier.

$$p(\mathbf{Y}|\mathbf{X}, \theta) = \prod_{i=1}^{M} \sum_{Z_i} p(\mathbf{y}_i, z_i|\mathbf{x}_i, \theta) = \prod_{i=1}^{M} \left(\frac{\gamma}{2\pi\sigma^2} e^{-\frac{\|\mathbf{y}_i - (\mathbf{x}_i + f(\mathbf{x}_i))\|^2}{2\sigma^2}} + \frac{1 - \gamma}{b} \right) \quad (1)$$

Thus, by using the i.i.d. data assumption, we have the likelihood function a mixture model like (1), where $\mathbf{X} = (\mathbf{x}_1, \mathbf{x}_2, ..., \mathbf{x}_M)^T$ and $\mathbf{Y} = (\mathbf{y}_1, \mathbf{y}_2, ..., \mathbf{y}_M)^T$ are the feature matrixs, $\theta = \{f, \sigma, \gamma\}$ is a set of unknown parameters. γ is the mixing coefficient refers to the marginal distribution of the latent variable. The parameter set θ can be estimated by a maximum likelihood method, that is $\theta^* = \arg \max_{\theta} p(\mathbf{Y}|\mathbf{X}, \theta)$. This is equivalent to minimizing its negative log-likelihood function:

$$E(\theta) = -\ln p(\mathbf{Y}|\mathbf{X}, \theta) = -\ln \prod_{i=1}^{M} \sum_{Z_i} p(\mathbf{y}_i, z_i|\mathbf{x}_i, \theta) \tag{2}$$

We use the EM algorithm to optimize this objective function. The complete-data log posterior is:

$$\wp(\theta, \theta^{old}) = -\frac{1}{2\sigma^2} \sum_{i=1}^{M} p_i \|\mathbf{y}_i - \mathbf{x}_i - f(\mathbf{x}_i)\|^2 + (\ln \gamma - \ln \sigma^2) \sum_{i=1}^{M} p_i + \ln(1-\gamma) \sum_{i=1}^{M} (1-p_i) \tag{3}$$

where $p_i = P(z_i = 1|\mathbf{x}_i, \mathbf{y}_i, \theta^{old})$ is a posterior probability. It reflects what extent $(\mathbf{x}_i, \mathbf{y}_i)$ is an inlier. The EM algorithm alternates with two steps:

E-step: the posterior distribution of the latent variables is calculated by using the current parameter values θ^{old}. Then, the diagonal matrix $\mathbf{P} = diag(p_1, p_2, ..., p_M)$ can be acquired via Bayes rules:

$$p_i = \frac{\gamma \exp\left(-\frac{\|\mathbf{y}_i - \mathbf{x}_i - f(\mathbf{x}_i)\|^2}{2\sigma^2}\right)}{\gamma \exp\left(-\frac{\|\mathbf{y}_i - \mathbf{x}_i - f(\mathbf{x}_i)\|^2}{2\sigma^2}\right) + \frac{2\pi\sigma^2(1-\gamma)}{b}} \tag{4}$$

M-step: the revised parameters θ^{new} is determined by: $\theta^{new} = \arg \max_{\theta} \wp(\theta, \theta^{old})$. We take derivative of $\wp(\theta)$ respecting to σ^2 and γ and set the derivative to zero. Then we get:

$$\sigma^2 = \frac{tr((\mathbf{Y} - \mathbf{X} - \mathbf{F})^T \mathbf{P}(\mathbf{Y} - \mathbf{X} - \mathbf{F}))}{2 \cdot tr(\mathbf{P})} \tag{5}$$

$$\gamma = tr(\mathbf{P})/M \tag{6}$$

where $\mathbf{F} = (f(\mathbf{x}_1), f(\mathbf{x}_2), ..., f(\mathbf{x}_M))^T$, $tr(\bullet)$ denotes the trace. After the EM iteration converges, we use the criterion (7):

$$\Theta = \{(\mathbf{x}_i, \mathbf{y}_i) : p_i \geq \hbar, i \in \mathbb{N}_M\} \tag{7}$$

with the predefined threshold $\hbar$ to acquire the inlier set Θ. It is worth noting that, to finish the EM algorithm, the transformation f should continue to be deeper analyzed in the M-step as follows. Extract the terms in (3) relating f, here comes

$$\wp(f) = -\frac{1}{2\sigma^2} \sum_{i=1}^{M} p_i \|\mathbf{y}_i - \mathbf{x}_i - f(\mathbf{x}_i)\|^2 \tag{8}$$

Generally, the above problem is an ill-posed problem, because f is non-rigid, which will results in infinite solutions. To alleviate the problem and to obtain a meaningful result, the regularization techniques is a better option. We suggest to operate in a Reproducing Kernel Hilbert Space (RKHS) [10]. Thus, f has the following explicit representation:

$$f(\mathbf{x}) = \sum_{i}^{N} \mathbf{\Phi}(\mathbf{x}, \mathbf{x}_i)\mathbf{k}_i \qquad (9)$$

where $\mathbf{\Phi}$ points to a positive definite matrix-valued kernel defining the RKHS χ, e.g., $\mathbf{\Phi}(\mathbf{x}_i, \mathbf{x}_j) = \delta(\mathbf{x}_i, \mathbf{x}_j) \bullet \mathbf{I} = e^{-\xi\|\mathbf{x}_i - \mathbf{x}_j\|^2} \bullet \mathbf{I}$ with $\mathbf{I}$ being an identity matrix. ξ refers to the range of samples interaction, namely, neighborhood size. $\{\mathbf{k}_i\}_{i=1}^{N}$ is the coefficients that need to be obtained.

Like [11], we introduce two constraints, i.e. the global (10) and local constraints (11), to control the complexity of the hypothesis space, and to preserve local structures among feature points respectively.

$$\kappa_1(f) = \|f\|_v^2 = tr(\mathbf{K}^T \mathbf{\Phi} \mathbf{K}) \qquad (10)$$

where $\mathbf{K} = (k_1, k_2, ..., k_N)^T$ is the coefficient set of f.

$$\wp(\mathbf{K}) = \sum_{i=1}^{M} p_i \left\| (\mathbf{x}_i^T + \mathbf{\Phi}_{i,.}\mathbf{K}) - \sum_{j=1}^{M} \mathbf{W}_{ij}(\mathbf{x}_j^T + \mathbf{\Phi}_{j,.}\mathbf{K}) \right\|^2 \qquad (11)$$

$\mathbf{\Phi}_{i,.}$ is the i-th row of $\mathbf{\Phi}$, and p_i is a posterior probability which plays a role of soft distribution to determine the inliers.

What's more, to excavate intrinsic structure information of the input data, we introduce the manifold regularization. The graph Laplacian [12] is chosen to define the manifold regularization term. The weighted neighborhood graph G can be established by acquiring the graph from vertex set $\bar{A} = \{\mathbf{x}_i\}_{i=1}^{N}$ with edges $(\mathbf{x}_i, \mathbf{x}_j)$ if and only if $\|\mathbf{x}_i - \mathbf{x}_j\|^2 \leq \zeta$, and assigning to edge $(\mathbf{x}_i, \mathbf{x}_j)$ the weight:

$$\mathbf{W}_{ij} = e^{-\frac{1}{\zeta}\|\mathbf{x}_i - \mathbf{x}_j\|^2} \qquad (12)$$

Then, the graph Laplacian of G represented by the matrix $\mathbf{\Omega}$ is given by (13).

$$\mathbf{\Omega}_{ij} = \mathbf{\Lambda}_{ij} - \mathbf{W}_{ij} \qquad (13)$$

where $\mathbf{\Lambda} = diag\left(\sum_{j=1}^{N} \mathbf{G}_{ij}\right)_{i=1}^{N}$. Let $\tilde{\mathbf{F}} = (f(\mathbf{x}_1), f(\mathbf{x}_2), ..., f(\mathbf{x}_N))^T$, the manifold term is finally derived as:

$$\kappa_2(f) = \|f\|_\omega^2 = \sum_{i=1}^{N}\sum_{j=1}^{N} \mathbf{\Omega}_{ij}(\tilde{\mathbf{F}}_i - \tilde{\mathbf{F}}_j)^2 = tr(\tilde{\mathbf{F}}^T \mathbf{\Omega} \tilde{\mathbf{F}}) = tr(\mathbf{K}^T \mathbf{\Phi} \mathbf{\Omega} \mathbf{\Phi} \mathbf{K}) \qquad (14)$$

Combining (10), (11) and (14), the coefficient set $\mathbf{K}$ of the transformation f can be solved by minimizing (using matrix patterns):

$$\wp(\mathbf{K}) = \frac{1}{2\sigma^2}(\mathbf{Y} - \mathbf{X} - \mathbf{J}\mathbf{\Phi}\mathbf{K})^T\mathbf{P}(\mathbf{Y} - \mathbf{X} - \mathbf{J}\mathbf{\Phi}\mathbf{K}) + \frac{\alpha}{2}tr(\mathbf{K}^T\mathbf{\Phi}\mathbf{K}) + \frac{\beta}{2}tr(\mathbf{K}^T\mathbf{\Phi}\mathbf{\Omega}\mathbf{\Phi}\mathbf{K})$$

$$+ \frac{\lambda}{2}\{[(\mathbf{X} + \mathbf{J}\mathbf{\Phi}\mathbf{K}) - \mathbf{W}(\mathbf{X} + \mathbf{J}\mathbf{\Phi}\mathbf{K})]^T\mathbf{P}[(\mathbf{X} + \mathbf{J}\mathbf{\Phi}\mathbf{K}) - \mathbf{W}(\mathbf{X} + \mathbf{J}\mathbf{\Phi}\mathbf{K})]\} \quad (15)$$

where the positive numbers α, β and λ keep the balance between data item, smooth item and the regularization items. $\mathbf{J} = (\mathbf{I}_{M \times M}, \mathbf{0}_{M \times (N-M)})$ with $\mathbf{I}$ being an identity matrix and $\mathbf{0}$ being a zero matrix. We follow the solution of seeking extreme value for a function making the derivative of (15) equals zero with respect to $\mathbf{K}$. Then we obtain an Eq. (16) where $\mathbf{K}$ can be easily solved.

$$(\mathbf{J}^T\mathbf{P}\mathbf{J}\mathbf{\Phi} + \alpha\sigma^2\mathbf{I} + \beta\sigma^2\mathbf{\Omega}\mathbf{\Phi} + \lambda\sigma^2\mathbf{J}^T\mathbf{S}\mathbf{J}\mathbf{\Phi})\mathbf{K} = \mathbf{J}^T[\mathbf{P}\mathbf{Y} - (\mathbf{P} + \lambda\sigma^2\mathbf{S})\mathbf{X}] \quad (16)$$

where $\mathbf{S} = (\mathbf{I} - \mathbf{W})^T\mathbf{P}(\mathbf{I} - \mathbf{W})$. Until now, all the parameters have been solved.

3   Experimental Results

In order to verify the availability of our method, we compare it with other three state-of-the-art feature matching methods, i.e. RANSAC [7], GS [8] and ICF [9].

3.1   Data Acquisition and Parameters Settings

The test data consist of 100 image pairs with size of 600×337 pixels captured by UAV parrot AR. Drone 2.0 in different height and direction. Among those, no. 1–no. 50 are image pairs of UAV with stable attitude, and no. 51–no. 100 are image pairs of UAV with motion (as shown in Fig. 1a). The experimental field is located in Zhuhai, Guangdong province, China ($22°02'59.2''$N $113°21'30.0''$E), where different types of crops such as lettuce, chives, etc. was implanted. Parameters setting: $C = 15$, $\alpha = 0.1$, $\beta = 0.001$, $\lambda = 3$, $\zeta = 0.1$, $\gamma = 0.9$ and $\hbar = 0.75$. We use the open source VLFEAT toolbox [11] to determine the initial set of matching points. For the ground truth, we adopted a method combining subjectivity and objectivity. First, the matching algorithm is used to establish the rough correspondences. Then, the final results are manually confirmed.

3.2   Results on UAV Images

This section gives quantitative comparisons of our method with RANSAC, ICF and GS. We use the precision and recall to evaluate the performance of the algorithm, wherein the precision is defined as the percentage of all the matching points retained by the algorithm, and the recall is defined as the percentage of all inlier points of the image retained by the algorithm.

The statistic results of the four methods are summarized in Fig. 1b. The cumulative distribution function of the initial inlier rate is shown above, from which the average inlier rate of the sample set is 46.34% and the inlier rate increases from 13.3 to 69.17%.

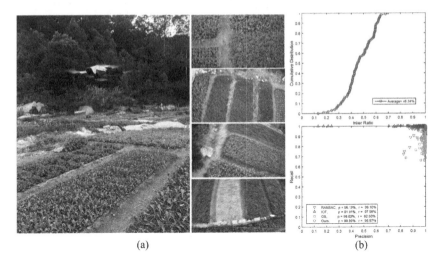

(a)   (b)

Fig. 1. (a) the UAV and some sample images included in the dataset. (b) quantitative comparisons of our method with RANSAC, GS and ICF.

Each discrete point in the bottom of Fig. 1b represents a precision-recall pair. The average precision and recall of four algorithms are (99.99%, 98.97%), (95.19%, 98.10%), (91.91%, 97.98%), (98.83%, 82.63%) respectively. For RANSAC, since it chooses the fundamental matrix as its parameter model, it can produce satisfactory results when the inlier ratio of the initial set is high, but it degrade dramatically when there exist too much mismatch points. The results showed that ICF either had a high precision or a high recall, but could not reach a higher value simultaneously. GS has a higher accuracy rate but the recall rate is not very satisfactory, this is mainly because the algorithm cannot automatically estimate the coefficient of affinity matrix and the algorithm does not have affine deformation. In contrast, as shown in the upper right-hand corner, our method has the highest precision and recall at the same time.

4   Conclusion

This paper presents a non-rigid image feature matching algorithm which is suitable for UAV in precision agriculture. It can determine the correct matching point set while estimating the transformation. The EM algorithm is adopted to complete the estimation in a maximum likelihood framework combining the global, local and manifold regularization constraints. The comparison test results confirm the effectiveness of our method, which is more superior to other state-of-the-art methods.

Acknowledgments. This work was supported by the Ph.D. Start-up Fund of Guangdong Natural Science Foundation (Grant No. 2016A030310306), the National Natural Science Foundation of China (Grant No. 41501505), Zhuhai Key Laboratory of Advanced Equipment Manufacturing and Material Processing Technology (Grant No. 201601A) and the Natural Science Foundation of Guangdong Polytechnic of Science and Technology (Grant No. XJPY2016003).

References

1. Yu, Z., et al.: Automatic image-based detection technology for two critical growth stages of maize: Emergence and three-leaf stage. Agric. For. Meteorol. **174**, 65–84 (2013)
2. Zhang, C., Kovacs, J.M.: The application of small unmanned aerial systems for precision agriculture: a review. Precis. Agric. **13**(6), 693–712 (2012)
3. Zhou, H. et al.: Seamless stitching of large area UAV images using modified camera matrix. In: IEEE International Conference on Real-Time Computing and Robotics (RCAR). IEEE (2016)
4. Courbon, J., et al.: Vision-based navigation of unmanned aerial vehicles. Control Eng. Pract. **18**(7), 789–799 (2010)
5. Se, S., Jasiobedzki, P.: Stereo-vision based 3D modeling and localization for unmanned vehicles. Int. J. Intell. Control Syst. **13**(1), 47–58 (2008)
6. Lowe, D.G.: Distinctive image features from scale-invariant keypoints. Int. J. Comput. Vis. **60**(2), 91–110 (2004)
7. Fischler, M.A., Bolles, R.C.: Random sample consensus: a paradigm for model fitting with applications to image analysis and automated cartography. Commun. ACM **24**(6), 381–395 (1981)
8. Liu, H., Yan, S.: Common visual pattern discovery via spatially coherent correspondences. In: 2010 IEEE Conference on Computer Vision and Pattern Recognition (CVPR). IEEE (2010)
9. Li, X., Zhanyi, H.: Rejecting mismatches by correspondence function. Int. J. Comput. Vis. **89**(1), 1–17 (2010)
10. Aronszajn, N.: Theory of reproducing kernels. Trans. Am. Math. Soc. **68**(3), 337–404 (1950)
11. Huabing, Z., et al.: Nonrigid feature matching for remote sensing images via probabilistic inference with global and local regularizations. IEEE Geosci. Remote Sens. Lett. **13**(3), 374–378 (2016)
12. Belkin, M., Niyogi, P., Sindhwani, V.: Manifold regularization: a geometric framework for learning from labeled and unlabeled examples. J. Mach. Learn. Res. **7**, 2399–2434 (2006)

A Comprehensive Evaluation Approach to Aviation Maintenance Support Ability Based on PCA

Yanming Yang[✉], Weituan Wu, and Ping Fang

Naval Aeronautical University Qingdao Campus, Qingdao, China
yymqd@126.com

Abstract. Principal component analysis (PCA) is a multivariate statistical method for reducing dimensionality by using the correlation between the original variables and interpreting the original variables through a few linear combinations of the original variables. As one of most important statistical methods, PCA can not only turn many index questions into less overall targets, but also provide the comparatively objective weight. The aircraft state index is the important parameter to measure the aviation maintenance support capability. It is indispensable to scientifically analyze aircraft state index data and to make scientific decisions on aviation maintenance to improve maintenance support capability. This paper discusses on multivariate evaluation method of aviation maintenance support ability using PCA. The mathematic model and algorithm steps of PCA method are given in detail. The application of PCA in aircraft integrated condition assessment is analyzed by an example. And the results show that the PCA-based method is effective for the comprehensive evaluation of the aviation maintenance support ability.

Keywords: Principal component analysis · Aviation maintenance support · Comprehensive evaluation · Multivariate statistics

1 Introduction

With the rapid development of aviation equipment, there are a lot of comprehensive evaluation issues in the aviation maintenance support. To carry out effective comprehensive evaluation of aviation maintenance support capacity, for the promotion of aviation maintenance work targeted, predictive, and promote scientific maintenance of aviation equipment plays a very important role. Aviation maintenance index data is becoming more and more complicated. Therefore, it is of great significance to establish an effective evaluation method of aviation maintenance support capability [1]. Principal component analysis (PCA) is a linear dimensionality reduction technique [2]. PCA has been widely used to evaluate aviation maintenance support ability. In general, there are the following basic relationships between the principal component obtained by PCA and the original variable: each principal component is a linear combination of the original variables; the number of final extracted principal components is less than the number of original variables; each principal component is not related to each other [3–5].

© Springer International Publishing AG 2018
F. Qiao et al. (eds.), *Recent Developments in Mechatronics and Intelligent Robotics,*
Advances in Intelligent Systems and Computing 690, DOI 10.1007/978-3-319-65978-7_89

2 The Mathematical Model and Algorithm of PCA

2.1 Mathematical Model and Calculation

Suppose the study object is n samples, p variable data. We can represent the original data as the following matrix:

$$X = \begin{bmatrix} x_{11} & x_{12} & \cdots & x_{1p} \\ x_{21} & x_{22} & \cdots & x_{2p} \\ \vdots & \vdots & & \vdots \\ x_{n1} & x_{n2} & \cdots & x_{np} \end{bmatrix} = (X_1, X_2, \cdots, X_p) \tag{1}$$

Linear transformation of X can form a new integrated variable:

$$\begin{cases} y_{i1} = u_{11}x_{i1} + u_{12}x_{i2} + \cdots + u_{1p}x_{ip} \\ y_{i2} = u_{21}x_{i1} + u_{22}x_{i2} + \cdots + u_{2p}x_{ip} \\ \vdots \\ y_{ip} = u_{p1}x_{i1} + u_{p2}x_{i2} + \cdots + u_{pp}x_{ip} \end{cases}, \ i = 1, 2, \cdots, n \tag{2}$$

Expressed as a matrix:

$$Y = XU \tag{3}$$

where $Y = \begin{bmatrix} y_{11} & y_{12} & \cdots & y_{1p} \\ y_{21} & y_{22} & \cdots & y_{2p} \\ \vdots & \vdots & & \vdots \\ y_{n1} & y_{n2} & \cdots & y_{np} \end{bmatrix}$, $U = \begin{bmatrix} u_{11} & u_{12} & \cdots & u_{1p} \\ u_{21} & u_{22} & \cdots & u_{2p} \\ \vdots & \vdots & & \vdots \\ u_{n1} & u_{n2} & \cdots & u_{np} \end{bmatrix}$

The k-th integrated variable is:

$$Y_k = Xu_k \tag{4}$$

where u_k is the coefficient of linear transformation. According to the requirements of the PCA, the following constraints are applied to the linear transformation: (1) Y_i is not related to Y_j; (2) Y_p is the largest variance of all linear combinations of $X_1, X_2, \cdots, X_p$ that are not related to $Y_1, Y_2, \cdots, Y_p$; (3) $u'_k u_k = 1$.

2.2 Algorithm Steps

Standardize the Source Data. Using Z-score method to standardize the data transformation:

$$z_{ij} = \frac{x_{ij} - \bar{x}_j}{s_j} \quad , i = 1, 2, \cdots, n; j = 1, 2, \cdots, p \tag{5}$$

where $\bar{x}_j$ and s_j are the mean value and standard deviation of the j-th variable, respectively.

Calculate the Correlation Matrix. To calculate the correlation matrix of the standardized matrix R:

$$R = \left(r_{ij}\right)_{p \times p} = \frac{Z'Z}{n-1} = \frac{1}{n-1} \sum_{k=1}^{n} z_{ki} z_{kj} \quad , i,j = 1, 2, \cdots, p \tag{6}$$

Calculate Eigenvectors and Eigenvalues to Determine the PCs. Using Jacobi method to solve the characteristic equation of the correlation coefficient matrix. Then its eigenvalues λ_i and corresponding eigenvector e_i of coefficient matrix R, can be also computed respectively as follows [6]:

$$|R - \lambda I| = 0 \tag{7}$$

$$[\lambda_i I - R] e_i = 0 \tag{8}$$

where R is the Correlation coefficient matrix I is the p-order unit matrix, $e_i = [e_i(1), \ e_i(2), \ \cdots, \ e_i(n)]'$. With e_i from Eq. (8), we have $v_i = e'_i X$ and select m $(m < n)$ vectors $v_1, \ v_2, \ \cdots, \ v_m$ to be principal components.

Calculate Variance Contribution Rate to Determine the Number of PCs. The variance contribution rate of the k-th principal component Y_k:

$$\alpha_k = \lambda_k \left/ \sum_{i=1}^{p} \lambda_i \right. \quad , i = 1, 2, \cdots, p \tag{9}$$

The cumulative variance contribution of the former m principal components $Y_1, Y_2, \cdots, Y_m$:

$$\sum_{i=1}^{m} \alpha_i = \sum_{i=1}^{m} \lambda_i \left/ \sum_{i=1}^{p} \lambda_i \right. \quad , m < p \tag{10}$$

In general, the number of principal components k value is determined by the variance contribution rate $\sum_{i=1}^{m} \alpha_i \geq 85\%$.

3 An Example of PCA

3.1 Problem Description

The six maintenance index data of 24 aircrafts in five years: intact rate, maintenance grounding rate, failure rate, air failure rate, flight sorties, flight time. Try to evaluate maintenance support capability for each aircraft using PCA. The data is standardized according to Eq. (6), and the standardized data are shown in Table 1.

Table 1. Aircraft maintenance support index data in 5 years (after standardized processing)

Aircraft ID	Intact rate (Z1)	Grounding rate (Z2)	Failure rate (Z3)	Air failure Rate (Z4)	Flight sorties (Z5)	Flight time (Z6)
1	0.7242	−0.3587	−0.3126	−0.4173	0.1326	0.0006
2	−0.7787	0.5223	0.6547	−0.4691	−0.0758	−0.4926
3	−1.3360	1.6659	0.3612	−0.8464	0.9095	0.9157
4	0.5782	−0.0760	1.5725	0.4763	0.1137	0.7400
5	0.4172	−0.5447	−0.0658	−0.1974	0.2842	0.5418
6	0.7189	−0.2815	0.9015	0.5485	−0.8906	−1.0653
7	1.2837	−1.1003	0.7202	0.1599	0.5874	0.5842
8	−0.4989	0.1563	1.9176	0.7365	−1.2506	−1.5239
9	−1.2070	1.4889	1.2232	0.8321	−0.7769	−0.7397
10	0.1577	−0.4844	0.4554	0.3974	−0.3790	−0.1209
11	−2.7980	1.9871	1.1131	0.8543	0.2463	−0.2940
12	−0.2998	−0.2140	−2.3092	−2.9913	0.1895	0.6731
13	−0.7594	1.2740	−0.5810	1.1661	−2.1601	−2.4531
14	1.2174	−1.0664	−1.5392	−0.3526	1.0800	0.4353
15	−0.8981	1.4113	−0.8401	−0.6959	0.5495	1.0348
16	−0.2919	0.8983	−0.4464	−1.6028	−0.0379	0.1375
17	1.1274	−1.2001	0.1104	0.6838	0.7390	0.6702
18	−0.5179	−0.3354	−0.0236	−0.4855	0.5116	0.0000
19	−0.3791	0.4703	−0.6744	−0.2236	0.4548	0.6312
20	0.4741	−0.9253	0.0345	0.9242	−1.2316	−0.4686
21	0.5407	−0.5304	−0.9616	1.6501	−0.4926	−0.7408
22	1.3634	−1.4469	0.0642	−0.3815	−1.9137	−1.4392
23	0.7773	−1.0663	−0.1268	0.8871	1.9516	2.1082
24	0.3846	−0.2442	−1.2478	−0.6531	1.4590	0.8656

3.2 Analysis Process

Calculate the Correlation Coefficient Matrix. According to the correlation between the two index data, calculate the correlation coefficient of the index data, the results shown in Table 2. For the same index data, the correlation coefficient is 1, with 1 as the boundary, the diagonal elements take the same value. According to Table 2 can be calculated correlation matrix R.

Table 2. Correlation coefficient between the indexes

	Z1	Z2	Z3	Z4	Z5	Z6
Intact rate	1.000					
Grounding rate	−0.913	1.000				
Failure rate	−0.204	0.184	1.000			
Air failure rate	0.057	−0.068	0.517	1.000		
Flight sorties	0.064	−0.092	−0.283	−0.294	1.000	
Flight time	0.147	−0.152	−0.280	−0.360	0.920	1.000

Calculate the Eigenvectors and Eigenvalues. The eigenvalues and eigenvectors of the correlation matrix R are calculated. According to the Eq. (7) to calculate the eigenvalue: $\lambda = (2.4845, 1.8393, 1.0793, 0.4397, 0.0923, 0.0648)'$. According to the Eq. (8) to calculate the corresponding eigenvectors as shown in Table 3.

Table 3. Eigenvectors of the correlation matrix

Variables	PC1	PC2	PC3	PC4	PC5	PC6
Intact rate	0.305	−0.626	−0.027	0.117	−0.609	0.361
Grounding rate	−0.307	0.623	−0.005	−0.126	−0.645	0.292
Failure rate	−0.396	−0.067	0.607	0.683	0.015	0.068
Air failure rate	−0.341	−0.318	0.536	−0.696	−0.053	−0.087
Flight sorties	0.506	0.258	0.432	−0.133	0.309	0.615
Flight time	0.533	0.219	0.397	0.048	−0.339	−0.628

Determine the Principal Components. In order to determine the number of principal components, the cumulative variance contribution rate should be calculated first. According to the Eqs. (9) and (10) to calculate the variance contribution rate and cumulative variance contribution as shown in Fig. 1.

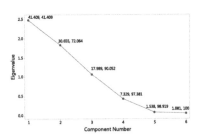

Fig. 1. Scree plot of PCA

It can be seen from Fig. 1 that the first three eigenvalues are greater than 1 and the cumulative contribution rate is over 85%, which indicates that the information reflected in the six aircraft maintenance index data can be reflected by three PCs, which explain 90.052% of total variation of variables in PCA. Thus, we only used the first three PCs to obtain the variables.

3.3   Result Analysis

From the analysis of Fig. 2 that: The first principal component includes flight sorties and flight time; the second principal component includes the intact rate and the maintenance grounding rate; the third principal component includes the failure rate and the air failure rate. The analysis shows that the first principal component is expressed as the strength of the aircraft, the second principal component is the troop maintenance capability, and the third principal component is the inherent reliability of the aircraft.

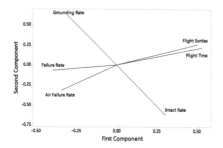

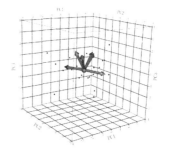

Fig. 2. Loading plot of PCA

Fig. 3. 3D scatter of PCA

According to the scores of the three principal components, three-dimensional scatterplots can be drawn to visually show how they explain the original variables, as shown in Fig. 3. It is also possible to see the main correspondence between the principal component and the original variable.

4   Conclusion

An effective comprehensive evaluation approach for the aviation maintenance support ability based on PCA is put forward. Based on the principle of principal component analysis, the mathematical model and algorithm are given in detail. In order to improve the operability of the method, taking the aviation maintenance support index data of 24 aircraft as an example, this paper discusses how to use the PCA-based method to evaluate the aviation maintenance support capability comprehensively. And the visual analysis results are presented graphically. The results show that the method is an effective comprehensive evaluation approach for aviation maintenance support ability. The method can also be applied to other multivariate evaluation issues in the field of aeronautical equipment maintenance support.

References

1. He, Q., He, X., Zhu, J.: Fault detection of excavator's hydraulic system based on dynamic principal component analysis. J. Central South Univ. **15**(5), 700–705 (2008)
2. Rezghi, M., Obulkasim, A.: Noise-free principal component analysis: an efficient dimension reduction technique for high dimensional molecular data. Expert Syst. Appl. **41**(17), 7797–7804 (2014)
3. Abdi, H., Williams, L.J.: Principal component analysis. Wiley Interdisciplinary Rev. Comput. Stat. **2**(4), 433–459 (2010)
4. Jolliffe, I.T., Cadima, J.: Principal component analysis: a review and recent developments. Phil. Trans. R. Soc. A **374**, 20150202 (2016)
5. Wang, R., et al.: Fault detection of flywheel system based on clustering and principal component analysis. Chin. J. Aeronaut. **28**(6), 1676–1688 (2015)
6. Wen, C., Wang, T.Z., Hu, J.: Relative principle component and relative principle component analysis algorithm. J. Electron. (China) **24**(1), 108–111 (2007)

Design and Implementation
of an Underwater Manipulator

Qin Wang[⊠], Junwei Tian, Yanfei Zhao, and Zhiyi Jiang

School of Mechatronic Engineering,
Xi'an Technological University, Xi'an 710021, China
wq992514@163.com

Abstract. In order to replace the human being under water, a Manipulator with three degrees of freedom is designed to work under water for 200 m. The Manipulator made of aluminum alloy as the body material, and the shell is designed as a cylinder. So as to make full use of the internal space and reducing the movement resistance, At the same time, the high pressures resistance of the whole structure is improved. The static analysis and calculation on the shoulder, elbow and hand are carried out and the driving module is selected; Using Solidworks to create 3D solid model, and the finite element simulation analysis of the claw is carried out. Manipulator seal is divided into two kinds: Static seal using O ring seal, dynamic seal using magnetic seal. The simulation results show that the design meets the requirements and the structure is reliable.

Keywords: Underwater manipulator · Solidworks modeling · Finite element analysis · Waterproof and sealing

1 Introduction

Underwater manipulators are an important tool that can replace mankind in underwater operations. Underwater has always been a general operating tool for the preferred configuration of underwater robots. The types of underwater robotic arms are also varied. Underwater robots could finish numbers of tasks through their robotic arms, largely replaced the staff to do complex, dangerous tasks [1–5]. Manipulators in the industrial, military, medical treatment, as well as space exploration and other fields have been widely applied and developed t. Although they have different forms, they can still finish different works of various areas, it is because they have a common feature is able to accept the order, and can accurately locate a point of a three-dimensional space to start its work. Compared with the Onshore robots, the underwater manipulator should have the following characteristics: due to long-term soaking in the sea, the robots requiring corrosion-resistant materials; meanwhile, according to the restrictions of underwater robots' equipment, small and lightweight are demanded, and the emergency device as well [6]; due to the tasks contents are different and complex, so it should be flexible to operate.

© Springer International Publishing AG 2018
F. Qiao et al. (eds.), *Recent Developments in Mechatronics and Intelligent Robotics*,
Advances in Intelligent Systems and Computing 690, DOI 10.1007/978-3-319-65978-7_90

2   The Design Program of Manipulator

The underwater manipulator as well as the traditional mechanical arm, mainly including the shoulder joint, elbow joint, wrist joint and paw [7] etc. The design is intended to that, when the underwater manipulator working in a three-dimensional space coordinate system, the shoulder joint could achieve pitching action in the X-Y plane, the elbow joint could achieve rotational motion in the Y-Z plane, the paw could achieve Grasping motion under the driving of wrist joint, to complete the task through clamping objects. The schematic diagram as shown in Fig. 1.

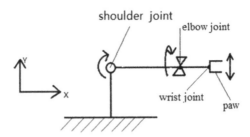

Fig. 1. The schematic diagram of underwater manipulator

In the actual industrial production, the shape of the manipulator is also diversified, the drive mode is also different, there are two drive modes: motor drive and servo drive. For this design, the principle scheme of how the paw clamp objects is: with electromagnet as the drive, according to the principle of magnetic flux leakage of spiral pipe, it could force gripper to open or close through switching the power of electromagnet to pull or release the rod constantly to grip objects. Elbow joint adopts stepper motor to drive, complete the rotary motion in Y-Z plane; Shoulder joint can be driven by the steering gear, complete pitching motion in X-Y plane.

3   The Structure Design of Underwater Manipulator

Because underwater manipulators need to work in the deep sea for a long term, have to meet the design requirements like high pressure resistance, corrosion resistance and light weight, so this article uses hard aluminum alloy as the material for underwater manipulator's main body. In addition, the shell is cylindrical, so that can both take the full use of the internal space, and also reduces the underwater movement resistance, meanwhile, improving the overall structure's high pressure resistance.

According to the structure of the underwater manipulator, we have designed the two-dimensional graphics of it, its overall forces as shown in Fig. 2:

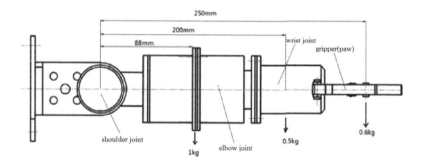

250mm

200mm

88mm

wrist joint

gripper(paw)

shoulder joint

1kg

elbow joint

0.5kg

0.6kg

Fig. 2. Overall force diagram

3.1 Static Analysis and Calculation of Gripper

The gripper is mainly calculate its clamping force: through the mechanical analysis we can know that the total clamping force of the gripper and the friction force generated by the object (μF) must be bigger than the gravity force of the object (mg), taking into account the acceleration and the impact force caused by underwater robot, so we must set a safety factor a, so need to meet these requirements:

$$\mu F \geq \alpha \bullet mg \tag{1}$$

That is $F \geq \alpha \bullet mg/\mu$

In the formula, μ is the coefficient of friction, the friction coefficient between the rubber and the cast iron is 0.8, and the safety factor is 3, m = 0.5 kg,

$$\text{so}: \ F \geq 3 \times 0.5 \times 10/0.8 = 18.75N \tag{2}$$

Therefore, by comparison, the choice of electromagnet model JP1564H-01, it is a push-pull electromagnet, the maximum suction force of 5 kg, or 50 N, because F > 18.75 N, so selected electromagnet is in line with the requirements.

3.2 Analysis and Calculation of Elbow Joints

The rotation movement of the elbow is driven by the stepper motor, assume the maximum equivalent radius of rotation of the gripper and the object is R = 50 mm, the total weight of the gripper and the object is 0.6 kg, the moment of inertia is:

$$J_1 = m_1 \bullet R^2/2 = 0.6 \times 0.052/2 = 0.00075kg \cdot m^2 \tag{3}$$

Set the arm speed ω 1 of the manipulator from 0 to 500 r/s, the required time (t) is 0.5 s, then the angular acceleration is:

$$\alpha_1 = \omega_1 \bullet \pi/(180 \bullet t) = 500\pi/180 \times 0.5 = 17.44 \ \text{rad/s}^2 \tag{4}$$

Load start moment of inertia (excluding static friction torque) is:

$$T_1 = J_1 \bullet \alpha_1 = 17.44 \times 0.00075 = 0.01308 \text{ N} \cdot \text{m} \tag{5}$$

As the stepper motor doesn't have instantaneous overload capacity, so take the safety factor of 2, the stepper motor output starting torque is:

$$T_{out} = 2T_1 = 2 \times 0.01308 = 0.02616 \text{ N} \cdot \text{m} \tag{6}$$

Since T_{out} must be less than the maximum torque of the stepper motor, so we select the 42BYGH33 gearmotor as we need. The main function parameters are shown in Table 1.

Table 1. Main performance parameters of Model:42BYGH33 gearmotor

Model	Max holding torque of step angle	Step angle	Weight	Level.1 reduction ratio
42BYGH33	1.58 N*m	1.8°	0.22 kg	1:5.18

3.3   Analysis and Calculation of Shoulder Joint

The pitching motion of the shoulder is driven directly by the steering gear. When the whole arm is in the horizontal state, the static torque of each part to the rotation center of the shoulder is the largest, and the algebra is:

$$\begin{aligned}
M_{肩} &= (m_1 \bullet l_1 + m_2 \bullet l_2 + m_3 \bullet l_3) \bullet g \\
&= (0.6 \times 0.25 + 0.5 \times 0.2 + 0.8 \times 0.88) \times 9.8 \\
&= (0.15 + 0.1 + 0.0704) \times 9.8 \\
&= 3.14 N \cdot m
\end{aligned} \tag{7}$$

In the formula, $m1$ is the total mass of the gripper and the object, l_1 is the distance between the gripper and the object's gravity center to the center of rotation of the shoulder, and l_1 is about 250 mm;

$m2$ is the mass of the wrist rotation center, about 0.5 kg, l_2 is the distance from the center of rotation of the wrist to the center of the elbow, and it's about 200 mm;

$m3$ is the mass of the elbow rotation center, about 0.8 kg, l_3 is the distance between elbow rotation center to the shoulder rotation center, about 88 mm;

The output torque of the steering gear is $3.14 N \cdot m$, according to the output torque, we choose the steering gear, the function parameters shown in Table 2.

Table 2. Function parameters of underwater steering gear

Maximum torque	Angle stroke	Weight	Rated voltage
3.5 N*m	300°	0.24 kg	6 V

The steering gear is a special underwater steering gear produced by Cehai Technology, with good water resistance, its maximum working depth of 200 meters, can meet the design requirements, but also reduces the design difficulty of manipulator's waterproofness and leakproofness.

4 Modeling and Simulation of 3D Entity

According to the structure and function of underwater manipulator, we designed and completed three-dimensional solid modeling, and it's shown in Fig. 3, the main performance parameters shown in Table 3.

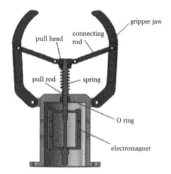

Fig. 3. Three-dimensional solid model of the underwater manipulator

Fig. 4. The gripper's fully opened and completely closed condition's three-dimensional simulation

Table 3. Main performance parameters of underwater manipulator

Total length	Maximum joint diameter	Shoulder joint pitch angle	Elbow joint rotation angle
397 mm	70 mm	±90°C	±90°C

What we designed shown on this paper is a gripper composed by simple connecting rods, mainly use the electromagnet's adsorption force and the spring's restoring force to achieve the closing and opening of the gripper. However, if the design of the gripper structure is not appropriate, will make gripper's opening-closing range couldn't meet the requirements, so the need to make a reasonable design. The gripper's fully opened and completely closed condition's three-dimensional simulation (in ideal condition) shown in Fig. 4.

It can be seen from the figure, in the case of not holding the object, when the electromagnet powers on, the rod will overcome the spring force, the lower surface of the lower end of the rod will contact with the suction surface of the electromagnet, gripper completely closed; After the iron is turned off, the gripper will fully opened under the action of the spring force, but the diameter of the rod head is larger than the diameter of the rod hole, and the lever is stuck so that the gripper has a limit tension.

Manipulator in the work, mainly rely on the grippers to hold the object, If the gripper structure is reasonable, or will the gripper be out of shape when clamping objects, which is very critical, will have a direct impact on the efficiency of the manipulator, we can do static simulation analysis towards a single gripper jaw [8, 9].

(1) The material properties are shown in Tables 4 and 5

Table 4. The material properties of gripper (1)

Number	Entity name	Material	Weight	Volume
1	SolidBody 1(Cut - stretch 3)	[SW]1060 alloy	0.0211755 kg	7.84277×10^{-6} m3

Table 5. The material properties of gripper (2)

Attribute name	Value	Unit	Value type
Elastic modulus	6.9×10^{10}	N/m2	Constant
Poisson's ratio	0.33	NA	Constant
Shear modulus	2.7×10^{10}	N/m2	Constant
Mass density	2700	kg/m3	Constant
Tension strength	6.8936×10^{7}	N/m2	Constant
Yield Strength	2.7574×10^{7}	N/m2	Constant
Thermal expansion coefficient	2.4×10^{-5}	/Kelvin	Constant
Thermal conductivity	200	W/(m.K)	Constant
Specific heat	900	J/(kg.K)	Constant

(2) Simulation results

The stress maps, deformation figure and strain diagrams of the grippers are shown in Figs. 5, 6 and 7.

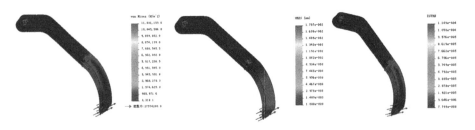

Fig. 5. Stress maps   **Fig. 6.** Deformation figure   **Fig. 7.** Strain diagrams

5 Waterproof and Seal of Underwater Manipulator

One of the key issues of underwater manipulator is the sealing problem [10, 11]. Meanwhile, the deeper the robot dives, the higher requirements of sealing performance need to meet. It's also an important indicator of underwater manipulator. The Underwater manipulator's sealing types are divided into static seal and dynamic seal.

(1) static seal

One of the commonly used static sealing ways is the O-ring seal, O-ring has advantages of simple sealing mechanism, compact installation, lighter weight and self-sealing effect, sealing performance is good. And O-rings and their trenches are standardized, so the cost is low, and easy to use or outsource. For these reasons, we selected the O-ring for underwater manipulator's static seal.

(2) dynamic seal

Some of the commonly used dynamic sealing ways are: Y-ring, mechanical seal, magnetic seal. The magnetic seal drive is a non-contact seal. Magnetic coupling transmission (shown in Fig. 8) mainly composed of three parts: the outer rotor, the inner rotor and isolation sleeve. As the magnetic seal's excellent sealing performance, can achieve zero leakage; do not need to add lubrication device, with self-sealing effect; and simple structure, small size. Therefore, in this article, the underwater manipulator wrist rotation center using magnetic coupling technology to seal, thus solving the waterproof seal problem of the motor case.

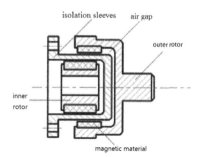

Fig. 8. Magnetic coupling transmission

6 Conclusion

This paper had designed a three-degree-of-freedom manipulator that can operate in 200 meters underwater. As a result of the cylindrical design in the underwater working case, the movement resistance is reduced and the high-pressure resistance of the whole structure is improved. The shoulder is driven by the underwater special steering gear, it can achieve 90 °C pitching motion, elbow joint is driven by the stepper motor can be achieve 90 °C rotation, is with the used of electromagnet adsorption force and the spring Restoring force, the gripper paw can grab 5 kg of objects. Through the simulation of the gripper paw, the results could meet the design requirements.

Funds. The department of project in Shanxi province. (NO. 2015KTZDGY-02-01).

References

1. An, J., Sun, C., Ling, H.: Underwater manipulator structure design and research. Mech. Eng. Autom. 4(2), 90–95 (2009)
2. Design and research of six-degree-of-freedom underwater manipulator. J. Hydraul. Pneum. 1, 10–12 (2004)
3. Zhang, L.: Study on structure and impedance control of three-degree-of-freedom underwater manipulator. Master thesis, Harbin Institute of Technology, Harbin (2008)
4. Zhang, M.: Underwater manipulator design and research. Master's thesis, Northwestern Polytechnical University (2005)
5. Ryu, J.H., Kwon, J.H., Lee, P.M.: Control of underwater manipulators mounted on an ROV using base force information. In: Proceedings of the 2001 IEEE International Conference on Robotics and Automation, Seoul, Korea (2001)
6. Meng, Q., Zhang, M., Zhang, J., et al.: Research and development of underwater manipulator. Appl. Technol. 4 (1996)
7. Liu, Z., Zhu, G., Wang, Y., et al.: Design of mechanical gripper structure based on typical underwater operating tasks. Comb. Mach. Tool Autom. Process. Technol. 8, 60–61 (2005)
8. Zhao, W., Su, D.: Finite element analysis of O-ring sealing performance of underwater robot compression shell. Mech. Eng. 7, 66–68 (2009)
9. Zhou, L., Du, Q., Zuo, D.: Palletizing robot arm finite element analysis based on ABAQUS. Mech. Eng. 9, 120–121 (2013)
10. Li, F., Chen, J.: Design and discussion based on underwater motion seal component. Water Leap Warf. Ship Prot. 3 (2007)
11. Zhong, X.: Dynamic sealing technology of underwater robot. Robot Technol. 1, 40–41 (2006)

Author Index

© Springer International Publishing AG 2018
F. Qiao et al. (eds.), *Recent Developments in Mechatronics and Intelligent Robotics*,
Advances in Intelligent Systems and Computing 690, DOI 10.1007/978-3-319-65978-7

Printed in the United States
By Bookmasters